BASIC GRAPHICS AND CARTOGRAPHY

Claude Z. Westfall

University of Maine at Orono

University of Maine at Orono Press

Cover: Orthophotograph with contours for a portion of Balden-Wurttemberg, Germany (Courtesy Carl Zeiss, Inc).

Copyright © 1984 by University of Maine at Orono Press, Orono, Maine, 04469. All rights reserved. No part of this book may be reproduced or ultilized in any form or by any means, whatsoever, whether electronic or mechanical, including photocoyping, recording, or by any information storage and retrieval system, without written permission from the publisher.

First edition
Printed in the United States of America

Library of Congress Catalog Card Number: 84-51063
ISBN 0-89101-061-0

0 9 8 7 6 5 4 3 2 1

PREFACE

This text has evolved from the teaching of graphics and mapping techniques to primarily forestry students at the University of Maine, Orono. It has been written for those who would like to learn more about the basics of graphic presentation and those whose interest is in cartography or for other individuals where this area is directly related to their field of activity.

The book is designed to be used as a basic reference work or as a text for a beginning course. It is realized that the field of graphics is enormous and a complete book covering the entire area would be too large. Sufficient material, however, is presented for either a one or two semester course at the introductory level.

All graphics, e.g., mechanical, electrical, architectural, cartographic, etc., requires a basic understanding of certain graphic principles or techniques for these activities to be carried out. These are instruments and their use, lettering, data plotting, projection theory, and scales, discussed in the first part of the book. The second part of the book uses this information as a background for cartographic considerations and applications. A general survey of cartographic techniques, principles and concepts is presented to help individuals plan, design, and present information in map form. These aspects are of importance and should be considered to increase one's ability to effectively use, read, and prepare a professional appearing drawing. The production of any graphic material, as the final step, requires that the information presented be pleasing and easily understood. Using the information contained in this book, individuals will find it possible to prepare drawings that will meet this requirement.

In graphic communication, a number of significant changes have occurred. New materials and the automation of graphic processes have had their effect on how we do things and will exert still greater influences in the future. Perhaps the most important development is the increased awareness of the significance of graphics in carrying out successful communications. The ability to communicate these ideas is enhanced by understanding and sharpening skills associated with drawing methods and techniques.

The unique features of this book include the following:

1. Basic graphics and cartography are both discussed for learning, developing, and applying skills.

2. New developments, materials, and equipment are surveyed.

3. More than 570 illustrations, using a wide variety of techniques, have been incorporated into the text to aid the reader to better understand the information presented.

4. Metric units as well as customary units have been incorporated into the text.

5. Each chapter is self-contained and presents a different topic. Chapters may be presented in order or rearranged to meet individual needs.

The field of cartography is dynamic and it is hoped that the text will enhance the reader's appreciation of maps and arouse an interest to pursue the field in more depth. Many articles and books on graphics and mapping have been published. At the end of the book is a bibliography of further readings for a more detailed treatment to areas of interest.

The author wishes to acknowledge and to thank the many students and colleagues who contributed to the development of the text through discussions in the classroom, offices, and professional meetings. An undertaking of this kind depends on the contributions of many people and I am greatly indebted to:

- the University of Maine at Orono (UMO) for support in the form of a leave of absence to work on the text.

- the following individuals for reviewing parts of the manuscript.

    Professors Marshal Ashley, Lawrence T. Fisher, Gary Furbish, Peter M. Messier, Dave Tyler, and Norman J. Viger (UMO).

    Terrance J. Keating, Kork Systems, Inc., Bangor, Maine.

    Terry Kelly, Public Information and Central Services (UMO).

    Stanley J. Plisga, Plisga and Day Land Surveyors, Bangor, Maine.

- the School of Engineering Technology secretarial staff (UMO) for typing the manuscript; Rosemary LaMountain and Sharon Steele.

Special appreciation is expressed to Professor Wallace C. Robbins, College of Forest Resources (UMO) for his help in preparing the chapter on Aerial Photogrammetry and Remote Sensing.

The author appreciates the generosity of the many organizations and firms for supplying appropriate illustrations.

Finally, I am indebted to the University of Maine at Orono Press for its help and assistance for this book to be produced.

The author would be most interested to receive suggestions for the improvement of the text at any time.

Claude Z. Westfall
Orono, Maine

CONTENTS

chapter

1	Introduction to Graphics and Cartography	1
2	Graphic Instruments-Equipment and Their Use	21
3	Scales and Measurement	39
4	Graphic Geometry and Construction	49
5	Drawing Media and Techniques	65
6	Lettering	89
7	Graphical Representation of Data	111
8	Orthographic Projection	127
9	Basic Dimensioning	157
10	Pictorial Systems	169
11	Sketching	193
12	Conventions and Symbols	199
13	Map Projections	219
14	Map Orientation	241
15	Land Location and Description	249
16	Area and Volume	273
17	Representation of Relief	285
18	Remote Sensing and Aerial Photography	319
19	Computer-Assisted Cartography	349
20	Printing Processes and Reproduction Systems	371
21	Map Information and Distribution	385
	Appendixes	401
	Index	413

INTRODUCTION TO GRAPHICS AND CARTOGRAPHY

1.1 THE GRAPHIC LANGUAGE

Throughout recorded history people have relied on pictures and drawings of many types to communicate and record ideas. These drawings have developed from crude pictorials of prehistoric man, through a period of highly artistic drawings and paintings, to the present well developed technical types of drawings required by society today. The word *graphic* means drawing to express ideas by means of lines, marks or characters impressed on a surface. This universal language system is used as one means to express ideas and thoughts. Several early examples of this form of communication are the Egyptian hieroglyphics (sacred writings), the South American Indian symbols and glyphs, and the Chinese characters, Fig. 1.1. The Chinese

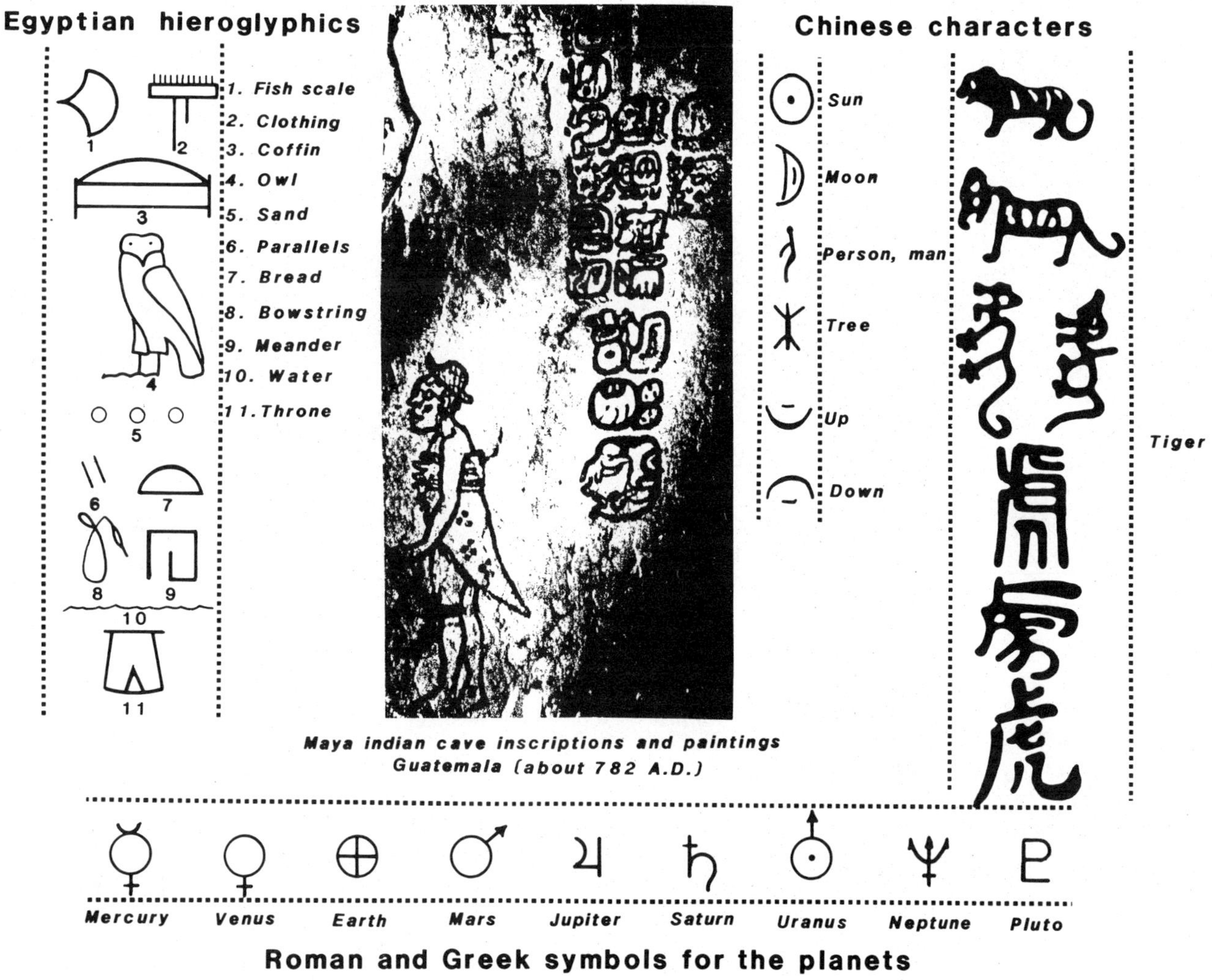

Egyptian hieroglyphics

1. Fish scale
2. Clothing
3. Coffin
4. Owl
5. Sand
6. Parallels
7. Bread
8. Bowstring
9. Meander
10. Water
11. Throne

Maya indian cave inscriptions and paintings Guatemala (about 782 A.D.)

Chinese characters

- Sun
- Moon
- Person, man
- Tree
- Up
- Down

Tiger

Mercury Venus Earth Mars Jupiter Saturn Uranus Neptune Pluto

Roman and Greek symbols for the planets

Fig. 1.1 Early communication forms.

characters date back thousands of years to inscriptions found on ancient oracle bones and other archaeological relics. As the Chinese society became more sophisticated many characters gradually became more abstract than pictorial, as illustrated for the tiger.

The only means for adequately describing the size, shape, and relationship of physical objects and features on the earth's surface is a drawing. A drawing contains many graphic symbols and abbreviations, and professions such as engineering, cartography, architectural, electrical, mechanical, etc. has its own unique code of symbols. The form and use of some of these symbols is discussed in subsequent chapters of this book as well as other reference books and manuals. Since almost everyone encounters numerous drawings in all areas of human endeavor, graphics along with reading and writing for communicating is of importance. In addition, well-trained professionals must have the ability to interpret or make drawings of many kinds, Fig. 1.2. Almost everything that is used requires a drawing or a series of drawings for the object to be produced. The graphic language for expressing engineering designs, data relationships, maps, and other forms of information is now more vital than ever before in this technology oriented world.

It is necessary to provide the basic elements of the graphic language for individuals to acquire a useful working knowledge of graphical principles and a modest competence in drawing skill. The values to be gained from undertaking a study of the graphic language are: (1) the ability to think in three dimensions, (2) learning basic traits required of all work, e.g., neatness, speed, and accuracy, (3) the ability to use constructive imagination and to be creative, (4) the development of the learning process through original thinking, and finally (5) the ability to synthesize, analyze, record, and communicate ideas graphically and professionally.

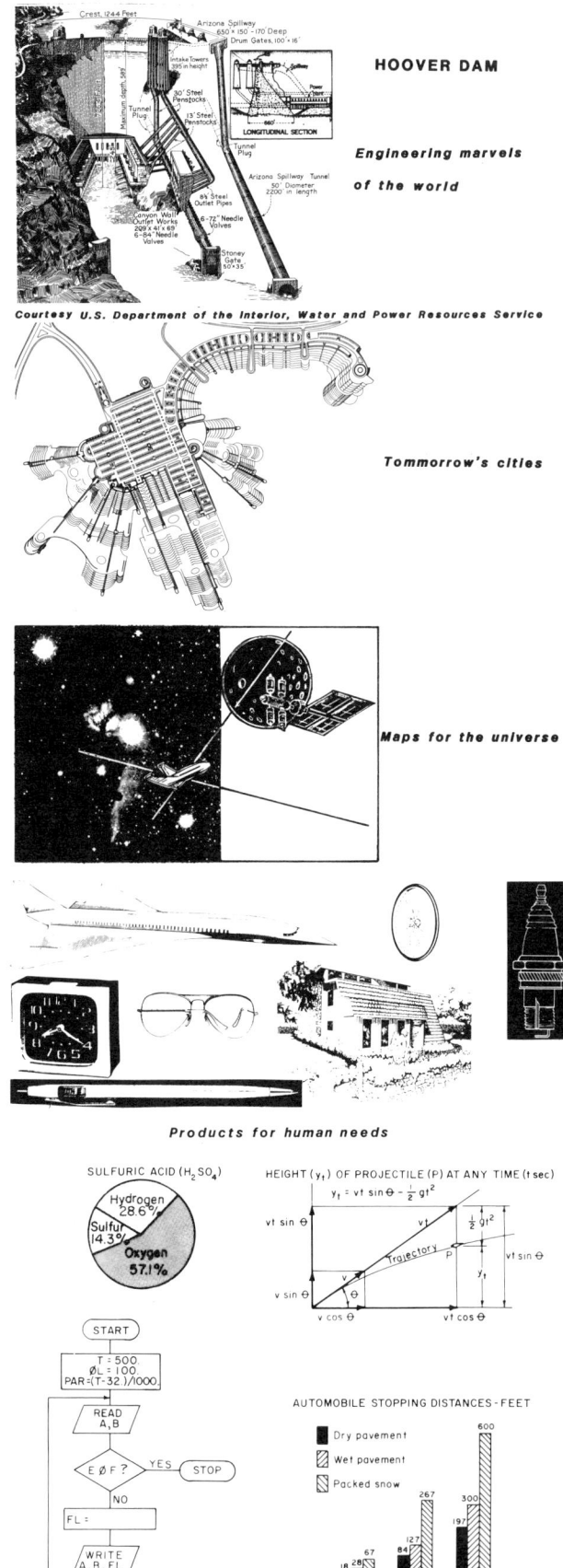

Fig. 1.2 Graphics encompasses all areas.

1.2 MAP CLASSIFICATION

A *map* is the graphic representation of all or a portion of the earth's surface or other celestial body, by means of signs and symbols or photographic imagery, at some given scale or projection. Maps may emphasize, generalize or omit certain items to meet specific needs. Basically maps can be classified as: (1) planimetric, (2) topographic, (3) thematic, (4) computer, and (5) photomaps. In each of these categories the scale, content, and requirements of the map may vary to allow for further classification. A word or phrase is usually employed to additionally describe the map or the type of information that is presented, e.g., property map, timberlands inventory, and land use map.

Planimetric maps. A *planimetric* map shows the correct horizontal position of natural or cultural features in plan only, Fig. 1.3. These features include water, land cover,

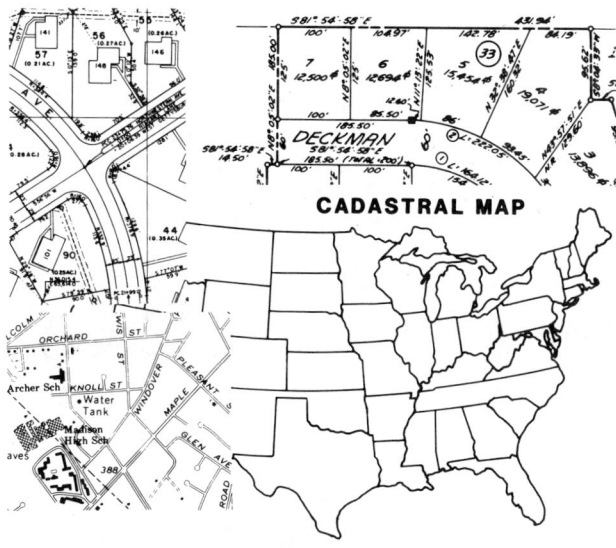

BASE MAP
Fig. 1.3 Planimetric maps.

transportation routes, buildings, utilities, and boundaries. Relief features may be shown in symbolic form but not in measurable form. Typical planimetric maps are the small scale maps of large regions that appear in atlasses. *Cadastral* maps are used to describe land and ownership. The boundaries, usually described with bearings and lengths, are given along with culture, drainage or other features relating to the value and use of the land. Other examples are base maps used for compiling various kinds of data and outline maps that typically show only boundaries and major water features.

Topographic maps. A *topographic* map shows the horizontal position of natural and cultural features (planimetry) along with the shape and elevation of the terrain, Fig. 1.4. The various methods that can be used to represent the terrain are discussed in chapter 17. Many special types of topographic maps are produced for planning engineering projects, recreational needs, and military requirements. Bathymetric maps are used to show water depths and underwater topography.

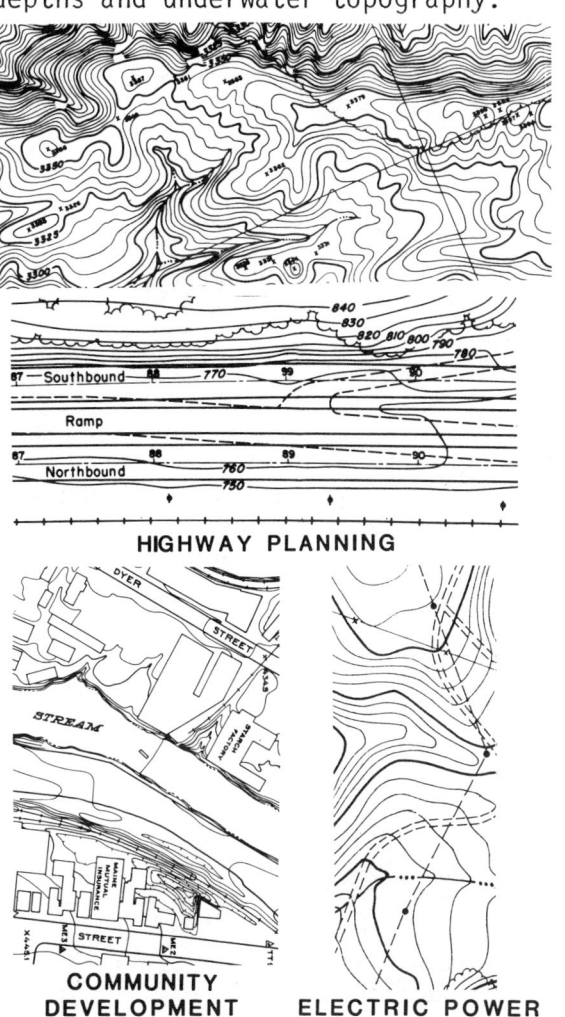

Fig. 1.4 Topographic maps.

Thematic maps. A *thematic* map, Fig. 1.5, is a special purpose or distribution type map that is used to emphasize a single topic such as population, meteorological data, land use, geology, crop production, and vegetation. The data is frequently in the form of statistical information and can be plotted in map form rather than graph form. Several methods for showing distributions include the use of choropleths, isopleths, and dots discussed in chapter 12.

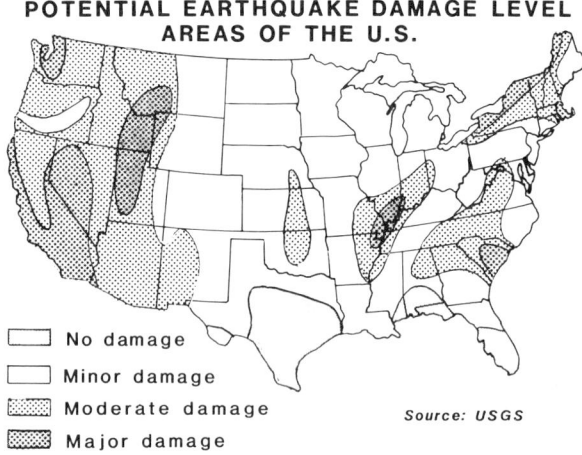

Fig. 1.5 Thematic map.

Computer maps. Computers are used to store cartographic information in digital form that can be processed and retrieved in graphic form using automated graphics systems, Fig. 1.6. Input to the computer for generating a map includes control survey data, required detail information scale, and contour interval. A program is necessary to direct the computer to solve for the position of points using the input data and plot the features. The main advantages of using a computer graphics system are speed, accuracy, the ability to select information to be presented, the ability to revise data more easily, and a finished map that is uniform and of high quality.

Photomaps. Any picture of the earth's surface in photographic form can be broadly considered a *photomap*, Fig. 1.7. The use of photographs as map substitutes obviously creates problems in that details require photo-interpretation, scale distortions occur, and all the information presented is frequently not required thus leading to confusion. Table 1.1 compares several factors related to conventional line maps and photographs.

The geometry of photographs, however, can be corrected to conform to that of a map as seen in orthographic projection. A corrected photograph used as a map is known as an *orthophotograph*. Orthophotographs can be further enhanced by modifying the image structure to improve the identification and classification of features. In addition, information in the form of contours, boundaries, and names can be overprinted on the photograph so as to more closely resemble a map.

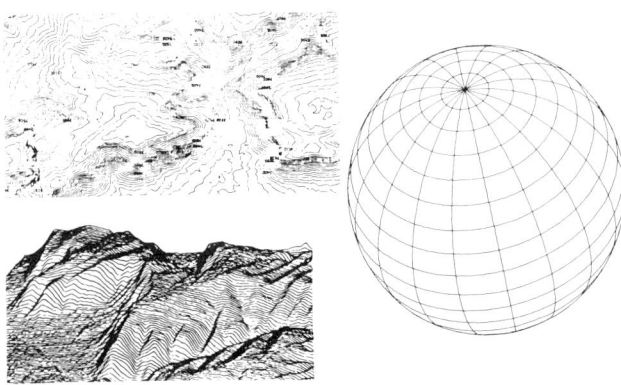

Fig. 1.6 Computer maps.

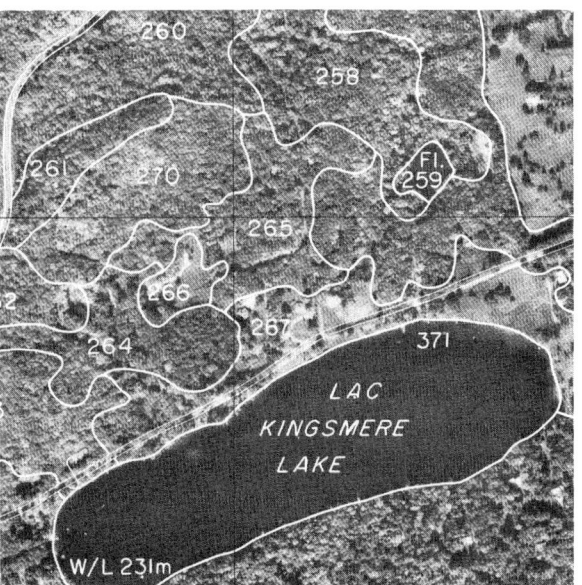

Fig. 1.7 Orthophoto map of Kingsmere Lake area in Gatineau Park, Quebec *(Courtesy Canadian Forestry Service).*

Table 1.1 Line map and photograph comparison.

LINE MAP	PHOTOGRAPH
Constant scale	Variable scale due to flying height and tilt displacements
Detail usually exaggerated and generalized	All detail shown (same ground areas may be hidden by vegetation or intervening terrain)
Names and classifications appear	No names and classifications
Relief may be shown	No relief
Information shown easy to interpret	Assessment of information requires experience in photointerpretation
Map is frequently out of date requiring changes and much time	Up-to-date photography fast and easy to obtain

1.3 CARTOGRAPHY AND MAP USE

The term *cartography* as defined by the International Cartographic Association is as follows:

"The art, science, and technology of making maps, together with their study as scientific documents and works of art. In this context maps may be regarded as including all types of maps, plans, charts and sections, three-dimensioned models and globes representing the earth or any heavenly body at any scale."

Every year new maps are published and distributed in a variety of forms. Individuals using these maps, in many instances, give very little thought to their preparation. Before the actual drawing can be started, considerable planning and synthesis must be initiated. Questions must be answered on what scale will it be?, what kinds of people will use the map?, what types of information will it include?, what size sheet of drawing media will be used?, and many others. After these questions have been reviewed, the production of a map involves a sequence of operations typified in Fig. 1.8.

These activities are carried out by cartographers, engineers, the cartographic drafter who physically draws the map, artists, lithographers, and printers. The work performed requires that individuals have a knowledge or be proficient in many areas that include geography, surveying, photogrammetry, mathematics, geomorphology, historial evolution, and cartographic techniques and production. Basic graphics must be understood and the necessary skills developed that go into making a map. A famous art critic has said that a real appreciation of a great work of art can be realized only if one knows and has some comprehension of the skills and techniques involved in the creation of a great painting. Maps are also the result of many specialized skills and precise techniques. A thorough understanding of map properties, their limitations, and interpretation is essential for evaluating, compiling, designing, and drawing maps using the latest technological methods available.

The educational importance of maps is recognized in subject areas such as literature, mathematics, for-

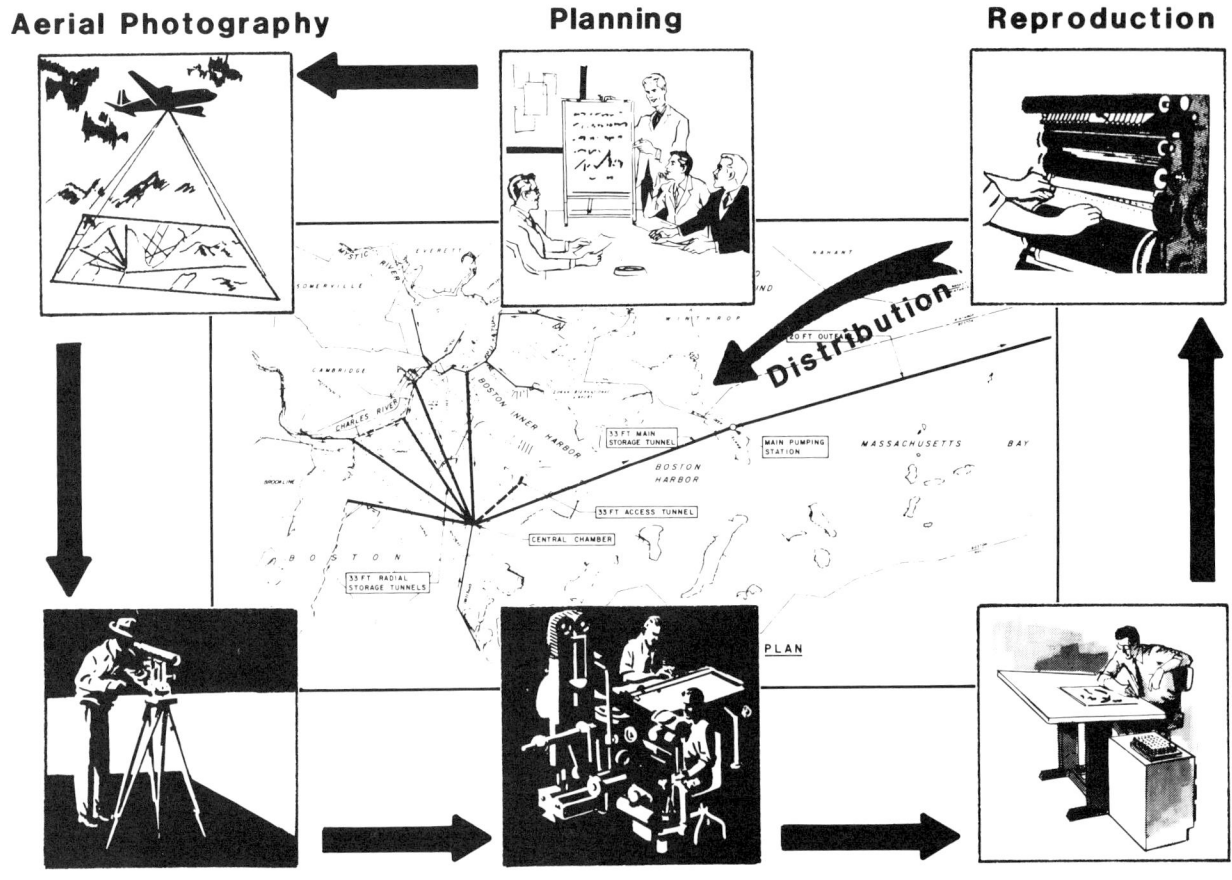

Fig. 1.8 The mapping process.

eign languages, history, business, and economics. At the same time, however, little academic time is devoted to learning about this subject area in the United States as compared to European schools. It is covered to a certain extent in program areas like geography, catography, and survey engineering. An exposure to understanding maps by all groups is greater now than ever before with the increasing pressure being placed upon the earth's land and water resources.

Map making is a dynamic ongoing activity. It provides the basic data for planning all types of essential services and form the essential base for the collection, recording, and investigation of hydrology, geology, soils, vegetation, etc. All of these facts and the results of various kinds of surveys are part of the required basic information for the planning, design and development of our natural resources which are rapidly being depleted. The demand for map products is staggering. For the year 1981 alone, the Defense Mapping Agency developed some 700 new maps and charts and updated 2100, as well as reprinting another 3400 to keep stockpiles current. Over 43 million copies of maps and charts were printed. In addition to this traditional production, digital data was developed for more than 5 million square miles on the face of the Earth. Other government agencies issue thousands of new maps annually. This does not include the thousands of maps published in books, newspapers, state, and private agencies. Since many maps rarely exceed a life span of five to seven years, cartographic information will continue to be mass produced to meet human needs. As one author writes, "a reason for the popularity of maps is that people are probably map conscious or perhaps lost."

Many different types of maps are necessary to meet the requirements of government, private enterprise, and individuals. A listing of the various uses of maps is endless and it is not possible to cover all of them. The following examples will serve to illustrate several ways accurate up-to-date map information is utilized.

Individuals engaged in various fields of conservation find maps indispensable. Complex problems in the management of public and private lands may be ascertained through the use of maps. Maps are used in the planning and development of roads, trails, buildings, and telephone lines. They are invaluable for planning fire prevention and fire control. Maps are useful for forestry cruising studies, type classification, drainage studies, marsh management, silvicultural work, volume estimation, evaluation of timber, and soil mapping.

In geology, maps form the most acceptable base for the construction of maps for both scientific and practical applications. Geologists find maps useful in showing volcanic and glacial areas, eroded regions, outcrops, rock formations, and resources. Maps are being used extensively for geophysical exploration, oil exploration, and production, geological interpretation, and mineral deposit studies.

Individuals engaged in town management find maps invaluable in the efficient planning of a community. Maps serve as a planning nucleus around which improvements, extensions, and other projects having to do with modern administration problems are conceived and executed. Services such as sanitary sewers, storm drainage, and street improvements, when based on modern maps, can be designed with both efficiency and the prevention of costly errors. In addition, maps are invaluable in the site planning of public buildings, recreational parks, private dwellings, zoning and assessing.

Engineers are concerned with a broad field of factual information which embraces all kinds of basic data concerning the land. Field operations can be cut to a minimum when properly prepared maps are available for location and construction pruposes. The final location may be drawn directly on a topographic map, the accurate contour intervals of which then permit immediate preparation of profiles and cross-sections. The extent of the necessary cuts and fills and the required allowance for watershed area may be quickly determined. Maps are part of the required basic information for the planning, design and construction of engineering projects such as highways, airports, water supply, irrigation, dams, pipelines, airport locations, defense location studies, traffic studies, river and harbor development, and industrial area studies.

Maps are also required and used by individuals in other activities and disciplines indicated in Table 1.2.

Table 1.2 Typical map uses.

Surveying	Mass communication
Education	Social Sciences
Geography	Aeronautical charts
Biology	Aerial photography
Irrigation	Geophysics
Zoning	Industry
Graphic arts	Military

1.4 HISTORICAL EVOLUTION OF MAPPING

The history of map making can be traced to every culture of the world. Recorded maps of various kinds provide a chronicle of civilizations achievements, travels, and explorations. The maps vary in form from crude maps carved on stone weapons, utensils, and the walls of caves to the accurate sophisticated maps produced today, Fig. 1.9. Primitive peoples such as the Eskimos have drawn maps of their own coasts with many of them considered superior to those of the same regions made by white men. Crude maps drawn on the ground or a hide were used by the American Indian. The aborigines of Australia used stones and sand as geographical symbols in their initiations ceremonies. Marshall Islanders showed wind and wave patterns for navigational purposes using stick charts fastened together with fibers, and shells or

Marshall Islands stick charts (Reprinted by permission of the Royal Geographical Society, London, England).

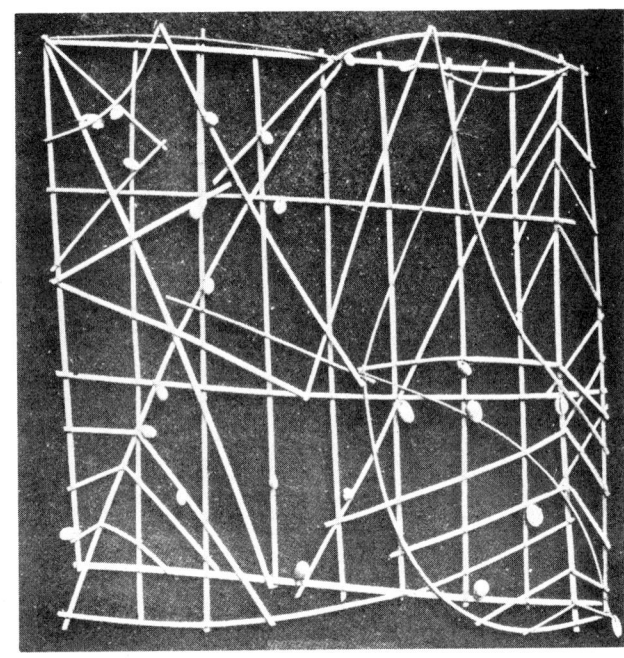

Fig. 1.9 Primitive maps.

coral inserted to represent islands. The decorative maps prepared by the Aztecs emphasized the recording of historic events rather than details of topography.

Innumerable accounts are cited in history books on the use of maps for travel and exploration of the entire cosmos. In the thirteenth century Marco Polo of Italy journeyed to China and back; Christopher Columbus, known as the discoverer of America, crossed the Atlantic in 1492; Robert E. Peary made several trips to the Artic and reached the pole in 1909; and Richard E. Byrd explored Anartica in the 1930's and 1940's. The first crossing of North American was made across Canada by Alexander MacKenzie, who reached the Pacific in 1793. Further development in the United States Northwest was made possible by the journey of Lewis and Clark in 1804-06. Extra-terrestrial travel became a reality when on July 20, 1969 the Apollo 11 landing craft, Fig. 1.10, landed on the Moon with Neil Armstrong announcing "Houston, Tranquility Base here. The Eagle has landed."

A detailed historical review of cartography, as a field of study has been well documented by other writers. Only a brief outline is presented in the following sections to allow the reader to savor this exciting and interesting topic. For the reader interested in this aspect of our proud heritage, the bibliography lists additional reading. Table 1.3 summarizes several mapping milestones that have had profound effects on the field of mapping.

Fig. 1.10 Apollo 11 on the moon (Courtesy NASA).

Table 1.3 Mapping milestones.

Time	Event
2500 B.C.	Circle divided into 360° by Babylonians.
6th Century B.C.	Spherical shape of the earth first described by Phythagoras the Ionian (Greek).
2nd Century B.C.	Earth's circumference measured by Eratosthenes of Cyrene, Greece.
105 A.D.	Invention and first use of paper by the Chinese.
1448	Moveable type developed by Johann Gutenberg, Germany.
	Survey tools and techniques developed.
1571	• Theodolite by Leonard Digges, England
1617	• Triangulation, Dutch
1620	• Surveyor's chain by Edmund Gunther, England
About 1700	• Spirit or differential leveling
1765	• Chronometer, England
	Important projections developed.
1569	• Mercator by Gerard Mercator, Dutch
1770's	• Transverse Mercator and Lambert's Conformal Conic by Johann H. Lambert, France
1805	• Alber's Conic Equal Area by H. C. Albers
1820	• Polyconic by Ferdinand R. Hassler, U.S.
1900's	• Miller Cylindrical by Osborn M. Miller, U.S.
18th Century	Photography developed.
1805	Lithography developed.
1875	Representatives of 20 nations met in Paris to adopt the convention of the metre.
1879	U.S. Geological Survey established.
1884	First international meridian conference held in Washington, D.C. to adopt the Greenwich, England longitude as the prime meridian.
19th Century	Color-separation developed.
1903	First offset press developed.
20th Century	Aerial photography mushrooms prior and during World War II. Basic principles of stereoscopic instruments worked out by Karl Pulfrich, a German inventor and Erich von Orel, an Austrian.
1947	Geodimeter developed by Erik Bergstrand, a Swedish scientist.
1950's	Scribing developed.
1956	Orthophoto instrument invented by Russell Bean, USGS.
1960's	Remote sensing, satellite, and laser technology developed.
1970's and 1980's	Computer generated maps make their impact. Spacecraft development and extraterrestrial mapping.

Early maps. The oldest known map in existence is probably of Babylonian origin from around 2500 B.C., Fig. 1.11 (a). The earth is shown on a clay tablet as a flat circular disk surrounded by ocean and several mythical islands. Early maps of this period portrayed items concerning basic topography and philosophical theories of the nature of the world. Other clay tablets in the form of cadastral maps have also been discovered in ancient Mesopotamia, Fig. 1.11 (b). Other than the great antiquity of the map-making art, the principal Babylonian contribution to cartography was the introduction of the 360-degree circle, with 60-minute degrees, and 60-second minutes.

 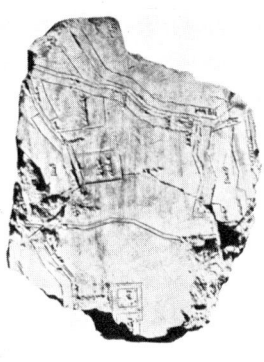

(a) World map (b) City plan

Fig. 1.11 Early Mesopotamia clay tablet maps *(Reprinted by permission Antiquity Publications Ltd., Cambridge, England)*.

The primary reasons for making maps in ancient times were for taxation, travel or exploration, and military operations. It is interesting to note that maps produced today continue to meet these major needs. Early mapmakers held to several theories concerning the origin of the earth. Homer thought the earth was a flat disc surrounded by a constantly moving ocean river. Another early philosopher described the earth as a flat rectangle, buoyed up in the sky by air that blew from beneath it. Still others believed the earth was shaped like a cylinder and that it was suspended in the heavens.

The Greeks. The true shape of the earth came first from Pythagoras the Greek who founded a school of philosophy in Crotona about 523 B.C. The Greeks continued to add to our present system of cartography through the work of Aristotle (384-322 B.C.) who is credited with the founding of scientific geography and demonstrated the theory of a spherical earth with its poles, equator, and tropics. The first real attempt to determine the circumference of the earth was undertaken a century later by Erastosthenes (276-196 B.C.), keeper of the library at Alexandria, Egypt. Early pioneering in the application of mathematics to cartography was largely due to the mathematician and astronomer Hipparchus (160-120 B.C.). Hipparchus aplied astronomic methods to mark the position of places on the earth's surface and devised the stereographic and orthographic projections for maps. The height of Greek cartography and geography reached its zenith with Ptolemy (87-150 A.D.) who contributed an outstanding piece of work in Geographia, Fig. 1.12. This is an eight volume set of books summarizing all geographic learning and the principles for the scientific construction of maps.

The Romans and the Dark Ages. Geography and cartography development declined sharply at the time the Romans were a great empire. The Romans were not interested in Greek speculations about the size and shape of the earth and reverted to maps primarily for military and administrative purposes. A further decline in mapping occurred during the period Europe was in the Dark Ages (500-1000 A.D.). Contributions to scientific map making continued to decline also during the Middle Ages (1000-1400 A.D.) since the official church doctrine held to the belief that the world was flat surrounded by the world ocean and divided into the three continents of Europe, Asia, and Africa.

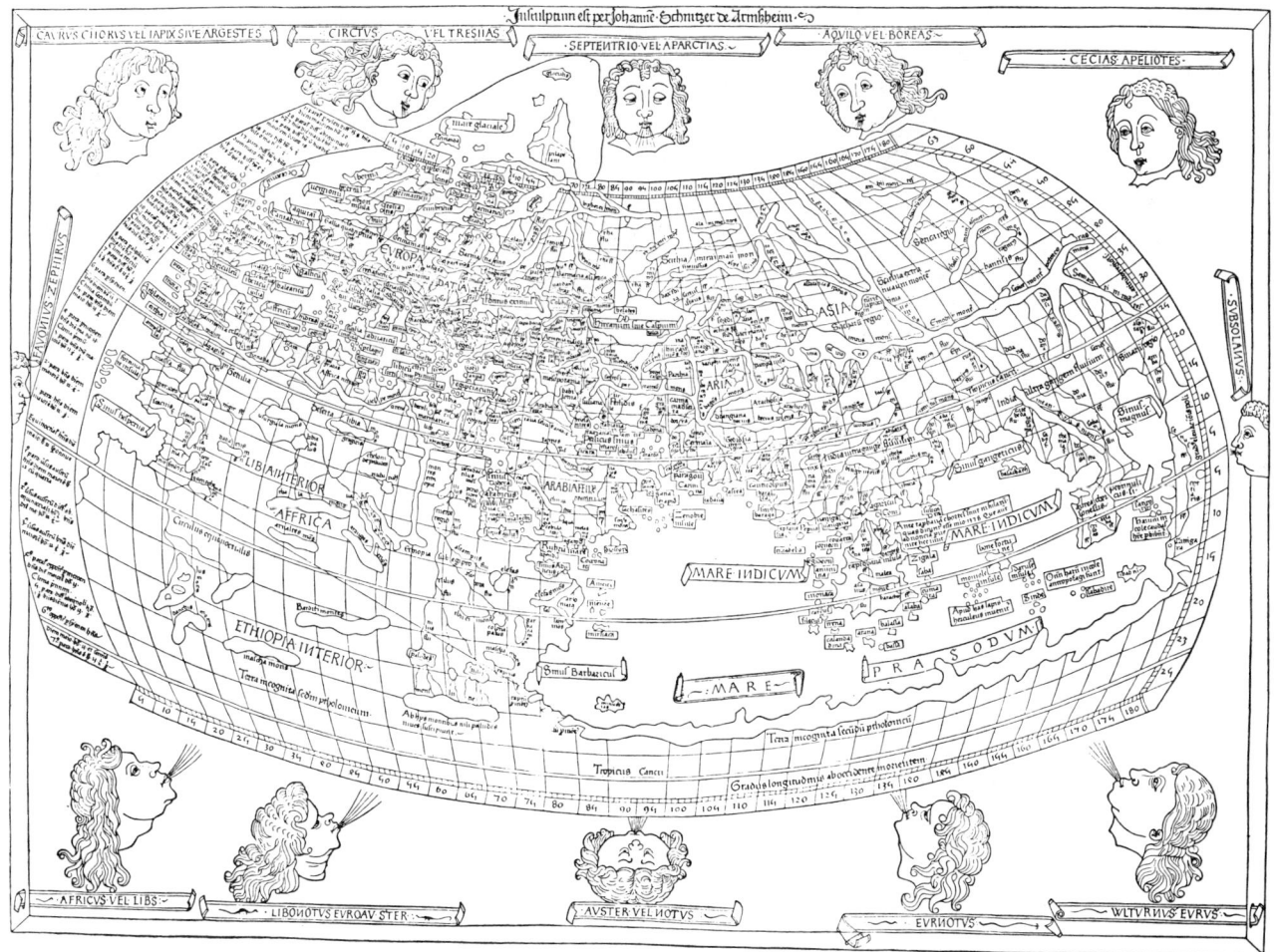

Fig. 1.12 The World, in Ptolemy's *Geographia* printed at Ulm, Germany in 1482 (Reprinted from *Decorative Printed Maps of the 15th to 18th Centuries*, R.A. Skelton).

The Renaissance and the Dutch. A rebirth of scientific cartography occurred during the Renaissance Period (approximately 1400-1700 A.D.). The rapidly increasing interest in cartography led to the establishment of many businesses publishing maps, atlases, and globes. From the sixteenth to the eighteenth century, the centers of cartographic research were the cities of Amsterdam, Antwerp, and Venice. Dutch cartography flourished in the sixteenth century and many high grade maps were produced. The father of Dutch cartography was Geradus Mercator who is best known for the Mercator projection system (1569) which is still in wide use today. In 1570 Abraham Ortelius from his workshop in Antwerp published the first modern atlas of the world, Theatrum Orbis Terrarum (Theater of the World), Fig. 1.13. Many other important map projections that were developed through the years include: Johann Heinrich Lambert, an eighteenth century Alsatian (French) mathematician, physicist, and astronomer developed formulas for a half-dozen new projections that comprise the Transverse Mercator and Lambert's Conformal Conic. H.C. Albers of Germany in 1805 developed Alber's Conic Equal - Area projection. Ferdinand B. Hassler in 1820, the organizer and first superintendent of the old Coast and Geodetic Survey, developed Polyconic projection that is used for many United States maps.

The invention of printing, by Alois Senelder (about 1805), and engraving greatly stimulated the progress of cartography. Prior to this time, all maps were laboriously drawn by hand and their expense limited their use.

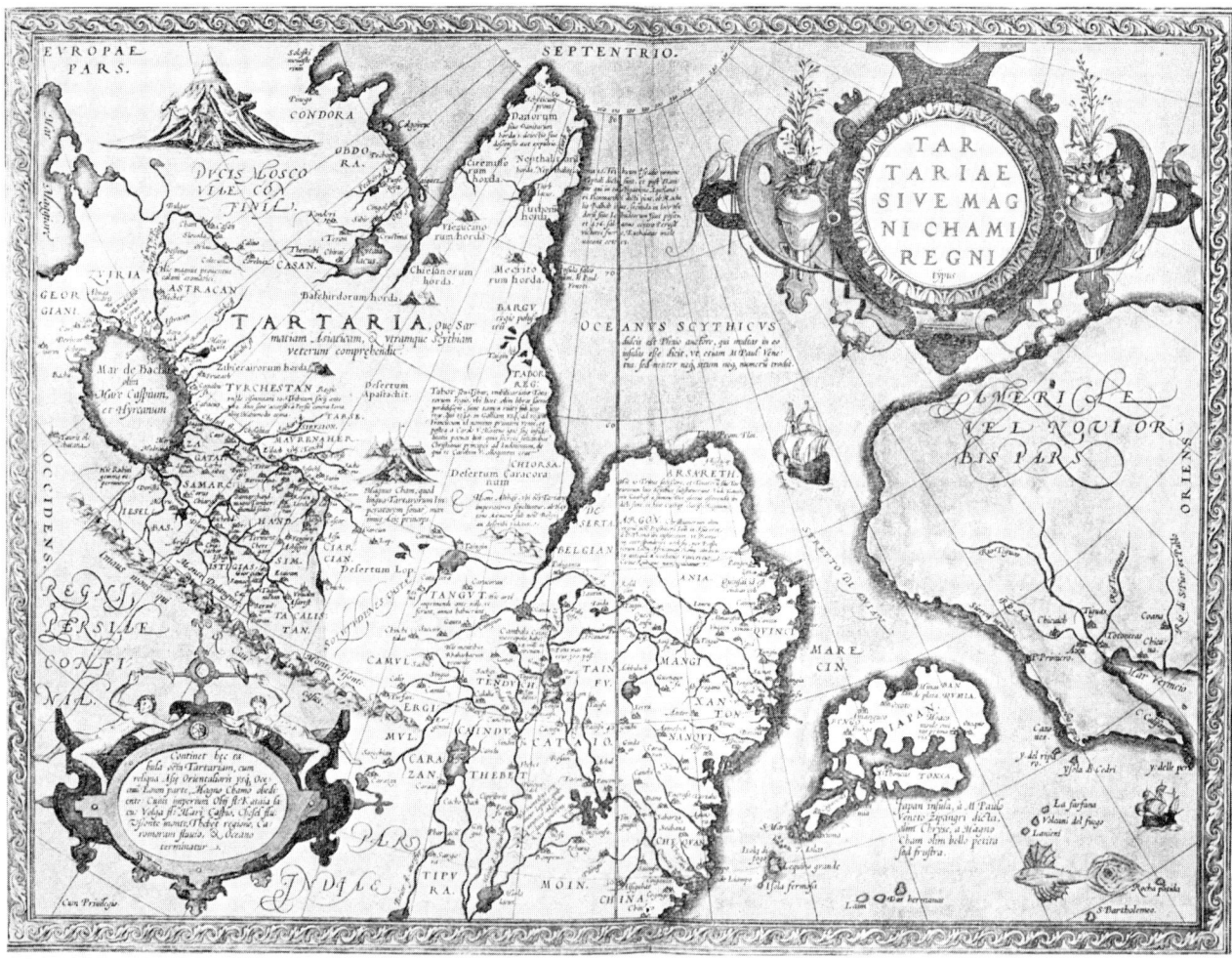

Fig. 1.13 Tartary, or the Kingdom of the Great Khan in Ortelius's *Theatrum*, 1570 (Reprinted from *Decorative Printed Maps of the 15th to 18th Centuries*, R.A. Skelton).

<u>The French</u>. The fundamental changes between the seventeenth century Dutch school and the eighteenth century French school of cartography are most notable. The Dutch decorated the empty spaces on the map with swash lines, animals, products of the country, and used decorative frames called cartouches to enclose titles, scales, and other descriptive material. Reformation in cartography led to a more scientific accuracy stressed in maps produced. William Blaeu (1571-1638), a master surveyor, globe maker and publisher, produced many high quality products. The Blaeu <u>Atlas Major</u> in twelve volumes is considered one of the most outstanding geographical works ever published. Leadership in many areas was provided by Nicholas Sanson (1600-1667) and Alexis Jaillot (1632-1712) who published hundreds of maps and atlases. Scientists such as physicist Christian Huygens and astronomers Jean Picard and Jean Cassini worked on methods to determine longitude on land.

Remarkable advances were also made by England, Germany, Italy, Russia, and the Scandinavian countries. Part of this was due to the rise of the great powers of Europe amidst almost constant large-scale warfare. An English contribution to cartography is the work of Edmond Halley in 1866, the originator of meterological and magnetic charts.

<u>The Americans</u>. The name America probably appeared for the first time on the maps of Martin Waldseemüler who suggested the name for the New World in honor of Amerigo Vespucci who explored and named the Venezuelan

coast in 1499. As further explorations of the east coast of the New World were searched and settled many maps began to appear. Several of these early maps include: a map of Cape Hatteras by John White in 1585, a map of New England by Captain John Smith in 1614, Fig. 1.14: and a map of Old and New Virginia in Sir Robert Dudley's Dell'Arcano del Mare, 1646-47, Fig. 1.15. Other important map makers, surveyors, and explorers were Joshua Fry, Augustine Hermann, George Washington, Thomas Jefferson, and John Mitchell.

American cartography during the early growth of the country was regarded primarily as a task of the Army and Navy. Mapping is still a big occupation in the military and is currently carried out under one command by the Defense Mapping Agency. After the establishment of the new republic, a slow emancipation of American cartography from European influence emerged. Early explorations and surveys of the West were carried out under four Federal territorial surveys authorized by Congress. These were known as the King, Hayden, Powell, and Wheeler Surveys, each named after its leader. The westward expansion in the early 1800's led to a great demand for maps. Organized state surveys were started and the beginning of private cartography was noted. Philadelphia, Boston, New York, Hartford, and Baltimore were the centers of map making for this period.

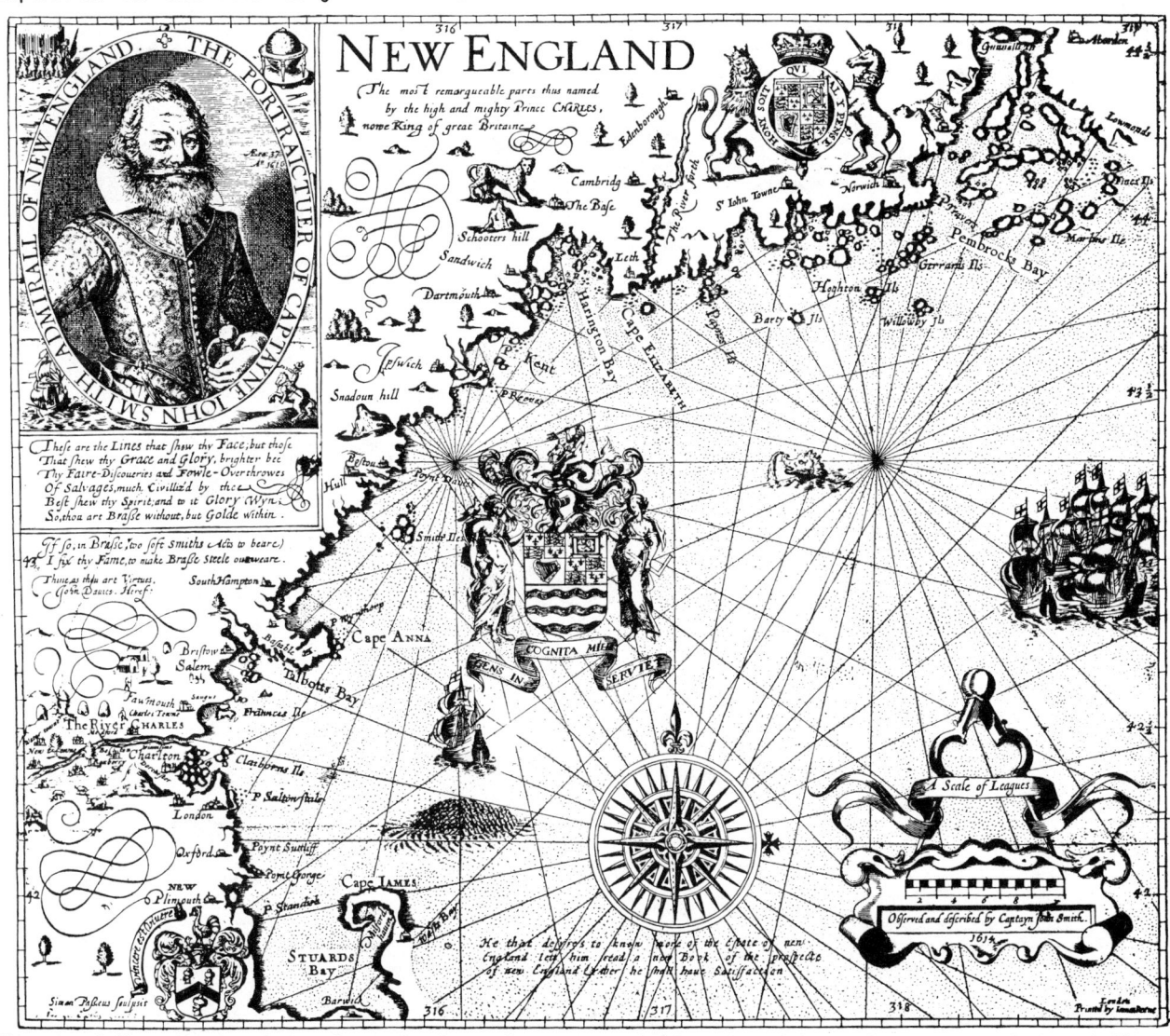

Fig. 1.14 New England, by Captain John Smith, 1614 (Reprinted from *Decorative Printed Maps of the 15th to 18th Centuries*, R.A. Skelton).

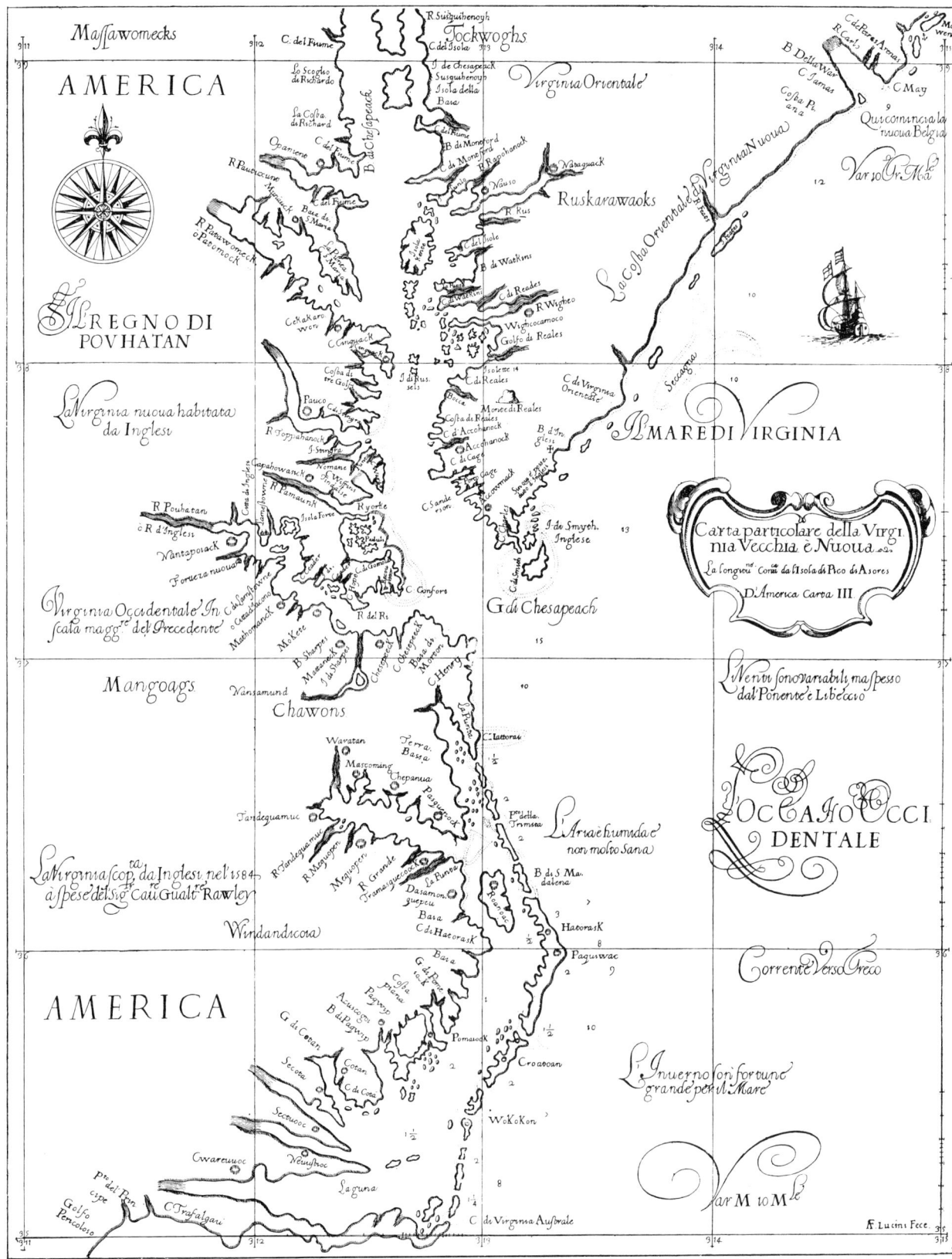

Fig. 1.15 Old and New Virginia, in Sir Robert Dudley's *Dell'Arcano del Mare*, 1646-47 (Reprinted from *Decorative Printed Maps of the 15th to 18th Centuries*, R.A. Skelton).

The National Ocean Survey (formerly the U.S. Coast and Geodetic Survey) came into existence in 1807 with responsibility for providing maps, coastal charts, and geodetic surveys, Fig. 1.16. The American Geographical Society, a national association, was founded in 1852 and produces a variety of maps and related materials. The National Geographical Society was incorporated in 1888 and supports exploration and research projects. Many outstanding maps appear in its journal, books, and atlases.

During the middle of the nineteenth century enormous tracts of land were acquired by the United States. The acquisition of huge territories included the Louisiana Purchase of 1803 and the buying of Alaska in 1867. Almost a third of the nation, nearly 740 million acres make-up the public domain managed by various agencies and bureaus in the Federal government. The survey of the public lands has been conducted since 1785 under several Congressional Acts that placed the responsibility for this work to be carried out. The Bureau of Land Management, established in 1946 in the Department of the Interior, is responsible for supervision of surveying the public lands through State and service center directors.

In 1879 the U.S. Geological Survey, the nations largest earth science research agency, came into existence. The USGS is responsible for producing maps of the lands of the national domain for general information and map series upon which professionals can base and plan activities. In 1882, in cooperation with the various states, the USGS began producing a series of topographic maps of the United States in quadrangle form. The principal map series and their essential characteristics are given in Table 1.4. These well-known maps furnish a vast field of information valuable to the map user for carrying out all kinds of detailed or preliminary studies. The prevailing scale until the 1950's was 1:62 500 used on the 15-minute (15 minutes of latitude and 15 minutes of longitude) maps, Fig. 1.17. With an increased need for greater accuracy and detail, the scale shifted to 1:24 000 that is used on the 7.5-minutes (7.5-minutes of latitude and 7.5 minutes of longitude) maps. The area covered on each sheet varies since the meridians converge at the poles.

Worldwide cartography. It has only been during the last two hundred years that the expansion of the science of cartography has been so phenomenal. During the nineteenth century, the period of the industrial revolution, map making worldwide was affected in many ways with the development of lithography, photoengraving, wax-engraving, and color printing. The middle of the nineteenth century has seen most nations completing either total or partial surveys of the land. The early twentieth century witnessed further technological developments particularly in transportation which created a new demand for maps. Since World War II the increased capacity and demand for cartographic information has in all probability seen more maps produced and distributed than in all the years prior to this time.

The accuracy and rapidity for obtaining data has increased with the development of aerial photography and computers. Radar has allowed inaccessible cloud-covered regions like the Amazon in Brazil, the rain forests of Guatemala, and New Guinea to be mapped. The seafloor has been mapped using sonar and airborne radio echo sounding has been used for sublacial maps at the North and South Poles. Underground mapping in search of natural resources is being carried out with instruments like the magnetometer and seismic methods. Remote-sensing of the earth's resources from space satellites is being used by land-use planners, cartographers, and agriculturists.

The political turmoil in the world has created a cartographers paradise. Political boundaries have changed in many areas and others are still in question. Cities and towns are changing rapidly and extensive mapping programs will always be needed.

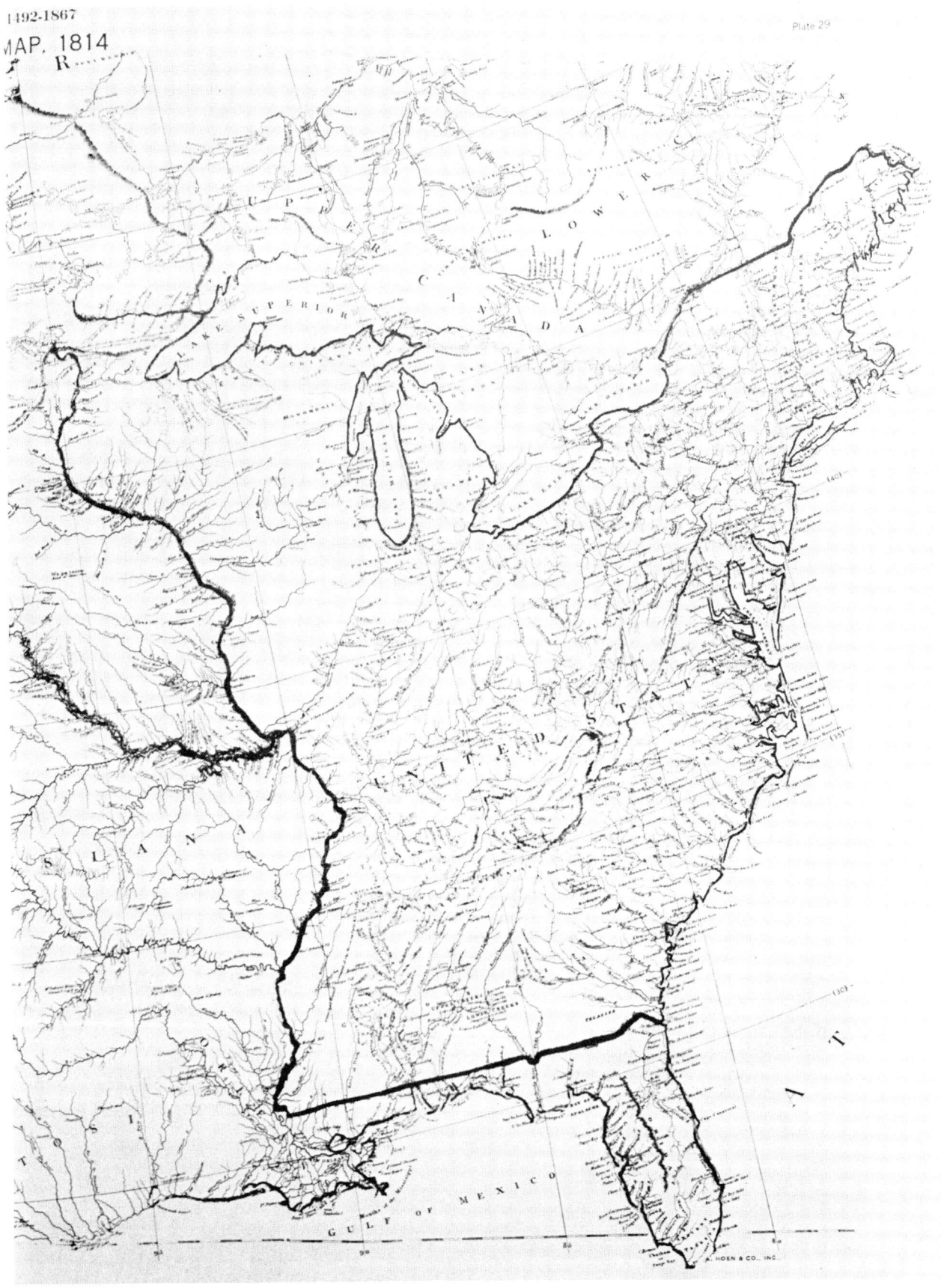

Fig. 1.16 A map exhibiting all the new discoveries in the interior part of North America by Aaron Arrowsmith, 1814 *(Courtesy Carnegie Institution of Washington)*.

Table 1.4 USGS Topographic map series.*

SERIES	SCALE		SIZE (latitude and longitude)	AREA		TYPICAL APPLICATION
	1 cm represents	1 inch represents		Square km	Square miles	
7.5-minute	1:24 000 0.24 km	2,000 feet	7.5x7.5 minutes	127-184	49-71	Detailed information and planning.
15-minute	1:62 500 0.625 km	1 mile approximately	15 x 15 minutes	510-730	197-282	
Intermediate Scale	1:100 000 1.0 km	8,333 feet	30 minutes x 1 degree	2964-5610	1,145-2,167	Land management and planning.
U.S.	1:250 000 2.5 km	4 miles approximately	1 x 2 degrees	11 857-22 443	4,580-8669	Comprehensive views of extensive projects and regional planning.
International map of the world	1:1 000 000 10.0 km	16 miles approximately	4 x 6 degrees	190 888-266 031	73,934-102,759	

*Some maps of Alaska, Hawaii, and Puerto Rico vary from these standards.

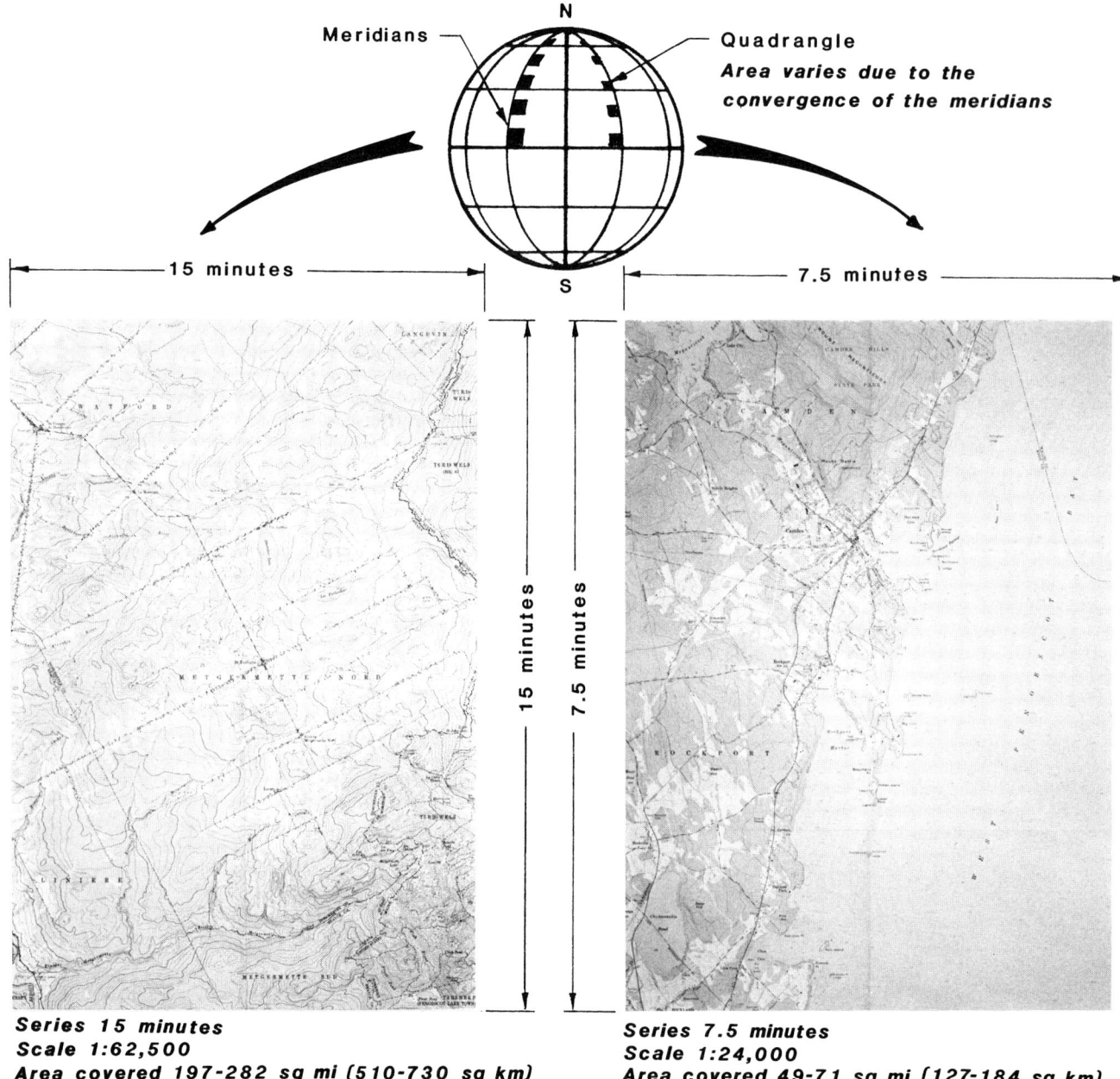

Fig. 1.17 USGS quadrangle sheets.

The flight into space has opened up new horizons for cartographers and thousands of spectacular photographs are now available, Fig. 1.18. This new technology has allowed a more precise knowledge of the size and shape of the earth to be found. The program has allowed the positions of Alaska and the Aleutian Islands to be precisely located. More accurate information of areas in Asia, Australia, Europe, and South America has also been obtained.

Extraterrestrial mapping has in an incredibly short time passed everyone's expectations. The outstanding maps of the moons lunar surface, Fig. 1.19 (a), were made possible by unmanned spacecraft that transmitted pictures to Earth. Maps have been produced of Mars, Fig. 1.19 (b), and valuable information is currently being gathered on other extraterrestrial bodies such as Jupiter, Saturn, and its moons. All of these exploits are exciting and unbelievable at times. However, cartographers can be assured that new problems will be encountered and that the future of maps is unlimited.

Earth Apollo-15

Gulf of Mexico and Florida

Earth and Mediterranean

U.S. East Coast

Fig. 1.18 Space photographs *(Courtesy NASA).*

(a) Moon

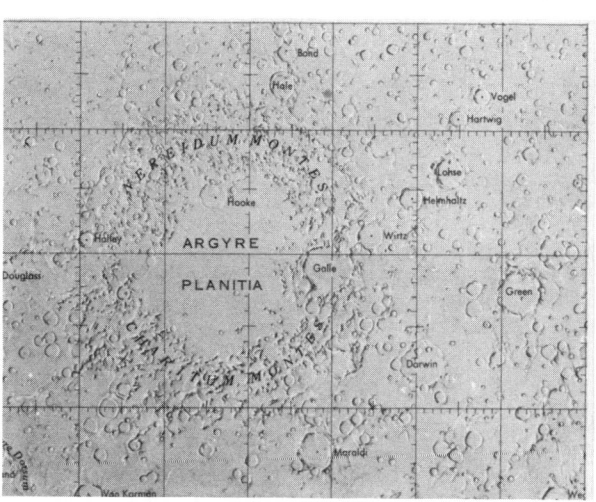
(b) Portion of Mars Argyre area

Fig. 1.19 Extraterrestrial maps *(Courtesy USGS).*

1.5 FUTURE MAPPING TRENDS

The techniques for surveying and mapping have changed with new technology and new scientific tools to meet changing requirements for map data. Several of these new tools include computers, automation techniques, digital applications, electronic distance measurement procedures, inertial navigation systems, remote-sensing technology, and applications of space science. Accurate maps will be produced utilizing sensors and all forms of photogrammetric data without further field work. The control for these maps will be established by, and referenced to, space satellite stations following a specific orbit. Ground orientation will take place by simultaneous satellite and ground observation electronically relayed to a computer by micro-wave transmission. Improvements in systems of this kind will allow for simultaneous up-to-date orthophotographs, digital terrain models, and other maps to be produced. Systems will be improved for theme extraction, automatic classification, and derivation of specialized information by manipulation of images from various sensors. Specific information of general interest and use will also be created on TV screens in the home, on display consoles in the automobile, and personal mini-computer systems.

Other cartographic changes and trends discussed by M. M. Thompson, 1979 include:
1. Photoimage or sensor-image forms of maps instead of conventional line and symbol maps will become more prevalent. The user will need to become more adept in photo-interpretation.
2. Additional cartographic products in graphic, digital or image form will need to be developed.
3. Standards and map contents including symbols will be modified to make them more compatible with a completely automated mapping process.
4. Automated techniques will increasingly be used for updating map contents.
5. A universally accepted map reference system or rectangular coordinates will eventually be used to express survey and map data. The United States Geological Survey (USGS) encourages use or the Universal Transverse Mercator grid (UTM) as the basic reference for use with the products of the National Mapping Program.
6. Complete conversion of map products to the metric system will occur to avoid being considered non-standard and outdated. The USGS in 1977 adopted a policy for the systematic conversion to the metric system as part of the National Mapping Program.
7. Central data banks containing all kinds of cartographic information in both digital and graphic form will be established to meet individual, private, state and government needs.
8. Map making from space will become more prevalent.

1.6 NATIONAL MAPPING PROGRAM

The drawing of maps is continually changing and new needs and uses have placed an increased demand for new kinds of maps to meet the requirements for the country. To help coordinate Federal mapping activities to better meet the basic cartographic data needs for the country, a National Mapping Program was adopted in 1975. The U.S. Geological Survey was named by the Department of the Interior as the Federal agency to administer the program. In this capacity the Geological Survey coordinates, defines and approves the program categories to be incorporated into the National Mapping Program.

GRAPHIC INSTRUMENTS EQUIPMENT AND THEIR USE

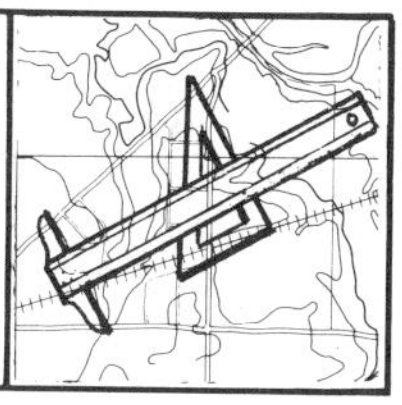

2.1 INTRODUCTION

An understanding of the basic equipment used for graphics is useful in order to produce a neat, legible drawing. An effective finished drawing will depend on the skill, clarity, and overall quality of the work performed by the individual with the equipment. The equipment purchased should be of good quality and will last years if given proper care. For most purposes the basic equipment, Fig. 2.1, should include:

Additional equipment is available and may be necessary for some types of work. In the field of cartography, e.g., a number of time-saving materials and instruments have been developed. The student should be familiar with these items and their proper use in order to attain professional quality and accuracy in this type of work.

2.2 DRAWING SETS

Drawing sets may contain numerous items, Fig. 2.2, and are manufactured by several firms. The average individual, however, seldom has need for a set with many instruments. A compass and a pair of dividers will normally be all that is required for most drawings.

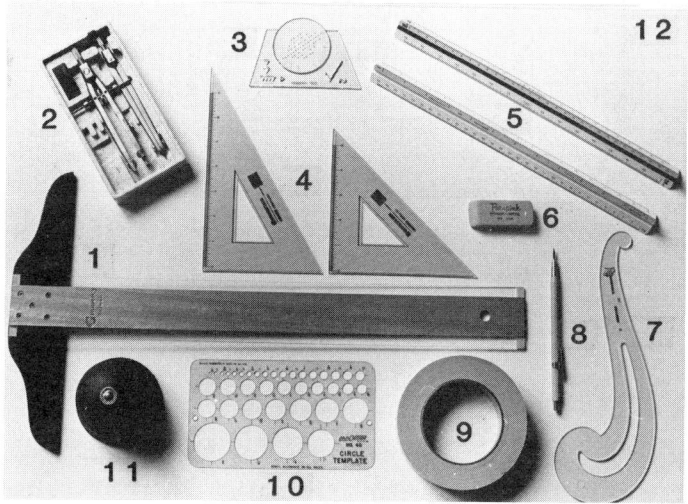

Fig. 2.1 Basic graphic equipment.

1. T-square
2. Compass and divider set
3. Lettering guide
4. Triangles- 30°- 60° and 45°
5. Scales- English and metric
6. Eraser
7. Irregular curve
8. Lead holder with H and 2H lead
9. Drafting tape
10. Template- circle or general use
11. Lead pointer- sandpaper pad or file
12. Drawing board or table

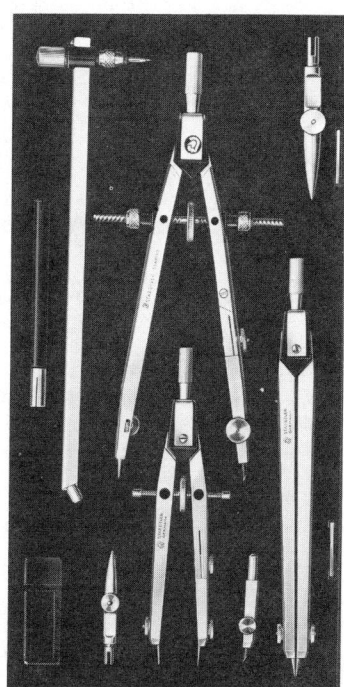

Fig 2.2 Typical drawing set *(Courtesy J.S. Staedler, Inc).*

2.3 COMPASSES

The *compass* is used for drawing circles and circular arcs. A large 15 cm (6"), rigid type compass, Fig. 2.3 (a) and (b), is recommended for general use. Since most drawings are in pencil, friction head compasses, Fig. 2.3 (c), are not recommended. To obtain dense black lines considerable pressure must be applied to the pencil lead and the legs tend to spread apart with compasses of this type.

To draw a circle or an arc, the procedure in Fig. 2.4 should be followed. For accuracy it is recommended that scale measurements be laid off on the drawing and not taken directly from the scale with the compass.

Circles up to approximately 12-15 cm (5-6") diameter may be drawn with a large compass. To increase the diameter of a circle an *extension bar* insert, Fig. 2.3 (a), is used. For small circles great care must be exercised in adjusting the shoulder needle point so that it projects about 1 mm (1/64") beyond the pencil lead, Fig. 2.5. Excessive pressure on the needle point should be avoided or a large hole will be drilled in the paper. Small circles may usually be drawn more easily with a template (see section 2.15) than with a compass.

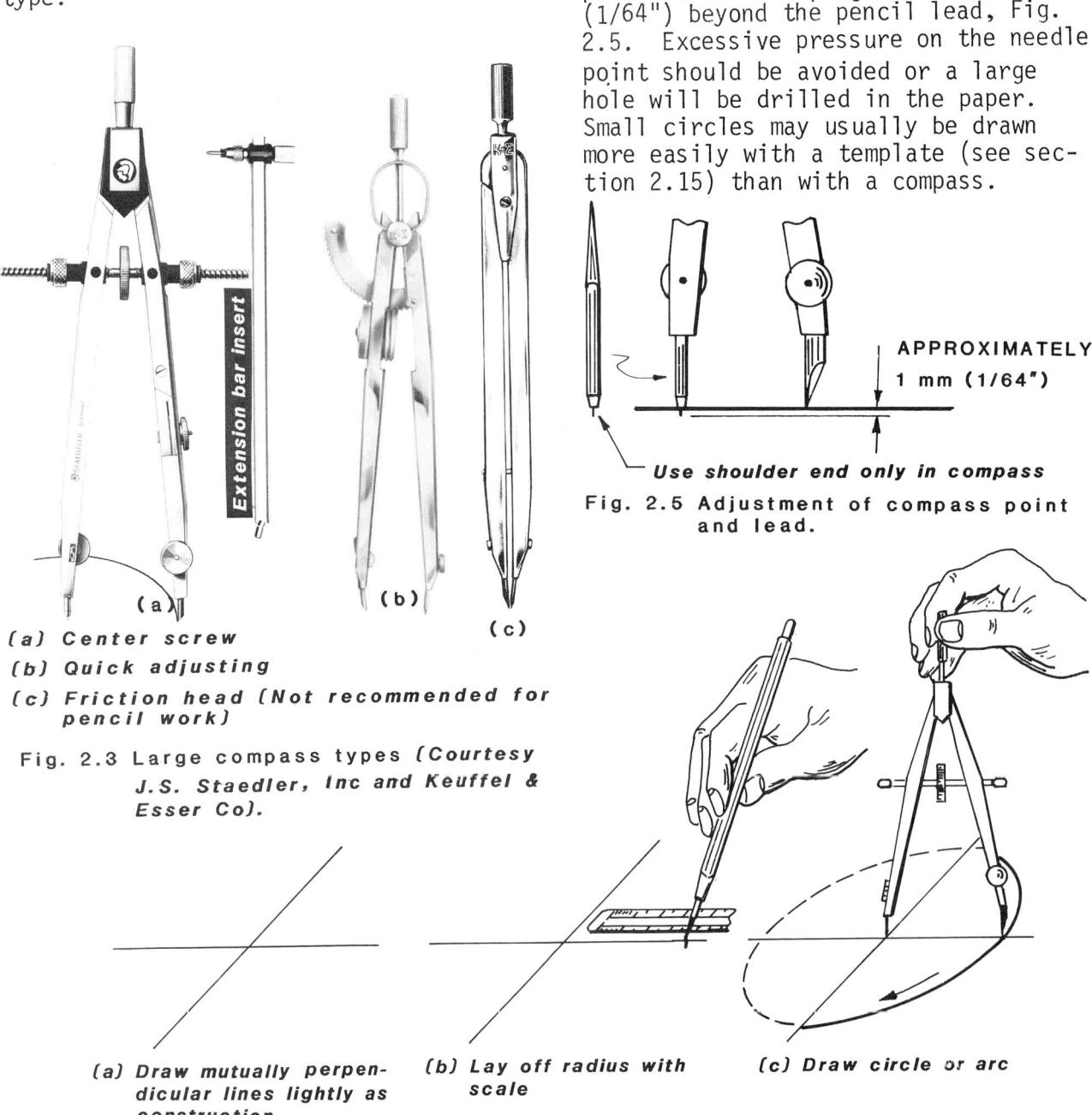

(a) Center screw
(b) Quick adjusting
(c) Friction head (Not recommended for pencil work)

Fig. 2.3 Large compass types (Courtesy J.S. Staedler, Inc and Keuffel & Esser Co).

Fig. 2.5 Adjustment of compass point and lead.

(a) Draw mutually perpendicular lines lightly as construction
(b) Lay off radius with scale
(c) Draw circle or arc

Fig. 2.4 Layout procedure for drawing circles or arcs.

A knee joint, Fig. 2.6, in one or both legs of the compass is helpful in drawing large circles, especially in ink. The legs should be bent so that they are always perpendicular to the paper. For inking work a compass attachment and resevoir pen is usually used, see section 2.16.

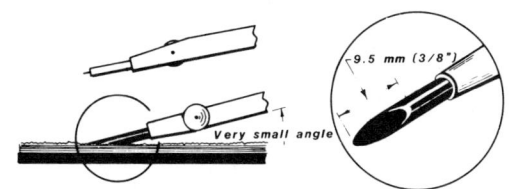

Fig. 2.7 Sharpening the compass lead.

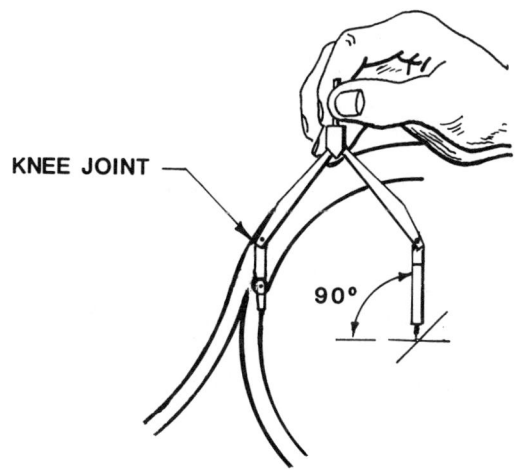

Fig. 2.6 Drawing with compass knee joint.

The lead used in the compass should be on the *soft* side (H) and should be properly sharpened at all times. This may be accomplished by rubbing the lead at a very small angle to the file or sand paper pad, Fig. 2.7. The lead should be sharpened flat to about 9.5 mm (3/8") and may be beveled slightly on the sides.

Slidecomp compass. The *slide-comp compass*, Fig. 2.8 is unique in that it allows circles and arcs to be drawn accurately and more rapidly than with a standard compass and scale combination. A sliding scale allows measurements to be made in one of three scales; fractional, decimal inch, and metric. The sliding scale may be interchanged with longer scales of the same type or with custom scales for cartography or other work. As the rule passes under a hairline index attached to the body of the compass, the user directly reads a radius setting in increments of 1/16 inch, 0.050 inch, or 1 mm.

For inking the slide scale bar may be removed and replaced with a bar that is designed to hold a reservoir pen. The instrument may be used as dividers by replacing the lead with a needle.

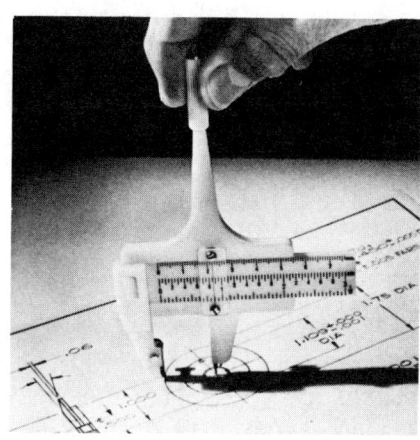

PENCIL

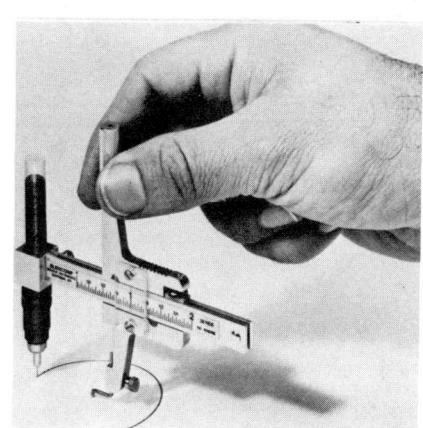

INK

Fig. 2.8 Slidecomp compass *(Courtesy Creative Instruments).*

Drop-bow compass. The *drop-bow compass*, Fig. 2.9, is convenient for the rapid inking of a large number of small, identical circles. The top is held between the thumb and index finger and the nib revolved around the stationary needle point. A typical use is the drawing of plotted points on graphs and map symbols.

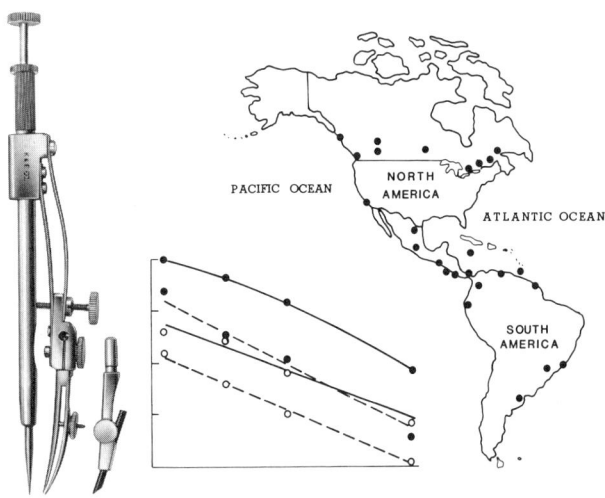

Fig. 2.9 Drop-bow compass *Courtesy Keuffel & Esser Co).*

2.4 DIVIDERS

The *dividers* have a needle point in each leg, Fig. 2.10, and are used to transfer measurements from one location to another.

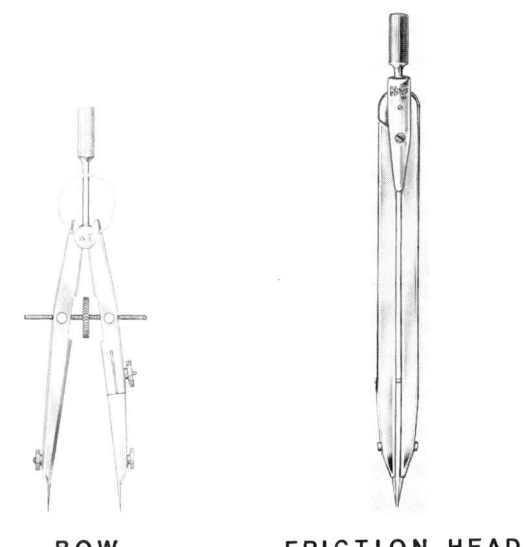

BOW **FRICTION HEAD**

Fig. 2.10 Divider types *(Courtesy Keuffel & Esser Co).*

Proportional dividers. *Proportional dividers*, Fig. 2.11, are used to enlarge or reduce drawings. The dividers are set by closing the legs, loosening the slider screw, and moving the slider up or down in the longitudinal slot. The index mark on the slider is then set on the number of the desired ratio. If the instrument is provided with rack and pinion, sliding is done by turning the pinion screw. The divisions marked "lines" are linear proportions and those marked "circles" is the setting used for dividing a circle into a desired number of equal parts when the large end is opened to the diameter of the circle. The lines on a drawing may be reproduced to any desired ratio between 1:1 and 1:10. Also the area of a plane figure or volume of a solid will be reproduced proportional to the original.

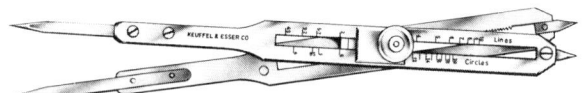

Fig. 2.11 Proportional dividers *(Courtesy Keuffel & Esser Co).*

2.5 PENCILS, LEADS, AND SHARPENERS

Pencils. Mechanical *lead holders*, Fig. 2.12 (a) and (b), are replacing the wood pencil, Fig. 2.12 (c), and are recommended for all drawing work. The fine line lead holder has the advantage in that the leads do not need sharpening. Wood pencils require that wood be removed to expose the lead for sharpening. This time-consuming procedure is eliminated by using a mechanical type lead holder.

Leads. Lead grades, Fig. 2.13, are designated by a number and letter, and range from 7B, very soft, to 9H, very hard. The quality of these leads has been improved by using finer grinds and mixtures of graphite and clay. High quality lines may now be drawn more easily that at anytime in the past. The lead selected should allow a crisp black line to be drawn on the particular surface used for the drawing. In general, the softer the lead, the black-

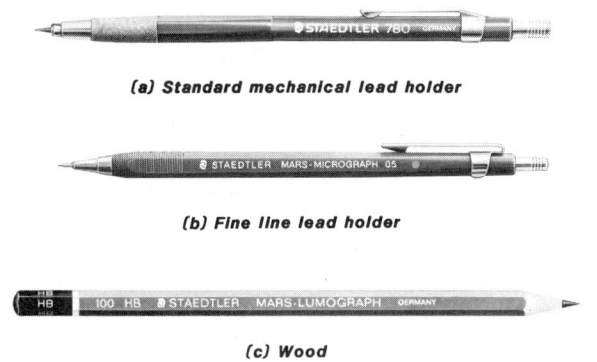

Fig. 2.12 Pencil types (Courtesy J.S. Staedler, Inc).

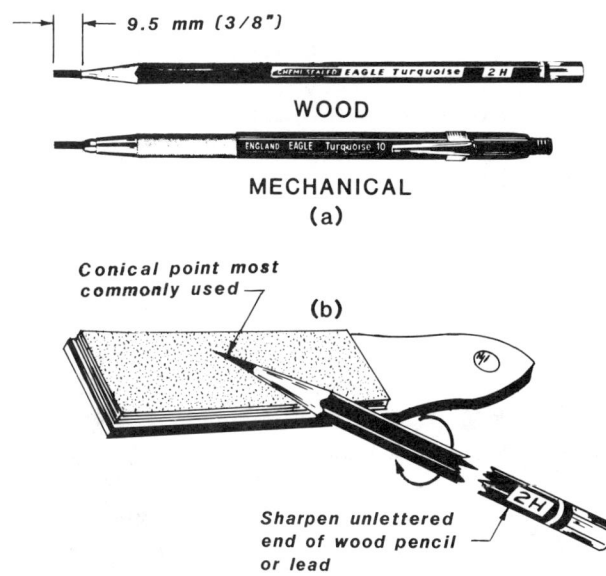

Fig. 2.14 Sharpening pencil leads.

er the line. For most pencil work the average individual will find that a 2H and H lead will give satisfactory results. The 2H lead may be used for layout or construction work and for darkening linework on the finished drawing. With practice and by varying the pressure exerted on the lead, both light construction lines and dense black finished lines may be drawn with this one lead. Lettering and sketching may be done with the softer H lead.

The fine line leads are available in four diameters; 0.3 mm, 0.5 mm, 0.7 mm, and 0.9 mm. These leads are economical, require no sharpening, and allow uniform lines to be drawn with little fluctuation in line width.

Drawing on polyester films of different kinds is common-place and leads have been developed especially suitable for surfaces of this kind. Leads are formulated of a special plastic substance instead of a mixture of graphite and clay. Some leads are smearproof and washproof allowing a soiled drawing to be washed with plain soap and water without loss of detail.

Sharpeners. Line quality and contrast is directly influenced by the sharpness of the lead point. In sharpening the wood pencil the wood first must be removed until about 9.5 mm (3/8") of lead is exposed, Fig. 2.14 (a). The lead may automatically be extended to any desired length by using mechanical lead holders. The exposed end of the lead may then be sharpened to a conical point, Fig. 2.14 (b), by rotating the pencil as the lead is rubbed on the file or sandpaper pad. To obtain good line technique revolve the

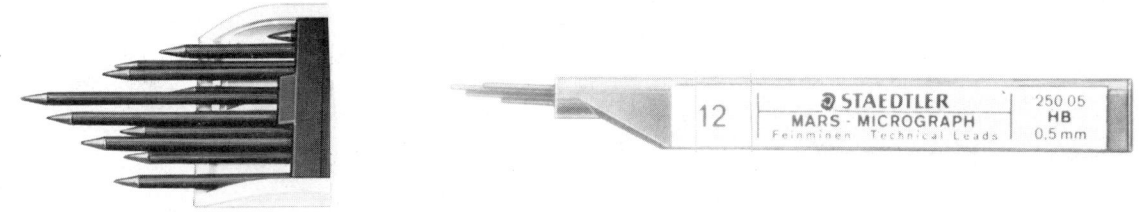

Standard leads   Fine line leads

9H 8H 7H 6H 5H **4H 3H 2H H** F HB B 2B 3B 4B 5B 6B 7B

← HARD   *Recommended for most drawing*   SOFT →

Fig. 2.13 Grades of drawing leads (Courtesy J.S. Staedler, Inc).

pencil slightly when drawing lines and resharpen the lead frequently to maintain uniform lines.

A variety of mechanical lead pointers, Fig. 2.15, is available. These are great time savers and eliminate the problem of disposing of graphite particles.

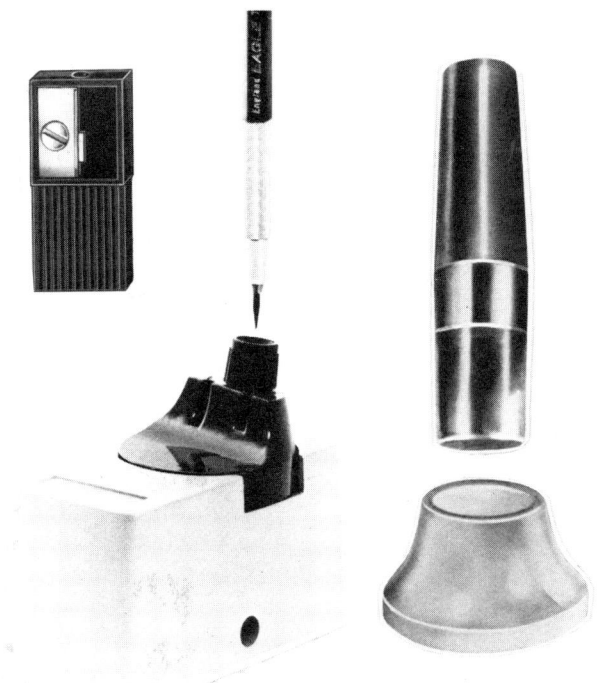

Fig. 2.15 Lead pointers (*Courtesy J.S. Staedler, Inc and Keuffel & Esser Co*).

2.6 LINE CONVENTIONS

Various types of lines are recommended by the American National Standards Institute (ANSI) to convey various information on drawings, Fig. 2.16. All lines (except construction) should be drawn as sharp crisp black lines. This gives the drawing a professional appearance that is easier to read, interpret, and reproduce. There should be a distinct contrast in line width for the types of lines drawn. Three basic widths are recommended - *thick*, *medium*, and *thin*. Depending on the purpose and size of the drawing the width of each line type may vary slightly. The lead should be properly pointed at all times to insure that lines remain of uniform width throughout their length. For thicker lines it may be necessary to re-

trace the lines several times in order to obtain the correct width contrast.

It is sometimes difficult to draw lines where circles or arcs and straight lines are tangent to each other. It is easier to draw a straight line tangent to a curved line than to draw a curved line tangent to a straight line. Therefore, circles and arcs should be drawn first and the straight lines allowed to blend smoothly into the arcs.

2.7 ERASERS AND ERASING TOOLS

A variety of *erasers*, Fig. 2.17 (a), has been formulated for use on various drawing surfaces such as film, tracing cloth, and paper. Non-abrasive plastic or vinyl erasers are rapidly replacing other kinds of erasers since they do not damage the drawing surface. They are excellent for erasing pencil lines and, when slightly moistened, erase ink lines easily from film, Fig. 2.17 (b). Avoid rubbing over finished pencil lines with any type of eraser so as not to affect the sharpness or blackness of the lines. A dusting brush, Fig. 2.17 (c), is generally used to remove eraser particles from the drawing after each erasure.

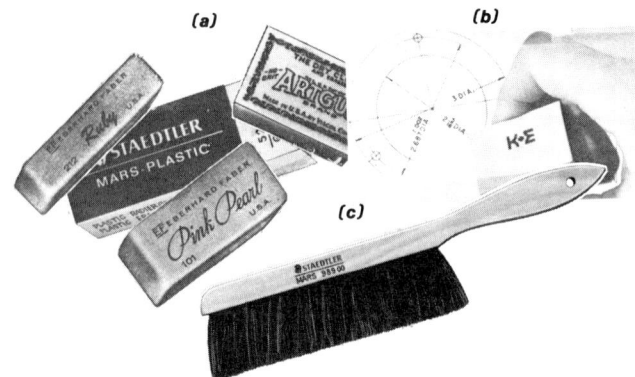

Fig. 2.17 Erasers and dusting brush (*Courtesy J.S. Staedler, Inc and Keuffel and Esser Co*).

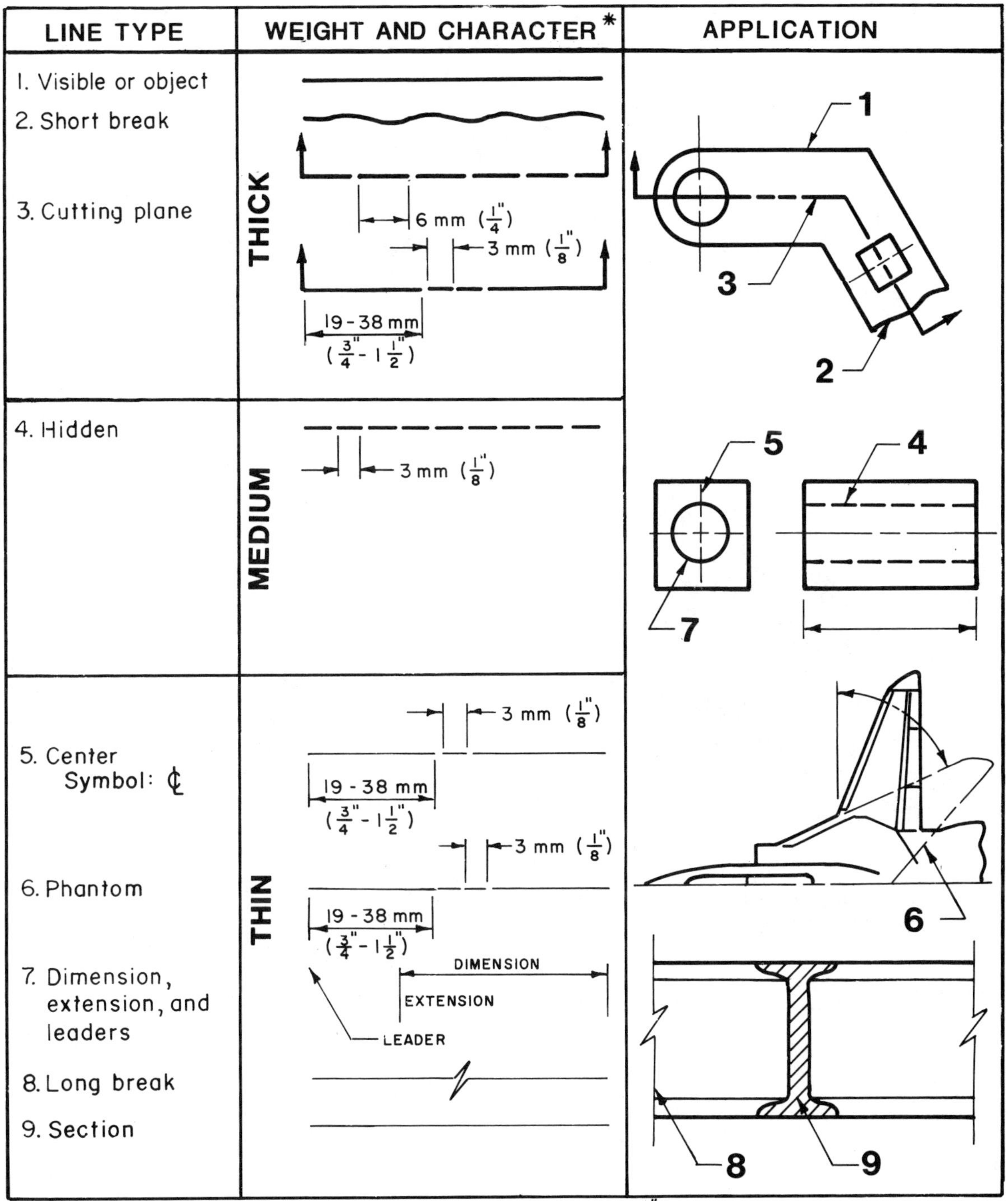

LINE TYPE	WEIGHT AND CHARACTER*		APPLICATION
1. Visible or object 2. Short break 3. Cutting plane	THICK	——————— ∼∼∼∼∼∼∼ ⊢ 6 mm (¼") ⊢ 3 mm (⅛") ⊢ 19–38 mm (¾"–1½")	
4. Hidden	MEDIUM	– – – – – – ⊢ 3 mm (⅛")	
5. Center Symbol: ℄ 6. Phantom 7. Dimension, extension, and leaders 8. Long break 9. Section	THIN	⊢ 3 mm (⅛") ⊢ 19–38 mm (¾"–1½") ⊢ 3 mm (⅛") ⊢ 19–38 mm (¾"–1½") DIMENSION / EXTENSION / LEADER	

Spaces between dashes for all lines are about 0.8 mm (1/32")

Fig. 2.16 Line characteristics and conventions.

Erasing shield. The *erasing shield*, Fig. 2.18, is made of thin metal with holes of various shapes and is useful for removing portions of a line without disturbing other lines. By selecting an appropriate opening and holding the shield down firmly over the line, erasures may be made neatly.

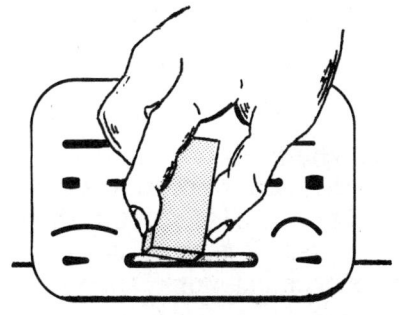

Fig. 2.18 Erasing shield.

Electric erasers. An electric *erasing machine*, Fig. 2.19, is a time saver for erasing pencil and ink lines. The eraser tip should be sharpened to a conical shape (using a file) in order to obtain the best results. To use the machine, touch it gently to the line to be erased. The machine revolves very rapidly and should be kept constantly moving to avoid damage to the drawing surface. Use the machine with the eraser point cutting against the metal edges of the erasing shield. This will prevent graphite or ink from accumulating on the eraser and will give a clean erased line.

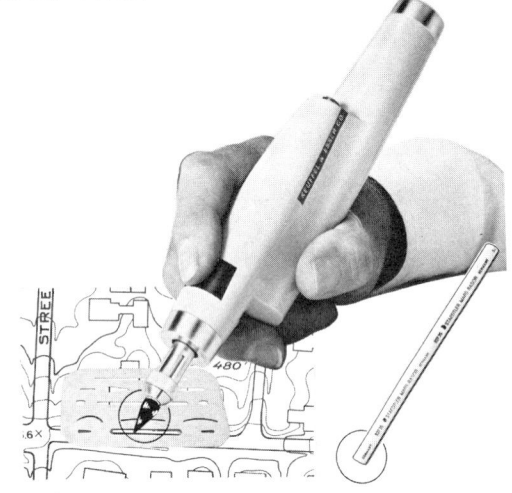

Fig. 2.19 Electric erasing machine (*Courtesy Keuffel & Esser Co*).

2.8 KEEPING THE DRAWING CLEAN

In order to maintain a clean professional appearing drawing, care must be exercised to keep loose graphite from contacting the drawing surface. Drawing leads should not be sharpened over the drawing or desk. The hands and instruments should be clean. Washing the triangles and T-square occasionally with soap and warm water is helpful. The triangles and T-square should be lifted slightly, instead of sliding over the drawing, to prevent line smudging. To protect portions of the drawing when not working on it, cover the finished portions with a sheet of paper. Moisture on the hands sometimes causes lines to be smudged when doing freehand work such as lettering. This may be alleviated by resting the hand on a piece of paper.

Dry cleaning pads, Fig. 2.20, containing pulverized gum eraser granules are sometimes used to help keep the drawing clean. The pad should not be used to scrub the drawing since pencil lines will loose their crisp black appearance. Since this is too frequently done, cleaning powders and pads are not recommended for general use. Cleaning materials of this type should also not be used when inking.

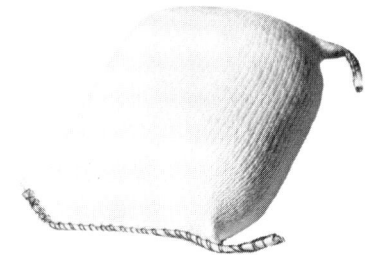

Fig. 2.20 Dry cleaning pad.

2.9 DRAWING SURFACE AND T-SQUARE

Drawing surface. Drawing boards of various sizes and composition and commercial tables are used for fastening and supporting the work material. Drawing surfaces should be smooth and have a straight working side edge for the head of the T-square. The drawing material is taped by the top two corners, Fig. 2.21, approximately 5 cm (2") from the edge and high enough on the board to allow the head of the T-square to remain in contact with the edge. The upper edge of the drawing material should be parallel to the upper edge of the T-square.

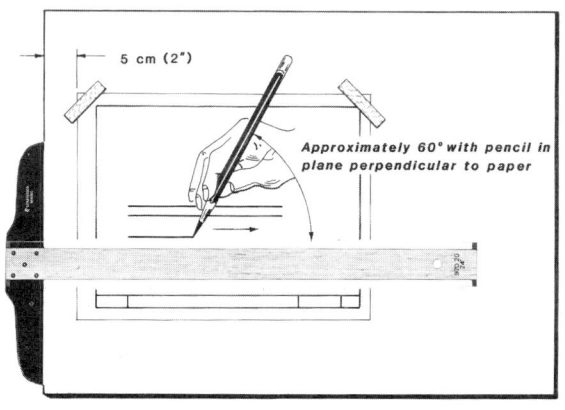

Fig. 2.21 Drawing board and T-square.

T-square. The *T-square*, Fig. 2.21, is available in a variety of materials, sizes, and qualities. Most T-squares consist of a long straight blade with transparent plastic edges attached to a hardwood head. The T-square is used for drawing horizontal lines (left to right for a right-handed person) and as a base for the triangles when drawing vertical or inclined lines. Lines should be drawn along the upper or working edge of the blade from left to right with the pencil held in a plane perpendicular to the drawing surface and inclined approximately $60°$. Darken horizontal lines by starting with the lines at the top of sheet to help keep work clean.

A parallel straightedge, Fig. 2.22, may be used in place of the T-square for drawing horizontal lines. The straightedge is permanently attached to the drawing surface by a system of cords and pulleys so as to give an exact parallel motion to the straightedge as it is moved over the drawing surface. Various lengths are available and are preferable to the T-square for drawing long horizontal lines.

Fig. 2.22 Parallel straightedge
(*Courtesy Keuffel & Esser Co*).

2.10 DRAFTING MACHINE

Drafting machines, Fig. 2.23 are helpful in producing finished drawings in less time, since the functions of a T-square or straightedge, triangle, scales, and protractor are all contained in one unit. Inter-

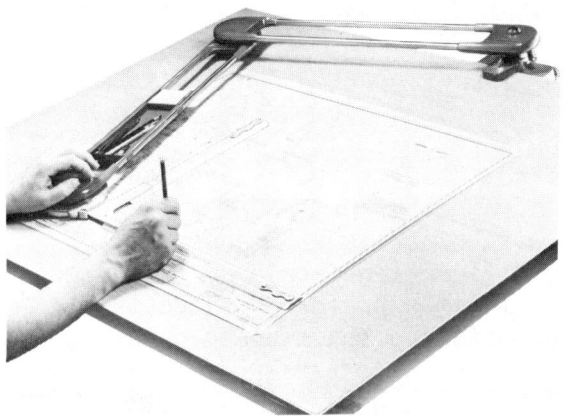

Fig. 2.23 Drafting machine (*Courtesy Keuffel & Esser Co*).

changeable scales are available in a wide variety of graduations suitable for any type of work. The scale assembly moves in a parallel direction so that one scale is always horizontal and the other vertical. This arrangement allows horizontal and vertical lines to be drawn from any point on the drawing. Any angle may also be drawn by shifting the scale assembly into place at any unit multiple of $15°$. A thumb lever releases the protractor ring to set in-between angles.

2.11 PARALLEL RULES

A variety of *parallel rules*, Fig. 2.24 has been devised for drawing vertical, horizontal, and angular lines. Lines may be drawn at various distances or evenly spaced with ease. A built-in metal roller allows the instrument to be smoothly moved up or down for drawing parallel lines.

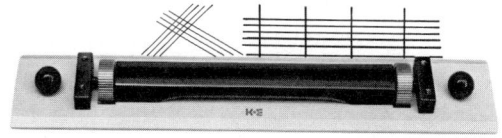

Fig. 2.24 Rolling parallel rule
(*Courtesy Keuffel & Esser Co*).

2.12 TRIANGLES

The standard 45° and 30°-60° plastic triangles, Fig. 2.25 (a), are available in several sizes, qualities, and colors. Good triangles are usually made from clear or tinted acrylic plastic with precision cut edges. The tinted fluorescent colored triangles are sometimes preferred since they reduce confusing shadows near the edge.

A number of triangles with special features are also available. The adjustable triangle, Fig. 2.25 (b), is convenient for drawing lines at any angle. Triangles with bevel edges or continuous finger lifts, Fig. 2.25 (c), are helpful when inking since the ink is less likely to run under the edge.

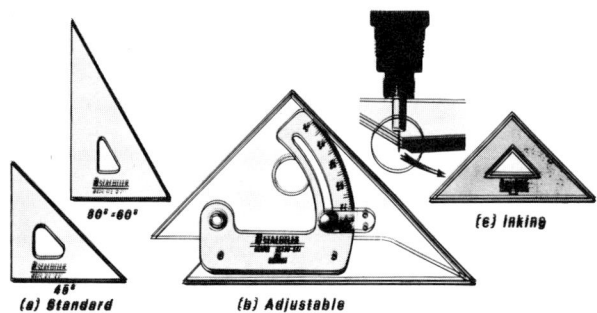

Fig. 2.25 Triangles (*Courtesy J.S. Staedler, Inc*).

Vertical lines may be drawn with either the 45° or 30°-60° triangle. For right-handed individuals the triangle should be placed on the top of the T-square with the vertical edge on the left, Fig. 2.26 (a). Lines are best drawn moving from the lower end of the triangle towards the top along the vertical edge. Drawing vertical lines from left to right will also help to keep the drawing clean. It is good practice when drawing a line to draw away from or across the body. The same procedure should be used to draw lines at 30°, 45°, and 60° with the horizontal, Fig. 2.26 (b). Reverse the triangles and the same angles may be drawn in the opposite direction, Fig. 2.26 (c).

Any angle of a multiple of 15° with the horizontal or vertical may be drawn with the 45° and 30°-60° triangles. The 15° and 75° angles inclined either to the right or left may be drawn as shown in Fig. 2.27. The triangles may be manipulated in various ways and the desired angles drawn with a little practice.

Parallel lines, Fig. 2.28 (a), may be drawn by aligning the edge of one triangle with a given line. The triangle is then moved along the T-square to draw a parallel line through any given point. Perpendicular lines may be drawn by using the adjacent side of the 90° angle of the 45° or 30°-60° triangles, Fig. 2.28 (b). If one triangle leg (not the hypotenuse) is parallel with a line, the side adjacent to the 90° angle will be perpendicular to it.

In place of the T-square another triangle may be used as an edge along which to slide the triangle for drawing parallel lines, Fig. 2.28 (c). The triangles should slide along their long edges (their hypotenuses) when drawing. The adjacent side of one of the 90° triangles again will allow perpendicular lines to be drawn, Fig. 2.28 (d).

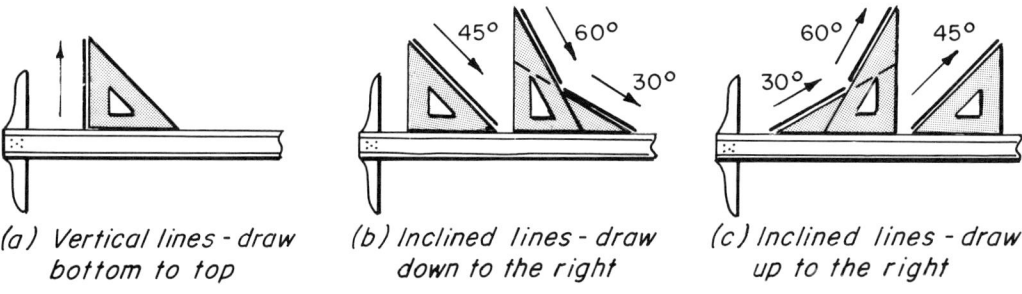

Fig. 2.26 Drawing vertical and inclined lines.

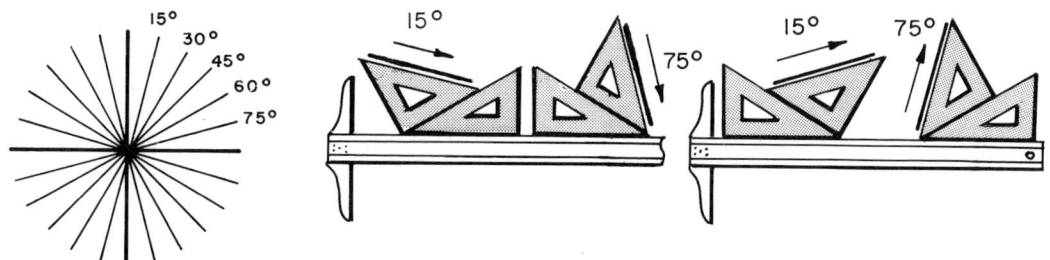

Fig. 2.27 Drawing angles of 15 and 75 degrees.

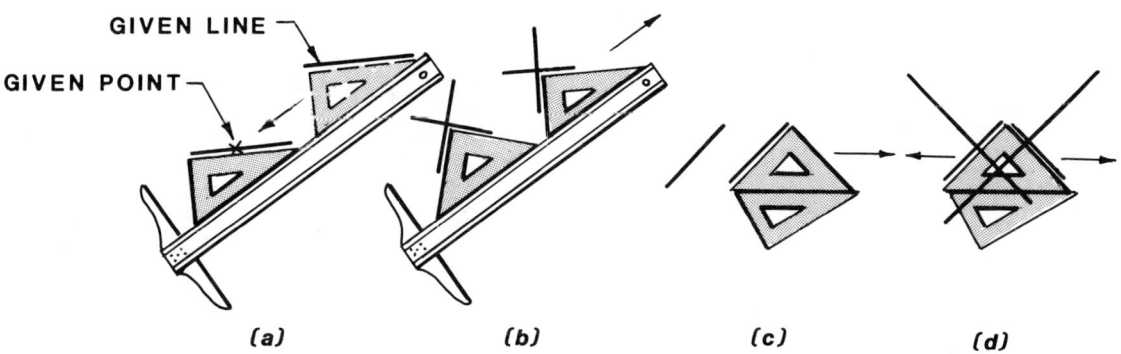

Fig. 2.28 Drawing parallel and perpendicular lines.

2.13 PROTRACTOR

The *protractor*, Fig. 2.29, is used for measuring angles. It may be circular or semicircular; made from metal, plastic, or paper; graduated in degrees or fractions of a degree. Special protractors are available with a moveable arm to allow for a line to be drawn at a specified angle in one operation. The protractor is used by placing the base line, indicated by the $0°$ and $180°$ marks, on one side of an angle with its center at the vertex of the angle. Any angle may then be measured by reading the appropriate degrees from the graduations.

2.14 IRREGULAR CURVES

The *irregular* or *French curve* is used for drawing non-circular curves. Many sizes and shapes of plastic curves are manufactured, Fig. 2.30 (a), with the shape based upon combinations of geometric curves like the ellipse, parabola, and hyperbola. An irregular curve that has a number of small curvatures as well as a few long slightly curved edges will usually work best. Adjustable curves, Fig. 2.30 (b), may be formed to fit any contour and are useful for drawing long lines of large curvature.

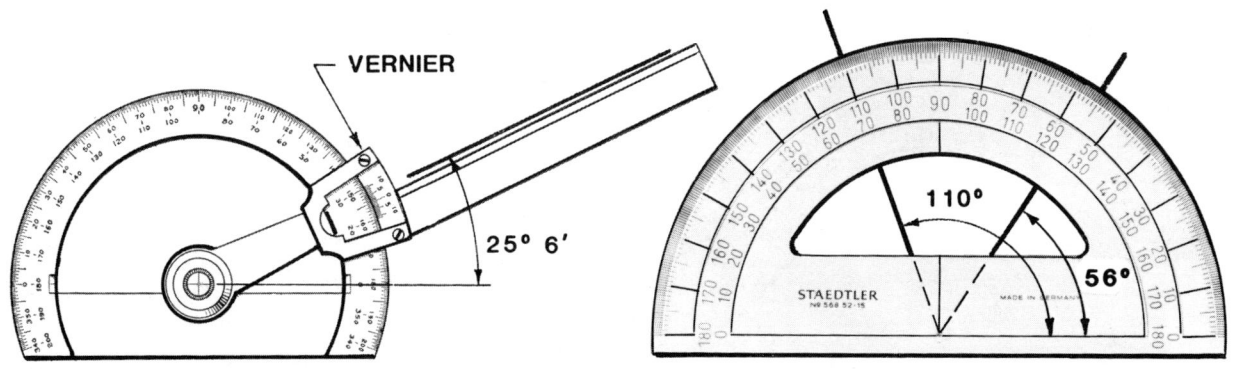

Fig. 2.29 Protractors *(Courtesy J.S. Staedler, Inc)*.

(a) Irregular curves

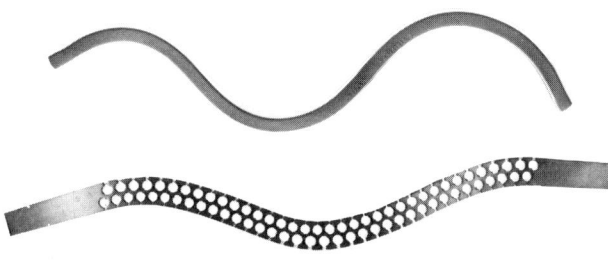

(b) Adjustable curves

Fig. 2.30 Irregular and adjustable curves *(Courtesy J.S. Staedler, Inc and Keuffel & Esser Co).*

To draw an irregular curved line through a series of plotted points, follow the procedure shown in Fig. 2.31. It is sometimes helpful to first sketch lightly a freehand smooth line through the points. A section of the irregular curve is then selected that coincides with the most points possible. Always make sure that the line of increasing curvature as indicated by the plotted points and the increasing curvature of the irregular curve correspond. Short segments of the curve are drawn with the irregular curve and then moved to the next series of points to extend the curve. To insure a smooth finished curve the irregular curve should overlap a portion of the preceding line.

2.15 TEMPLATES

Many plastic *templates* with machine cut symbols are available for a variety of drawing purposes, Fig. 2.32 (a). For cartography work, templates with mapping symbols, Fig. 2.32 (b), are useful. Where different shapes are frequently needed, the template is convenient and saves time for drawing an almost unlimited range of standard symbols and figures.

To obtain good results in drawing circles or arcs with a template, the required center lines should first be drawn. The template opening of the appropriate size is placed over the center lines, lining up the four marks on the template with the center lines, Fig. 2.32 (c). All cut-outs have a pencil allowance to provide accuracy in drawing.

2.16 INKING

It used to be common practice to lay out drawings in pencil and then to prepare a final drawing in india ink on tracing paper or cloth to insure that good reproduction results could be obtained. Many kinds of drawings are no longer inked since pencil line drawings may now be copied to produce results that will satisfy client requirements. Drawings frequently need to be copied by some process and the copy can be no better than the original. Ink still remains superior to pencil for obtaining high quality drawings and reproductions.

Inking has not gone out of style and is used for a number of different kinds of drawing. Its use in industry

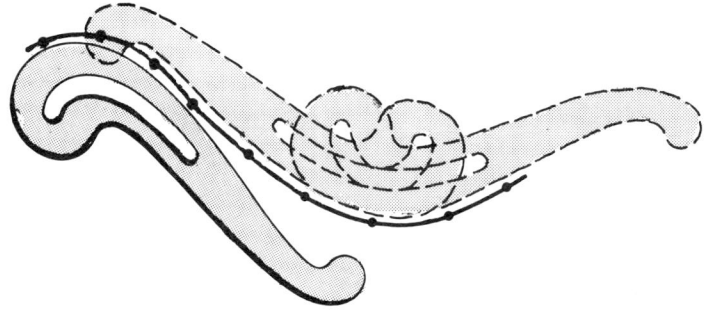

Fig. 2.31 Use of the irregular curve.

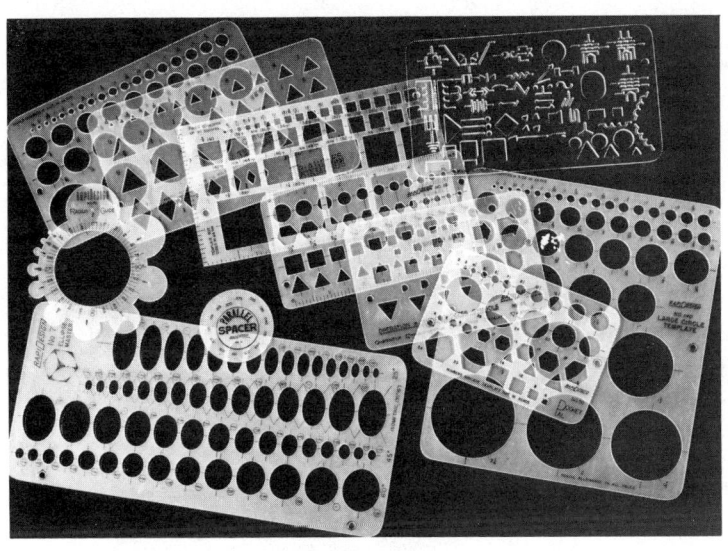
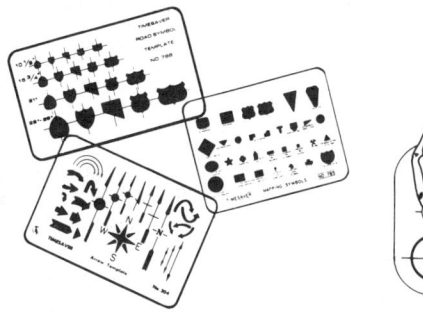
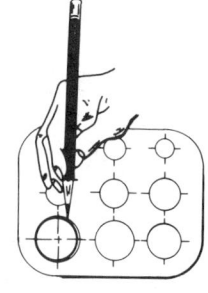

*(a) Template shapes
Courtesy Keuffel & Esser Co*

(b) Mapping templates

(c) Circle template use

Fig. 2.32 Templates.

is increasing since legibility in all reproductions, and consistent line width and density is obtained. Industries that are microfilming for the storage, retrieval, and reproduction of drawings have found that pencil drawings create problems due to varying line weights and gray tones. Drawings intended for publication are normally drawn in ink as well as presentation and display drawings. In cartography, ink is frequently used for the drawing of maps.

Although inking may still require more time, the margin has been cut considerably by using special drawing media (polyester film), improved fast drying inks, and reservoir-type pens. By drawing in ink, imprinted symbols on clear film with a self adhesive backing may advantageously be used on the drawing to also save time. The use of this material on pencil drawings is not advisable since the blackline imprint is incompatible with gray pencil tones. The use of new equipment of this type allows anyone without experience to obtain professional results in a short period of time.

Reservoir pens. The inking of both freehand and mechanical lines has been revolutionized with the development of *reservoir* type inking pens, Fig. 2.33. They may be used for all kinds of drawings, maps, lettering, graphs, etc. Most pens have an air tight plastic cartridge that may easily be filled with ink. An outstanding feature of reservoir pens is that consistent line widths and line densities may be obtained by using different pen point sizes that are available in standard American or metric sizes, Fig. 2.34. The nine metric pen sizes designated by an asterick conform to the International Standard ISO 3098/1 for line widths. The line widths are in a geometric ratio of 1.414 or $\sqrt{2}$ between progressive line widths. Each line width is 1.414 times larger than the preceding line width, following the same geometric progression as the five

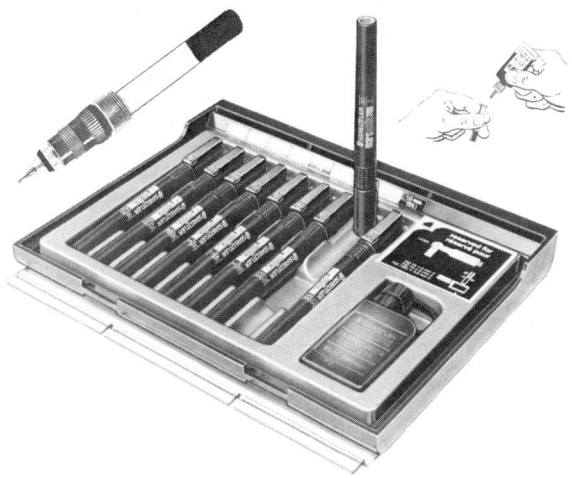

Fig. 2.33 Typical reservoir pens
(Courtesy J.S. Staedler, Inc).

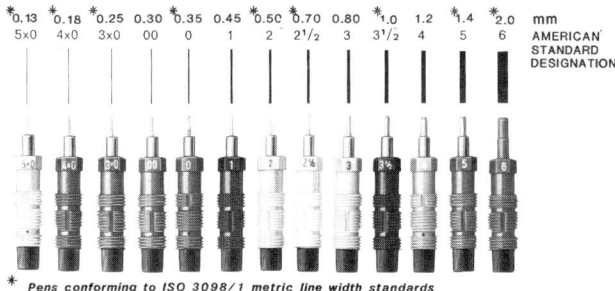

Fig. 2.34 Reservoir pen point sizes
(Courtesy J.S. Staedler, Inc).

The tip of the pen point is designed with a shoulder to help eliminate ink from running under straightedges, triangles, and templates providing it is held at the proper angle. Triangles are available with raised edges that will help eliminate ink from running under the edges if the pen is held at a slight angle, see section 2.12.

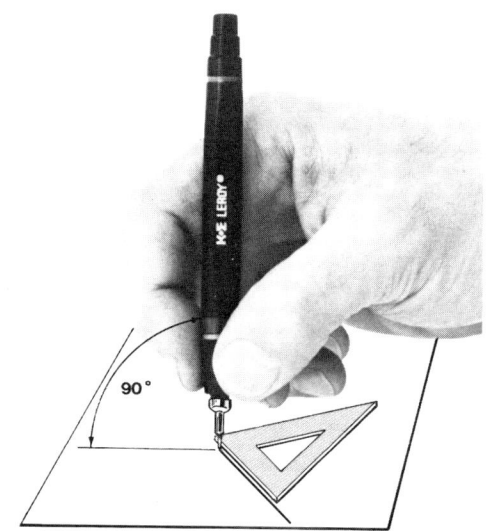

Fig. 2.35 Using the reservoir pen.

standard American paper drawing sizes, A through E. In addition, the line width ratios correlate directly to the reproduced line widths in microfilm reductions and blowbacks. If an original drawing (using pen number 0.50) is reduced on microfilm, for example, and blown back to half size, revisions may be made directly on the blowback with a metric point two sizes smaller (pen number 0.25). For a blowback one size smaller, revisions will be made with a metric point one size smaller than used on the original. The minimum width of lines for microfilming drawings prepared on A4, A3, and A2, paper sizes is 0.25 mm, and 0.50 mm for A1 paper size. A minimum distance of 1.5 mm should be kept between parallel lines.

Pen point tubes made from stainless steel are satisfactory for occasional use on most kinds of drawing media. Film is highly abrasive and will cause stainless steel pens to wear excessively so as not to function properly. Points made of synthetic jewels, sapphire or tungsten carbide are recommended for film and heavy use.

Reservoir pens are easy to use and the time span required to complete a drawing is substantially less than with other kinds of pens. If the ink does not flow immediately, gently shake the pen to force the ink into the point tube. The weighted cleaning wire will make a slight noise and must be free to move to regulate the flow. Drawing the pen point across a moistened finger will also help to start the ink to flow in a stubborn pen. The pen should be held close to perpendicular for inking smooth uniform lines, Fig. 2.35.

Ink lines must be dry before drawing intersecting lines. Considerable time may be saved by inking horizontal lines from left to right, starting at the top of the drawing and working down. Vertical lines are usually drawn in an upward direction and from left to right. Many of the directions discussed, however, may have to be reversed for left-handed individuals. For inking curved lines such as contours on a map, hold the pen vertically between the thumb and index finger with the heel of the hand resting on a piece of paper, Fig. 2.36. The pen is easier to move, for following a curved line, without having to turn the hand since there will be little friction as the paper slides smoothly over the drawing media.

For large circles and arcs that cannot be drawn easily with a template a compass attachment is used to hold the reservoir pen, Fig. 2.37. Some attachments have an adjustment pivot bearing that will allow the pen to be angled in any plane for drawing small circles.

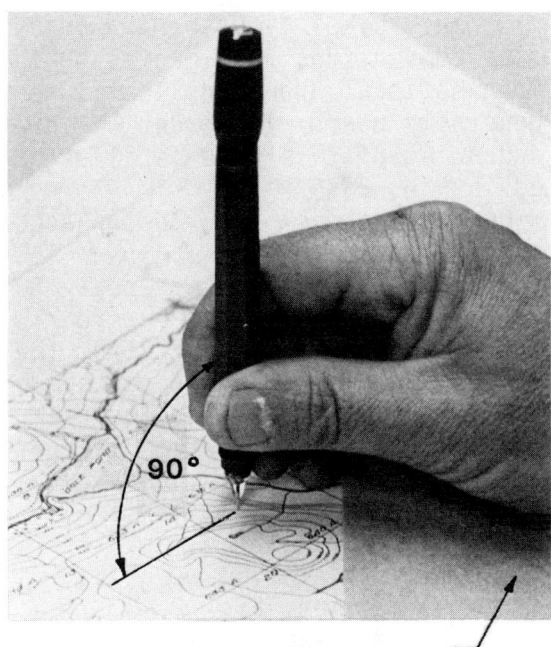

Fig. 2.36 Drawing freehand curves with a reservoir pen.

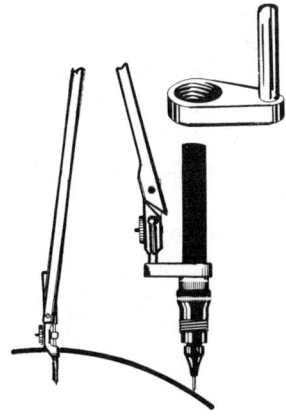

Fig. 2.37 Compass attachment for reservoir pens.

Inks. Drawing ink should be fast-drying, free flowing, and opaque to give good results. All inks are not compatible with different drawing media and the correct ink should always be used. Some inks that will not chip or crack have been developed for use on film while other special inks have been formulated with micro-fine carbon particles that are ideal for use in reservoir pens. Inks with the carbon held in a colloidal solution should not be used in reservoir pens since the tubes of the pens may become blocked.

Cleaning the reservoir pen. Inking pens must be kept clean to insure proper operation. Pens not used frequently should be carefully disassembled, Fig. 2.38 (a), and cleaned with lukewarm water or special cleaning solvents. When the pen is not in use, always replace the cap securely to avoid drying of ink. The pens should not be left lying in a horizontal position so that the ink remaining inside the point section will drain back into the ink cartridge to help reduce clogging problems. It is recommended that the manufacturer's instructions be followed concerning preventive maintenance of reservoir pens. Access to an ultrasonic cleaner, Fig. 2.38 (b), increases efficiency since it minimizes correc-

tive and preventive pen cleaning time. To start the ink flow, the tip of the pen should be immersed into the cleaner for a few seconds. If the pen is dried up and clogged from long disuse, it may be immersed in the cleaner to free the parts. The pen may then be disassembled without damage for a thorough cleaning of the individual parts. A commercial pen cleaning fluid is recommended for use with units of this type.

Planning to draw in ink. The use of polyester film, when inking, is recommended since corrections or revisions

Fig. 2.38 Reservoir pen, exploded view, and ultrasonic cleaner (Courtesy Keuffel & Esser Co).

are easy to make using vinyl erasers that have been slightly moistened. New eradicators having special formulations of eraser and liquid are also available. In using any product, the manufacturer's recommendations should be followed.

The inking of a drawing should be done carefully with the work planned so that erasures are minimized. This may be accomplished by having some definite order of precedence in inking the various classes of information after the pencil layout has been completed. A good rule to follow is to select the most important feature to ink first. Remaining or secondary features are then inked in the order of their importance, toning down the weight of the lines in proportion. Care must be exercised since information often must be superimposed upon another feature to complete the drawing.

On mechanical types of drawings, the line conventions recommended by the American National Standards Institute are generally used. The order of inking a drawing of this kind is illustrated in Fig. 2.39.

One of the most important phases of map drawing is that of *contrast*. Contrast is important and necessary to give a map "life and vigor" in order to make each feature distinct from the next. This one cartographic technique is important since most maps include several kinds of lines to symbolize features. The weights of lines may be varied in accordance with the size, purpose, and nature of the map. At least three distinct line weights should be maintained, Fig. 2.40. The order of inking a map is illustrated in Fig. 2.41.

ORDER OF INKING *

1. Circles and arcs
2. Irregular curves
3. Horizontal lines (top to bottom)
4. Vertical lines (left to right)
5. Inclined lines
6. Dimensions (dimension and extension lines, leaders)
7. Freehand work (arrowheads, lettering, etc)

* *Ink all lines having the same thickness at one time, e.g., visible, hidden, etc.*

APPLICATION

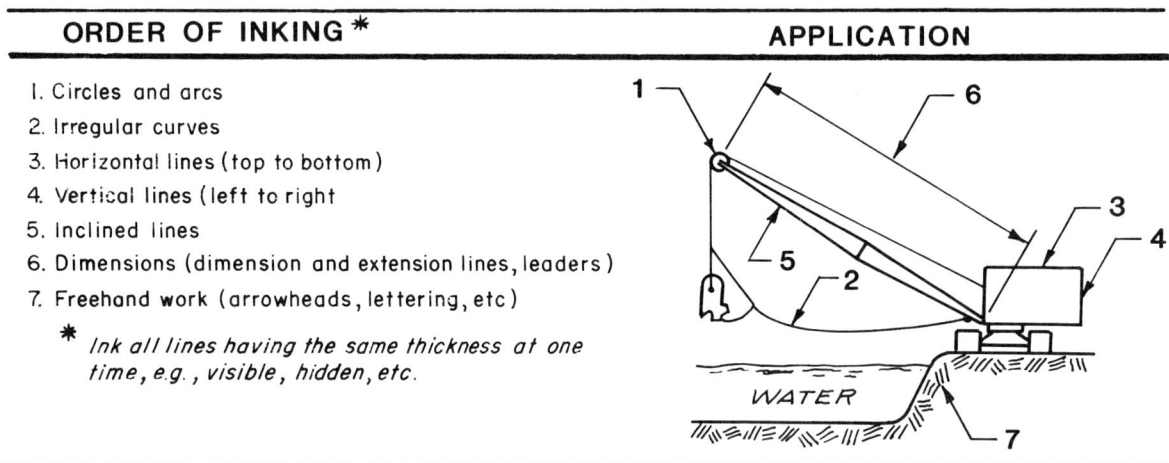

Fig. 2.39 Inking mechanical type drawings.

0.13 mm (.005")

0.30 mm (.012")

0.46 mm (.018")

Fig. 2.40 Line gage.

ORDER OF INKING	APPLICATION
1. Buildings 2. Roads, railroads, and communication lines 3. Boundaries and fences 4. Lakes, rivers, and streams 5. Index and intermediate contours (do not draw contours through buildings) 6. Marshes, swamps, and bogs 7. Vegetation 8. Map lettering 9. Marginal information and lettering	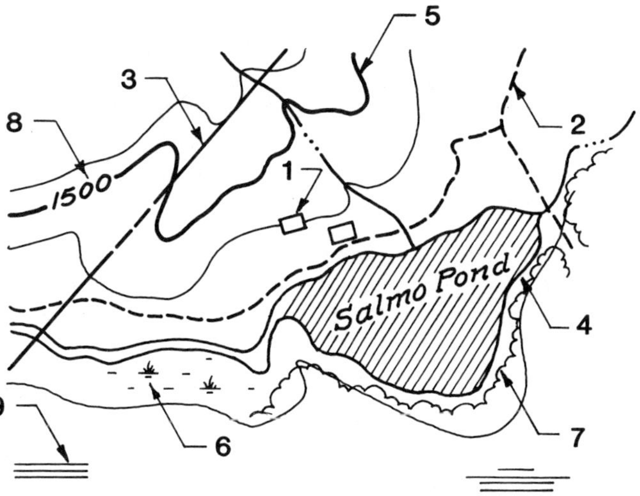

Fig. 2.41 Order of inking a map.

SCALES AND MEASUREMENTS

3.1 INTRODUCTION

Scales of different types are necessary for drawing objects to fit various sheet sizes. It is not possible to draw everything full size and the ratio of the measurements on a drawing of an object to the same measurements on the actual object must be maintained. Drawings are frequently rendered worthless through inaccurate measurements and improper use of the scale. The correct scale must be used for laying off measurements on the drawing. Calculations are not necessary and should not be used to convert reduced or enlarged measurements. For example, a full scale measurement of 3/4" should not be used to lay off a measurement of 1½" where the scale is ½"=1". A scale of ½" represents 1" should be used to measure the 1½" directly in its reduced size. To avoid cumulative errors, a series of measurements should be marked off, where possible, without moving the scale.

Standard scales may be obtained in either triangular or flat shapes, Fig. 3.1. Full-divided scales have the smallest unit represented throughout the length of the entire scale. They are sometimes double numbered to provide both left-to-right and right-to-left reading in order to provide two different scales on the same face. Open-divided scales have the smallest units only at one end of the scale.

3.2 MECHANICAL ENGINEER'S SCALE

For objects which need not be greatly reduced, either the *mechanical engineer's* scale or the civil engineer's scale may be used, Fig. 3.2 (a) and (b). The mechanical engineer's scales are marked 3/4"=1" or 3/4 size, ½"=1" or

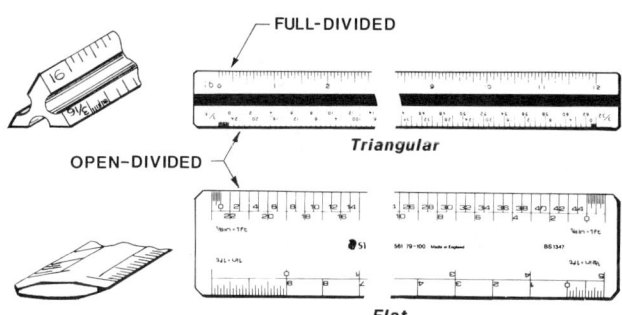

Fig. 3.1 Types of scales (*Courtesy J.S. Staedler, Inc*).

½ size, 3/8"=1" or 3/8 size, and ¼"=1" or ¼ size. On these scales, a fraction of an inch on the drawing represents a full inch on the object to be drawn. The subdivisions at the end of the scale represent the commonly used fractions of an inch, 1/2, 1/4, 1/8, 1/16, etc.

3.3 CIVIL ENGINEER'S SCALE

In measuring distances in surveying, highway, and map work the *civil engineer's* scale or more simply: the engineer's scale, is used, Fig. 3.2 (b). This scale is fully divided with each inch division divided into 10 (1"=10 units), 20 (1"=20 units), 30 (1"=30 units), 40 (1"=40 units), 50 (1"=50 units), or 60 (1"=60 units) parts. The whole number at the end of the scale, e.g. 10, identifies the scale. When using the 10 or 50 scale, full size values in decimal fractions of an inch may be laid out conveniently. On the 50 scale each of the 50 subdivisions represents .02 of an inch. The 10 scale could be used for 1"=10', 1"=100', 1"=1000', etc. Since the units may vary, the scale may represent any quantity, e.g., 1"=10 miles, 1"=10 kilometre, 1"=100 chains, 1"=1000

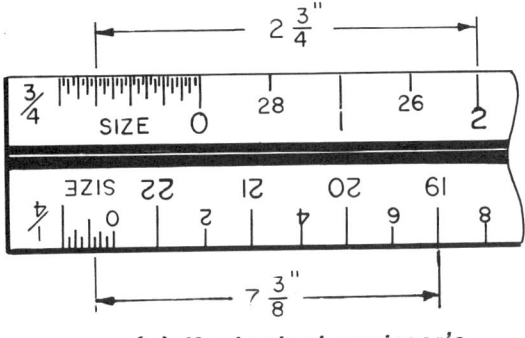

(a) Mechanical engineer's

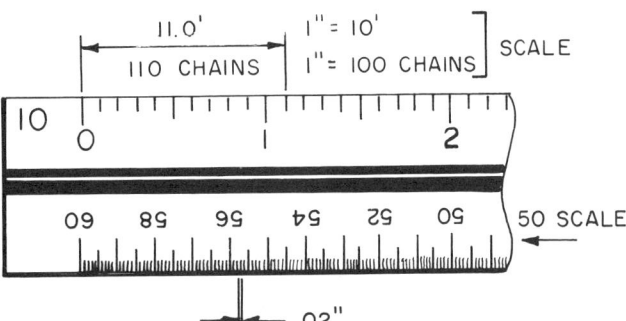

(b) Civil engineer's

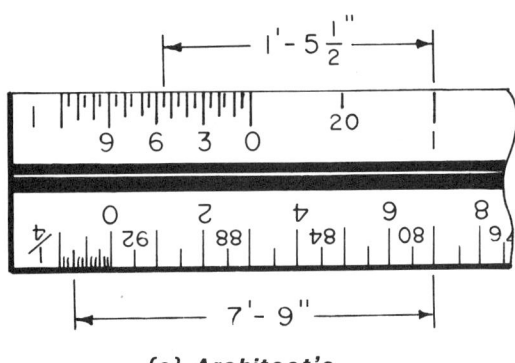

(c) Architect's

Fig. 3.2 Mechanical engineer's, civil engineer's, and architect's scale.

3.4 ARCHITECT'S SCALE

The *architect's* scale, Fig. 3.2 (c), is used for many types of drawings (structural, construction, landscaping, etc.) where it is necessary to reduce measurements given in feet and inches. Architect's scales are open-divided for laying out units representing a foot with the end subdivided to measure inches or fractions of an inch. Listed below are the different scales in use.

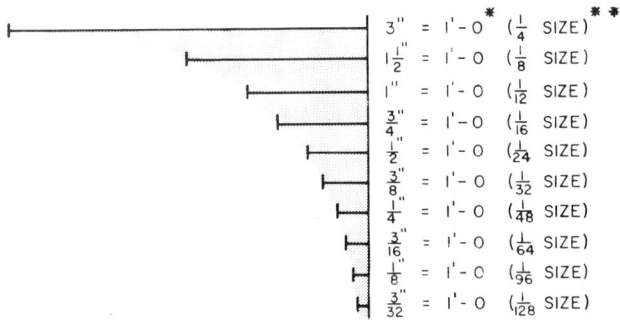

* *The inch marks are normally omitted when expressing zero inches*
** *Ratio of drawing size to object size*

Considerable care must be exercised when selecting the correct architect's scale if it is expressed as a ratio. The scale marked 3"=1'-0 would have to be used for a drawing of ¼ size since 3" is ¼ of 1'-0. On any of the architect's scales, it must be remembered that an open unit on the scale represents one linear foot on the object. Two compatible scales are sometimes combined and the graduation marks must be read in the direction of increasing scale value.

3.5 METRIC SCALE

The *International System of Units* (Système International d'Unités") or SI is a modernized version of the metric system established by international agreement in 1960. It is used by most countries today and provides a logical and interconnected framework for most measurements in science, industry, and commerce.

The use of the metric scale means using millimetres and metres instead of inches, feet, and yards. Measurements may be made easily because relation-

pounds, 1"=10 yards. Similarly, an inch on the 20 scale may represent 1"=200', 1"=2000 ton, etc. This method applies to the other civil engineer's scales as well. The civil engineer's scale also may be used for small measurements where it is desirable to represent measurements in decimal inch fractions. For instance the 40 scale may be used for quarter-size measurements.

ships between a unit and its multiples and sub-multiples are based on tens instead of numbers such as two, three, four, and twelve which are used in the English system. To convert from one unit of metric measurement to another is just a simple matter of moving the decimal point either to the right or left since the units employed are almost always related by powers of 10. The *metre* (common international spelling) is the basic unit of length in the metric system and is equal to 10 decimetres, or 100 centimetres, or 1000 millimetres. To convert 24.76 metres to centimetres (2476 cm) requires only shifting the decimal point two places to the right.

Decimal multiples and division of SI units may be expressed by adding prefixes to the unit names. The prefix symbol is applied to the unit symbol to form a new symbol, e.g., millimetre for mm and kilometre for km. Some common prefixes are shown in Table 3.1.

The common metric units, Fig. 3.3, used for linear measurements are:

1. millimetre (mm)

   Used for all dimensions on engineering drawings and for measuring very small lengths.

2. centimetre (cm)

   Used for measuring common everyday lengths (instead of using inches), e.g., body measurements.

3. metre (m)

   Used for measuring intermediate lengths or distances, e.g., lumber length, sporting events, and surveying.

4. kilometre (km)

   Used in measuring long distances, e.g., maps.

Table 3.1 SI metric unit prefixes.

Prefix	Symbol	Multiplication Factor	
tera	T	1 000 000 000 000	$=10^{12}$
giga	G	1 000 000 000	$=10^{9}$
mega*	M	1 000 000	$=10^{6}$
kilo*	k	1 000	$=10^{3}$
hecto	h	100	$=10^{2}$
deca	da	10	$=10^{1}$
deci	d	0.1	$=10^{-1}$
centi*	c	0.01	$=10^{-2}$
milli*	m	0.001	$=10^{-3}$
micro	μ	0.000 001	$=10^{-6}$
nano	n	0.000 000 001	$=10^{-9}$
pico	p	0.000 000 000 001	$=10^{-12}$

*Most commonly used.

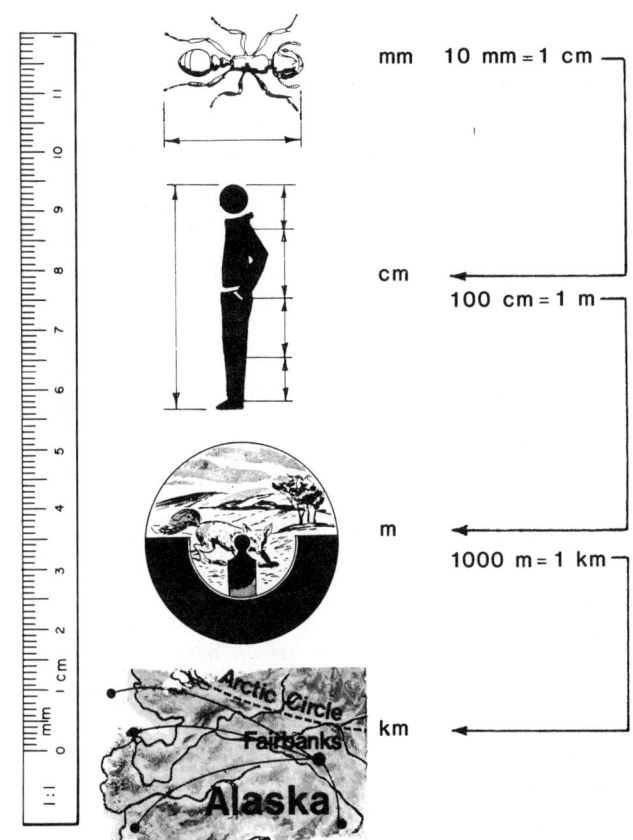

Fig. 3.3 Common linear metric units.

In accordance with SI Standards, the correct style usage should be adhered to. The basic rules are summarized:

1. Prefix symbols are lower case except, tera (T), giga (G), and mega (M).

2. Prefixes, when written out in full, are lower case.
   kilo not Kilo
   centi not Centi

3. Units, when written out in full, are lower case except at the beginning of a sentence.
   millimetre not Millimetre
   centimetre not Centimetre

4. Symbols are written with lower case letter; except for units named after persons
   m (metre) not M (metre)
   N (Newton) not n (newton)

5. Leave a space between the figure and the symbol.
   1.83 cm not 1.83cm

6. Periods are not used with any units, except at the end of a sentence.
   2.76 mm not 2.76 mm.

7. A zero must precede the decimal point for numerical values less than one.
   0.4 not .4

8. When using five or more digits, space them in groups of three counting from the decimal point toward the left and right. Do not use commas since in some countries the comma is used as a decimal marker.
   62 471 not 62471
   108 417 not 108,417
   4.141 672 not 4.14 167 2
   3000 or 3 000 (A narrow space may be provided in numbers of four digits.)

The selection of a metric scale to replace the inch scale will be determined by the kind of drawing to be done. Several kinds of metric and English or straight metric scales are available. A metric scale containing ratios of 1:1, 1:2, 1:3, 1:4, 1:5, and 1:6 is recommended. It should be emphasized again that the decimal point may be shifted to create several different common scales. The metric scale, Fig. 3.4, identified as 1:1 may be used to represent scales of 1:10, 1:100, 1:1000, etc. The same procedure may be applied to the other scales as well. In addition to the English and metric scales described, a number of other scales, such as 1"=66' and 1"=660' are available that have been designed for use on maps. A number of other units involving length are often encountered in surveying and mapping. Appendix 6 should be referred to for additional values useful in mapping.

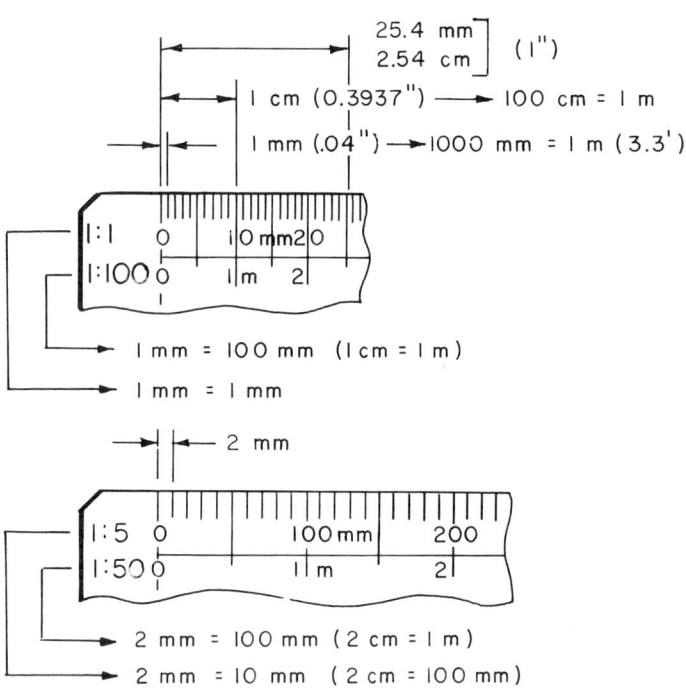

Fig. 3.4 Metric scale.

3.6 U.S. SURVEY FOOT

In the United States the *U.S. Survey Foot* is used as a fundamental unit for surveying, mapping, charting, and associated activities. The length of the U.S. Survey Foot is based on the legal definition of a metre established by Congress in 1866 and is exactly 39.37 inches being equal to one metre. Since surveys involve long distances and many old surveys are still encountered, this definition to define the U.S. Survey Foot is used to avoid inconsistencies if a different equivalence were used. The *International Foot* is still another unit of linear measurement and is based on one inch equals 25.4 millimetres exactly.

3.7 MAP SCALES

Map scales are dependent on the area mapped, the amount of detail required, accuracy desired, and the size of the map sheet. The trend is toward the use of larger scales reflecting the need by map users for more detailed information about the land surface. This need may be attributed to an increase in the world's population, construction projects of various kinds, political turmoil, more intensive and scientific use of the land in addition to many other factors.

The scale of a map is defined as the ratio between a distance as it is represented on the map and as it is measured horizontally on the ground. The formula for expressing the map scale is:

$$\text{Map Scale} = \frac{\text{Map Distance}}{\text{Ground Distance}}$$

It must be remembered that distances measured on a map are *horizontal* and true ground distances are seldom measured. The scale may be expressed as a representative fraction, verbally or in graphical form.

Representative Fraction. A *representative fraction* (RF for short) shows one type unit on the map, e.g., 1 cm or 1 inch, equal to a certain number of equal type units on the ground. A scale of 1:24,000 means that 1 unit on the map represents 24,000 of the same units on the ground. Thus, 1 cm on the map represents 24 000 cm (240 m) on the ground or 1 inch on the map represents 24,000 inches on the ground. The unit of distance on both sides of the ratio must be the same. The representative fraction may be written as a fraction or ratio.

$$RF = \frac{\text{Map Distance}}{\text{Ground Distance}} = \frac{1}{24\ 000} \text{ or } 1:24\ 000$$

Table 3.2 lists several representative fractions and their equivalent in metric and English units.

Verbal statement. Expressed verbally, the scale is a statement of some map distance to some ground distance. It is common to see 1 cm or 1 inch on the map representing whole numbers of tens, hundreds or thousands of metres, kilometres, feet or miles on the ground. Typical examples are: 1 cm = 50 metres, 1 cm = 10 kilometres, 1 inch = 100 feet, and 1 inch = 16 miles.

Table 3.2 Common map scales and their equivalents.

	Metric	English
Scale	1 cm represents (km)	1 inch represents (feet or miles)
1:20 000	0.20	1,666 feet
1:24 000	0.24	2,000 feet
1:25 000	0.25	2,083 feet
1:30 000	0.30	2,500 feet
1:50 000	0.50	4,167 feet
1:62 500	0.625	1 mile approximately
1:75 000	0.75	6,250 feet
1:100 000	1.0	8,333 feet
1:125 000	1.25	2 miles approximately
1:250 000	2.5	4 miles approximately
1:500 000	5.0	8 miles approximately
1:1 000 000	10.0	16 miles approximately

Graphic scale. A *graphic* or *bar scale* is a line subdivided into map distances corresponding to convenient units of length on the ground. It may be drawn as a single line or with two closely spaced parallel lines, shaded or unshaded as illustrated in Fig. 3.5. The length of the graphic scale varies depending on the size and purpose of the map. It is recommended that they be drawn with at least three whole units to the right of the zero mark where possible. The units of measurement should always be stated. Ordinarily the units begin at zero with an extension to the left of zero. This is a convenience for anyone who must figure out small distances which are fractions of the unit. For purposes of allowing different units of measurement to be used on a map, two scales are used, Fig. 3.5 (f) and (g). The lines are divided and labelled with the distances generally in metres and feet or kilometres and miles. The zeros should be lined up under one another when two or more units are represented in order that comparisons between ground distances in the different units may be made quickly.

A graphic scale should always be placed on a map. There is a danger that the material on which the map is drawn may stretch, shrink or that the usefulness will be diminished when reproduced to a larger or smaller scale. Since the scale in printed form always remains the same, it is important that maps intended for reproduction show the final scale and not the drafting scale.

The drawing of a graphic scale may be confusing when the scale of the map is expressed as an RF. For example, it may be desirable to prepare a graphic scale of 4,000 feet where the map scale is 1:15,000. The map distance which is equal to 4,000 feet on the ground is as follows:

$$MD = \frac{\text{Ground Distance}}{\text{Denominator of RF}} = \frac{4,000}{15,000} =$$

$$\frac{4,000 \times 12}{15,000} = \underline{3.24 \text{ inches}} \text{ (map distance equal to 4,000 feet)}$$

A graphic scale, Fig. 3.5 (h), may be prepared with this information. A line is drawn 3.24 inches long and line AB is drawn at any convenient angle from one end of the line. Line AB is then divided into four equal parts using the method depicted in Fig. 4.5. Next draw a line from the last divi-

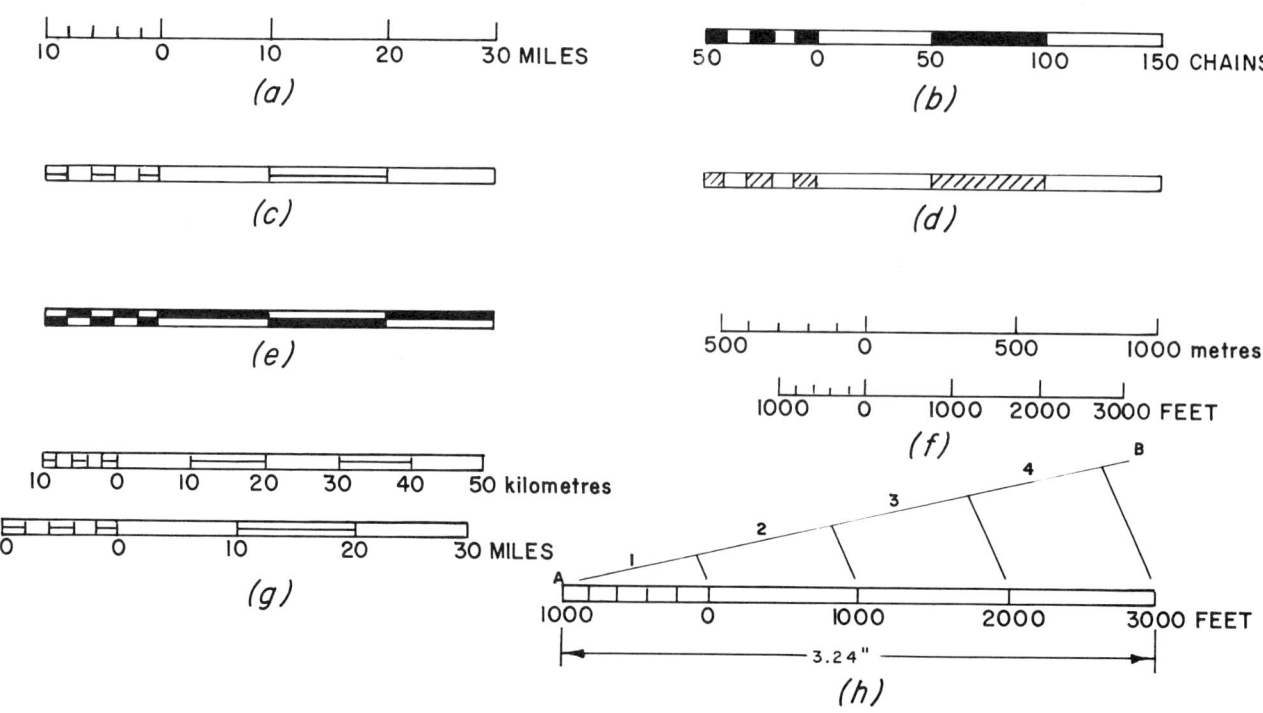

Fig. 3.5 Types of graphic scales.

sion point on line AB to the other end of the line. Parallel to this line draw lines through the other division points to divide the line into four equal parts. The scale may be labeled and minor divisions indicated at one end.

Large vs. small scale maps. When one map is said to be of a larger or smaller scale, Fig. 3.6, the meaning of the statement may not be immediately apparent. An easy way to remember whether a map is large or small scale is to think of the scale as a fraction; 1/25 th is larger than 1/50 th. The larger the denominator of the scale ratio, the smaller the scale of the map.

Large and small scale maps may be defined broadly as follows: This is an arbitrary classification and other references will indicate different limits.

Large scale maps - 1 cm = 5 m to 50 m (1" = 40' to 300'): Large scale maps portray areas with a relatively large amount of detail and are used for densely settled areas and other areas where more detailed map information is required.

Small scale maps - 1 cm = 50 m + (1" = 300' +): Maps at these scales cover large areas and the features shown must be generalized. Small scale maps are useful in many ways, such as planning state-wide and nation-wide projects and other mapping projects.

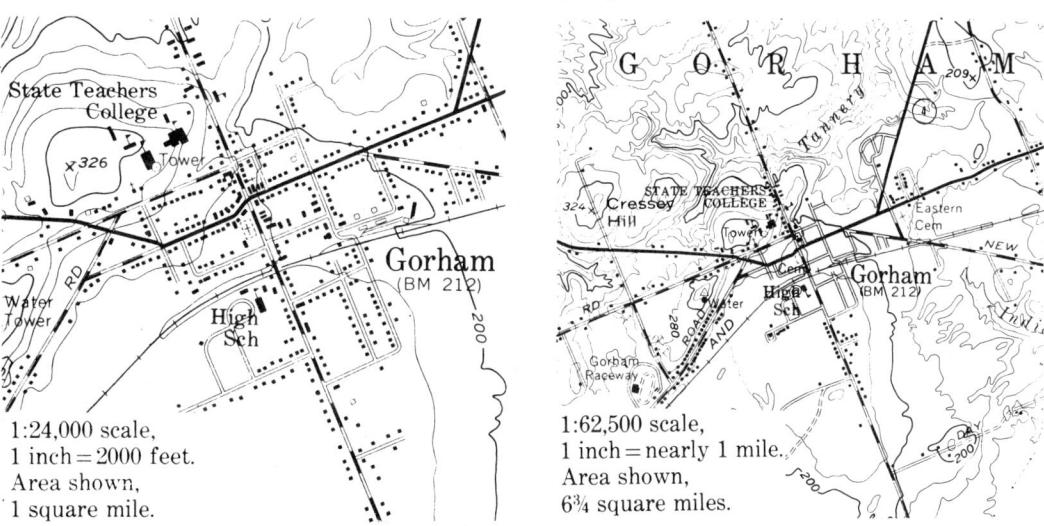

Fig. 3.6 Comparison of map scales (Courtesy USGS).

Scale and area relationship. The three maps illustrated in Fig. 3.7 have the same dimensions, but are drawn at different scales. A different area will be represented on each map, since a relationship exists between the map scale and the area represented. For any map of given outside dimensions the ground area represented will vary inversely with the square of the change in scale. Since the map at (b) is twice the scale of map (a), the area is only one-fourth as great. The scale of map (c) is four times larger and will cover one-sixteenth as much area.

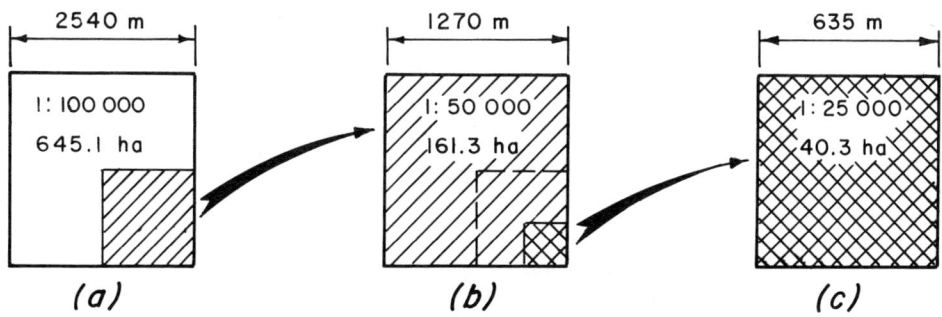

Fig. 3.7 Scale and area relationship.

3.8 SCALE TRANSFORMATION

A change or transformation in scale between an RF, a verbal statement, and graphic scale is often required to satisfy different purposes. At other times it may be necessary to convert ground distance to map distance and vice-versa. The following examples illustrate how these conversions can be made.

Problem 1: <u>Converting a verbal statement to an RF</u>

(a) Scale: 1" = 5 miles

$$\frac{MD}{GD} = \frac{1"}{5 \text{ miles}} = \frac{1"}{5 \times 63,360}^* = \frac{1}{316,800} \text{ or } \underline{1:316,800}$$

* Map and ground distances must be in the same units requiring that the denominator be multiplied by the number of inches per mile where 1" = 63,360"

(b) Scale: 4" = 1 mile

$$\frac{MD}{GD} = \frac{4"}{1 \text{ mile}} = \frac{4"}{1 \times 36,360} = \frac{4/4}{63,360/4}^* = \frac{1}{15,840} \text{ or } \underline{1:15,840}$$

* The numerator of an RF is always 1, requiring that both the numerator and denominator be divided by 4.

Problem 2: <u>Converting an RF to a verbal statement</u>

(a) Scale 1: 100,000

Number of inches/mile = number of inches in a mile/Denominator of the RF

$$= \frac{63,360}{100,000} = \underline{.64}$$

(b) Scale: 1:50,000

Number of miles/inch = Denominator of the RF/number of inches in a mile

$$= \frac{50,000}{63,360} = \underline{.79}$$

(c) Scale: 1:20 000

Number of cm/km = Number of cm in a km/Denominator of the RF

$$= \frac{100\ 000}{20\ 000} = \underline{5}$$

(d) Scale: 1:250 000

Number of km/cm = Denominator of the RF/Number of cm in a km

$$= \frac{250\ 000}{100\ 000} = \underline{2.5}$$

Problem 3:   Determining ground distance

  Map Scale: 1:62,500
  Map Distance: 8.5 inches
  Ground Distance: Map distance x Denominator of the RF

  GD = 8.5 x 62,500 = 531,250 inches

  GD (in feet) = $\frac{531,250}{12}$ = 44,270.8 feet

Problem 4:   Determining Map Distance

  Map Scale: 1:25 000
  Ground Distance: 150 m or 15 000 cm
  Map Distance: Ground distance/Denominator of the RF

  MD = $\frac{15\ 000}{25\ 000}$ = 0.6 centimetres or 0.006 metres

3.9 NATIONAL MAP ACCURACY STANDARDS

The location and elevation of points are of vital interest to map users. If maps are to be used where accuracy is essential, a dependable map with errors reduced to a practical minimum must be insured. The scale of a map and its accuracy, however, is closely related. To help individual Federal agencies producing maps, standards of accuracy for location (latitude and longitude) and elevation (height above sea level) have been established. The U.S. Bureau of the Budget is responsible for the standards that follow:

"With a view to the utmost economy and expedition in producing maps which fulfill not only the broad needs of standard or principal maps, but also the reasonable particular needs of individual agencies, standards of accuracy for published maps are defined as follows:

1. Horizontal accuracy. For maps on publication scales larger than 1:20,000, not more than 10 percent of the points tested shall be in error by more than 1/30 inch, measured on the publication scale; for maps on publication scales of 1:20,000 or smaller, 1/50 inch. These limits of accuracy shall apply in all cases to positions of well-defined points only. Well-defined points are those that are easily visible or recoverable on the ground, such as the following: monuments or markers, such as bench marks, property boundary monuments; intersections of roads, railroads, etc.; corners of large buildings or structures (or center points of small buildings); etc. In general what is well-defined will also be determined by what is plottable on the scale of the map within 1/100 inch. Thus while the intersection of two road or property lines meeting at right angles would come within a sensible interpretation, identification of the intersection of such lines meeting at an acute angle would obviously not be practicable within 1/100 inch. Similarly, features not identifiable upon the ground within close limits are not to be considered as test points within the limits quoted, even though their positions may be scaled closely upon the map. In this class would come timber lines, soil boundaries, etc.

2. Vertical accuracy, as applied to contour maps on all publication scales, shall be such that not more than 10 percent of the elevations tested shall be in error not more than one-half the contour interval. In checking elevations taken from the map, the apparent vertical error may be decreased by assuming a horizontal displacement within the permissible horizontal error for a map of that scale.

3. The accuracy of any map may be tested by comparing the positions of points whose locations or elevations are shown upon it with corresponding positions as determined by surveys of

a higher accuracy. Tests shall be made by the producing agency, which shall also determine which of its maps are to be tested and the extent of such testing.

4. Published maps meeting these accuracy requirements shall note this fact in their legends as follows: "This map complies with National Map Accuracy Standards."

5. Published maps whose errors exceed those aforestated shall omit from their legends all mention of standard accuracy.

6. When a published map is a considerable enlargement of a map drawing (manuscript) or of a published map, that fact shall be stated in the legend. For example, "This map is an enlargement of a 1:20,000-scale map drawing," or "This map is an enlargement of a 1:24,000-scale published map."

7. To facilitate ready interchange and use of basic information for map construction among all Federal mapmaking agencies, manuscript maps and published maps, wherever economically feasible and consistent with the use to which the map is to be put, shall conform to latitude and longitude boundaries, being 15 minutes of latitude and longitude, or 7½ minutes, or 3 3/4 minutes in size."

3.10 METRIC-ENGLISH SCALE EQUIVALENT

Drawings of different kinds and maps will be in English units of measurement for some time. Although metric conversions are presented in the appendices, it is best to avoid converting units and to think and work with the metric system. The following is a table of commonly used drawing and map scales in customary English units together with the probable nearest metric equivalent which may be used.

Table 3.3 English Metric Scale Equivalents.

CIVIL ENGINEER'S		MECHANICAL ENGINEER'S	
English	Metric*	English	Metric*
1 = 10	1:100	1/8 size	1:10
1 = 20	1:200	1/4 size	1:4 or 1:5
1 = 30	1:300	3/8 size	1:3
1 = 40	1:400	1/2 size	1:2
1 = 50	1:500	3/4 size	1:1 or 1:2 (as appropriate)
1 = 60	1:600	Full size	1:1

ARCHITECT'S		MAPPING	
English	Metric*	English	Metric*
1/32" = 1'-0	1:300	1" = 50'	1:500
1/16" = 1'-0	1:200	1" = 80' to 100'	1:1000
3/32" = 1'-0	1:100	1" = 200'	1:2000
3/16" = 1'-0	1:50	1" = 400' (1'-0 = 1 mile)	1:5000
1/4" = 1'-0	1:50	1" = 800' to 1000' (6" = 1 mile)	1:10 000
3/8" = 1'-0	1:30		
1/2" = 1'-0	1:20 or 1:25	1" = 2000' (2½" = 1 mile)	1:25 000]**
3/4" = 1'-0	1:20	1" = 1 mile	1:50 000
1" = 1'-0	1:10	1" = 2 miles	1:100 000
1 1/2"= 1'-0	1:10	1" = 4 miles	1:250 000
2" = 1'-0	1:5	1" = 8 miles	1:500 000
3" = 1'-0	1:4 or 1:5	1" = 16 miles	1:1 000 000
4" = 1'-0	1:3		

* The scales of the metric system are identified by the ratio of the size of the drawing to the size of the actual object drawn.
** USGS metric scales in the various topographic series.

GRAPHIC GEOMETRY AND CONSTRUCTION

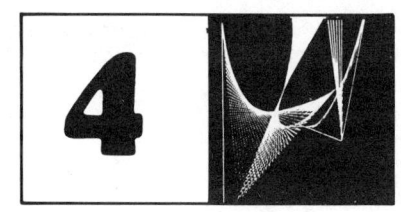

4.1 INTRODUCTION

In graphics it is necessary to understand geometric shapes and their construction for the drawing of objects and the solution of problems. Many of the geometric constructions may be drawn by a number of methods with the instruments discussed in chapter 2. The constructions presented represent several of the more useful methods and are illustrated with step by step instructions for easy understanding and drawing.

4.2 DEFINITION OF TERMS

The nomenclature to describe many common geometric shapes dealing with angles and surfaces is defined and shown in Fig. 4.1 through 4.4.

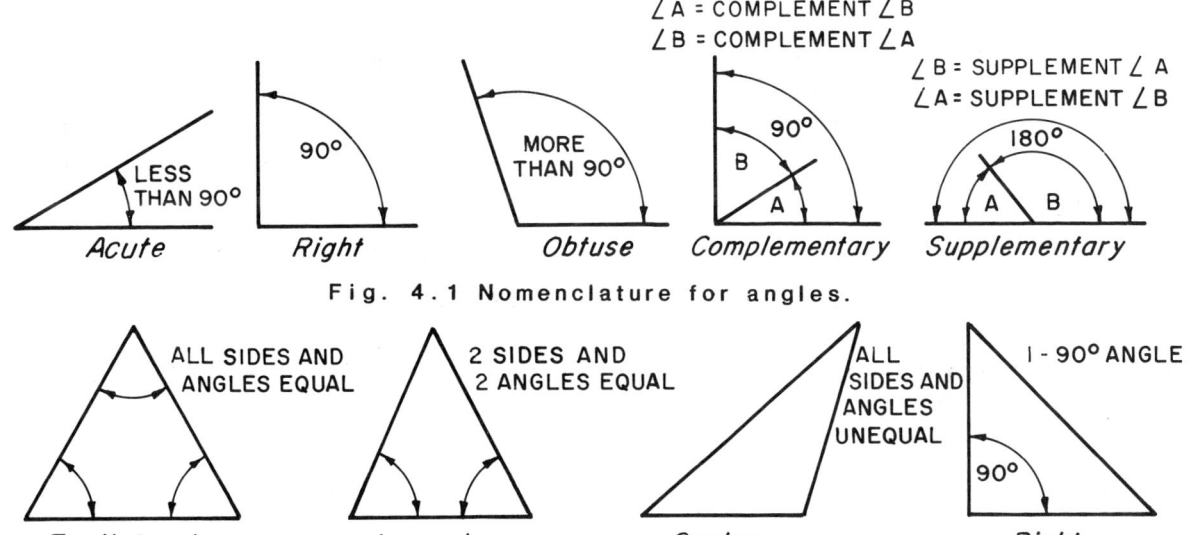

Fig. 4.1 Nomenclature for angles.

Fig. 4.2 Nomenclature for triangles.

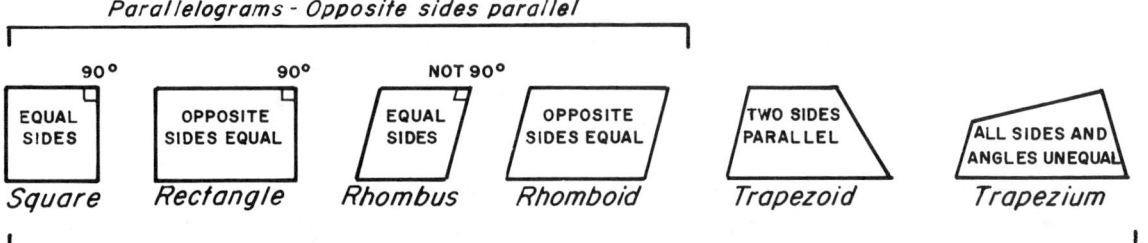

Fig 4.3 Nomenclature for quadrilaterals.

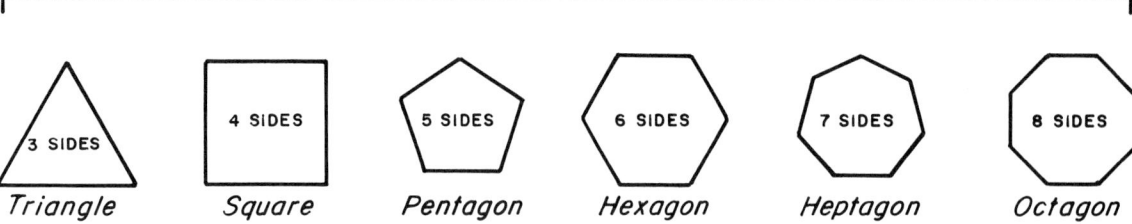

Fig. 4.4 Nomenclature for regular polygons.

4.3 DIVIDING A LINE INTO ANY NUMBER OF EQUAL PARTS (GIVEN LINE AB)

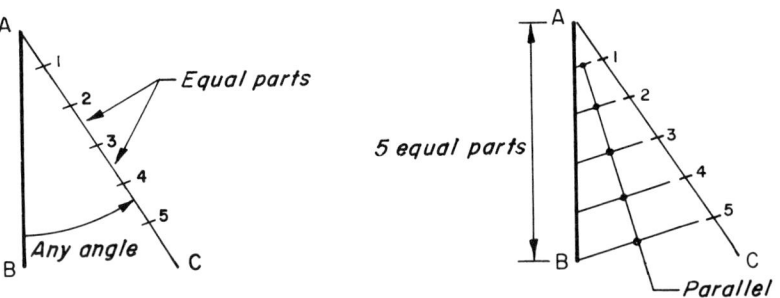

Draw line AC at any angle with line AB and lay off the number of required divisions using the scale or dividers.

Draw a line from point B to point 5. Lines drawn parallel to line B5 through the remaining points will divide line AB into five equal segments.

Fig 4.5 Dividing a line into any number of equal parts.

4.4 BISECTING A LINE OR CIRCULAR ARC (GIVEN LINE AB OR ARC AB)

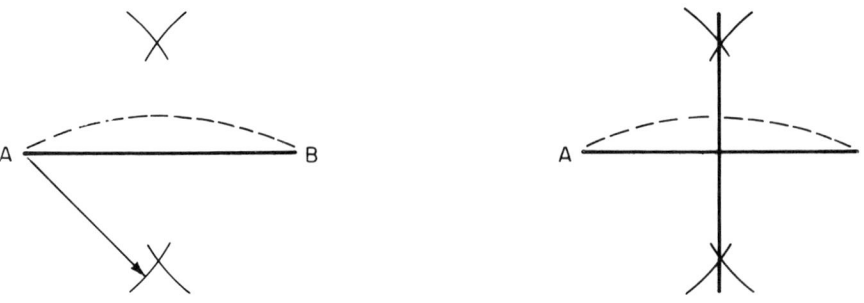

Draw arcs from points A and B using a radius larger than one-half of AB.

Draw perpendicular bisector through points where the arcs intersect.

Fig. 4.6 Bisecting a line or circular arc.

4.5 BISECTING AN ANGLE (GIVEN ANGLE ABC)

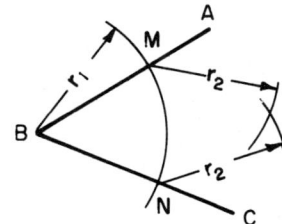 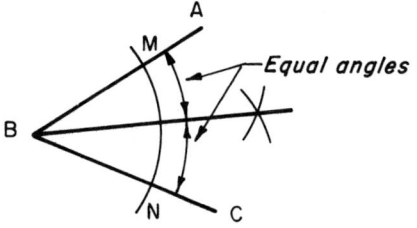

Draw an arc from point B using any radius, r_1. From points M and N where the arc intersects lines AB and BC draw two arcs using a radius, r_2, larger than one-half of MN.

Draw the bisecting line through the point where the arcs intersect.

Fig. 4.7 Bisecting an angle.

4.6 DRAWING AN ANGLE EQUAL TO A GIVEN ANGLE (GIVEN ANGLE ABC)

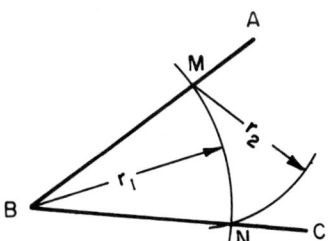 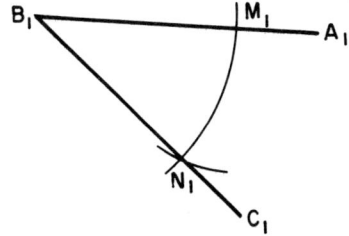

Draw an arc of any radius, r_1, intersecting line AB and BC at points M and N.

Establish the new position for the required angle and locate point B_1. Draw a line from point B_1 in the desired direction. Using radius, r_1, draw an arc to find point M_1. Draw a second arc of radius, r_2, to locate point N_1. Draw a line through points B_1 and N_1. Line segments AB and BC may be laid off to locate points A_1 and B_1.

Fig. 4.8 Drawing an angle equal to a given angle.

4.7 DRAWING AN ANGLE BY THE TANGENT METHOD (DRAW AN ANGLE OF 38°)

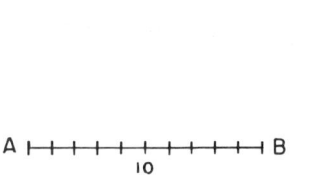
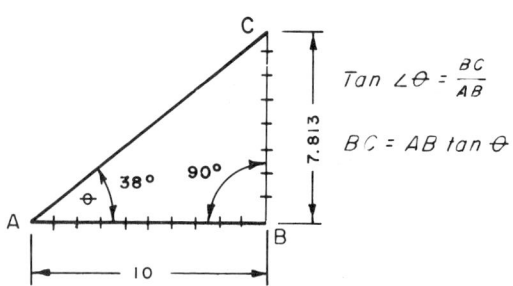

Multiply the tangent of 38° (0.7813) by 10.
0.7813 x 10 = 7.813
Draw a line AB and lay off 10 equal units.

The tangent of an angle is the ratio of the opposite side over the adjacent side in a right triangle. Draw a perpendicular at B and lay off 7.813 along the line to locate point C. A line from A to C will locate the required angle CAB equal to 38°.

Fig. 4.9 Drawing an angle by the tangent method.

4.8 DRAWING A CIRCLE THROUGH THREE GIVEN POINTS (GIVEN POINTS 1, 2, AND 3)

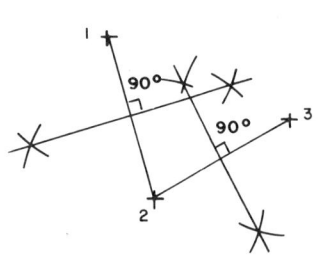
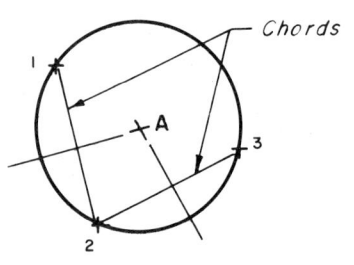

Draw lines between points and find the perpendicular bisector of each line.

Using point A as the center where the perpendicular bisectors intersect, draw the required circle through the three points.

Note: To find the center of a given circle or arc, reverse the above procedure. Draw any two chords, e.g., 1-2 and 2-3. Draw perpendicular bisectors. The bisectors intersection will be the center.

Fig. 4.10 Drawing a circle through three given points.

4.9 DRAWING A TRIANGLE (GIVEN THE THREE SIDES OF A TRIANGLE A, B, AND C)

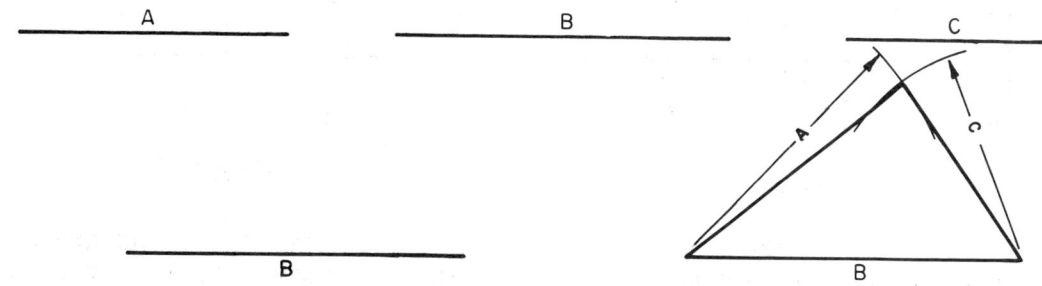

Draw any one side as the base.

Draw arcs having radii equal to the remaining two sides using the end points of line B as centers. Draw sides of triangle from point of intersection to the ends of the base line B.

Fig. 4.11 Drawing a triangle with three given sides.

4.10 DRAWING A REGULAR PENTAGON (GIVEN THE CIRCUMSCRIBED CIRCLE)

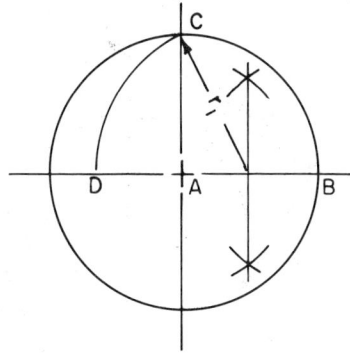
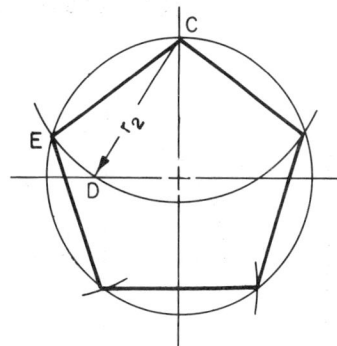

Bisect AB, and draw arc CD having a radius of r_1.

Draw an arc of radius, r_2, through point D, intersecting the circle at point E. Using CE as one equal side of the pentagon, step off this distance around the circumference to establish the five points.

Fig. 4.12 Drawing a regular pentagon in a circle.

4.11 DRAWING A REGULAR HEXAGON

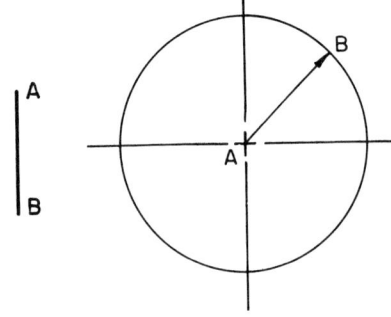
Draw a circle using the given side AB as the radius.

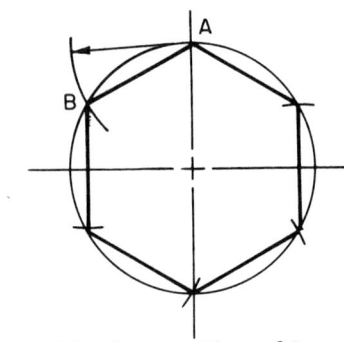
Step off the radius AB around the circumference to establish the six points.

(a) One side given (AB)

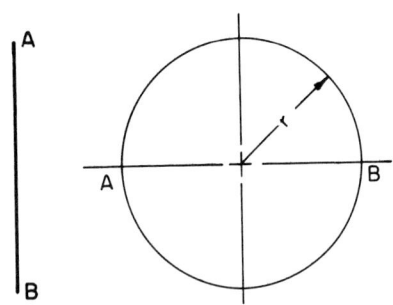
Draw a circle with the distance across corners, AB, as a diameter, r (d/2).

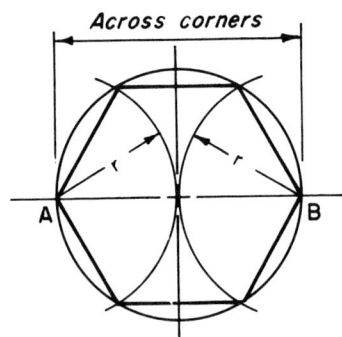
Using the same radius, r, used for the circle, draw two arcs with points A and B as centers to establish the six points for the *inscribed* hexagon.

(b) Distance across corners given (AB)

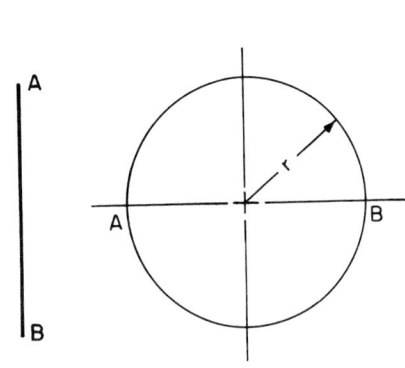
Draw a circle with the distance across flats, AB, as a diameter, r (d/2).

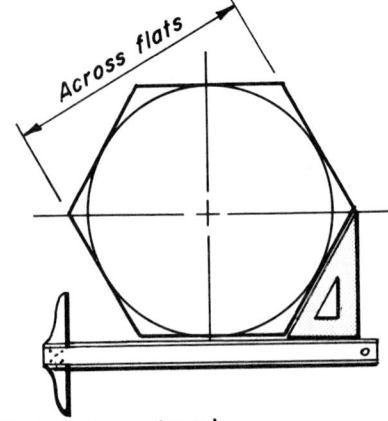
Using the T-square and 30°-60° triangle draw six lines tangent to the circle to establish the *circumscribed* hexagon.

(c) Distance across flats given (AB)

(c) Distance across flats given (AB)

Fig. 4.13 Drawing a regular hexagon.

4.12 DRAWING A REGULAR OCTAGON

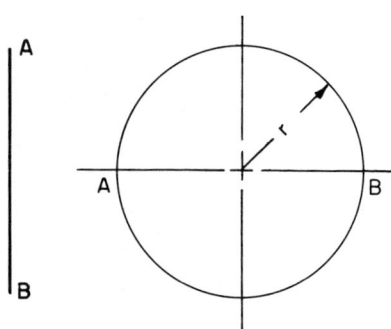

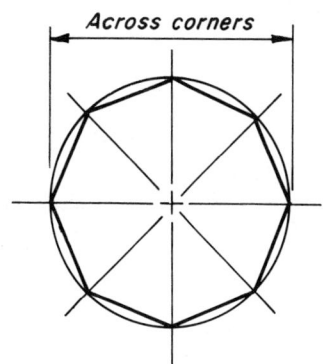

Draw a circle with the distance across corners, AB, as a diameter, r (d/2).

Draw two 45° center lines through the center of the circle. The eight points of the octagon are established where the center lines cross the circle.

(a) Distance across corners given (AB)

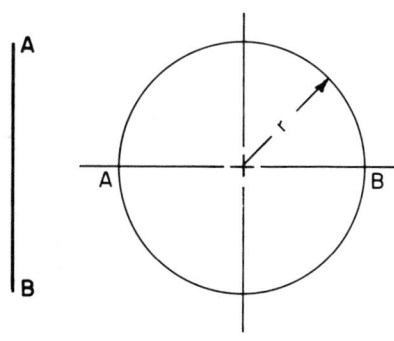

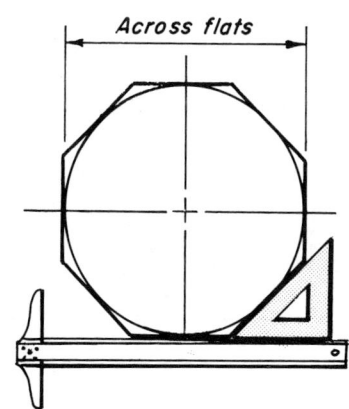

Draw a circle with the distance across flats, AB, as a diameter, r (d/2).

Using the T-square and 45° triangle, draw eight lines tangent to the circle.

(b) Distance across flats given (AB)

Fig. 4.14 Drawing a regular octagon.

4.13 DRAWING A CIRCULAR ARC OF A GIVEN RADIUS, r, TANGENT TO ANY TWO STRAIGHT LINES

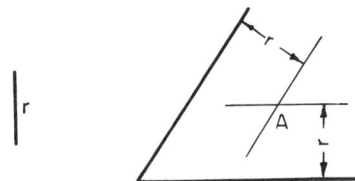

Draw two lines at distance, r, parallel to the given lines. The point of intersection, A, of these two lines will be the center to draw the required arc.

Find points of tangency (PT) by constructing perpendicular lines from point A to the given lines. Draw arc of given radius, r, from point A between tangent points.

Fig. 4.15 Drawing a circular arc of a given radius tangent to any any two straight lines.

4.14 DRAWING A CIRCULAR ARC OF A GIVEN RADIUS, r, TANGENT TO ANY STRAIGHT LINE AND ARC

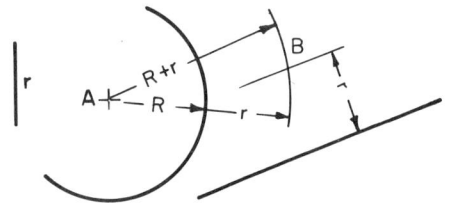

Draw a line at distance, r, parallel to the given line. Draw a *concentric* arc from point A using a radius of R + r. The intersection of the line and arc, point B, will be the center to draw the required arc.

Points of tangency (PT) are determined by drawing line AB and constructing a perpendicular bisector from point B to the given line. Draw arc of given radius, r, from point B between tangent points.

Fig. 4.16 Drawing a circular arc of a given radius tangent to any straight line and arc.

4.15 DRAWING A CIRCULAR ARC OF GIVEN RADIUS, r, TANGENT TO ANY TWO ARCS

Draw *concentric* arcs from points A and B using a radius of R + r. The intersection of the two arcs, point C, will be the center to draw the required arc.

Points of tangency (PT) are determined by drawing lines AC and BC. Draw arc of given radius, r, from point C between tangent points.

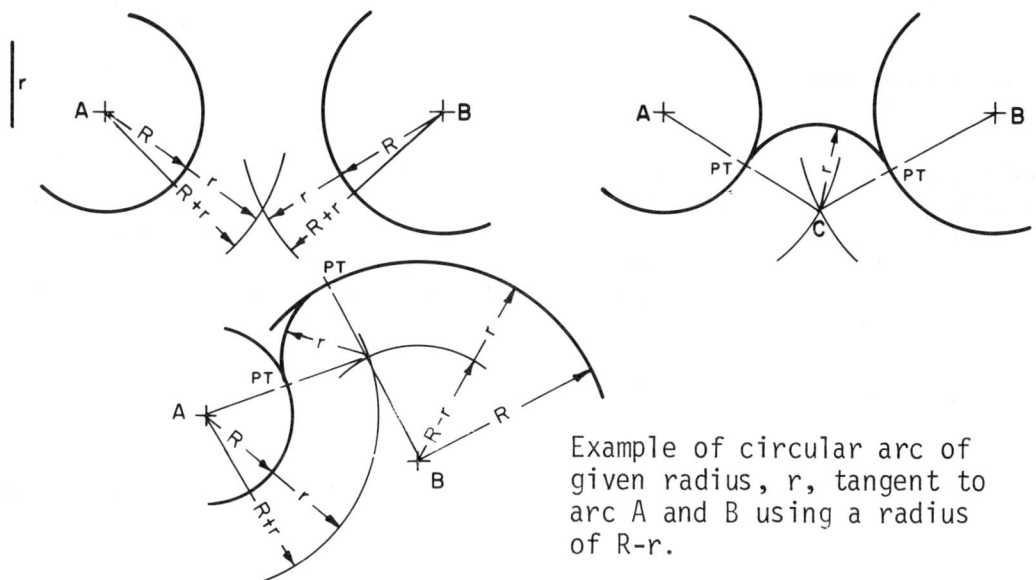

Example of circular arc of given radius, r, tangent to arc A and B using a radius of R-r.

Fig. 4.17 Drawing a circular arc of given radius tangent to any two arcs.

4.16 DRAWING A REVERSE (OGEE) CURVE TO TWO PARALLEL LINES USING EQUAL ARCS (GIVEN PARALLEL LINES AB AND CD)

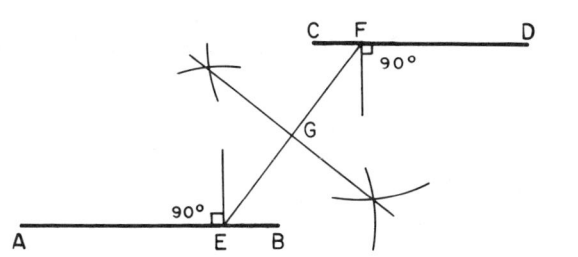
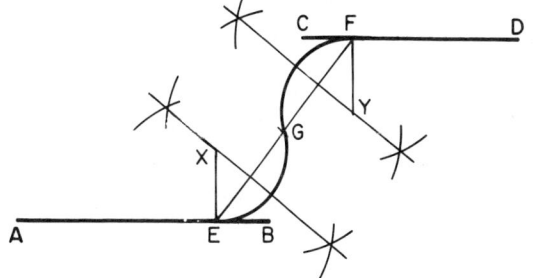

Draw any line, e.g., EF and divide into equal parts, EG and GF. Draw a perpendicular line to AB and CD at points E and F.

Draw perpendicular bisectors for lines EG and GF. The intersection of the bisectors with the perpendicular lines from E and F locates the centers, X and Y, to draw the required arcs of radius XB and YC. A line, XY, drawn between the centers of the arcs must always pass through the tangent point, G, of the arcs.

Fig. 4.18 Drawing a reverse (ogee) curve to two parallel lines using equal arcs.

4.17 DRAWING A REVERSE (OGEE) CURVE TO TWO PARALLEL LINES USING UNEQUAL ARCS (GIVEN PARALLEL LINES AB AND CD)

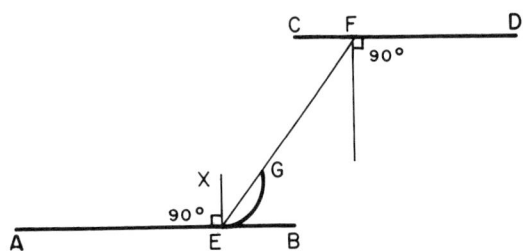 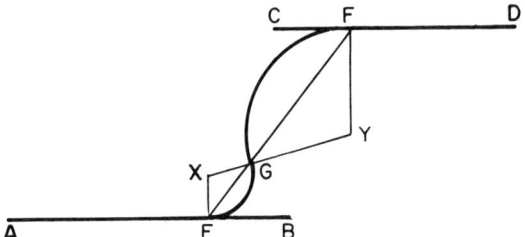

Draw any, e.g., line EF and construct a perpendicular line to AB and CD at points E and F. On the perpendicular line to AB lay off the radius, XE, of one of the arcs. Draw an arc of radius XE until it intersects EF at point G.

Draw line XG and extend until it intersects the perpendicular line drawn to line CD from point F. Point Y is the center to draw the remaining arc of radius YF. The two arcs will be tangent at point G.

Fig. 4.19 Drawing a reverse (ogee) curve to two parallel lines using unequal arcs.

4.18 DRAWING A REVERSE (OGEE) CURVE TO TWO NONPARALLEL LINES (GIVEN NONPARALLEL LINES AB AND CD)

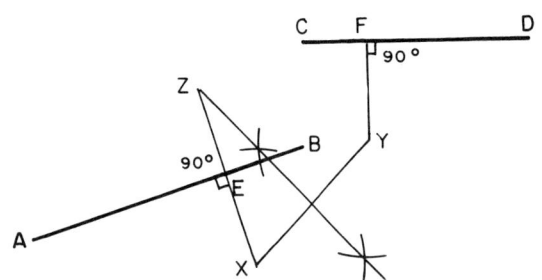 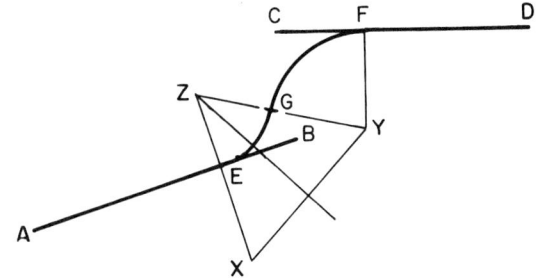

Draw a perpendicular line to AB and CD at any two points E and F. Lay off the required radius along both perpendicular lines to locate points X and Y. Draw line XY and bisect. Extend the bisector until it intersects the perpendicular from point E at point Z.

Using points Y and Z as centers, draw the required arcs of radius YF and ZE. The tangent point, G. for the two arcs is found by joining the centers with a straight line.

Fig. 4.20 Drawing a reverse (ogee) curve to two nonparallel lines.

4.19 CONIC SECTIONS

A cutting plane may be passed through a right circular cone of revolution to give four curves of intersection that are called *conic* sections, Fig. 4.21. An infinite variety of curves are possible, dependent on the slope of the elements of the cone, the slope of the cutting plane, and the relation between the cutting plane and the axis of the cone. Problems are frequently encountered in mathematics and graphics e.g., graph theory, roadways, structures, etc., that involve the conic sections. Several geometric methods for constructing the conic sections are presented as follows.

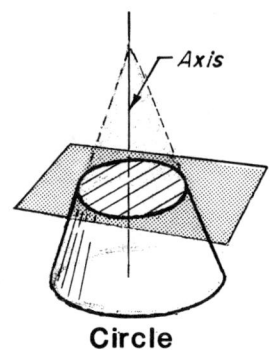
Circle

a. <u>Circle</u>. The section formed by a cutting plane that cuts all elements of the cone and is at right angles to the axis or is parallel to the base of the cone.

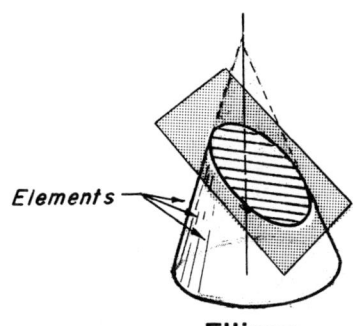
Ellipse

b. <u>Ellipse</u>. The section formed by a cutting plane making a greater angle with the axis than the elements of the cone. The plane is not parallel to any element of the cone, but cuts all the elements.

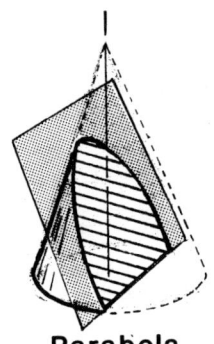
Parabola

c. <u>Parabola</u>. The section formed by a cutting plane making the same angle with the axis as the elements of the cone or is parallel to one of the elements.

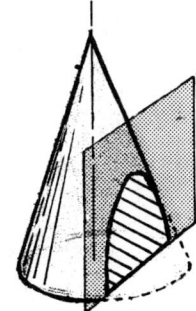
Hyperbola

d. <u>Hyperbola</u>. The section formed by a cutting plane making a smaller angle with the axis than do the elements of the cone or is parallel to the axis.

Fig. 4.21 The conic sections.

4.20 THE CIRCLE

The *circle* is a closed plane curve such that all of its points are equidistant from a point within, the center. Associated geometric terms for a circle are illustrated in Fig. 4.22.

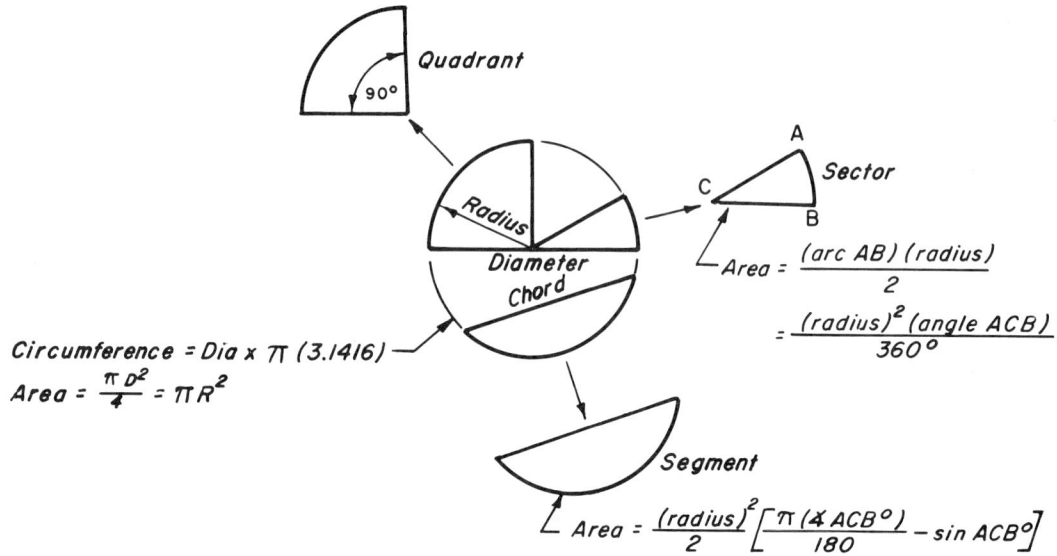

Fig. 4.22 Geometry of the circle.

4.21 DRAWING THE ELLIPSE

The *ellipse* is a curve generated by a point moving so that at any position the sum of its distances from two fixed points, the *foci*, is a constant, equal to the major axis (diameter). The long axis of an ellipsi is the *major axis*, and the short axis is the *minor axis* (diameter). In addition to the three methods shown here for drawing an ellipse, an approximate method which is sometimes easier is described in section 10.6

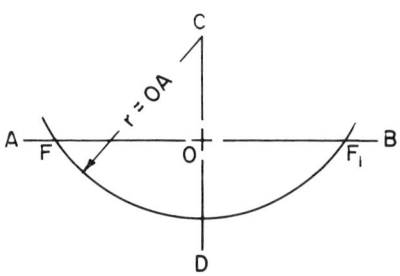
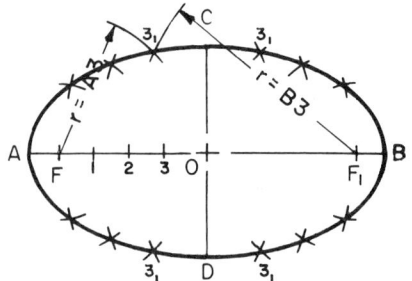

Draw major and minor axes, AB and CD. Locate foci, F and F_1, by drawing an arc from point C having a radius of OA (one-half the major axis.)

Divide FO into any number of convenient parts. Additional points will increase the accuracy. Draw arcs from points F and F_1 having radii of A3 and B3 respectively. The intersecting arcs locate point 3_1 on the ellipse in all four quadrants. The remaining points are found in the same way. Draw a smooth curve through the points using a French curve.

(a) Foci method.

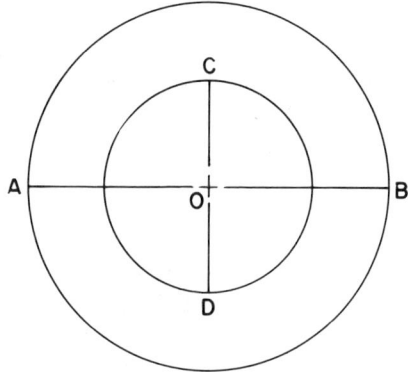
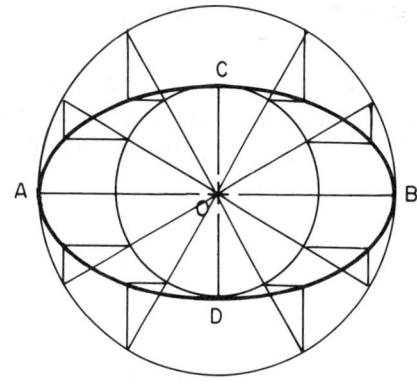

Draw major and minor axes, AB and CD. Using point O as the center draw two concentric circles having the major and minor axes as diameters.

Draw arbitrary radial lines through point O intersecting both circles. Points on the ellipse are found by drawing vertical lines from the points on the outer circle and horizontal lines from the points on the inner circle. The intersection of these lines determine the points on the ellipse. A sufficient number of points should be established to allow the ellipse to be drawn accurately as a smooth curve using the French curve.

(b) Concentric circle method

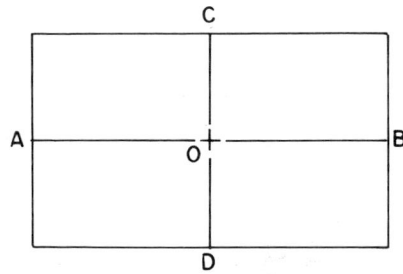
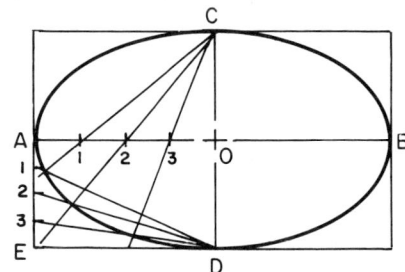

Draw major and minor axes, AB and CD. Construct the parallelogram (rectangle), using the major and minor axes as the center lines that will enclose the ellipse.

Divide AO and AE into the same number of equal parts. Additional parts will increase the accuracy. From points C and D draw lines through points 1, 2, and 3. The point of intersection of these lines, e.g., lines C1 and D1, will be on the required ellipse. Points in the remaining quadrants are found in the same manner. Draw a smooth curve through the points using a French curve.

(c) Parallelogram method

Fig. 4.23 Drawing an ellipse.

4.23 DRAWING A PARABOLA

The *parabola* is a curve generated by a point that moves so that at any position its distance from a fixed point, the *focus*, is always equal to its distance to a fixed line, the *directrix*.

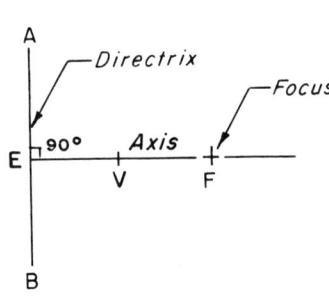 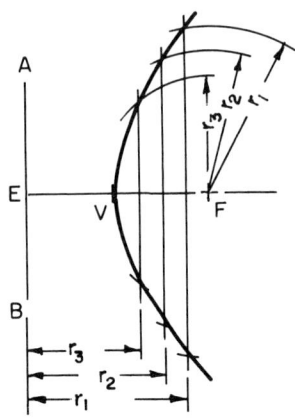

Draw the axis of the parabola perpendicular to the directrix AB. The vertex, V, of the parabola will occur halfway between points E and F.

Draw a series of lines through the axis parallel to the directrix AB. Using F as the center and the distances r_1, r_2, and r_3 as a radius, draw arcs intersecting the lines to locate points on the curve. A sufficient number of points should be found to draw a smooth curve.

(a) Directrix (AB) and focus (F) given.

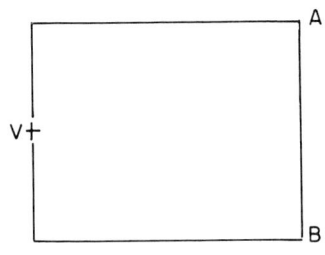 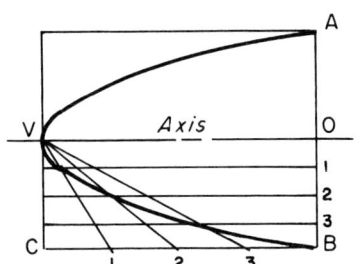

Construct a parallelogram (rectangle) with the sides passing through points V, A, and B.

Divide OB and CB into the same number of equal parts. Additional parts will increase the accuracy. Draw lines through the points on OB parallel to the axis. From point V draw lines to the points on CB. The point of intersection of the lines in pairs locate points on the parabola. Repeat the procedure for the remaining half and draw a smooth curve through the points.

(b) Parallelogram method. Given the vertex (V) and two symmetric points (A and B) on the curve of the parabola.

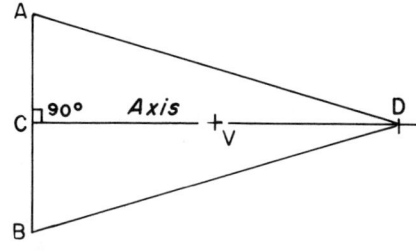 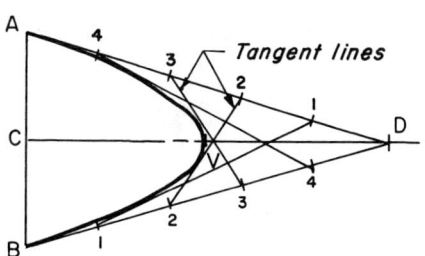

Draw the axis of the parabola CV perpendicular to AB. Extend the axis and lay off VD equal to CV. Draw lines AD and BD that will be tangent to the parabola at points A and B.

Divide line AD and BD into the same number of equal parts and label as illustrated. Draw lines between corresponding points. The required parabola is drawn as a smooth curve tangent to these lines.

    (c) Tangent method. Given the vertex (V) and two symmetric points (A and B) on the curve of the parabola.

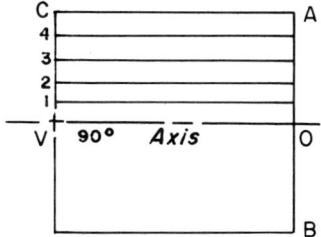 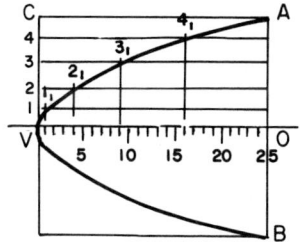

Construct a retangle with the sides passing through points V, A, and B. Divide VC into any number of equal parts (5) and draw lines through the points parallel to the axis VO. Additional parts will increase the accuracy.

The offsets vary as the square of their distances from point V. Therefore, divide line VO into 25 equal parts and mark the points 1, 4, 9, and 16 on the line VO. Draw perpendicular offsets to locate points 1_1, 2_1, 3_1, and 4_1, on the parabola. Repeat the procedure for the remaining half and draw a smooth curve through the points.

    (d) Offset method. Given the vertex (V) and two symmetric points (A and B) on the curve of the parabola.

Fig. 4.24 Drawing a parabola.

4.23 DRAWING A HYPERBOLA

The *hyperbola* is a curve generated by a point moving so that at any position the difference of its distances from two fixed points, foci, is a constant. Note that the constant difference between the radii is the distance AB in the example (a).

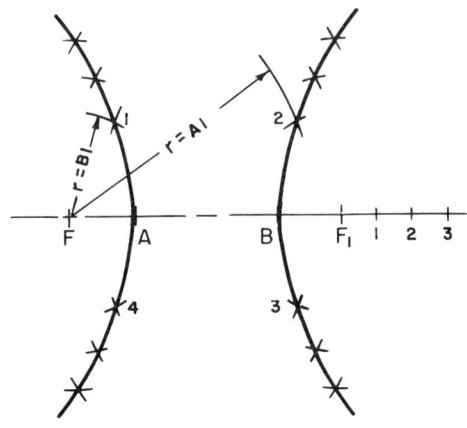

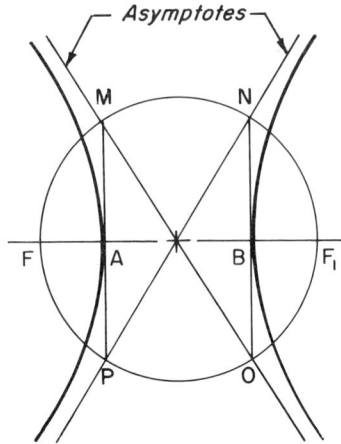

Lay off to the right of F_1 a series of points, 1, 2, 3, etc. Draw arcs of radius A_1 and B_1 using F and F_1 as centers. The intersecting arcs locate four symmetrical points, 1, 2, 3, 4, on the curve. Additional points are found by repeating the process. Draw a smooth curve through the points.

A hyperbolic curve extended toward infinity will gradually approach two straight lines known as *asymptotes*. These are located by drawing a circle having the diameter FF_1. Draw perpendicular lines through the vertices A and B that will intersect the circle at points M, N, O, and P. The asymptotes are located by drawing the lines as illustrated.

(a) Given the foci (F and F_1) and the vertices (A and B) on the curve of the parabola.

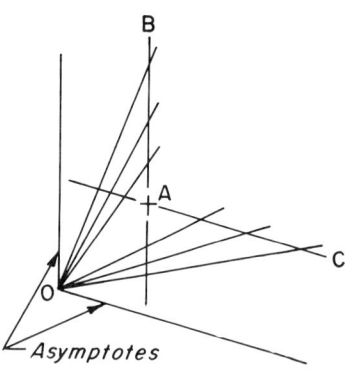

Draw two lines through point A parallel to the asymtotes. From point O draw a series of radial lines, on both sides of point A, intersecting line AB and AC.

Draw lines parallel to the asymptotes from every point where the radial lines intersect line AB and AC. The intersection of these lines are points, 1 through 6, on the curve through the points.

A rectangular or equilateral hyperbola is obtained when the asymptotes are at right angles to each other.

(b) Given the asymptotes and point A on the curve of the hyperbola.

Fig. 4.25 Drawing a hyperbola.

DRAWING MEDIA AND TECHNIQUES

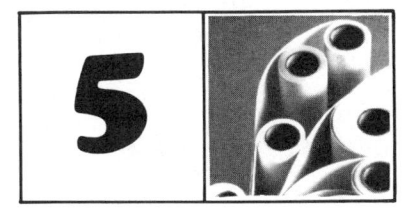

5.1 INTRODUCTION

The preparation of artwork is a principal activity for drawings to be reproduced and printed. Drawings are usually composed of line images, screens and tints, appliqué patterns, photographs, and lettering. The drawing may be reproduced to the same scale or copied to a different scale. It is a common practice to draw the original copy to a large scale and then reduce to the required size. This procedure allows a drawing to be prepared with greater precision and detail. Reducing a drawing will normally sharpen the linework. The specified reduction, however, must always insure that the drawing will be readable. Artwork must be prepared using line widths, appliqué patterns, and lettering that is large enough to maintain a good relationship when reduced.

A drawing may be positive or negative or either right-reading or wrong-reading, Fig. 5.1.

In preparing positive artwork, dense black sharp lines are necessary to obtain good reproduction. An ink original is the usual method for preparing positive copy on media that is white or clear. The original may be entirely drawn in ink or partly drawn and assembled from finished pieces of other artwork, preprinted symbols and patterns, and photographs, Fig. 5.2. This procedure is sometimes referred to as composite or paste-up drafting. A composite original produced in this way is easier, faster, and allows for interesting creative effects to be obtained.

The paste-up of artwork may be accomplished with rubber cement. Excess cement may be picked-up quickly after drying with a kneaded ball of dry rubber cement. Since this may be messy, it is easier and faster to use a waxer to lay down a coating of wax on the materials to be pasted together. Hand waxers, Fig. 5.3 or large roller type waxing machines that electrically heat the wax are ideal for this purpose. The adhesive is not tacky to the touch and permits papers, plastics, films, and photographs to be slid easily into position. No drying is required and materials may be peeled off safely for making modifications or corrections.

Artwork is prepared in right-reading form and will become wrong-reading when copied onto another material. This needs to be understood since in some printing processes (see chapter 20) the plate image must be

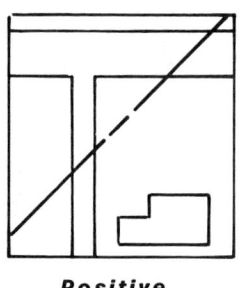
Positive

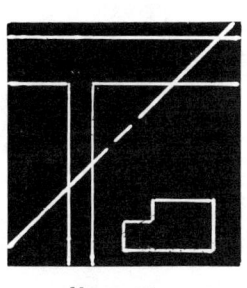
Negative

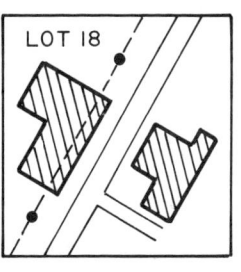
Right-reading

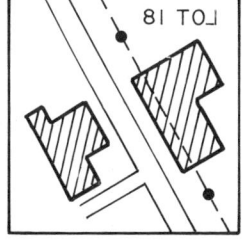
Wrong-reading

Fig. 5.1 Positive-negative and right-wrong reading artwork.

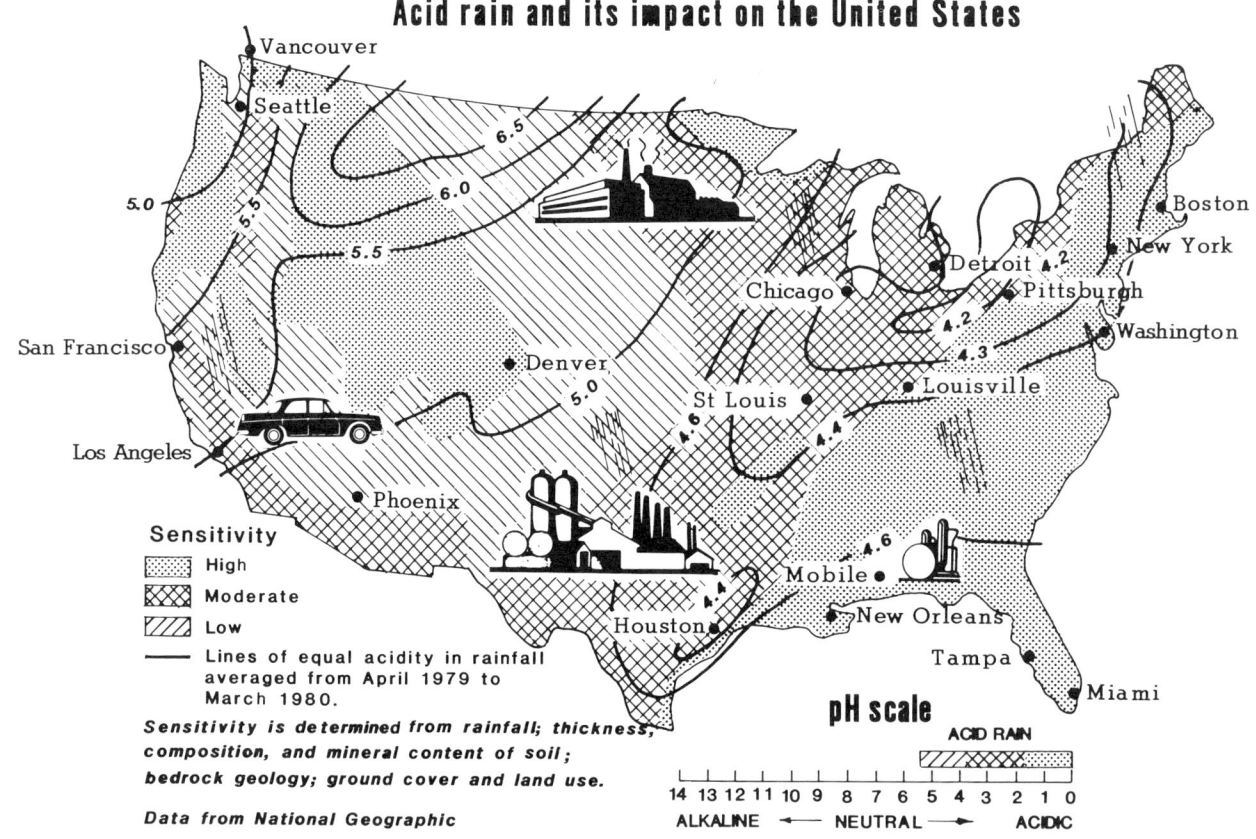

Fig. 5.2 Paste-up positive artwork.

wrong-reading to produce a print.

The final artwork for cartographic applications is frequently prepared in negative form such as scribing. Negative artwork is seen as white lines on a background that is black or opaque. Scribing is discussed in subsequent sections of the chapter along with other materials, methods, and considerations for the preparation of artwork.

5.2 APPLIQUÉ

Transparent micro-thin acetate sheets with a pressure-sensitive adhesive backing is used to print all kinds of patterns, Fig. 5.4, and graduated screen tints, Fig. 5.5. The application of this material instead of drafting is known as *appliqué* or *stickup*. Anyone preparing drawings can find many kinds of symbols, screens, colored films, shading patterns, and lettering to meet almost any requirement. Custom sheets are also made to order to serve special needs. It is virtually impossible to cover all of the materi-

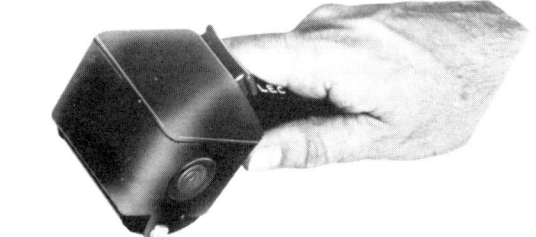

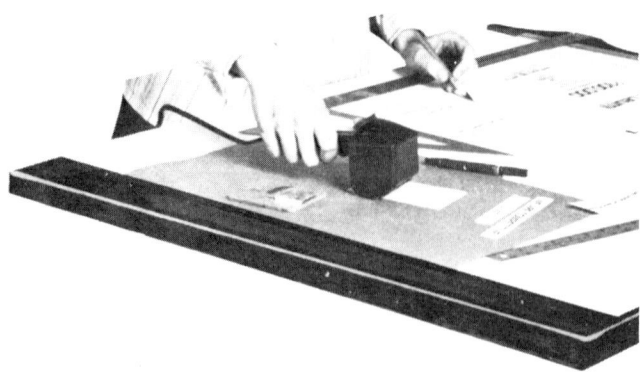

Fig 5.3 Lectro-Stik hand waxer
(*Courtesy Lectro-Stik Corp*).

Fig. 5.5 Graduated screen tints
(*Courtesy Chartpak*).

als that are available and manufacturers' catalogs should be referred to. Two different finishes are available - glossy and matte. The glossy sheets are most common and enhance colors particularly for direct visual presentation. Matte surface materials are used where original copy is to be reproduced or it is desired to type or draw on the material with pencil or ink. If changes are contemplated, lines drawn on the materials may be erased or the materials may be completely removed from the drawing. The non-reflective finish of matte surface materials makes it practical for copy shooting and direct visual communication in group presentations and exhibits. Light does not bounce back eliminating glare and "hot spots."

After the stickup is positioned on the drawing, it is cut out, Fig. 5.6, by using a sharp needle, razor blade,

Fig. 5.4 Examples of appliqué
screens and shading mediums
(*Courtesy Chartpak*).

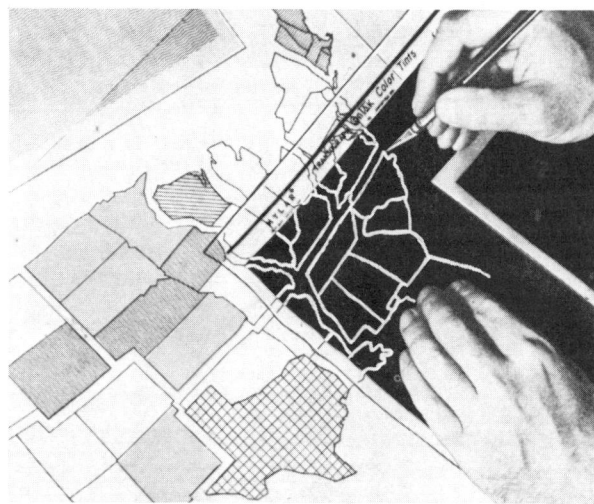

Fig. 5.6 Stickup application
(*Courtesy Chartpak*).

stylus or frisket knife and permanently affixed by *burnishing* (rubbing the cut out part with a burnishing tool or ball point pen). A sheet of paper should always be laid over the material before burnishing. Stickup coated with a wax adhesive should not be used on drawings intended for reproduction in thermal copy-type machines. The wax will melt and may cause the stickup to shift on the original copy. Maps and other kinds of drawings prepared with these materials may be executed with efficiency and a real savings in drafting costs may also be realized. It has a further advantage in that a more uniform appearance may be obtained. The problem is one of knowing when to use the materials. Final success in using this time-saving technique will depend on the awareness of the individual preparing the drawing. Shading films and color tints are ideal for indicating areal information. Patterns color or both, may also be used to highlight and emphasize information.

Pressure-sensitive tapes, 1/64" to 2" in width, printed on opaque or transparent surfaces are available in many patterns, and colors, Fig. 5.7. It is used for preparing graphs and drawings requiring straight or curved lines. Tapes are coated with a heat resistant, repositionable adhesive. It may be lifted from quality drawing surfaces and repositioned repeatedly. When properly burnished, the tape will not come loose during reproduction.

Tape holding devices, Fig. 5.8, have been designed to hold and unroll a roll of tape smoothly either freehand

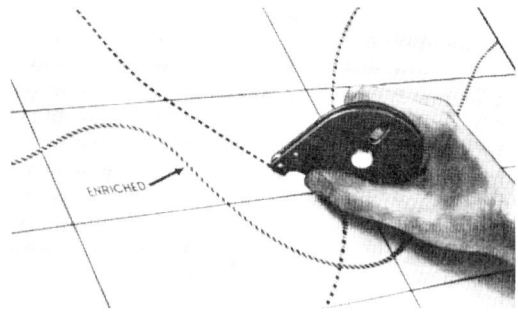

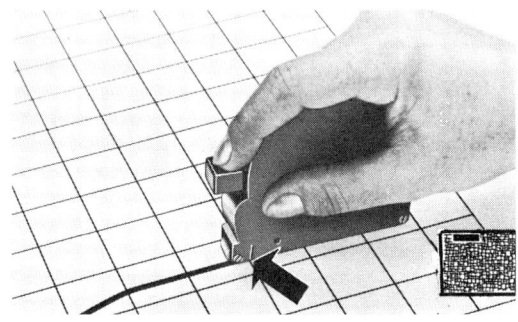

EXACT CUT-OFF INDICATION

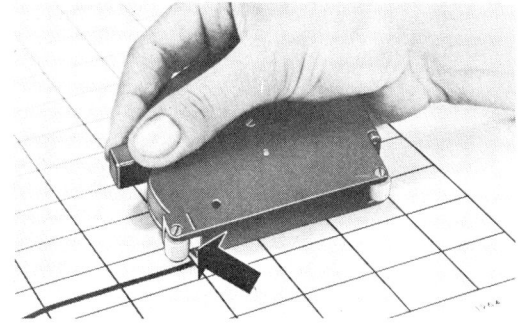

INTERNAL CUTTER BLADE

Fig. 5.8 Tape-holding devises *(Courtesy Para-Tone Inc)*.

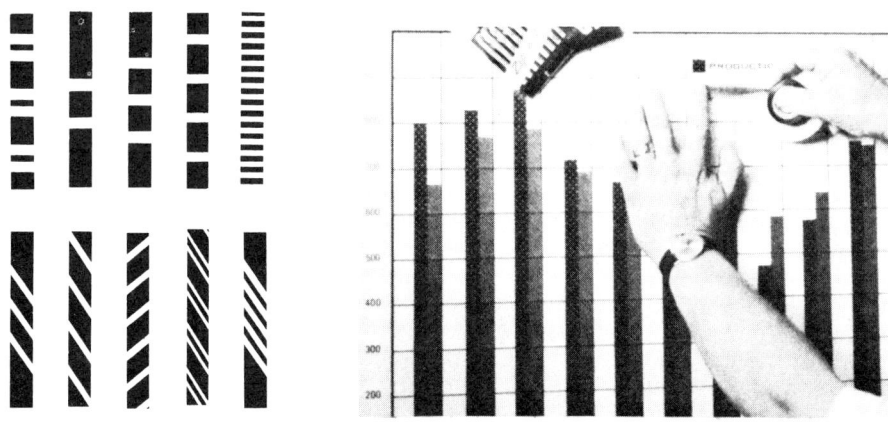

Fig. 5.7 Tape patterns *(Courtesy Chartpak)*.

or along the edge of a straightedge. Widths up to 1/4" may be used in many tape pens. They are particularly helpful for providing a quicker, easier, and more exacting way of positioning fine lines. An internal cutter blade, with a marker, allows the tape to be cut at any exact point.

Dry transfer letters, numbers, symbols, and repetitive data in sheets and rolls are available in virtually every style and size, Fig. 5.9. See chapter 6 on dry transfer lettering. The protective backing is removed and the symbol or letter is positioned on the drawing and rubbed lightly with a pencil, burnishing tool or ball point pen. This releases the image from the transparent carrier sheet and transfers it directly on the working surface. The backing sheet and not the carrier sheet should be laid over the image for final burnishing.

Transfer drafting films are available for applying preprinted repetitive symbols, letter type, diagrams, title blocks, typed data, and other information to drawings, Fig. 5.10. This procedure saves time and makes layout or design work easier. The material ac-

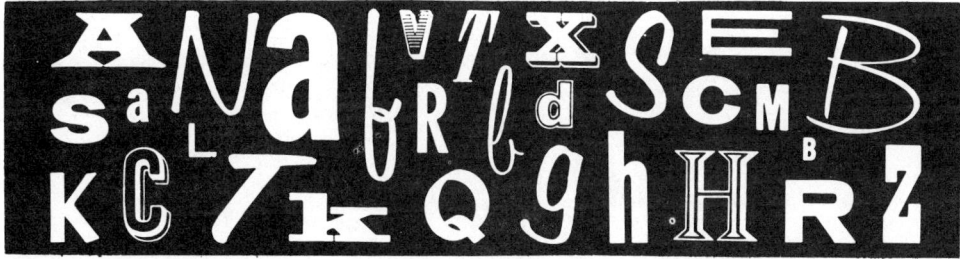

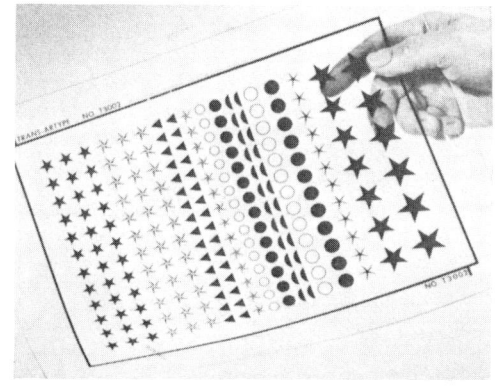

Fig. 5.9 Dry transfer appliqué *(Courtesy Artype Inc)*.

cepts pencil, ink or typing, and may be printed on using offset or letterpress methods. The film is coated on one side with an adhesive enabling the backing sheet to be easily removed for the film to be positioned on the drawing wherever required.

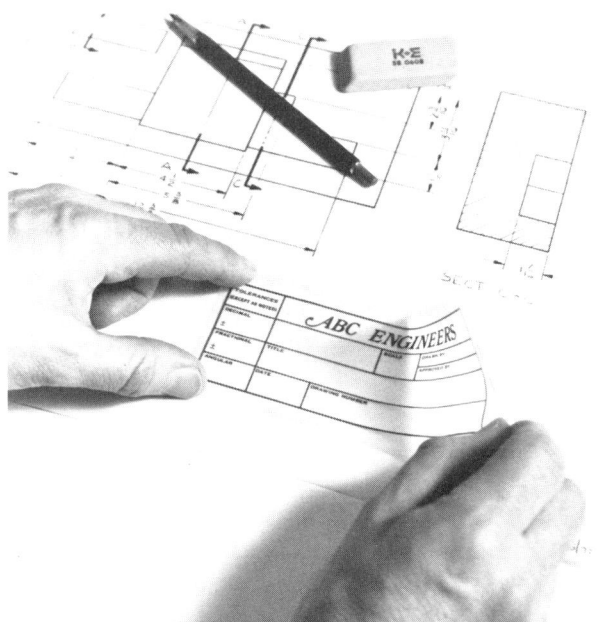

Fig. 5.10 Transfer drafting film (Courtesy Keuffel & Esser Co).

5.3 DRAWING MEDIA

Drawing media of different kinds are available in various size rolls and cut sheets of different sizes. Listed for reference are the dimensions of the metric International Standards Organization (ISO) paper sizes, Fig. 5.11.

The metric drawing sizes are based on the A0 size, having an area of one square metre and a length to width ratio of 1: $\sqrt{2}$. The smaller sheet sizes have an area half of the preceding size.

The dimensions in inches of the five Standard American Sheet sizes approved by the American National Standard Institute are shown in Fig. 5.12.

ANSI SIZE	ENGLISH-INCHES
A	8.5 x 11
B	11 x 17
C	17 x 22
D	22 x 34
E	34 x 44

Fig. 5.12 English paper sizes.

In determining the kind of material to use, the purpose of the drawing and how it will be reproduced should be considered. Characteristics to consider are transparency, strength, erasability, and permanence. Manufacturers will provide information on the qualities and characteristics of drawing media to help in selecting the best material.

Paper. For making original drawings in pencil or ink, opaque drawing paper is often used. High grade drawing papers are recommended where performance and erasability are important. Reproductions of a drawing often require that the original drawing be made on tracing paper or vellum. Two types of tracing paper, natural or transpar-

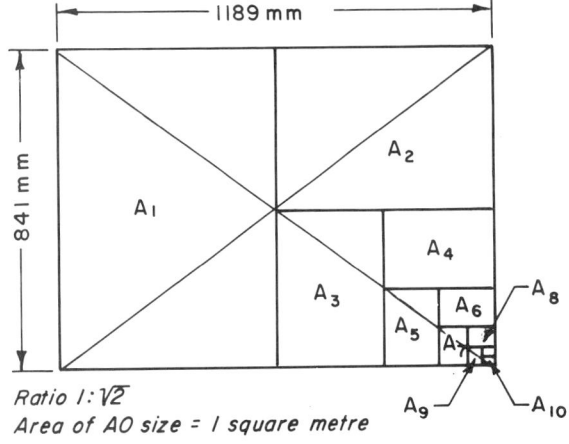

ISO SIZE	METRIC - mm
2A	1189 x 1682
A0	841 x 1189
A1	594 x 841
A2	420 x 594
A3	297 x 420
A4	210 x 297
A5	148 x 210
A6	105 x 148
A7	74 x 105
A8	52 x 74
A9	37 x 52
A10	26 x 37

Fig. 5.11 Metric paper sizes.

entized, should be considered depending upon the use and quality of paper required. Natural tracing paper is usually very transparent and possesses only moderate strength. It is excellent for sketching or other kinds of drawing where permanency is not a factor. Transparentized or prepared tracing papers are of high quality and will withstand repeated erasures and considerable handling.

Cloth. Tracing cloth is made from cotton fibers and treated to give a surface that is excellent for both pencil and ink work. Cloth is superior to paper and should be used where the combined advantages of transparency, surface quality, strength, and permanence are important. Moisture-resistant tracing cloth is recommended since ordinary tracing cloth treated with starch is water-soluble causing white spots to appear on the material when subjected to water or perspiration.

Gridded media. A vast assortment of sizes and shapes of gridded media is available to meet a variety of applications, Fig. 5.13. Grid patterns of different kinds have been produced to help save time and effort in the preparation of graphs, pictorials, sketches, maps, etc. Different grid colors and various kinds of base material, opaque or tracing paper, cloth or film, are also available. An orange grid should be used where it will be important to retain the grid detail when making multiple copies using some reproduction process. A, "no-print," blue grid will not reproduce and should be used when it is desirable to have only the drawn lines reproduced.

Film. A number of polyester base films have been introduced under trade names such as "Herculene," "Stabilene," and "Mylar." Characteristics that make film desirable for drawing are outstanding dimensional stability, resistance to warping and buckling, imperviousness to humidity, ability to withstand varying thermal conditions and great translucency. Dimensional stability is particularly important if maps and other types of drawings are to be drawn with one or more overlay sheets or if several colors are used. The overlays must

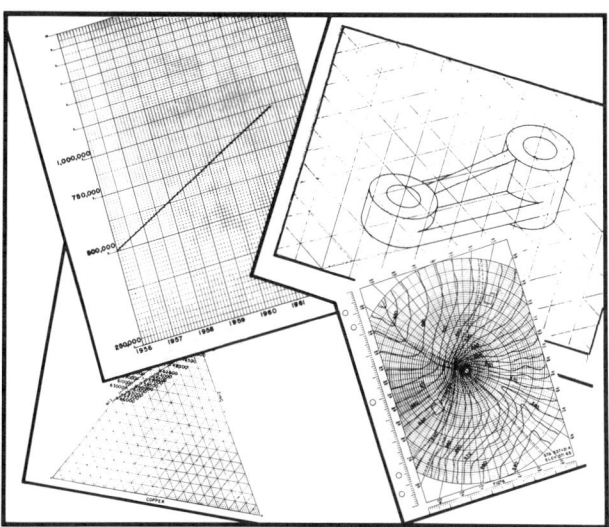

Fig. 5.13 Types of gridded media (Courtesy Keuffel & Esser Co).

register properly with the base map for reproduction and printing. In cartography most of the work is done on film.

For drawing in pencil or ink, film with a matte surface (light textured surface) on one or both sides is used. It is recommended that the film used be 2 (.002"), 3 (.003"), or 4 (.004") mil in thickness. Special pencils, inks, scribing tools, and erasers have been developed for use with many kinds of films. By using these products that are designed for compatibility with film, exceptional lines with a high degree of accuracy and clarity can be obtained. Films are very tough and will withstand repeated surface erasures without virtually having any effect on the material. Abrasive erasers should never be used since the matte coating may be worn away.

Special media. A number of advancements have been made in developing media for a variety of drawing functions. New materials are appearing on the market every day and manufacturers will furnish information on their use. Wash-off or sensitized materials that are compatible with many reproduction systems are used to restore drawings that become badly soiled, smudged, have lost their line density or legibility. Modifications in drawings may be made quickly by using positive to positive sensitized materials that will allow elements of the drawing to be eliminated with an erasing fluid prior to adding new lines to the drawing.

Drawings such as maps that are printed in color require a series of drawings that must be in perfect register. "No-print" sensitized film will allow any number of drawings to be prepared that will be in perfect registration.

Maps are frequently required for planning many kinds of projects and it may be helpful to stress certain aspects or features. Sensitized film may be used to prepare an intermediate print of a drawing in non-reproducible blue lines. The existing or new elements to be emphasized are then traced onto the surface of the film with pencil or ink. Reproductions made from the film will reproduce the pencil or ink lines, dropping out the non-reproducible blue lines. Other films will allow copies to be made that will show the drawn elements full strength with the rest of the drawing appearing in a subdued tone. Print copies may also be made that will show all traced or newly added elements in a blue color and the original elements in a red color.

5.4 OVERLAYS

Drawings do not have to be made on one piece of material. A series of *overlays* may be utilized to help transmit several types of information. Overlays are generally drawn on transparent or semitransparent sheets of polyester film with a matte surface on one side that can be registered to a common base drawing. Overlays are used in making architectural drawings, printed-circuit masters, overhead projection transparencies, and charts for displays. In cartography, the overlay method has been used for years, with separate sheets serving as the overlays for contour lines, grid lines, land characteristics, roads, waterways, lettering, etc.

The use of overlays offers many advantages. Clarity is improved for drawings so cluttered with detail information that their usefulness is limited. Considerable time may be saved and costs reduced for drawings that require frequent updating since the overlay "originals" are drawn separately. Different kinds of information may be emphasized where a significant relationship exists. Versatility may also be achieved by combining overlays with the basic drawing to make photo-reproductions. Colored maps are printed one color at a time. The colors must be separated with one overlay used for each different color. Press plates are made from each separate piece of artwork.

In preparing an overlay or artwork for color separation, an accurate dependable register system must be used to reliably position the base drawing with other pieces (flaps) of material. The registration marks are applied to both the base drawing and the overlay flaps, Fig. 5.14 (a). Usually the registration marks consist of small fine crosses drawn in the margins of the working area. A more precise method

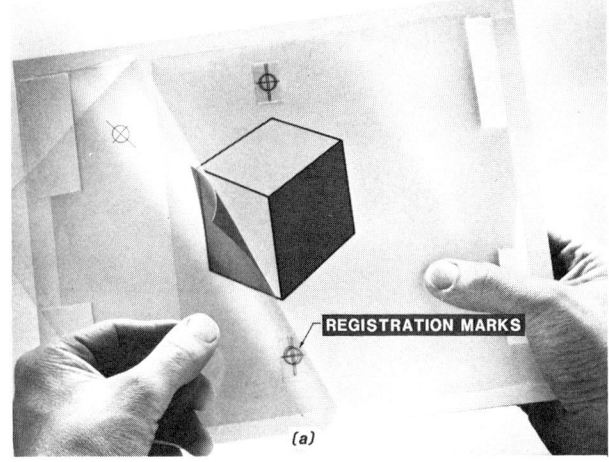

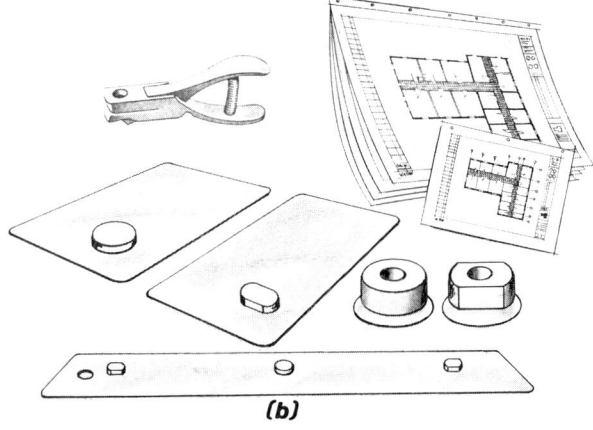

Fig. 5.14 Registration of overlays
(*Courtesy Keuffel & Esser Co and Bishop Graphics, Inc*).

of registering is to use some form of punch system in which holes are punched in the edges of the flaps. Registration pins with bases, individual registration pins or registration bars may be used to lock sheets together, Fig. 5.14 (b). The machined metal pins are inserted in the holes and held securely in place using double-sided tape.

To insure accuracy the basic drawing and all overlays should be prepared on dimensionally stable polyester film. Matte-surface films are appropriate for drawings made with pencil or ink. If two or more matte-surface overlays are going to be viewed simultaneously, back lighting is essential. For tape, appliqué materials, rub-on

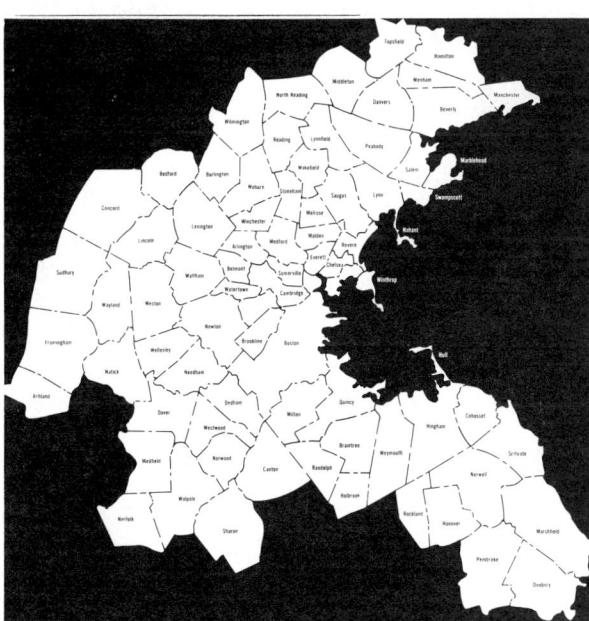

Base map showing boundary lines of the greater Boston area's cities and towns

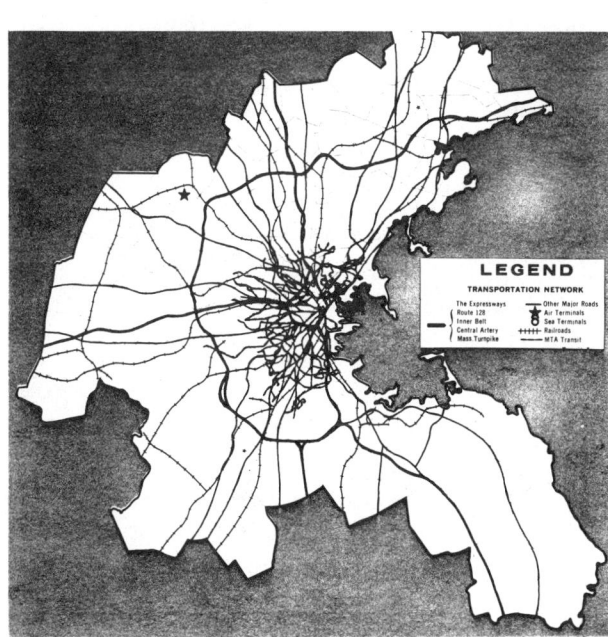

Base map and transportation network

Base map and major retail markets

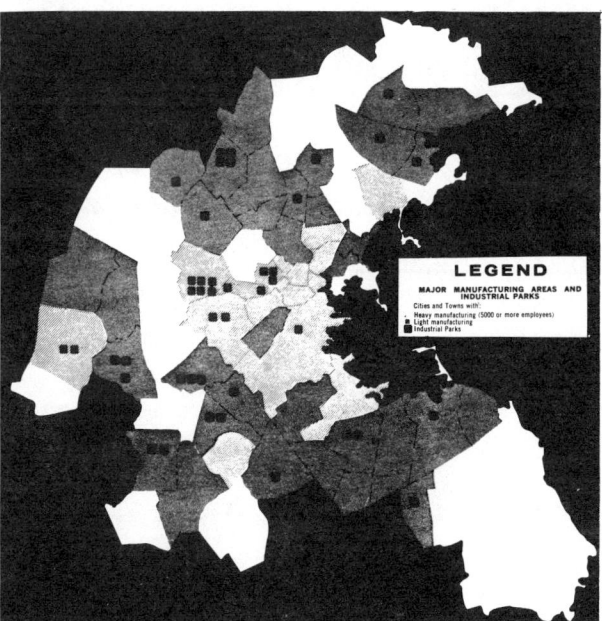

Base map and major manufacturing areas and industrial parks

Fig. 5.15 Use of base map and overlays (*Courtesy Greater Boston Chamber of Commerce*).

letters, and similar materials, either clear or matte-surface film may be used. It is helpful to apply appliqué materials to the back of the drawing so that it may be changed without affecting the lines drawn on the front.

An overlay may be limited to one attribute or type of information. In cartography, a basic drawing might be a topographical map. Information like timber types, geological outcrops, etc., would be an overlay. The basic map and overlays in Fig. 5.15 illustrate their use.

5.5 MASKS

In cartographic work, areas of the map such as water, vegetation, etc., are frequently printed in a color. A *mask* made of opaque material is used to prevent light from striking an area while other areas are exposed during plate making. The mask may be positive or negative, Fig. 5.16. A positive mask leaves the area of the drawing clear except the area to be treated which is opaque. Where a solid color is required to be printed, a positive mask is generally used since it may be exposed directly to the printing plate. Negative masks leave the area to be treated clear, while the remaining area is opaque. Areas that are to be screened require negative masks which may be exposed to the plate in combination with the screen. Screens are used in cameras to break continuous tone copy such as a photograph into a dot pattern, (see chapter 20). Both positive and negative masks may be converted by photographic contact printing.

The mask may be made by drawing the outline of an area and filling it in with an opaque substance on a clear material or by using plastic materials with a superimposed film and cutting away the film in the required areas. The film may be removed by cutting through the outlines by hand. In all cases, the outline must be followed exactly.

The use of sensitized masking film will eliminate hand cutting and help to insure perfect registration to produce several masks for color separation work. A photographic negative is made from the original line art and contact exposed to sheets of sensitized film. Any number of sheets may be exposed depending on the number of desired colors required for plate making and printing. The sheet may be developed in full daylight with every line of the original art photochemically etched into the film. The developed sheets are opaqued with a sponge brush to remove unwanted lines, and peeled to produce the negatives for plate making.

5.6 SCRIBING

A process referred to as negative *scribing* is widely used, in large scale production work, for producing line drawings and maps. Engineering tests and operational experience indicate a considerable time saving in using this method over the conventional pen and ink process. In scribing, a sheet of dimensionally stable polyester film that is coated with a thin film of opaque material is used. The surface of these materials is visually translucent for clearly scribing the image on a light table.

Materials. Scribe coat films are available with yellow, green, rust or white colored coatings. Since scribing is in the negative form, the white scribe coat film was developed to allow the drawing to be photographed as a positive by backing the scribe sheet with black paper. Outline Scribe Film with a double coating, white on one side with green or rust on the other side, is also used to obtain positive copy. A light table is not required

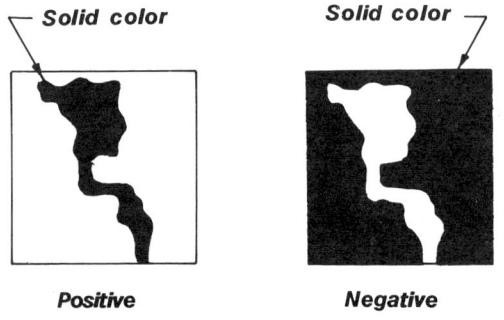

Fig. 5.16 Masks.

since the scribed lines will stand out clearly against the green or rust coating. Linework and lettering may be drawn or appliqué materials applied directly on the white side. When photographed, if backed up with black paper, all information including the scribed lines will be seen as positive copy. Since a positive (reflective copy) for camera sizing and a negative (contact) may be produced in a single step, the use of double-coated films is helpful in the reproduction process to save time.

When much scribing is necessary as in overlay scribing to produce map feature plates, Transparent Scribe Film is recommended. This material is highly transparent, allowing faster scribing. For scribing with high speed plotters, scribe films with a soft surface should be used. It is easier to scribe since less pressure is required.

The materials referred to are produced by the Keuffel and Esser Company. K & E has pioneered much of the development work on films and has published brochures dealing with the use of these materials in detail.

Instruments used for scribing are similar to those in conventional drafting, except that precision ground points (or blades) referred to as scribers (or gravers) are used in place of drawing pens and pencils, Fig. 5.17. Scribe points of various types are available for drawing fine, heavy, single or multiple, and straight or curved lines. The tips of the gravers may be either conical or spadetip like that of a chisel. They may be of steel, sapphire or other material which can be finely and accurately ground. The scribe points can be used in a pen type holder and used freehand for scribing. A variety of other type holders, having several legs mounted on ball feet, may be preferred for scribing large amounts of detail since they slide easily over the scribe material. Some models have a spring for adjusting the pressure on the graver point and a magnifying lens for detail scribing.

<u>How to scribe</u>. In scribing the manuscript is laid on a light table for tracing using one of the scribing points previously described, Fig. 5.18. For original drawings the design may be drawn lightly in pencil directly on the scribe surface. Changes and erasures may be made as the design progresses and develops. After the design has been scribed, linework and notes may all be wiped from the surface with a damp cloth. It is not necessary to remove pencil lines and notations since they will not

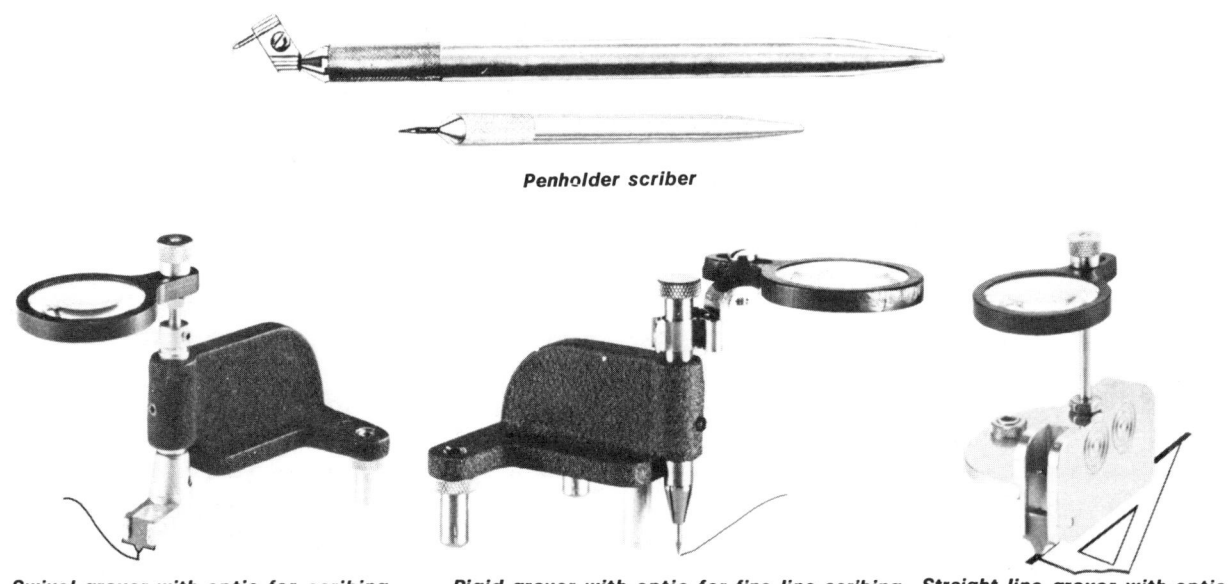

Penholder scriber

Swivel graver with optic for scribing where heavy or multiple lines are required.

Rigid graver with optic for fine line scribing.

Straight line graver with optic for scribing various width straight lines along a straightedge.

Fig. 5.17 Tools for precision scribing (*Courtesy Keuffel & Esser Co*).

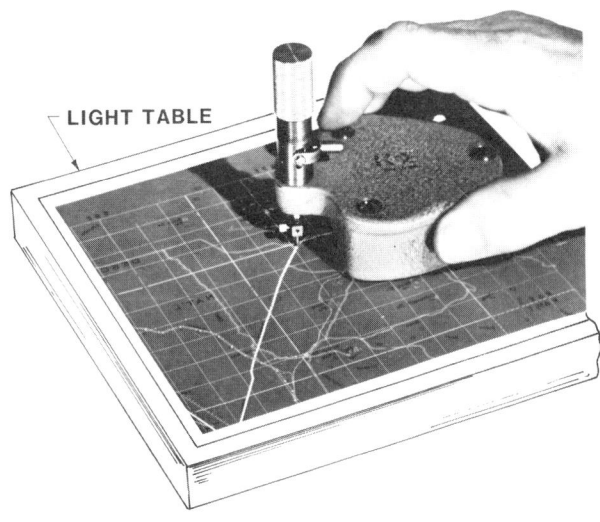

Fig 5.18 How to use scribe coat film
(Courtesy Keuffel & Esser Co).

show on a reprint. Lettering on scribe coat films may be facilitated by using a scribe point adapter with mechanical lettering devices such as LEROY. Several scribe points are available to scribe either fine-line or wide-line characters.

For making corrections and revisions several methods may be used. The quickest and easiest, for lines up to 0.25 mm (.010") wide, involves the use of an orange crayon pencil, Fig. 5.19 (a). Rubbing the pencil over the scribed line, at right angles, leaves a deposit in the lines. Any excess deposit should be removed with a tissue or cloth, using a light motion at right angles to the scribed lines. This will not rub off the deposit that is in the line. A touch-up liquid may also be used to paint over the scribed lines, Fig. 5.19 (b). It may be used for any line width. After the fluid has dried, the excess touch-up "build up" may be shaved from the surface using a single edge razor blade. Areas that are touched-up using crayon pencil or touch-up fluid may be rescribed. For areas that need to be blocked out and not rescribed, an opaque masking tape may be used, Fig. 5.19 (c).

<u>Sensitized materials</u>. In addition to unsensitized films, a number of sensitized films are available for a wide range of applications. Clear and matte sensitized films may be used to reproduce high-quality black line diazo prints of existing drawings. Addi-

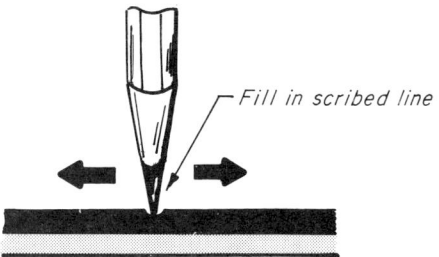

(a Crayon pencil

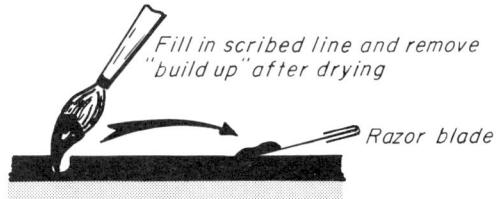

(b) Touch-up liquid

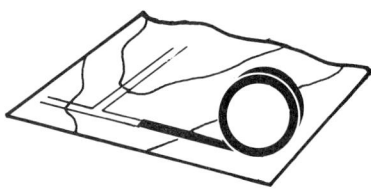

(c) Masking tape

Fig. 5.19 Touch-up techniques for scribing.

tions or corrections in pencil or ink may be drawn on the surface of these materials.

The image to be scribed may be transferred to sensitized diazo scribe films. A positive image is obtained by exposing a positive original, Fig. 5.20 (a), to the diazo sensitized scribe film, Fig. 5.20 (b), using either a Pulse Xeon or Carbon Arc light in a glass covered vacuum frame and developing with ammonia. Since the image is on the scribe surface, a light table although helpful is not required to scribe accurate lines. In multicolor map work several reproductions may be made from the original copy for scribing the number of color plates required.

Scribe film sensitized for photographic reproduction is used to produce duplicate line images and continuous tone positive reproductions by contact and projection printing, Fig. 5.21. The photographic emulsion on the scribe coat surface permits continuous imagery such as orthophotos, aerial or other continuous tone type negatives to be

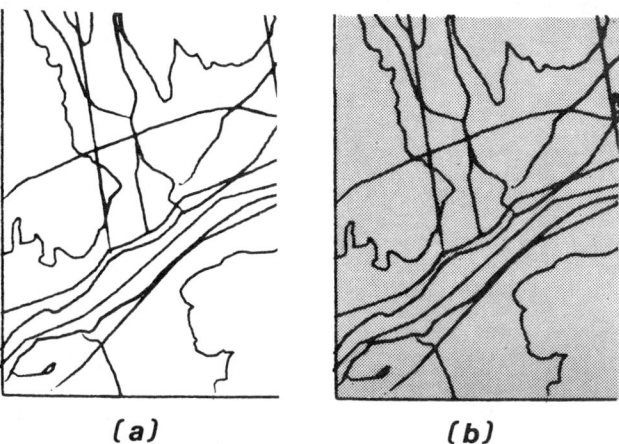

Fig. 5.20 Sensitized scribe film.

Fig. 5.21 Photographic guide line image for scribing (*Courtesy Keuffel & Esser Co*).

reproduced without loss of tonal characteristics on the scribe surface. The detail and tonal quality of the image allows lines to be scribed with speed and precision directly onto the photo image as a manuscript. The scribed lines are clean, sharp, and uniform.

Scribing may also be done photochemically, Fig. 5.22. Sensitized scribe coat film is used and contact-printed with a line drawing or film positive. The sheet is developed with an etching solution to remove the opaque coating wherever line detail prevents light from reaching the sheet. Etching is used if the reproduction contains a large quantity of numbers and symbols. The process is a quick and easy way to prepare copy when a map is being revised. All features may by retained except those that need to be rescribed.

Fig. 5.22 Scribing by etching.

Advantages. Scribed sheets resemble photographic negatives and will reproduce by contact printing, diazo methods, etc. Using the scribe film as a negative to produce printing plates eliminates the costly procedure of making a photographic negative of an inked original. Scribing offers advantages by insuring constant line widths, faster drafting, less time in training personnel, and absolute registration of several sheets for multicolor printing. By scribing line-work colors on separate scribe sheets, the expense of subsequent color separation is avoided and production time reduced. The scribed sheets, since they are in negative form, may be used for preparing multicolor proofs prior to printing (see chapter 20).

The scribing technique is used extensively by Federal agencies and private mapping organizations. An example of a map prepared by the scribing method is shown in Figure 5.23.

5.7 PEELABLE FILMS

Film of different kinds is available for stripping or peeling. It is extremely useful for preparing drop-out masks, and for producing large areas of exact dimension in a short period of time for map features like vegetation, soil types, and open water. After the surface is cut with a sharp needle or knife, it is peeled off leaving a perfect image, negative or positive as desired, Fig. 5.24.

Screens and tints for multicolor map production may be quickly and easily prepared with peel coat film. A separate piece of film is used to trace the information for each color, Fig. 5.25. All sheets must be in perfect registration with each other so that the final map printing will not show overlapping colors. Plates for printing each color may be made directly from the peel coat film overlays.

Photosensitized peelable film is recommended since it will allow the image of the map to be transformed to the film by photographic processing. The use of these films eliminates tracing and enables the cartographer to selectively and accurately remove areas that are to print on each sheet. Two types of photo-sensitive peelable film are available. Negative to negative peelable sheets are those made directly from scribed or photographic negatives. Positive to negative peelable sheets are obtained directly from line posi-

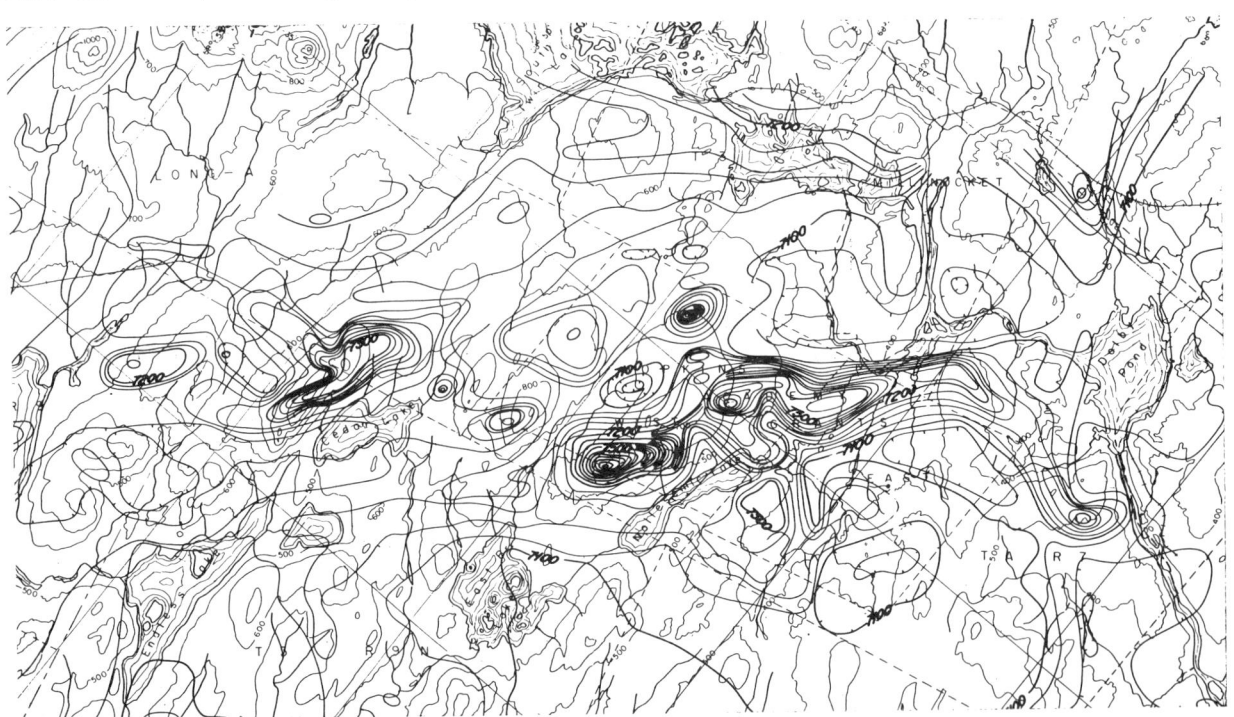

Fig. 5.23 Portion of a map prepared by scribing.

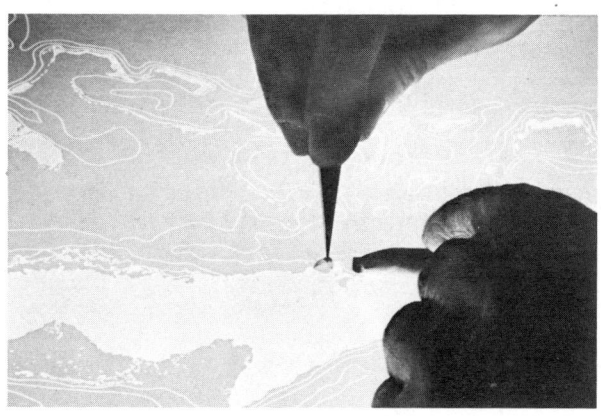 

Negative form   *Positive form*

Fig. 5.24 Strip or peel film application *(Courtesy Keuffel & Esser Co).*

tives, ink drawn or by photographic processes.

Figure 5.26 illustrates how peelable film open window negatives may be prepared from a single scribed original. The sensitized scribed film master, Fig. 5.26 (a), is obtained as explained in Fig. 5.20. Contact printing and etching the scribed master onto the peelable film, Fig. 5.26 (b), is accomplished as follows. Exposure is carried out by placing the peelable film in a vacuum frame, sensitized side out with the scribed original in contact so that the coated side of the scribed original or the emulsion side of the film negative is towards the operator. The vacuum frame is equipped with a rubber pad, a glass lid, and a pump to remove air from inside the frame. During the exposure, it is necessary that the film and scribed sheet be in tight contact. The exposure time, usually three or four minutes, will vary depending on light source, intensity, and distance. No darkroom facilities are required to develop the film. The sensitized surface (exposed side) is placed down in the developing solution for approximately three minutes. After developing, the film should be rinsed in water and thoroughly dried. Etching of the image requires etching solution and an etching block. The etching block is used to spread the etching solution evenly on the surface of the sensitized peel coat film. After 45 to 60 seconds the etching solution may be applied again. Applying light pressure with the etching block will remove the peelable coating not protected by the emulsion. The etched film should again be rinsed in water and dried. An opaquing liquid is applied to the film using a squeegee to mask out unwanted lines or areas. After drying (4 or 5 minutes) the film may be peeled, Fig. 5.26 (c), by lifting the edge of the section to be removed with a sharp instrument or a pair of tweezers. In Fig. 5.26 (b) a small section of film has been peeled to reveal the scribed road lines. Since the remaining features have been opaqued, no additional lines can be seen. The use of a light table will help to clearly see the areas that require peeling. Appliqué materials applied to the open window areas are excellent for creating different patterns, Fig. 5.26 (d). Any number of additional open window negatives (for tints, screened ruling, etc) may be prepared, using this procedure, to facilitate the preparation of artwork for color printing.

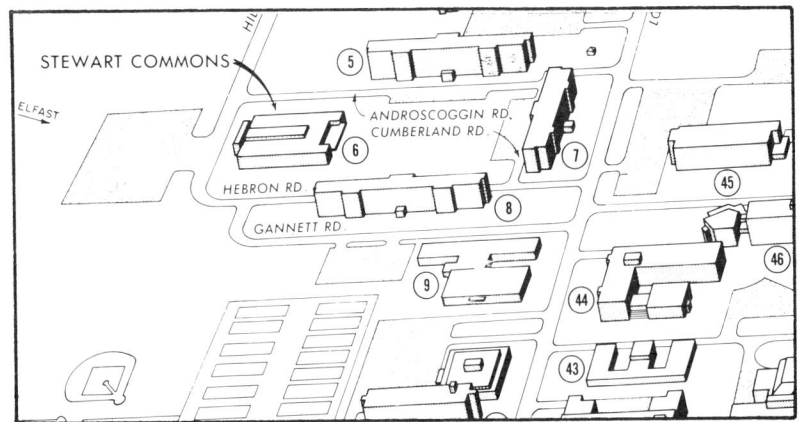

Base map drawn on Herculene film with black india ink.

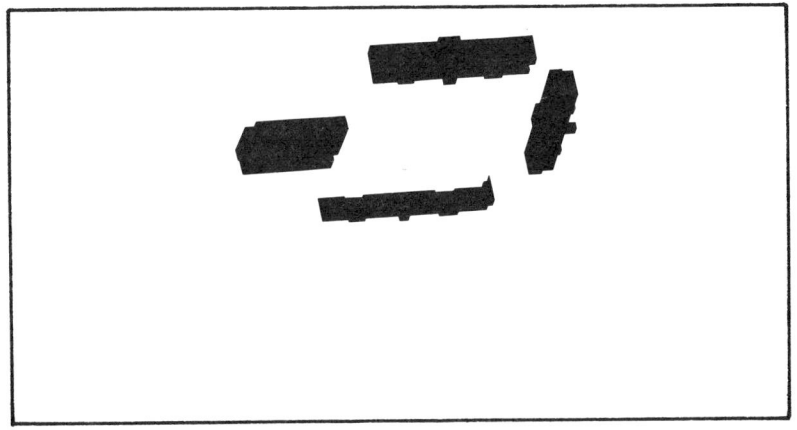

Peel coat positive film overlay for features to be printed in blue.

Peel coat positive film overlay for features to be printed in green.

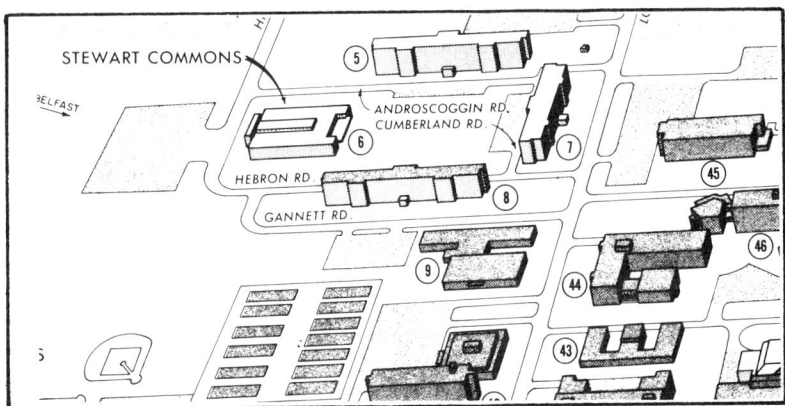

Finished multicolor map (colors omitted).

Fig. 5.26 Multicolor map production with peel coat film.

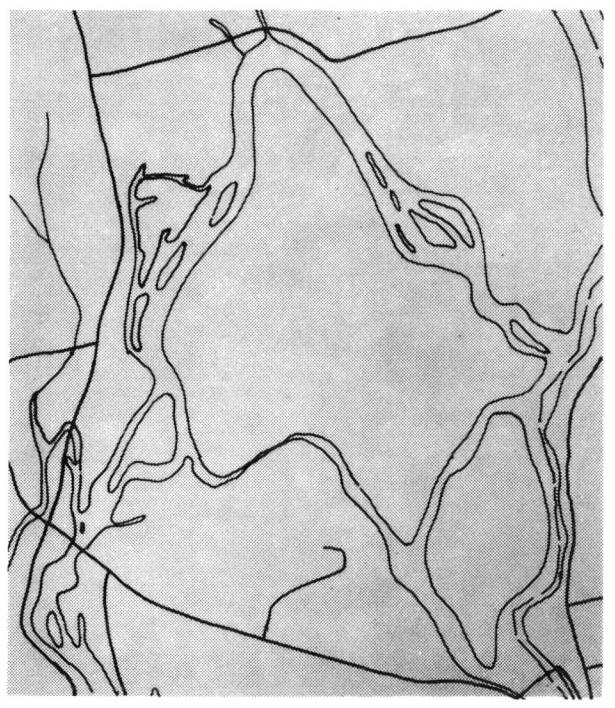

(a) Sensitized scribe film master

(b) Sensitized peelable film-Exposed, developed, and etched

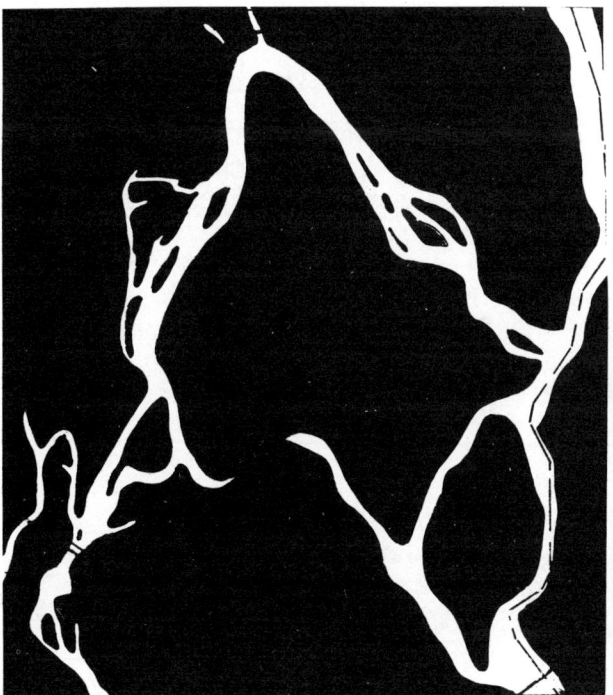

(c) Film masked with opaquing solution and peeled to produce an open window for water tint

(d) Application of applique symbol in open window for indicating a marsh

Fig. 5.26 Preparation of open window negatives using sensitized peelable film.

5.8 COLOR

The effectiveness of maps or other kinds of drawings may be improved by using various colors. A black-and-white map becomes confusing once details begin to overlap. Many features lose their definition and map reading becomes a chore. Various tints or solid color combinations may be used to identify and clearly distinguish specific features such as topography, culture, land classification, forest types, historical changes, etc. A multicolored map that is imaginatively conceived and skillfully produced has a psychological and graphic impact that is lacking in its black-and-white counterpart. Since the study of color is a complex subject, only major points are discussed to introduce the beginning student to the basic principles of color.

Color terminology. A number of elements should be considered to obtain good results. These include color selection, arrangement, and aesthetics. The effective use of color also requires a basic understanding of its various dimensions - hue, value, and chroma.

Hue is synonymous with color and is the quality which makes it possible to distinguish colors. The color wheel, Fig. 5.27, is an orderly arrangement of the basic colors and should be understood if pleasing combinations of colors are to be achieved. The hue of red, yellow, and blue are considered the *primary* colors. The *secondary* colors; orange, green, and violet are obtained by mixing adjacent primaries; e.g., green is obtained with blue and yellow. The *intermediate* or *tertiary* colors may be obtained by mixing a primary with a secondary color. Strong hues may be subdued by changing their value or chroma.

The colors frequently used in multicolor work for maps are black, dark blue, red, brown, and yellow. The color green is made from the combination of blue and yellow and purple from the combination of red and blue. Gray is normally produced by using a halftone of the black.

For preparing very presentable single maps and drawings, colored inks,

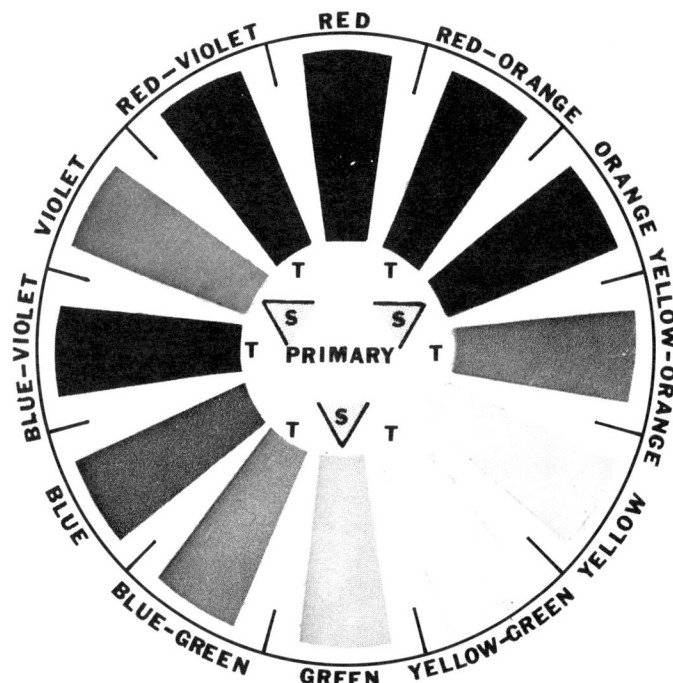

Colors frequently are reproduced and printed in black. This figure also illustrates how the various colors will reproduce. Many variations in value are possible due to varying exposures, use of filters, and different film.

Fig. 5.27 The color wheel.

flow pens, and pencils are manufactured in all the standard shades. Colored pencils are excellent for outline, tone, pastel, or wash effects, but have the disadvantage in that they smudge easily. An even tint is produced by rubbing the colors into the drawing surface with a pointed paper stomp. Contrasting colors should be used to distinguish features and the number of colors minimized to avoid an undesirable appearance of gaudiness and possible confusion. A drawing improperly colored cheapens it and may unjustly give it an appearance of poor drafting.

Value refers to the lightness or darkness of a hue. A color may be lightened, producing a *tint*, Fig. 5.28 (a), by mixing a lighter hue of the same color or by the addition of white. Solid colors may be modified by using halftone tints, which consist of very small dots or lines. A wide range of appliqué screens are also available which give a continuous tone appearance. Coordinated dot screens are available in intensity from 10 to 80 percent, and in screens of 27 to 100

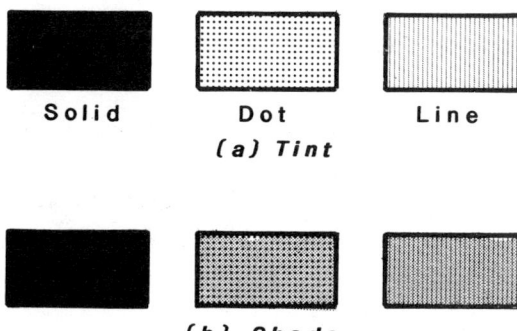

Fig. 5.28 Tints and shade.

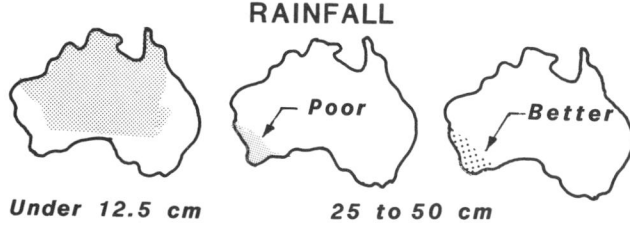

Fig. 5.29 Color saturation for large and small areas.

lines per inch. The problem in using this material is where the drawing is reduced or enlarged since patterns may become too small or too large for good reproduction. Dark values termed *shade*, Fig. 5.28 (b), are achieved by adding either a darker hue of the same color or black.

The use of tints is important in the preparation of maps. Color adds weight to the elements on the drawing and affects the balance. Solid colors do not work well for coloring large areas and should be reserved for emphasis on small areas. A color used with black should normally be relatively light, screened at a 30 or 40 percent level, so it will not draw undue attention from the black. The use of dark or light colors and dull (weak) or bright colors will provide a pleasing balance or relationship.

Perceptual differences in color will be affected by the size of the colored area. A very small area of a given tint will appear less saturated than a large area, and may appear to have a different tint, Fig. 5.29. The value of the tints used for small areas must be sufficient to allow two different area symbols to be differentiated. A light tint should be used for coloring large areas since the larger the area of a given hue, the more intense it will appear.

A fully saturated hue will appear dominant on a map and will "overprint" hues that are lighter and less saturated. A dark blue line will print clearly against yellow or pale green. Colors like red and blue-green, or browns and purple will always conflict. Pale yellow or yellow-green backgrounds will work well where the other elements are drawn in black, red, blue, and brown hues. The color environment should be light where the features of the drawing are dark, and vice versa, Fig. 5.30.

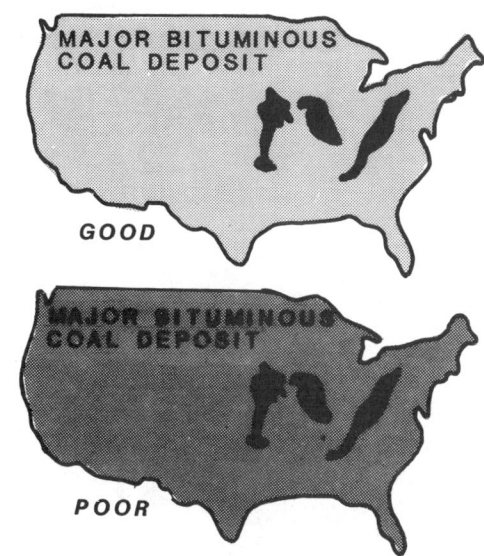

Fig. 5.30 Color environment-light and dark background.

Contrast in value, light versus dark, is more significant than contrast in hue, blue versus yellow for example. As a rule it is recommended that the colors applied to large areas be kept light and little saturated to maintain contrast and legibility of lettering and all symbols, Fig. 5.31.

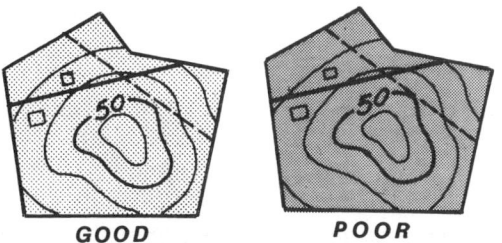

Fig. 5.31 Contrast-light and dark background.

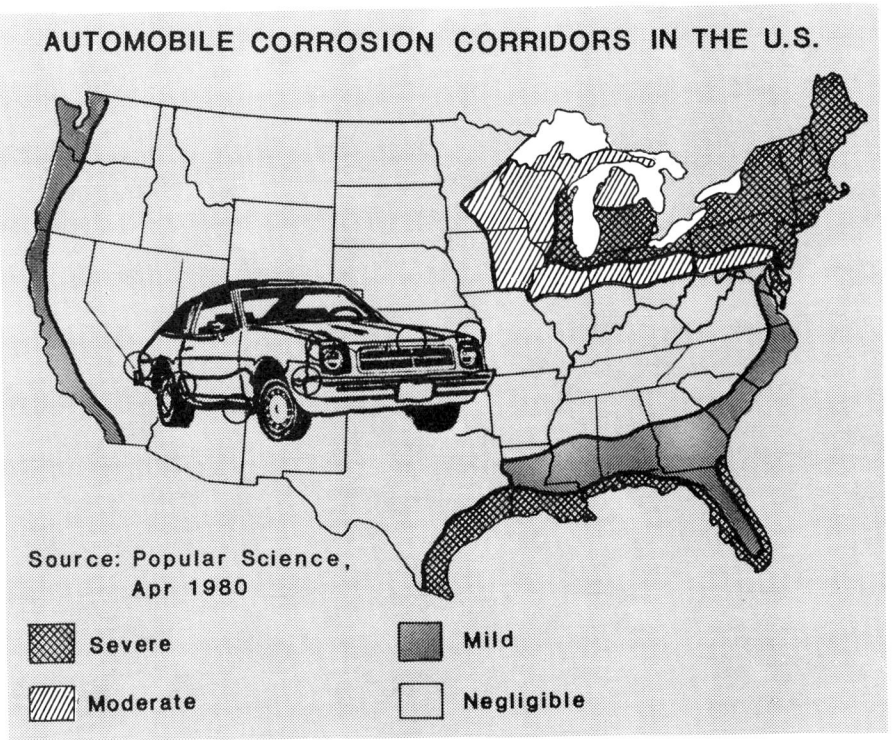

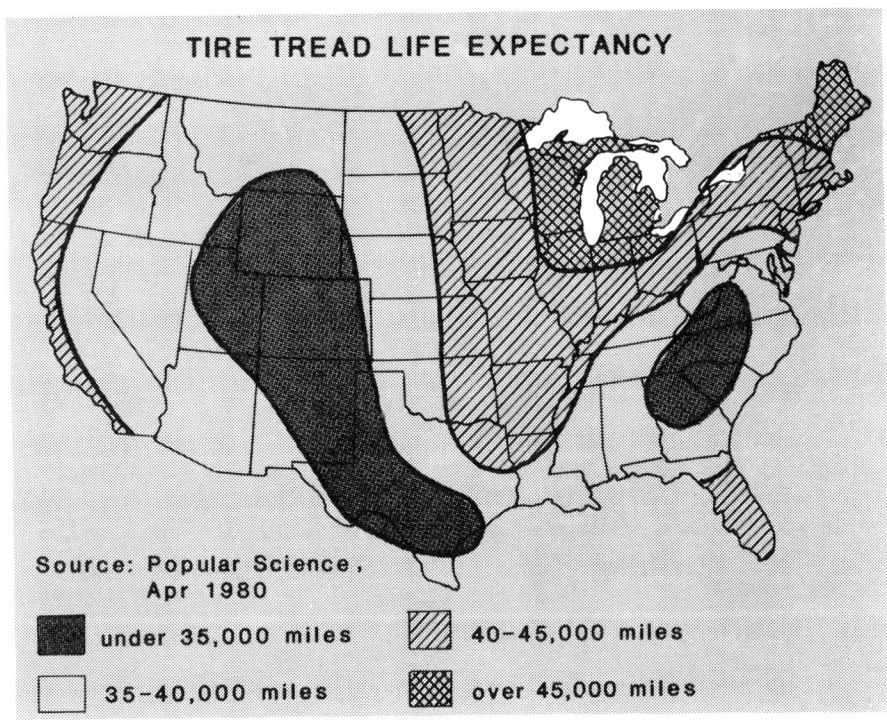

Fig. 5.32 Map background and shading techniques.

Maps and charts are frequently more effective if the background is of a neutral shade of gray, Fig. 5.32.

When two backgrounds are needed, e.g., a map showing land and sea masses, light and dark shades of gray in matte finish are satisfactory. The portion of the map of greatest concern should be represented using the lighter gray tone. Linework for the map itself should be drawn in black. A number of colors; yellow, blue, green or red to obtain a good visual display of information or conditions may be used to represent the data. If black is used, simple black patterns and symbols are recommended.

Black lettering may be used or white lettering is very effective by dropping out the background area where lettering occurs. The use of white lettering will help preserve the identity of the features drawn in other colors.

The preparation of the separation drawings for obtaining the map in the top illustration of Fig. 5.32 is explained in Fig. 5.33. All drawings must be carefully registered to insure accuracy of the final composite map. Registration marks are not shown since they fall outside the map boundary. The base map or black plate, Fig. 5.33 (a), was drawn using black india ink on film. All

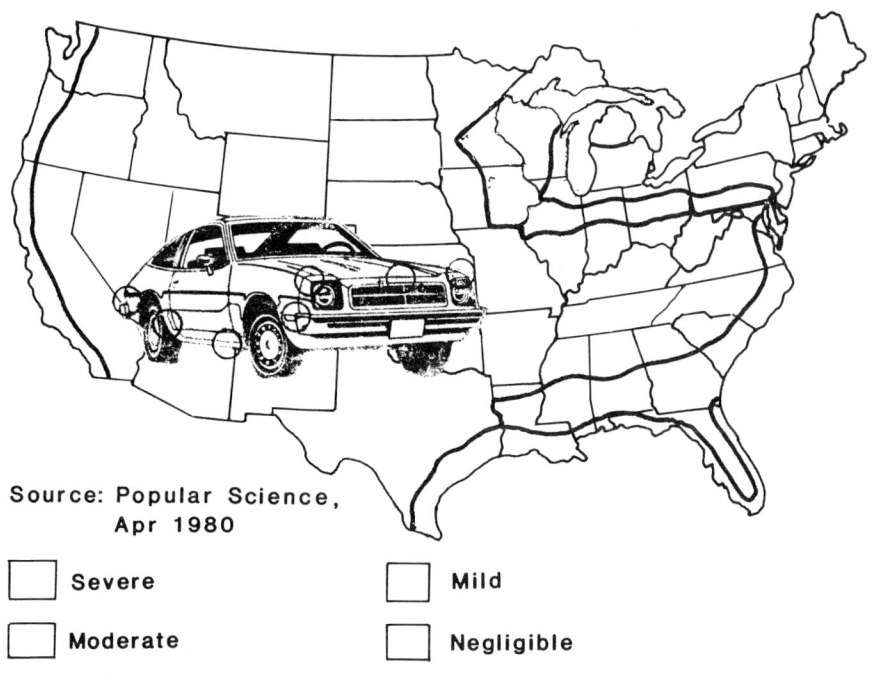

(a) Base map or black plate drawn on film with black india ink. Automobile cut out and pasted on drawing.

(b) Peel coat negative for 20% black screen.

Fig. 5.33 Preparation of color separation drawings.

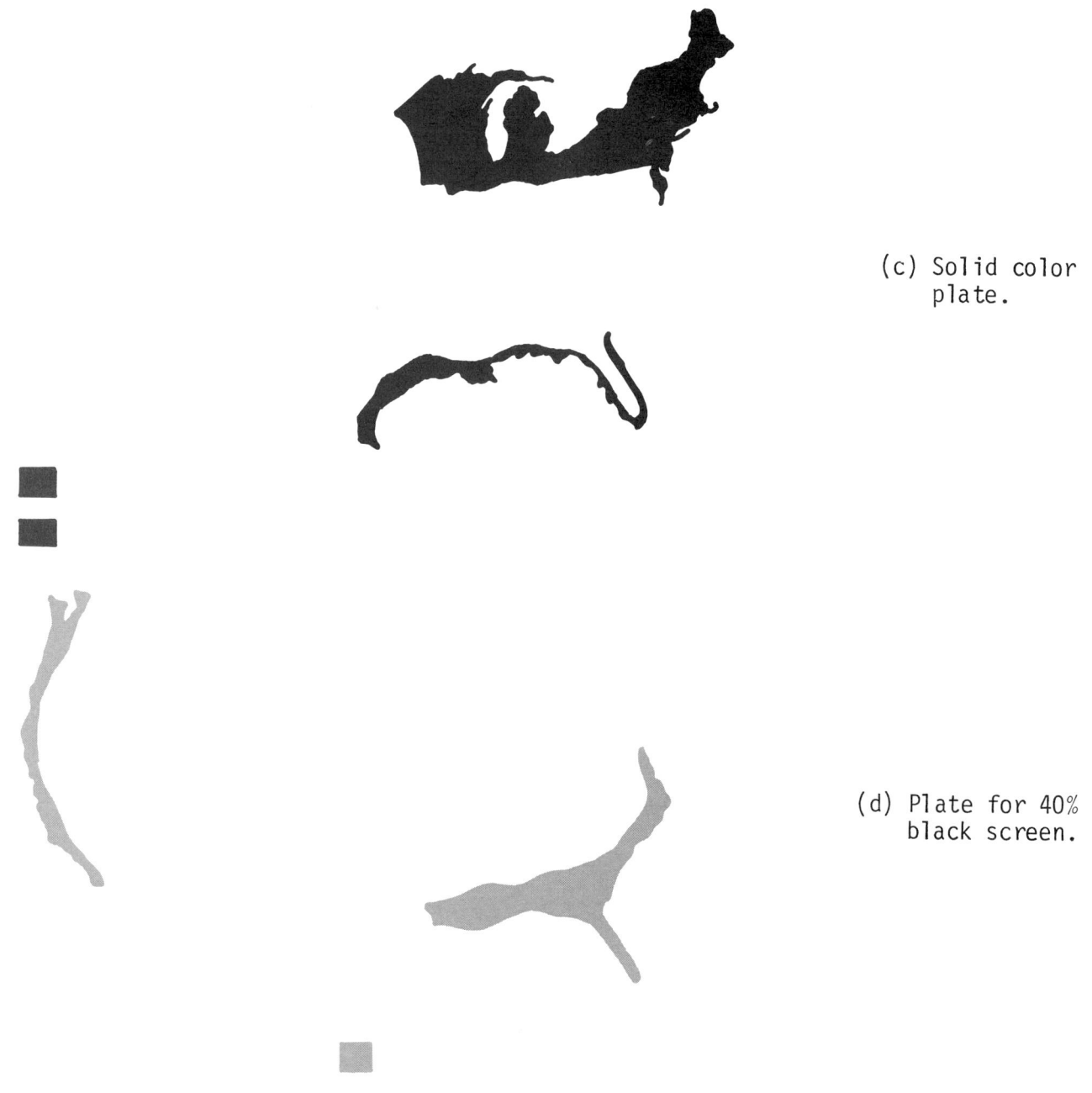

(c) Solid color plate.

(d) Plate for 40% black screen.

Fig. 5.33 Preparation of color separation drawings (continued).

lettering was produced using KROYTYPE and applied to the drawing in label form. The automobile was taken from a picture, cut out, and pasted on the base map. A 30% screen of black was used for the background area, Fig. 5.33 (b). This plate was prepared in negative form using peelable film. Areas where no screen should appear have been left open. The solid color plate, Fig. 5.33 (c), may be prepared using peelable film or drawing the outline of the areas on film to appear in color and filling in the areas. For small areas the latter method works well.

The solid color prints in the open areas (the cross-hatched area) left by the previous plate to delineate the moderate corridor. At the same time the color overprints areas (the double cross-hatched area) that were screened in black to portray the severe corridor. The final plate, Fig. 5.33 (d), may again be prepared using peelable film or drawn on film with black india ink and filled in. A 40% screen of black was used to overprint the 30% screen of black for representing the mild corridor.

The map at the bottom of Fig. 5.32 also requires three separation drawings: (1) base map drawn on film with black india ink with appliqué area symbols or drawn in ink, (2) peelable film overlay for 30% black screen, and (3) peelable film overlay for 50% black screen overprint. A fourth drawing containing a letter type dropout overlay could be produced to obtain white letters.

Chroma is the intensity of brightness inherent in a color. The alteration of chroma will result in a change of tone by dulling or neutralizing the color. This may be achieved by adding a complementary color or gray.

Functions of color. The harmonious use of color has many facets to make a drawing more effective. Its major use is to attract attention for items of greatest significance. One color plus black will offer the greatest contrast. Other schemes to obtain contrast in order of precedence are complementary, split-complementary, analogous, and monochromatic.

A complementary scheme uses colors opposite each other on the color wheel that are considered to be harmonious. These can be cool like blue and green or warm such as red and yellow. The cool colors on a drawing will cause elements to appear to recede while the warm colors will cause elements to appear to advance toward the viewer. Warm colors are also higher in visibility than cool colors. Green lies between the warms and cools and is considered to be relatively neutral.

Colors opposite each other on the wheel are referred to as complementary. The split-complementary scheme has three contrasting colors. A color like red may be used in contrast with the colors yellow-green and blue-green that adjoin its complementary, green, on either side.

The analogous scheme employs two or three adjacent colors on the color wheel such as red, red-orange, and orange, with the red predominating to create more interest. This scheme will not produce exciting results since contrast is missing.

The monochromatic scheme uses values of only one hue. In general, this scheme works well for large weak dull areas with bright colors used for small areas to provide contrast.

The psychological impact of color should receive careful consideration before colors are selected. Different moods are associated with certain colors and the colors that predominate on the drawing should fit the occasion. Warm colors imply ideas associated with life such as action and gaiety. Cool colors connote distinction, reserve, and serenity.

Drawings that have been colored help to increase the memory value by helping the viewer to remember what was seen. The use of colors like blue for water, green for vegetation, etc., is due to the ease with which the reader may identify with the feature.

Color procecesses. In the past, cost has often limited the use of color. Sophisticated new developments in reproduction methods such as color copying, color diazo printing, and the application of color technology to digitized plotting has now made the use of color feasible for all kinds of drawings, e.g., schematics, mechanical systems, electrical systems, structures, architectural drawings, maps, etc.

A full color illustration is usually printed using either pre-separated art for multicolor printing or the four-color process. For cartographic work pre-separated artwork is normally used. Film overlays or flaps are prepared using black india ink, one for each color, including the black flap. A map having three colors would need three color artwork and three plates for printing; one for black, one for red, and one for yellow. The base maps or the most complicated items are generally drawn on clear film. A clear sheet of film is then taped over the drawing and registration marks are positioned. Registration is critical so as not to destroy the accuracy of the drawing. The information to be shown in the second color is drawn on the clear film with the base map showing through as a guide. Additional colors may be added to the map by creating an in-register overlay illustrating other elements of the map. After all artwork has been completed, a photographic negative is made of each film overlay. For each negative, a plate is made for printing purposes.

All overlay drawings should be prepared on dimensionally stable base films by scribing or drawing with black india ink. The color of the ink on the overlay doesn't have to correspond to the eventual color of the plate which will be made from the photographed overlay. For each overlay, the desired hue should be noted.

The four-color process utilizes a single colored drawing for producing color plates, one for each of the primary colors and one for black. The process colors are yellow, red (magenta), and blue (cyan). This is accomplished by photographing the original artwork through a color filter to record each of the colors. Each separated negative is an image of the color to be reproduced. The printing of the process color plates in precise registration produces a color blended image having all the colors and values of the original. A black plate is added to emphasize detail, depth, and contrast using neutal shades of gray.

Colored prints in one pass may be made using a color diazo copier. The original copy is prepared in black ink or pencil on translucent material like tracing paper or film. The desired areas are blocked out on the back side of the drawing using special color pens that are available in red, blue, brown, yellow, and copier prints in gray. The original, together with a special sensitized paper, is run through a color diazo copier to produce copies up to 660.4 mm (26 inches) in width and indefinite length.

Digital plotting equipment may be used with a multipen assembly to draw lines having different thicknesses or different colors. Some plotters will handle as many as eight pens. The chief benefit of this type of unit is that it is a hands-off operation once the data necessary for the creation of a plot of a design is stored in the computer and readied for use.

Color separation. In cartography, the use of certain colors has been well established by convention. For the beginning student it is therefore suggested that the following colors be used.

Black (Culture)
1. Secondary roads, railroads
2. All place names (cities, towns, etc.)
3. Man-made features (buildings, bridges, etc.)
4. Political subdivisions
5. Bench marks
6. All marginal data
7. Spot elevations

Brown (Hypsography)
1. Relief (hills, mountains, valleys, slopes, etc.)
2. Contour numbers

Red
1. Main highways (Red and also yellow used as a color fill to classify roads)
2. Range, township, and section lines
3. Range and township numbers
4. Road classifications and route symbols

Blue (Hydrography)
1. Water features including lettering (rivers, ponds, lakes, marshes, wetlands, etc.)

Green (Surface cover)
1. Forests, meadows, crops, lowlands, etc.
2. Lettering for such features as vegetation areas

Pink
1. Large cities
2. Built-up areas

Gray
1. Sometimes used in place of brown

LETTERING

6.1 INTRODUCTION

Drawings of all kinds require lettering (calligraphy) of several types to convey essential information. Proper lettering may be achieved only when one studies and understands the proper form of various letter styles and keeps these in good proportion. Skill must be developed and meticulous care and judgment used in the artistic placement and proper positioning of all lettering. These factors are necessary to attain good lettering, and the additional labor to acquire this result is worthwhile. With a little patient practice anyone can obtain reasonable professional results. Attention to details like lettering will require little more time to do the job well than to be "sloppy."

6.2 LETTERING TYPES

The most common types of lettering used on working drawings and maps may be classified as being *Roman* or *Gothic*. The term Roman means that the letters have strokes consisting of both heavy and light lines, Fig. 6.1 (a). The ends of the strokes for some Roman types are terminated with short lines called *serifs*. The term Gothic means that all lines of the letters are composed of uniform width strokes, Fig. 6.1 (b). The Gothic alphabet that is generally used is classified as *sans serif* (without serifs). Single stroke Gothic lettering requires several strokes to draw most letters, but each line of a letter is usually made with only one stroke and will be only as wide as the line made by the pencil or pen being used. Letters that are inclined (slanted) are termed *italic*, Fig. 6.1 (c).

Pen points in many different styles and sizes (square, round, and oblong tips) have been designed to fit any lettering task, Fig. 6.1 (d). Style "B" speedball pens have a round tip and may be used for Gothic lettering. Style "C" pens may be used for Roman letters. The pens should not be dipped into ink,

EDUCATION ← SERIFS

(a) Roman

EDUCATION

(b) Gothic

Education

(c) Italic

A-STYLE — Designed to produce square Poster Type. Letters with single stroke — 6 sizes.

B-STYLE — For single stroke round Gothic letters — for drawing and sketching — 8 sizes.

C-STYLE — Will duplicate strokes of the flexible hand cut reed pen. Used for the Roman and Italic alphabets with "thick and thin" elements — 7 sizes.

D-STYLE — With oval marking tip, used for bold Poster Roman alphabet with "thick and thicker" elements — 7 sizes.

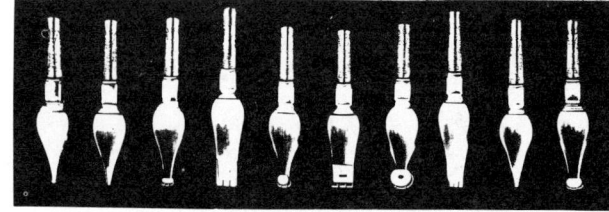

(d) Speedball pens

Fig. 6.1 Alphabet types *(Courtesy Hunt Manufacturing Co)*.

but filled with the quill or dropper from the ink bottle. After they have been used, the pens should be carefully cleaned.

Form. The first prerequisite for lettering is a knowledge of the correct form and shape of the individual letters and numbers. Vertical and inclined single-stroke Gothic letters and numbers are illustrated in Fig. 6.2 and Fig. 6.3. The letters are shown in a square to allow the width of each letter to be compared to its height. It should be noted that most letters are

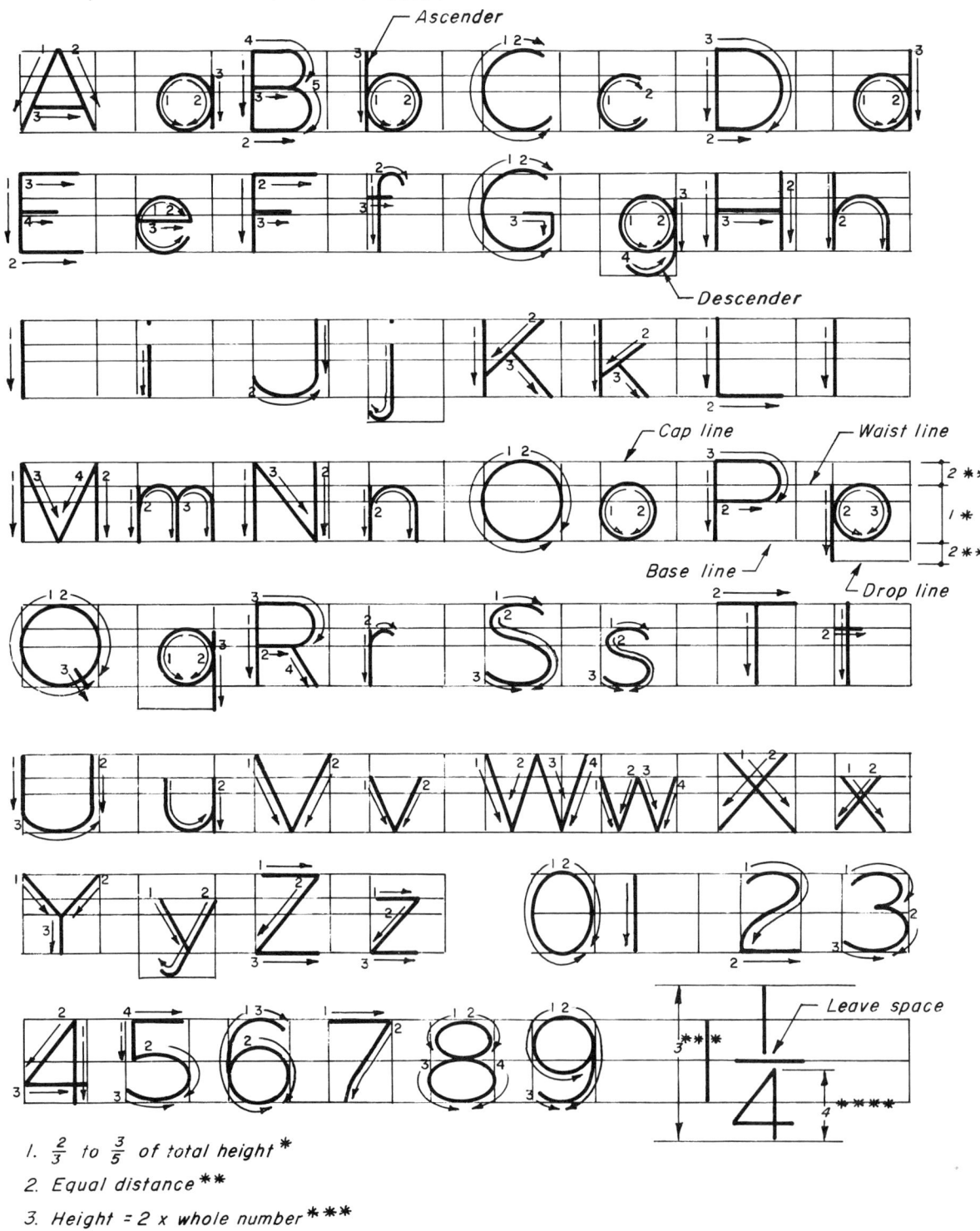

1. $\frac{2}{3}$ to $\frac{3}{5}$ of total height *
2. Equal distance **
3. Height = 2 x whole number ***
4. $\frac{2}{3}$ to $\frac{3}{4}$ height of whole number ****

Fig. 6.2 Vertical Gothic letters and numbers.

not as wide as they are high. The width of each number is shown in a rectangular box that is approximately five-sixths the height of the box. The arrows and numbers indicate the order and direction of the strokes. Vertical strokes are made downward and horizontal strokes from left to right.

Numerals and fractions must be legible and are made with a horizontal division line. The total height of the fraction is twice the height of a whole number with each figure in the fraction two-thirds to three-quarters the height of the whole number. A space must be left above and below the line to pre-

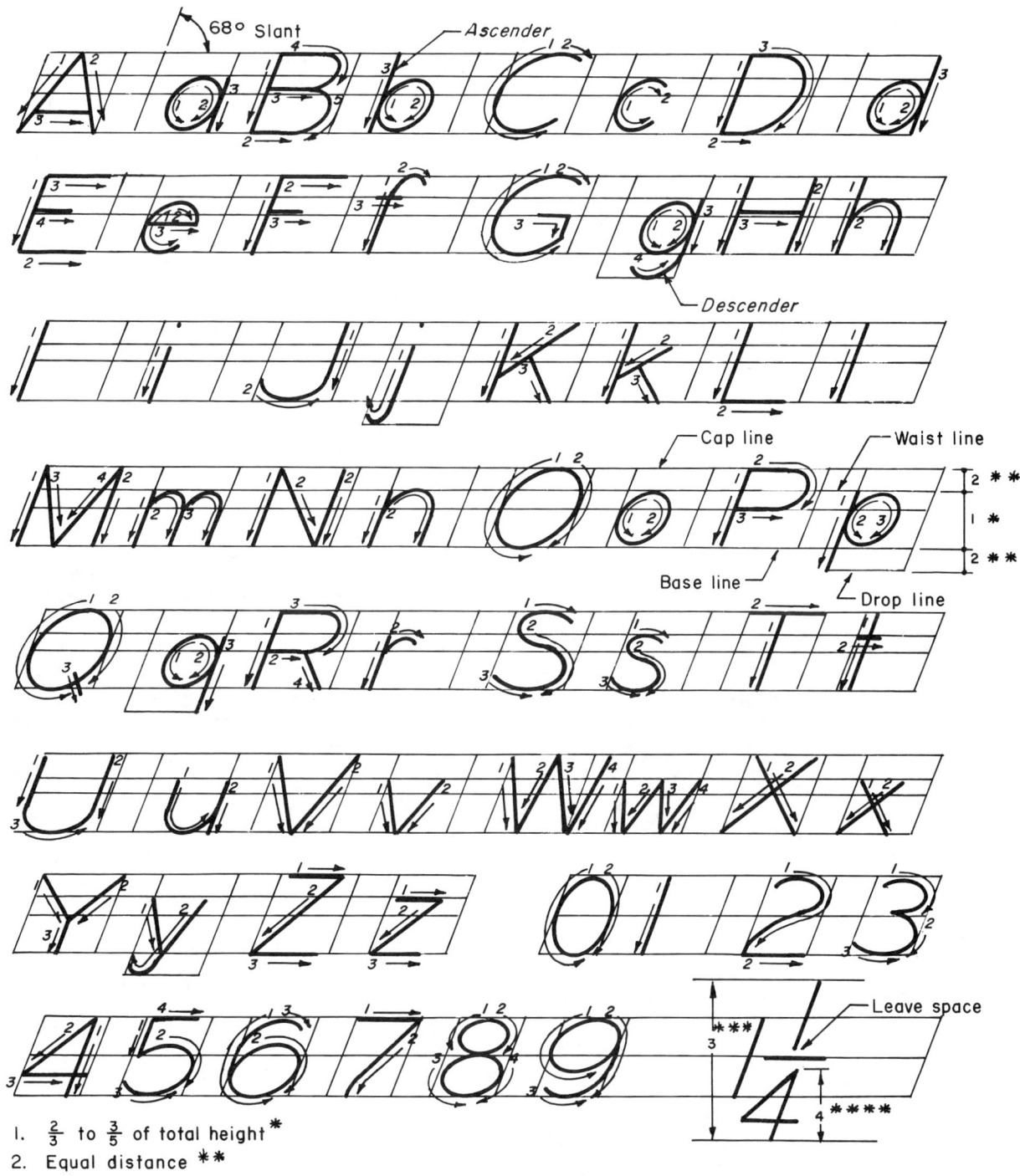

1. $\frac{2}{3}$ to $\frac{3}{5}$ of total height *
2. Equal distance **
3. Height = 2 x whole number ***
4. $\frac{2}{3}$ to $\frac{3}{4}$ height of whole number ****

Fig. 6.3 Inclined Gothic letters and numbers.

vent the figures of the fraction from touching the horizontal line.

The lower-case letters are usually made with the body two-thirds to three-fifths the height of the capitals. Ascenders should extend to the cap line and the descenders drop the same distance below the body of the letters.

Several styles of lettering are used on most finished maps to help designate features of different character and the cartographer should have a knowledge of two basic forms of letters, Roman and Gothic, well enough to use them when necessary. Since Roman letters are difficult to draw, it is recommended that where the lettering is to be executed freehand that the alphabet, serif, and Gothic, illustrated in Fig. 6.4, be used in place of the Roman type. The lettering is pleasing in appearance and easy to execute since all strokes are of uniform width. The form of both vertical and inclined letters is similar and proficiency for drawing both should be developed.

<u>Classification of letter styles for use on maps</u>. The lettering used on maps has undergone considerable change through the years. This is best seen by comparing many of the Old World maps, Fig. 6.5, to those that are being produced today. Most of the lettering on older maps was executed freehand using

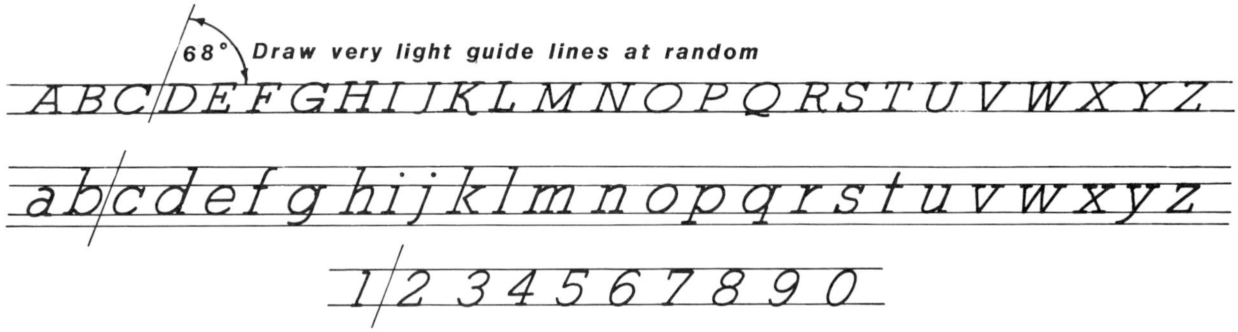

Fig. 6.4 Inclined serif and Gothic letters and numbers.

Fig. 6.5 Lettering on old world maps, portion of Fez and Morocco in the Mercator-Hondius Atlas-Amsterdam, 1606 (reprinted from *Decorative Printed Maps of the 15th to 18th Centuries*, R.A. Skelton).

pen and brush. One can only marvel at the pride these craftsmen must have realized in what is fast becoming a lost art.

The use of several different styles of lettering on a map will affect the overall appearance more than any other single factor. Examples of the styles of lettering recommended for freehand use are illustrated in Fig. 6.6. The use of capital versus lower case letters is one of judgment and will depend on the size of the feature and its importance. Studies show that, due to legibility and perceptibility, lower case letters are easier to recognize than capitals. The color of the lettering will also affect its legibility and perceptibility. Black lettering is generally used on maps having a light background with white or reversed lettering working well on a dark background to provide contrast. The size of the letter used should be proportional to the size of the drawing. Letter size will also be dependent on the density of detail and its purpose. The lettering must be large enough, however, to be read clearly if the map is to be reduced. Fig. 6.7 illustrates the effect of lettering when reduced at different scales. As a guide the letter size should probably not be any smaller than 1.5 mm (.05").

6.3 GUIDE LINES

A series of light guide lines should be drawn before any lettering is attempted. The Ames lettering guide, Fig. 6.8, has a series of holes for the drawing of guide lines. The number 2, 3, 4,....10 on the disc represent thirty-seconds of an inch. To use the instrument, the height of the lettering is determined, e.g. 1/4", and the number 8 (8/32" or 1/4") is set on the bottom index mark. The guide is placed on a

ORONO, MAINE University of Maine	*Place and Area Names* *States, counties, cities, etc* *Vertical Roman Letters*
MOOSEHEAD LAKE *Penobscot River*	**Hydrographic Names** *Oceans, lakes, rivers, etc* *Inclined Roman Letters*
MOUNT KATAHDIN Greenbush Ridge	**Hypsographic Names** *Mountains, valleys, hills, etc* *Vertical Gothic Letters*
COLLEGE AVENUE *Bangor & Aroostook R R*	**Public Works** *Railroads, highways, bridges, etc* *Inclined Gothic Letters*
BM x 1250 △ 306 x 472	*Control Data* *Benchmarks, spot elevations, etc* *Vertical Gothic Letters*
——500——	*Contour numbers-Inclined Gothic*

Fig. 6.6 Classification of letter styles for use on maps.

Fig. 6.7 Effect of reduction on the legibility of lettering for several selected scales.

straightedge, either a T-square or triangle, and a sharp pencil is inserted in one of the holes. As the guide is moved along the straightedge, a light line may be drawn. Using different holes in the disc, the process is repeated until the required number of lines has been drawn. The guide lines do not need to be erased if they are drawn lightly.

The holes in the disc are grouped in sets of three for drawing the top guide line, the center guide line for lower-case letters, and the bottom guide line for each line of lettering. The different spacing of the groups indicated by the fractions 3/5 and 2/3 provides for different sizes of lower-case letters.

Horizontal guide lines should be drawn to insure that lettering is the same height. The beginner will find it helpful to also draw vertical or inclined guide lines at random. The edge of the Ames guide is used to draw the vertical (90°) or sloping (68°) guide lines.

Metric guide lines. The series of holes and numbers at the extreme left of the disc are used for drawing metric guide lines in millimetres. Guide lines may be spaced equally using the holes marked with the right brackets or at half

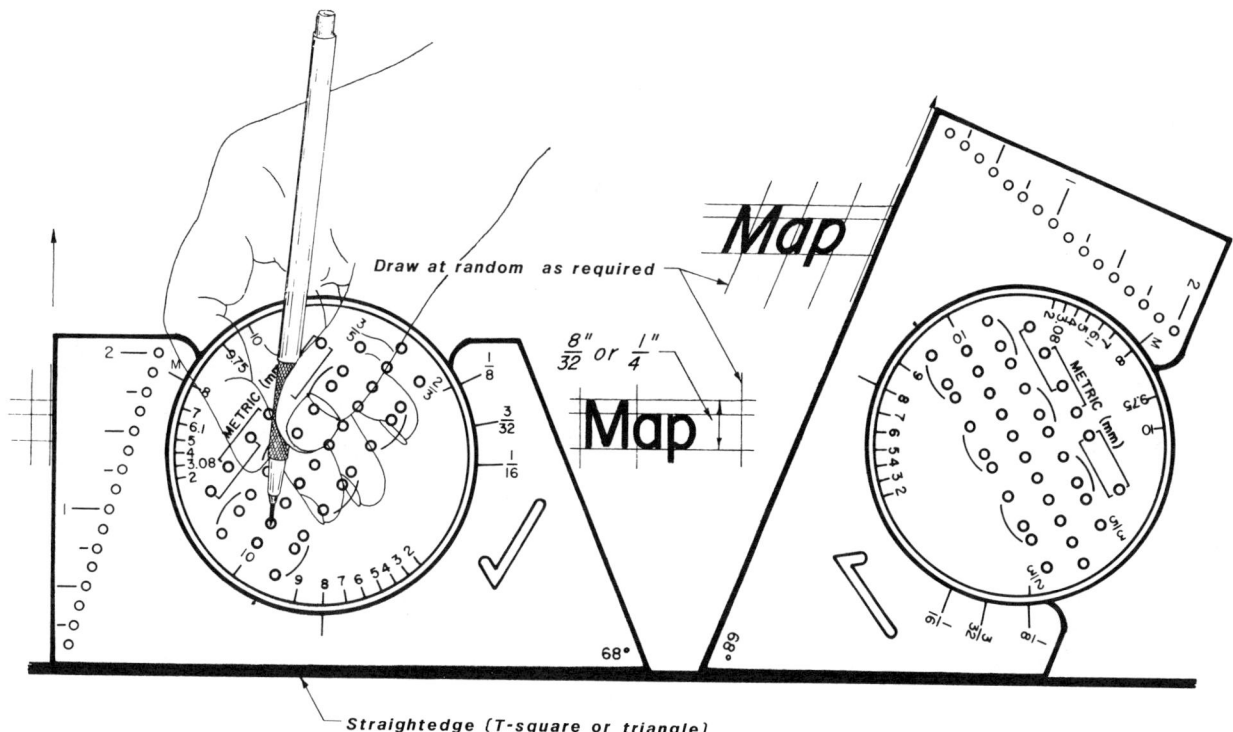

Fig. 6.8 Ames lettering guide.

space using the holes marked with the left brackets. To draw metric guide lines the height of the lettering is determined and the number set on the index marked "M." The numbers 3.08 (.12"), 6.1 (.24"), and 9.75 (.38") are for standard letter heights used in the United States.

6.4 LETTERING ESSENTIALS

To obtain good results, judgment should always be used when lettering. A number of essential techniques should be kept in mind that will be of considerable help when exercising this judgment.

Proportion. The correct relationship of width to height should be maintained for letters, Fig. 6.2, 6.3, and 6.4. In general, the lettering will look better if the width and the height are about the same. In some instances where a small or large space exists, it may be desirable to use a condensed or extended form of lettering, Fig. 6.9.

Stability. Top heavy or unstable letters and numbers should be avoided. The upper portions of most letters and numbers should generally be smaller in size to offset this effect. Examples of stability are illustrated in Fig. 6.10.

Spacing. The legibility of a line of lettering is as dependent on the spacing of letters and words as of letter shapes. Letters should be spaced with reference to their shape. The apparent areas between letters should be equal rather than space the letters an equal distance apart, Fig. 6.11. Adjacent letters having parallel strokes (e.g., the M, I, and N) are separated slightly more because of irregular spaces between other letters. Letters with curves like the O and G are brought closer together because of the large space between letters at the top and bottom. Curved and open-sided letters such as the L and O are also brought

CARTOGRAPHY | CARTOGRAPHY

Condensed   _Extended_

Fig. 6.9 Condensed and extended lettering.

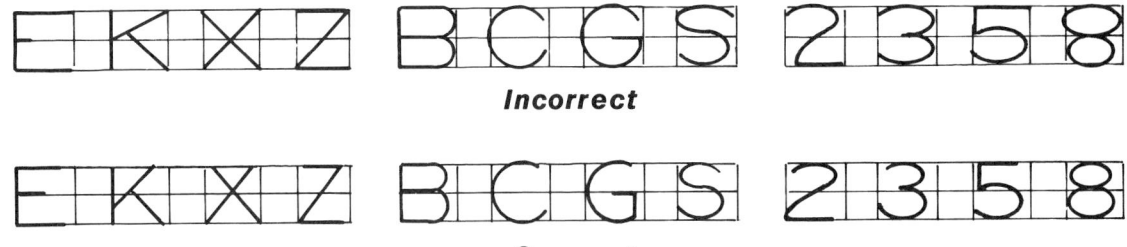

Incorrect

Correct

Fig. 6.10 Stability of letters.

closer together because of the large space in the open-sided letter.

The spacing between words, Fig. 6.11, should be about the height of the capital letters. A good rule to follow is to leave room for the letter "O" between words.

Fig. 6.11 Spacing of letters and words.

Lines of lettering may be spaced any distance apart as long as it is appropriate and in harmoney with the rest of the drawing. As a general rule, allow approximately one-half to one and one-half the height of capital letters between lines.

Where the letter spacing is to be expanded, the position of each letter may be indicated along the guide line by eye or with the aid of dividers, Fig. 6.12. For easy reading, spaces between letters should not exceed four times the height of the letters. Each letter of a name placed on a curve must be placed so as to follow the vertical or inclined guide lines. The spacing between words may also be increased and is generally uniform.

Consistency. Certain pitfalls should be avoided when lettering. The style of letters used should be consistent and alphabets should not be mixed, Fig. 6.13 (a). If serifs are used on some letters, e.g., the "I" and "J", they should be used on all letters, Fig. 6.13 (b). Capital and lower-case letters should not be mixed together, Fig. 6.13 (c).

Position for lettering. The immediate identification of features and complete legibility are the basic requirements for information appearing on a drawing. The lettering should not obscure detail. If necessary, the detail may be broken provided that it is not of vital importance. Guide lines should not touch the lines identifying any feature.

Ordinarily, the lettering is placed parallel to the bottom or sides of a drawing so that it may be easily read. The names of linear features should fol-

Fig. 6.12 Expanded letter and word spacing.

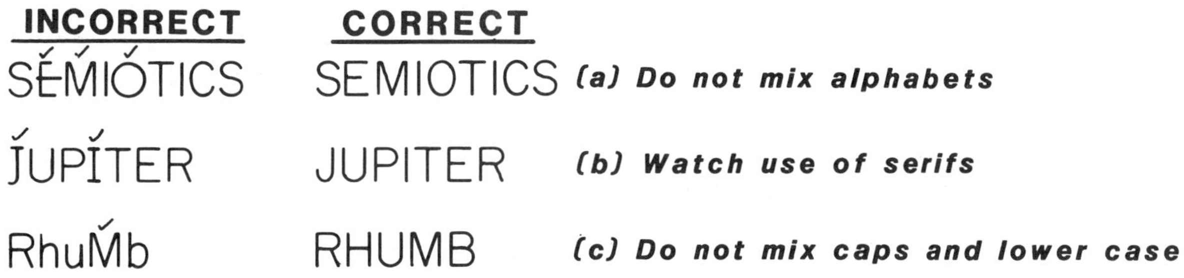

INCORRECT	CORRECT	
SÉMIÓTICS	SEMIOTICS	(a) *Do not mix alphabets*
JUPITER	JUPITER	(b) *Watch use of serifs*
RhuMb	RHUMB	(c) *Do not mix caps and lower case*

Fig. 6.13 Consistency of letters.

low the alignment of the feature. If this is not possible, the lettering should be placed to the right of the feature. Vertical features should be lettered from the bottom of the drawing to the top so as to be read from the right side. Where diagonal lettering is necessary, it should read from left to right with respect to the bottom of the sheet. Fig. 6.14 will serve as a reminder for the proper placement of lettering.

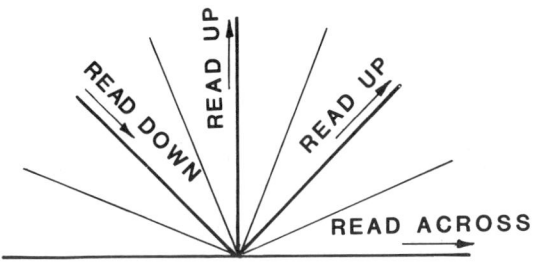

Fig. 6.14 Lettering direction.

Where a feature has adequate area, the lettering is placed entirely inside or entirely outside the boundary limits. It should be placed symmetrically in the feature or spaced to indicate the extent of the feature. In the case of a two- or three-word name, the lettering may be placed on two lines but not on three unless there is no alternative. The parts of a compound name may be divided and placed above and below the feature if necessary.

Abbreviations. It is sometimes convenient to use abbreviations on drawings to save time and space. Only common abbreviations that are widely understood should be used. The use of uncommon or unfamiliar abbreviations should be avoided since they may result in confusion and misinterpretation. It is recommended that abbreviations be used only when necessary. Periods after abbreviations are usually omitted unless they spell a word, e.g., dia not dia., BTU not B.T.U., fig. not fig, etc.

6.5 LETTERING MAP FEATURES

Name information. Map names are of universal interest and importance, and a name error may be more serious than an elevation error. Since map names have the same meaning for everyone, even to people who may not have an understanding of contours and map symbols, this information should be correct and complete. Information for names may be obtained from many different sources, varying widely in reliability and completeness. Sources of information are published maps, signs (road, building, etc.), bulletins, books, reports, local histories, local residents, and officials.

Letter alignment. Names shown in larger type, such as townships or large rivers, are usually placed in position first. If the name is to be placed on a curve, the guide lines are drawn with a smooth curve which closely follows the feature, Fig. 6.15. Lettering on reversed curves should be avoided if possible. Where a name is composed of more than one word, it is placed on a smooth continuous line and never in an angular or zig-zag arrangement. Because of the formality of a circular curve, a more pleasing effect may be obtained if the curve is slightly pronounced at one end. The name is more easily read if its beginning approximates a straight line.

Hydrography. Hydrographic features are lettered as indicated in Fig. 6.15. The name of a large area, such as a lake or bay, is located in the approximate center of the area and is placed to follow the general configuration of the feature.

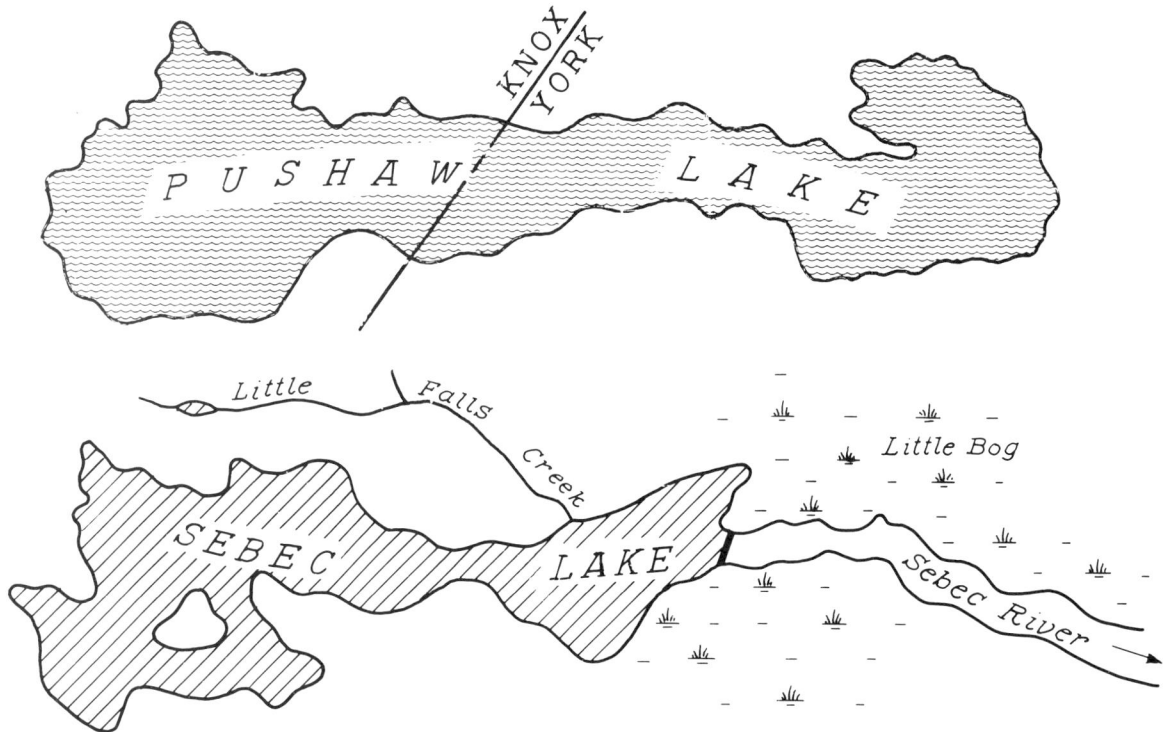

Fig. 6.15 Letter alignment and placement.

The name of a double-line stream (one of sufficient width at map scale to justify showing both shore lines) is placed within the outline where the space is wide enough to accommodate it. In placing stream names the type must be located entirely within or entirely outside the shore lines of the stream.

Where possible, the names of single-line streams, creeks, brooks, and branches are arranged in a smooth curve above the feature so that they identify the drain throughout its entire length. The name is repeated where necessary for identification on long streams. A general rule which has proven practical in the labeling of short or medium length features is to place the name along the middle third of the feature.

Names of ponds, lakes, reservoirs, and swamps are arranged horizontally and to the right of the feature unless the area is large enough to accommodate the name within its limits.

Hypsography. It is best to place names of large mountains and ridges so as to follow the feature where possible, Fig. 6.16. Names of small features, such as tops, hills and saddles, are lettered horizontally and located clear of their highest points for legibility. Where a spot elevation is shown, the name should clear the small cross used to locate the point to which the elevation applies. The names of islands, capes, necks, and shoals are located within the outline of the respective features, if possible, and placed to show their extent. The names of small islands, points, and rocks are placed horizontally in the water area and clear of the features.

Culture. The names of man-made features as railroads, highways, trails, canals, telephone, telegraph, and pipe lines are placed on the upper side of the feature and centered wherever practical, Fig. 6.17. If a name applied to the middle third of a feature will not identify it sufficiently, the name may be shown in two or more places. A name, however, should not cross the feature labeled.

Symbols for towers, pumping stations, tanks, etc., identify the exact location of the objects and seldom require lettering.

Elevations. The selection of an elevation for publication is based primarily on the potential importance of the elevation to the map user. Usually, el-

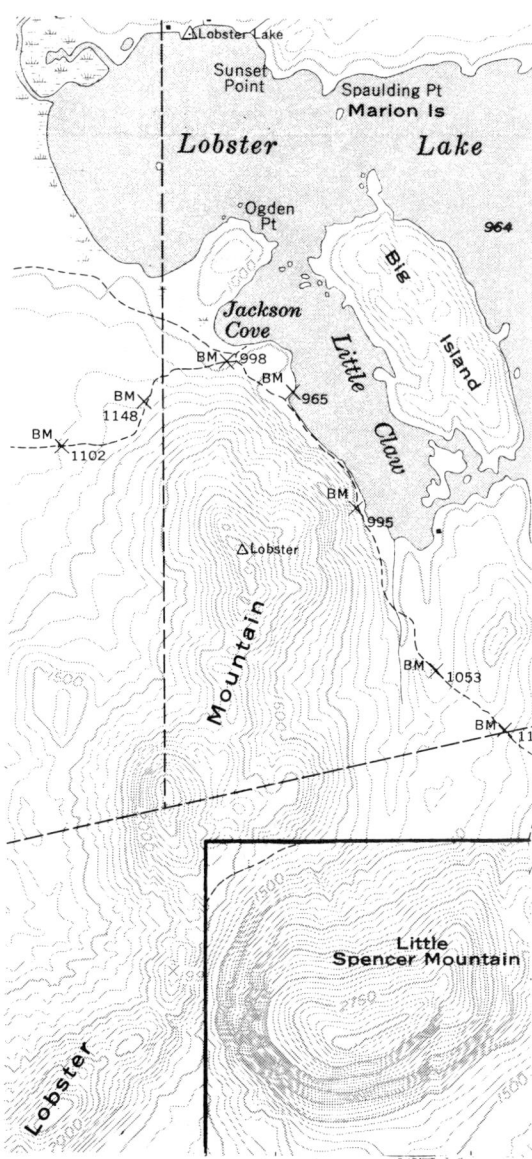

Fig. 6.16 Lettering hypsography (*Courtesy USGS*).

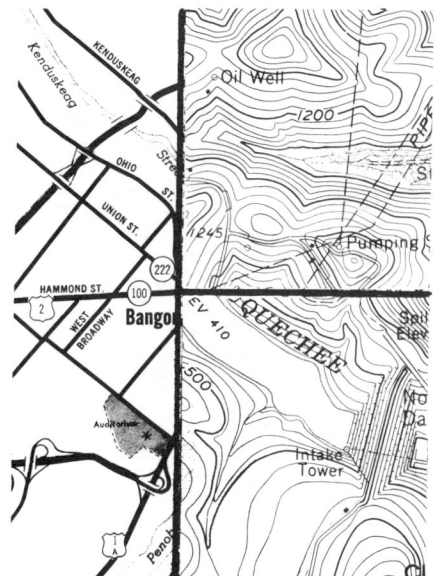

Fig. 6.17 Lettering culture (*Courtesy USGS*).

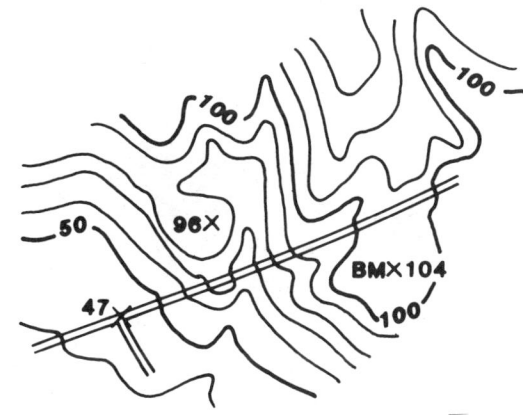

Fig. 6.18 Lettering elevation information.

evations of prominent road forks, summits of mountains, stable water surfaces, etc., are labeled, Fig. 6.18. Elevation numbers are placed parallel to the bottom of the map. Lettering over contour numbers, stream lines, contour lines, and cultural features should be avoided.

Contour numbers. Contour numbers are placed at strategic locations on the contour lines to facilitate the approximation of any elevation without a tedious search for a reference number. Numbers are usually placed on the index contours, by leaving a gap, although they may be placed on widely spaced intermediate contours. Locations for lettering the numbers usually are chosen before the contour lines are drawn in finished form.

6.5 DRY TRANSFER LETTERING

The use of dry transfer lettering has increased rapidly during the past few years. Many styles and sizes of letters are available from most drafting firms. The decision to use this material depends on the final results desired and where a savings is realized in both time and money. For large lettering or single items this type of lettering works quite well. Its use should be avoided on drawings requiring the lettering of many small size items. In defense of still being able to prepare a well-lettered drawing freehand, the stick-up process is essentially hand-done and may be very time-comsuming and costly. The application of dry transfer lettering is illustrated in Fig. 6.19.

1. Remove the protective backing sheet and position the desired letter, using the alignment marks located at the beginning and end of each row.

2. Shade lightly over the entire letter with a ball point pen, soft pencil or a burnisher such as the wooden Zipatone burnisher shown here.

   Do not use more pressure than is necessary to transfer the letter. Hold the sheet firmly in position during burnishing to avoid cracking the letter. A broad ended burnisher should be used for large letters.

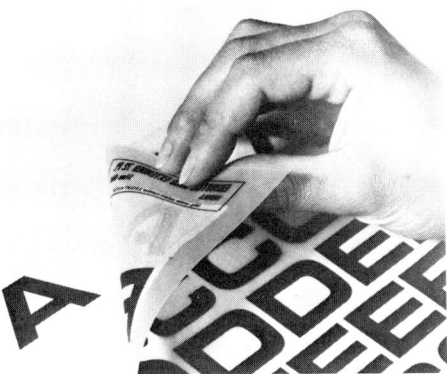

3. Carefully lift away the sheet, making certain the entire letter has transferred to the receiving surface. Repeat until your word or phrase is complete.

4. Place the protective backing sheet over the transferred letters, hold in position and burnish firmly to insure perfect adhesion. A blunt instrument such as the flat end of the Zipatone burnisher should be used.

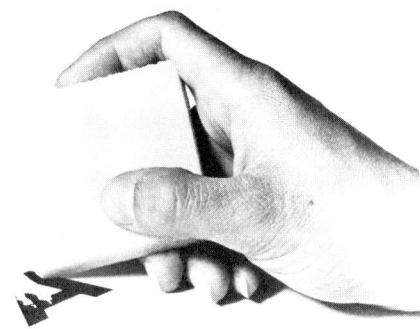

5. Corrections:
   Zipatone Transfer Lettering may be removed from hard surfaces such as metal, glass or acetate with adhesive tape. Use a rubber-cement pickup or tape to remove it from artboard or paper **before final backing sheet burnishing.**

6. Protection:
   Under normal handling and copying conditions, Zipatone Transfer Lettering needs no further protection if Step No. 4 of the Instructions is followed. Letters that are exposed to outside conditions such as auto or truck doors should be sprayed with several coats of clear acrylic or plastic spray. Spray lightly at first and let dry. Avoid heavy coating.

Fig. 6.19 Application of dry transfer lettering *(Courtesy Para-Tone Inc)*.

6.7 LETTERING EQUIPMENT

Many kinds of mechanical lettering devices are used extensively. The Varigraph, Fig. 6.20, is a versatile lettering device that may be used to produce letters from 150 different type styles. The alphabet and numbers of each type style are engraved metal matrixes. Each matrix is inserted into the Varigraph. Adjustment controls on the instrument allow the height, width, and slant of the type to be changed from only one type style matrix for lettering.

LEROY sets are available with metric pen point sizes and a matching series of fixed ratio templates, Fig. 6.20, that will give a constant ratio of 10:1 between letter size and line width. The lettering sizes and line widths follow the geometric progression of 1:414 or $\sqrt{2}$ and permits a direct correlation of character sizes and line widths to standard drawing sizes (see section 2.17, reservoir pens) to be maintained, Fig. 6.21. The lettering style is based on International Standard ISO 3098/1 for technical drawings and lettering. It concerns primarily the features of legibility, uniformity, and suitability for microfilming and other photographic reproduction.

The lettering guides or templates are generally made of a plastic material with a series of engraved letters, numbers or symbols. Templates are available in a wide variety of styles and sizes. Capital and lower-case let-

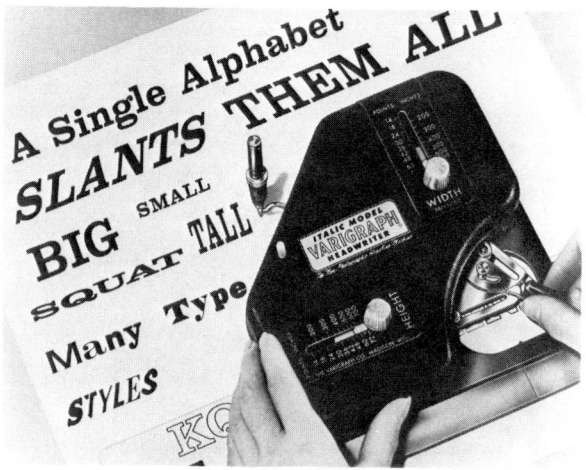

VARIGRAPH

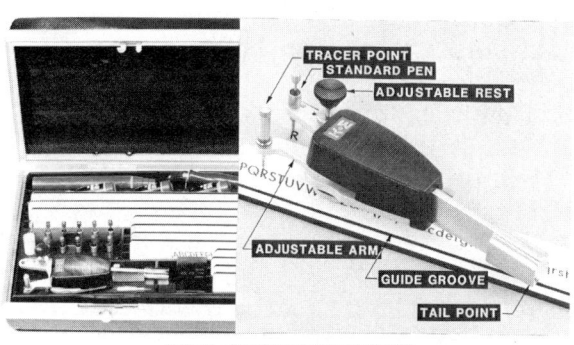

LEROY LETTERING INSTRUMENT

Fig. 6.20 Mechanical lettering equipment *(Courtesy Keuffel & Esser Co and Varigraph, Inc).*

ters and numbers are frequently combined on one template. To letter, place the template along a straightedge. The recommended pen size is engraved on each template to give the correct proportion of letter thickness and size. Set the pen in the socket in the upper arm of the scriber for inking or the lead

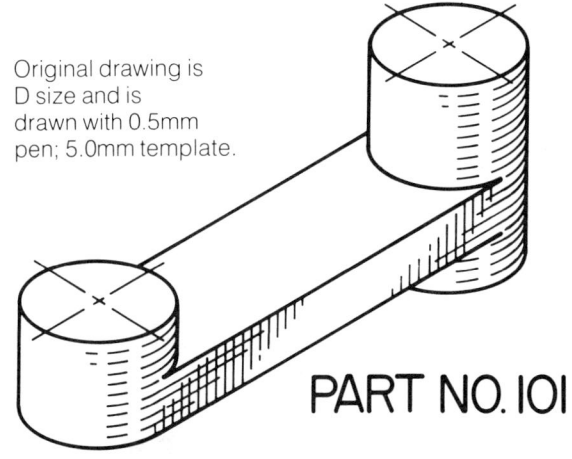

Original drawing is D size and is drawn with 0.5mm pen; 5.0mm template.

PART NO. 101

½ size blowback is B size. Changes made directly on blowback with .25mm pen; 2.5mm template.

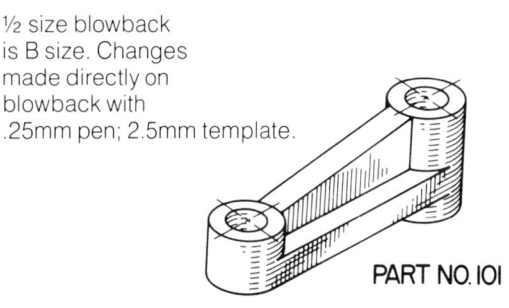

PART NO. 101

Fig. 6.21 Example of character sizes and line widths to standard drawing sizes *(Courtesy Keuffel & Esser Co).*

clutch in the socket for pencil lettering. The reservoir pens are also designed for use in the scriber in place of the standard pens. Vertical or inclined lettering, 0° to 22 1/2° forward slant, may be obtained by moving the adjustable arm on the scriber to the position desired. Slant lettering of 15° is recommended as the maximum slant to be used for maintaining legibility on drawings that are to be reduced in size. The tail point of the scriber should be set in the straight guide groove of the template. When the tracer point of the scriber is inserted into any of the engraved letters, the pen will reproduce the letter in full view above the template, as the tracer point is moved around the contour of the letter. The rest on the scriber should be adjusted because the pen will make a satisfactory letter only when it is in very light contact with the drawing.

The templates have a convenient scale on the bottom edge for quick centering and spacing of any line of lettering. Each space represents distance, center to center, between normal letters. Spacing between letters is determined by eye. All templates may be used for obtaining different size letters without moving the straightedge since there is a uniform distance between the bottom of each template and the bottom of the letters.

A lettering system similar to LEROY is Stano-script, Fig. 6.22. The advantage in using this scriber is that

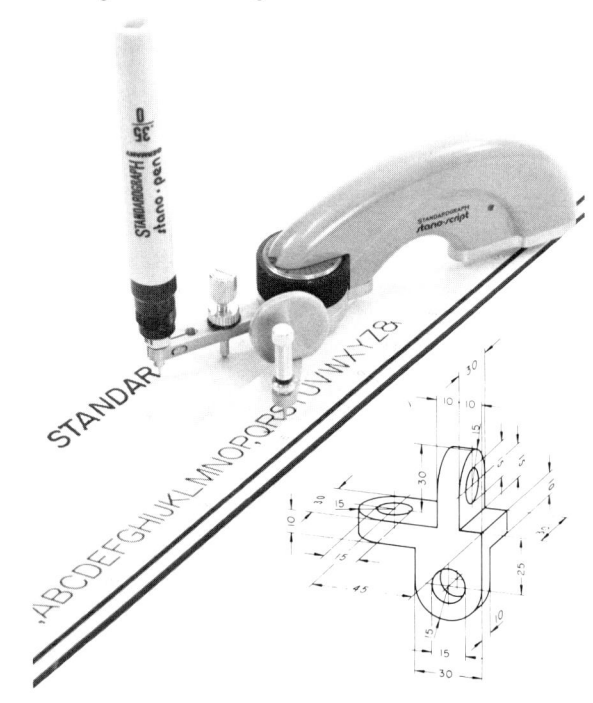

Fig. 6.22 Stano-script lettering system *(Courtesy Martin Instrument Co).*

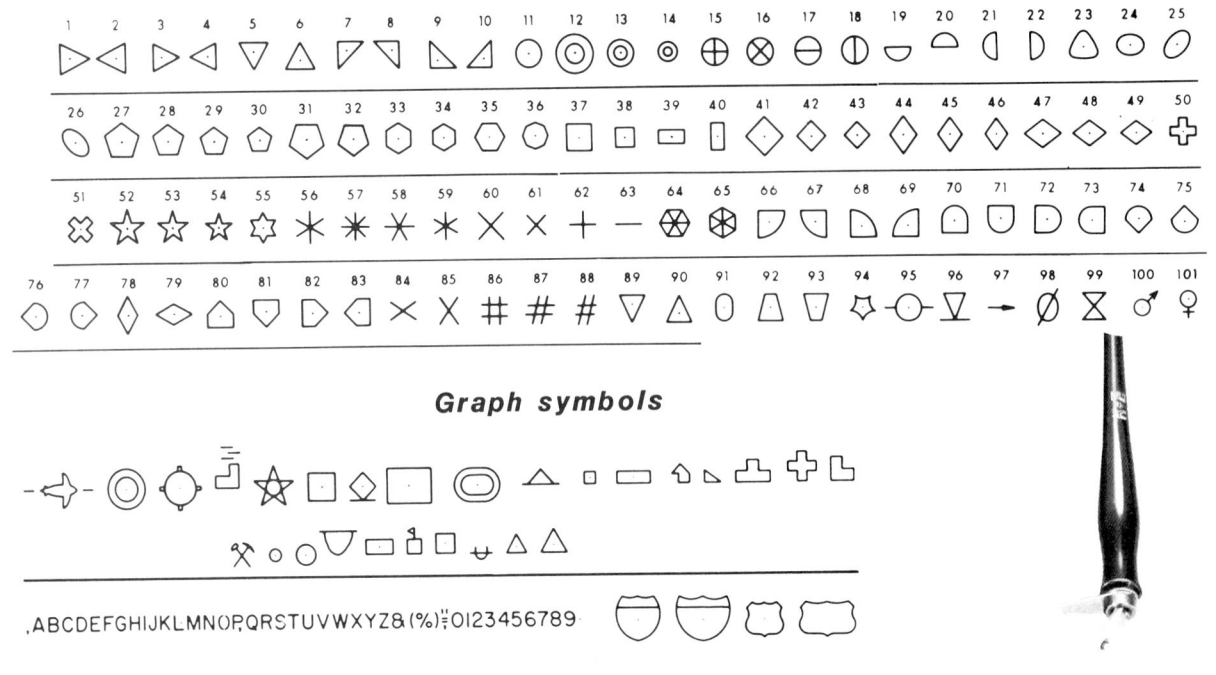

Fig. 6.23 Symbol templates *(Courtesy Keuffel & Esser Co).*

it can be adjusted to any desired slant to the left or right. This is useful where it is necessary for the lettering to exactly match the projection system employed, e.g., isometric or perspective.

Special templates, Fig. 6.23, are available containing the characters of map symbols, graph symbols, north arrows, etc. Many sets come with a penholder that may be used with these templates or for drawing other kinds of freehand work.

A variety of lettering guides with precision cut open style alphanumeric characters are also available for use with pencil or reservior type pens, Fig. 6.24. The guides are frequently made with elevated metal rails for fast lettering without danger of smearing.

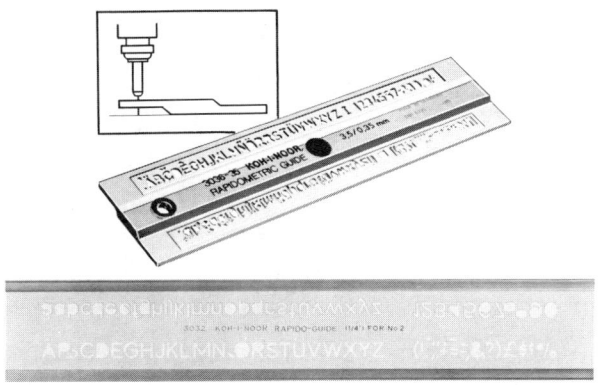

Fig. 6.24 Lettering guides *(Courtesy The Ben Meadows Co).*

On drawings where the equivalent of quality hand lettering is required, the Datascribe III, Fig. 6.25 (a), will produce legible uniform characters much faster than by hand. The machine is completely self-contained and is basically an X-Y plotter with a pen that may be positioned using either a joystick or a four directional key set. With the pen positioned, the user may key in 1000 character combinations to print. Letter size and width may be set automatically. Other features include the ability to set left and right margins with a single key and the drawing of horizontal or vertical lines.

The Datascribe III will permit up to 2000 characters to be stored in ten separate memories. This information may be repeated on command in any order. A micro cassette tape device permits storing up to 2000 keystrokes per cassette side for future use. Alternate character and symbol sets may be loaded into the memory using preprogrammed cassette tapes to meet special requirements. Several examples of available character sets and symbols are illustrated in Fig. 6.25 (b).

The KAD II Auto Draw is an electronically controlled lettering instrument for producing lettering and symbols with liquid ink, Fig. 6.26. A plotter is connected to a controller that accepts preprogramed standard or custom electronic template modules. The controller is activated by the plotter keyboard. Visual proofing of keyboard entries before any drawing is made possible by a

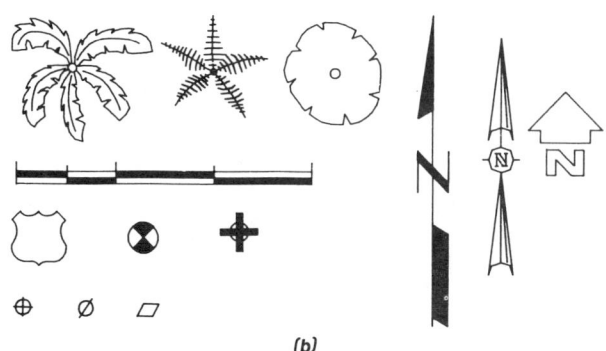

Fig. 6.25 Datascribe III lettering machine *(Courtesy Alpha Merics).*

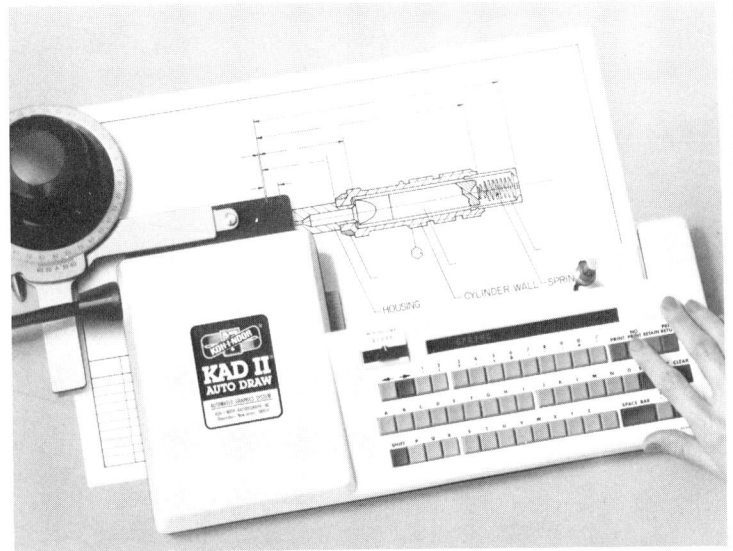

Plotter — Controller and template modules

Fig. 6.26 KAD II Auto Draw *(Courtesy Koh-I-Noor Rapidograph, Inc).*

light omitting diode red-light display. The instrument is completely portable and may be placed on the drawing surface at any angle or attached to almost every type of drafting machine.

6.8 LETTERING COMPOSITION

The use of hand lettering for single maps works well but is not feasible for production work. Where considerable lettering involving different styles and sizes is required, the lettering should be prepared by type-setting, either mechanically or optically. Regardless of the method used, the names to be produced should be collated in an orderly manner by type face and print sizes. The type of equipment used will govern how this information is to be listed.

Type measurement. It is necessary to be able to understand and specify type size when using printers type, hot or cold, and dry transfer kinds of lettering, Fig. 6.27. In measuring type, the smallest unit is a *point*. Each point measures 0.351 mm (0.0138"). There are approximately 72 points to 2.54 mm (1"), 12 points to one *pica*, and 6 picas to 25.4 mm (1"). The spacing or leading between lines is measured in points with line lengths expressed in picas.

Type sizes range from 4 to 120 points with sizes 8 to 24 point most commonly used. The point size applies to the body size of the metal block on which the type is cast and not to the type face itself. The true size of the letters will therefore be several points smaller than designated.

Fig. 6.27 The pica point.

Composition of type. The original method of setting type involved the selection of individual letters from a type case and arranging them in order by hand on a composing stick. This method is only suitable for small blocks of copy and is rarely used in modern printing practice or for producing map lettering. Mechanical methods are almost exclusively employed for the composition of type. Linotype equipment may be used to cast complete lines of justified metal type of any predetermined length. A keyboard similar to a typewriter allows an operator either manually or by using pre-coded tapes to quickly prepare many type faces from hot liquid metal. Line casting machines are widely used to set type for newspapers, magazines, books, and other material in small to medium sizes.

Fig. 6.28 Phototypesetter –Comp/Edit 6400 *(Courtesy Varatyper).*

For map production it is necessary that the lettering consist of type that can be composed quickly and positioned easily wherever required. Photographic type-setting equipment is replacing older methods for type composition. Photocomposing machines, hand operated and computer automated, in several forms are available for printing characters one after the other, Fig. 6.28. Direct-entry type phototypesetters allow lines of type to be produced on a keyboard, with errors noted and corrected. The copy may be automatically justified, hyphenated, and centered in any manner. Tape-operated phototypesetters, with some models having a mini-computer, that use punched paper or magnetic tape have the capability to record copy that may be re-entered, changed, added to, and re-set with a different type face, size or line measure. The use of cathode ray tubes (CRT's) and computers in typesetting is increasing. Their high speed allows complete pages of justified copy to be produced in a few seconds.

Many systems for producing phototype copy involve the imaging from a photomatrix negative. Only one letter image is needed in a particular style to produce the variety of sizes required. Expanded, condensed, backslanted, or italicized justified type may also be created from any one photo display film font. Type that is composed by photo display equipment offers a wide range of flexibilities in the horizontal spacing that may be obtained between individual letters and the vertical spacing between lines of lettering. The finished characters may be produced as film negatives, film positives, and positive or reverse type on paper. When printed on thin transparent sheets of film and coated with an adhesive, the lettering may easily be applied to the map original.

An inexpensive lettering machine is produced by Kroy that has the ability to produce professional looking type, Fig. 6.29. A variety of typesets and characters are available in sizes ranging from 8 to 36 point. Each typestyle/size is a circular disc. It is mounted on the machine and controlled manually. When a letter is selected, the print button is pushed causing colored dry ink to print on transparent or opaque tape. The machine automatically spaces and advances the tape as each character is dialed and printed. Upon completion of a word or line, the cut button is depressed to release the tape. The tape is removed from its backing and applied to the drawing.

Typewriters. Camera-ready copy of acceptable quality may be prepared with most typewriters. The letters, however, must be photographed to provide a film transparency since the characters are typed on an opaque medium. Justified margins on some machines are difficult to obtain and since the characters are all of the same width, standard typed copy does not have a desired printlike

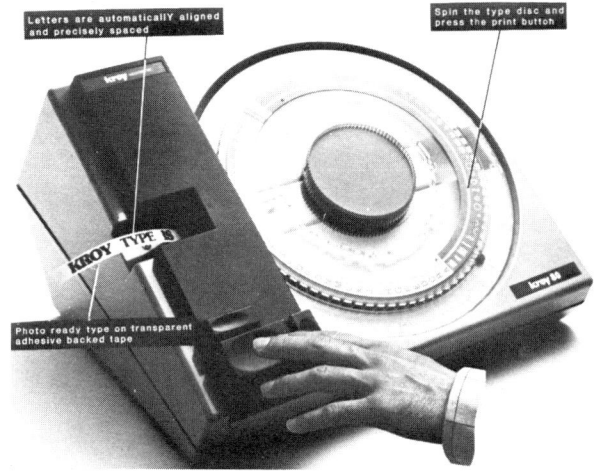

HELVETICA LIGHT
HELVETICA REGULAR
Helvetica Medium
Helvetica Bold
Helvetica Med. Cond.
Helvetica Med. Italic
STYMIE MEDIUM
STYMIE BOLD

Helvetica Med. Cond. Vertical

OPTIMA
OPTIMA SEMI BOLD
FLASH
GOTHIC
Century Schoolbook
𝕮𝖑𝖔𝖎𝖘𝖙𝖊𝖗 𝕿𝖊𝖝𝖙   Souvenir Medium
Commercial Script   **Souvenir Bold**

Typestyles

Fig. 6.29 Kroy lettering machine *(Courtesy Kroy Industries Inc).*

quality. There are a number of machines, however, like the VariTyper and IBM Selectric Composer that have special proportionately designed type face characters for producing good quality justified composition.

6.9 MARGINAL DATA

It is essential to include many kinds of information pertaining to drawings if they are to be used to their fullest extent. This information is usually located along the border of the drawing and is known as the marginal data. In order to attain uniformity in types of drawings, a standard style is usually adopted by agencies producing different kinds of working drawings and maps.

A drawing is identified by different kinds of marginal information, Fig. 6.30. For working type drawings a title box, frequently printed directly on the drawing media in the lower right-hand corner, is usually all that is required. The information that appears in the margin is frequently rearranged, limited, or added to since no definite set of rules can be given to fit all cases. It depends entirely upon the drawing, its subject, and purpose. On maps it may be necessary to relocate the infor-

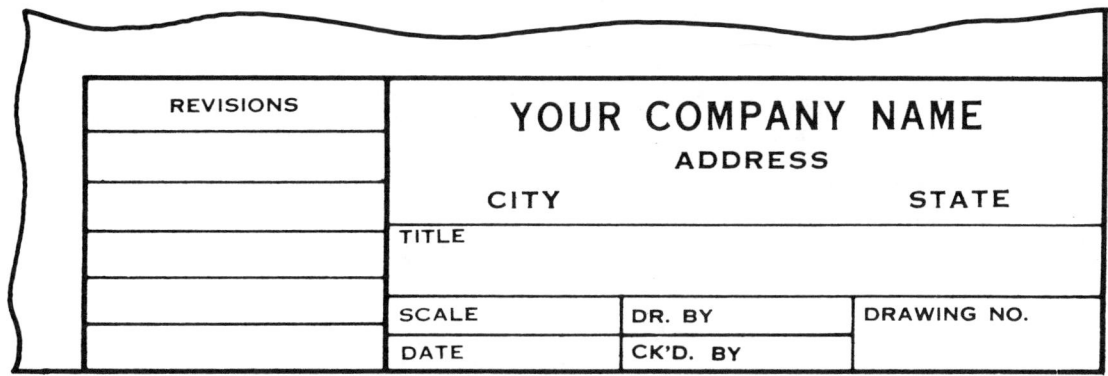

Title block for working type drawings

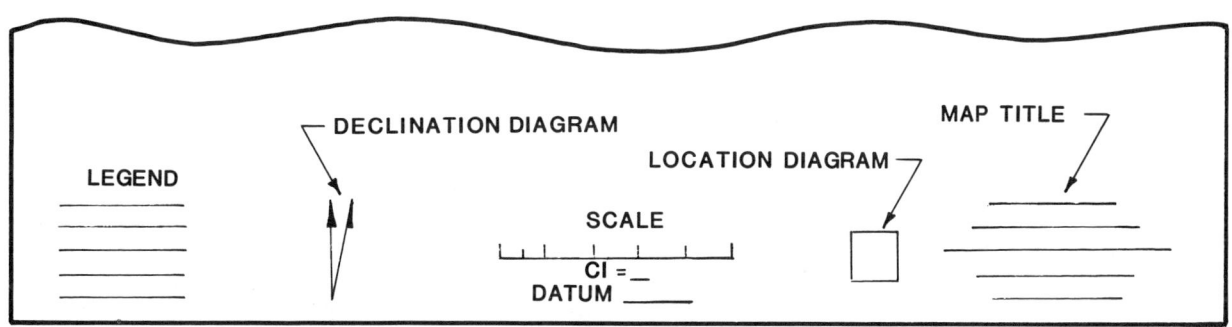

Title strip for maps

Fig. 6.30 Typical arrangement of marginal information.

mation, other than illustrated, to fit the sheet better. A pleasing well-balanced drawing should be the final objective. The lettering should be kept in proportion to the size of the drawing and is generally all vertical Gothic capitals. The marginal information corresponds somewhat to the table of contents and introduction of a book; it tells briefly how the drawing was made, the type, scale, and gives other information to make it more useful.

A format for the location of the marginal map information illustrated in Fig. 6.30 is as follows:

Southeast corner. Map title, Fig. 6.31, type of map, location, name, (usually taken from a prominent cultural or geographical feature that appears on the map), sheet number (number identifying sheet), date, name of mapping organization and draftsman. The most important information should be drawn in the largest letters with the remaining information smaller in proportion to its importance.

The title should be balanced on a center line. To balance a line of lettering, count the letters including word spaces. Begin the lettering by placing the middle letter or space on the center line of the title layout. By carefully lettering in both directions, a symmetrical arrangement may be obtained. It may be helpful to first sketch each line lightly in place and then to shift slightly left or right if the line is not balanced.

Southwest corner. Legend to illustrate symbols used for prominent features on maps. When multiple sheets are used to present map information, a legend is usually placed only on the index or title sheet. All symbols that are not self-explanatory should be explained and drawn in the same size and manner as on the map, e.g., names of organizations responsible for control, topography, photography, date of map survey, etc., Fig. 6.32.

Lower center. Scale, graphic scale, contour interval and datum note.

Lower right margin. Index to adjoining sheets, a diagram of the map showing the area drawn with a picture of all the other maps that border on the one in

```
                    TOPOGRAPHIC MAP
                   UNIVERSITY OF MAINE
                      ORONO, MAINE
    _____
    C.O. FORESTER CO            BANGOR, MAINE
    BY: B.E. NEAT              FEBRUARY 7, 1982
                   SHEET 1 OF 4 SHEETS
                        <--|-->
```
Letter symmetrically on center line

Fig. 6.31 Example of title composition.

LEGEND

PRIMARY ROAD	═══════
SECONDARY ROAD	=======
TRAIL	-------
PROPERTY BOUNDARY	— — —
BOUNDARY MARKER	△

Fig. 6.32 Legend.

question. This index is called a location diagram. Sheet one of four will have the location diagram as illustrated in Fig. 6.33 (a). The area cross-hatched represents the map information shown on sheet number one. For sheet number two, the location diagram is drawn the same but with area number two cross-hatched instead of area number one. The remaining sheets have the location diagram drawn in a similar way.

For maps drawn on single sheets a location diagram thoughtfully planned and designed is helpful to orient individuals who may not be familiar with the area represented, Fig. 6.33 (b).

Fig. 6.34 illustrates the arrangement of the marginal information appearing currently on many of the USGS metric maps.

Borders. The finished drawing should be framed using a simple rectangular border where pre-printed sheets are not used. A rectangle having a proportion of three to five is most pleasing and should be used where possible. The border should be drawn rather heavy with a thin line at least 12 mm (0.5") drawn around this to serve as a trim line.

On maps, fancy borders, titles, and legends should be avoided. These were commonly used by Dutch Renaissance cartographers and the Elizabethan map makers, but are out of place on modern maps.

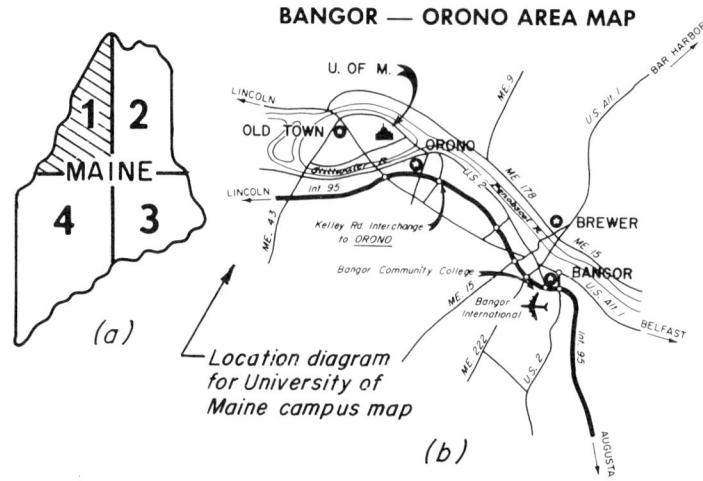

Fig. 6.33 Location diagram.

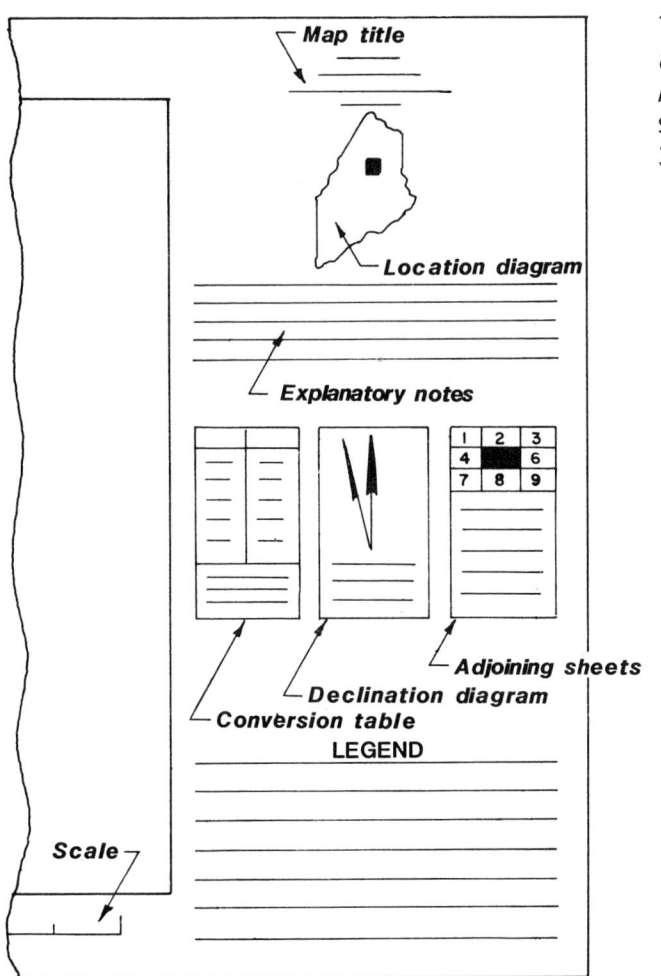

Fig. 6.34 USGS marginal information format for metric maps.

The minimum height of letters on A4, A3, and A2 paper sizes is 3 mm and 5 mm for A1 or larger paper sizes. A minimum spacing between lines of lettering of 3 mm should be maintained.

6.10 MICROFONT LETTERING

Drawings prepared for microfilming must be prepared carefully to allow for reductions and blowbacks. The linework and lettering must be of high quality to remain legible. Microfont lettering, Fig. 6.35, is recommended since it will maintain maximum clarity and legibility at extreme reduction and enlargement ratios. Standard alphanumeric characters are available in both LEROY lettering templates and dry transfer lettering. Individual LEROY templates are available in six sizes: 3, 4, 5, 6, 10, and 12 mm.

ABCDEFGHIJKLMNOPQRSTUVWXYZ.,:;1234567890
=÷+−± @&✻?#x"%'()[ ]°!¢$/_∠∞∆≈~⌀⊥<>µα√ớδΣ
ɣπβθωΩ∴‖

Fig. 6.35 Microfont lettering for microfilmed drawings.

GRAPHICAL REPRESENTATION OF DATA

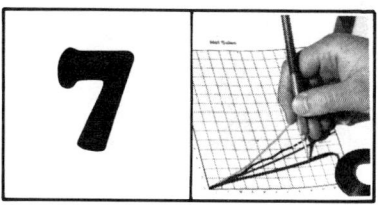

7.1 INTRODUCTION

The function of presenting information graphically is to translate data into a visual form that makes relationships easy to grasp, understand, and remember. A graphical presentation usually can be interpreted and analyzed much more quickly and clearly than a table of endless figures, Fig. 7.1. Data presented in graphical form frequently brings out hidden facts and relationships that stimulates as well as aids analytical thinking and investigation.

All segments of today's society cannot avoid an exposure to information presented in graphical form by the mass media in some creative manner. The newspapers, television, magazine advertisements, etc. frequently represent data of all types in a visual form which is captivating for lay people. Professional people, in addition to being able to understand and interpret many types of graphs, must be knowledgeable of the methods and techniques for presenting and analyzing data and solving problems. The principles of graphical construction, design, and limitations must be thoroughly understood for presenting data graphically in an effective way. The basic feature that should be kept in mind is simplicity. Several points to remember are: (1) try to convey one idea, (2) use well-defined graphic elements to make the point, (3) use color judiciously for emphasis, visibility, and representation, (4) keep lettering to a minimum.

7.2 CLASSIFICATION

With the exception of map charts (chapter 12) and with the omission of polar graphs, trilinear graphs, and alignment charts which are very specialized for presenting specific data required by industry and research, graphical information may be presented in a variety of ways and may be classified according to the following types. Each of these types is discussed in detail in subsequent articles.

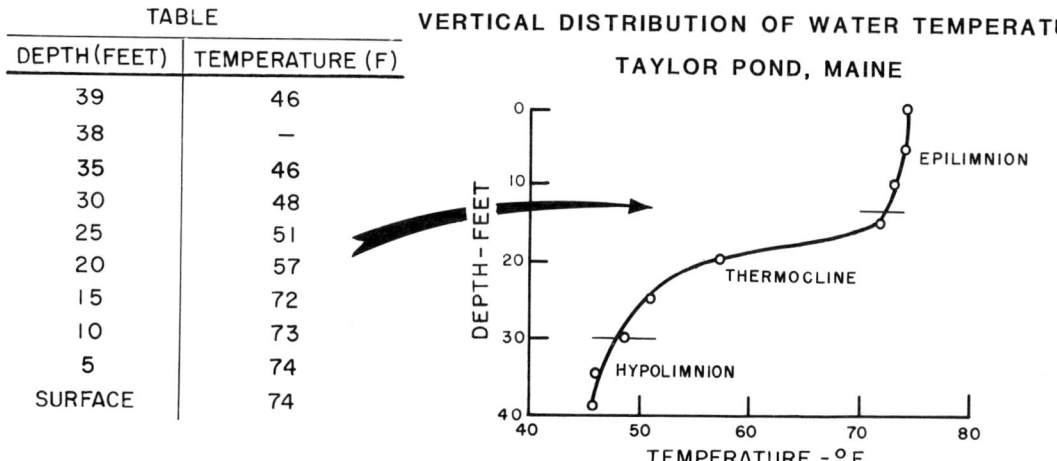

Fig. 7.1 Use of graph to translate data into visual form.

- Rectilinear graphs
- Semilogarithmic graphs
- Logarithmic graphs
- Bar charts
- Area charts
- Pictorial charts
- Organization and Flow diagrams
- Map charts
- Polar graphs
- Trilinear graphs
- Alignment charts

The terms chart, graph, and diagram may be confusing and their meaning should be clearly understood. Maps used for navigation in air or water or maps used to exhibit, in geographical aspect, temperature variations, a plan for military operations, or the like are referred to as charts. A *chart* is an inclusive term referring to any form of graphical representation of data. Tabular information and all graphs are also termed charts. A *graph* is a form of chart where the data symbolizes a system of interrelations that is plotted in reference to a grid system. *Diagrams* are line drawings of various forms for presenting data where no grid system is used.

7.3 BASIC GRAPH OR CHART COMPONENTS

A system of rectangular coordinates is used most frequently for plotting graphs or charts, Fig. 7.2 (a). The horizontal axis is referred to as the *X-axis* or *abscissa*, the vertical axis as the *Y-axis* or *ordinate*, and the point of intersection is called the *origin*. The two intersecting axes form the four quadrants shown by Roman numerals. Since most data involves positive values, only the first quadrant, Fig. 7.2 (b), is used when drawing a graph. If negative values are to be graphed, it will be necessary to use one or more of the other four quadrants, Fig. 7.2 (c). Conventionally positive values are plotted to the right of the origin along the X-axis and upward along the Y-axis from the origin. Negative values are plotted to the left of the origin along the X-axis and downward along the Y-axis from the origin.

In the rectangular coordinate system point P (a,b) is located from two intersecting axes (OX and OY) at 90° to each other. To locate point P, the X coordinate is always written first. In this example P (a,b) represents the point P whose X coordinate is a and whose Y coordinate is b, Fig. 7.2 (a).

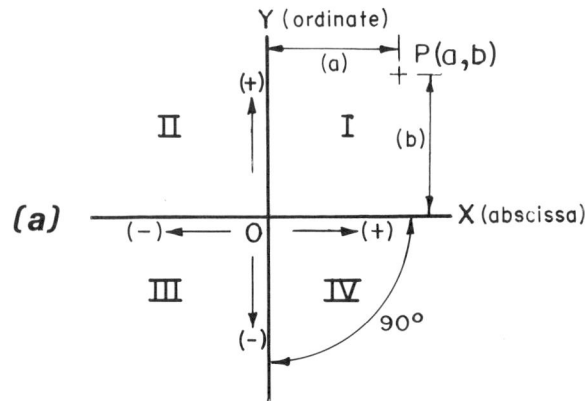

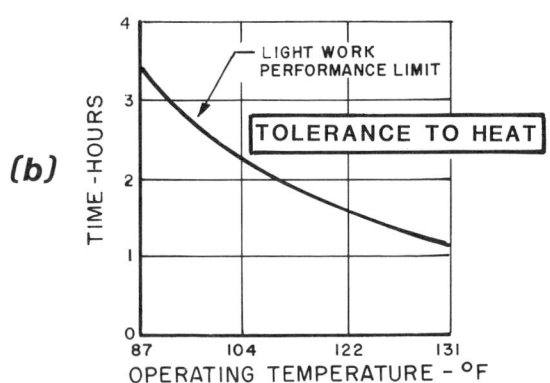

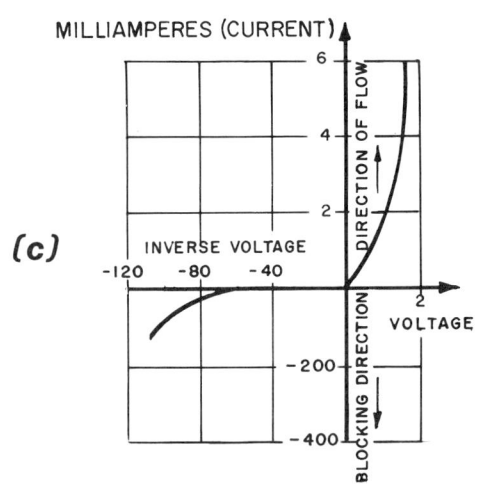

CHARACTERISTIC OF A SEMI-CONDUCTOR DIODE

Fig. 7.2 Rectangular coordinates.

7.4 RECTANGULAR COORDINATE GRAPHS

The most common form of graph is the rectangular coordinate or rectilinear graph. Graphs of this type are used extensively since they are easy to construct and read. The direction or relative trend of the data can be quickly analyzed for one or more curves. Their use should be limited where there are only a few plotted points or for showing variations in absolute amounts.

As a guide to effectively prepare a graph, the procedures that follow should be carefully considered and adhered to. The standards discussed are also applicable to graphs drawn on other types of gridded media other than rectangular coordinate paper.

1. <u>Grid selection</u>. Many kinds of commercially prepared coordinate paper are available in different spacings, sheet size, and color for plotting data (see chapter 5). The most generally used kinds of grid paper are rectangular, logarithmic, and semilogarithmic. Depending on the type of reader to be reached and the purpose of the graph, the printed grid lines may not be desirable on printed copies of the graph. Non-reproducible grid media is available in blue that will not print or photograph. Generally speaking, graphs intended for publication or the general public should have the grids entirely omitted, Fig. 7.2 (a), or spaced well apart, approximately 2.54 cm (1"), to make the graph easier to read, Fig. 7.3 (b). For technical and laboratory reports the closely spaced grid lines are desired to aid the reader to make computations accurately, to analyze, and to interpolate the data, Fig. 7.3 (c).

2. <u>Location and selection of axes</u>. The origin point of the axes is placed a minimum of 2.54 cm (1") from the edges of the sheet to allow sufficient space for lettering scale values and captions, Fig. 7.4. If values are plotted that are both positive and negative, e.g., an algebraic expression, the origin point will have to be shifted accordingly. Usually the *independent* or controlled variable, the quantity arbitrarily varied, is placed on the horizontal or X-axis. The *dependent* variable is normally placed on the vertical or Y-axis. Variables involving time are also placed on the X-axis.

Two different Y-axes are sometimes required for plotting two different but related dependent variables. The axes in general should be drawn hidden line weight with the grid lines, if shown, drawn center line weight.

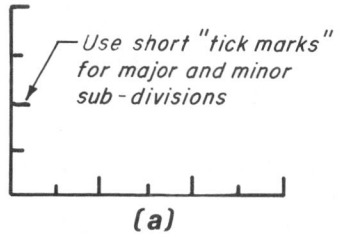

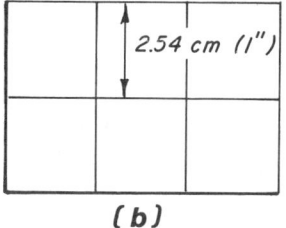

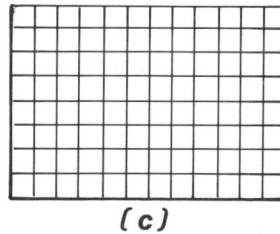

Fig. 7.3 Use of grid lines.

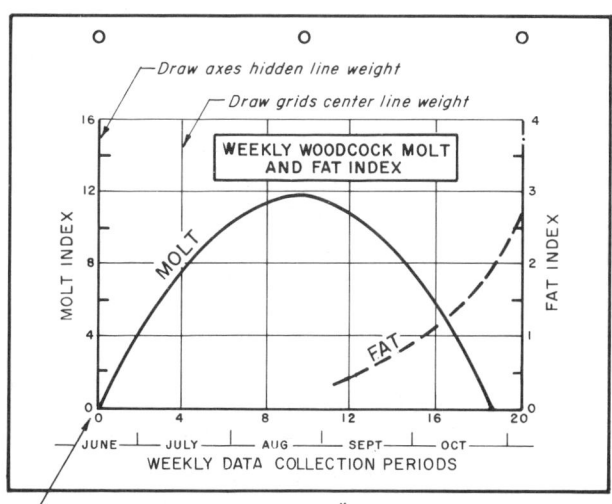

Fig. 7.4 Use of two Y-axes.

3. _Scale units._ The axes should be labeled carefully for easy readability. Generally, the intersection of the axes is the zero point for both the vertical and horizontal axes. The zero point is necessary where the graph is qualitative or where visual comparison of plotted magnitudes is desired, Fig. 7.5 (a). Where paper saving is important, the origin point can be shown with the unused portion of the grid broken out to save space, Fig. 7.5 (b).

Zero values may be omitted, with the origin starting at a value other than zero, if the graph depicts absolute values or is quantitative in nature, Fig. 7.5 (c).

In determining the scale for both axes, an orderly progression should be used for lettering the major grid lines. Multiples of 1, 2, 4, 5, 10, etc. (multiplied or divided by 1, 10, 100, etc.) work well and the units chosen must be compatible with the graph paper being used if the data is to be easily interpolated, Fig. 7.6 (a). Multiples of 3, 6, 7, 9, etc. should be avoided since each of the smaller grids would no longer represent a value that would be easy to work with, Fig. 7.6 (b).

Small or exaggerated scales should not be used in order to avoid a possible distortion in the data, Fig. 7.7 (a). The scale units used should be chosen so as to utilize the available grid area and at the same time give a slope to the curve that will represent a true picture of the data. Curves of steep slope

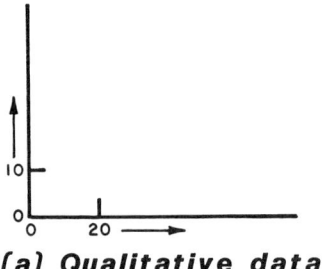
(a) Qualitative data

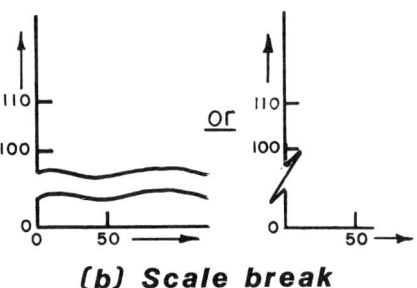
(b) Scale break

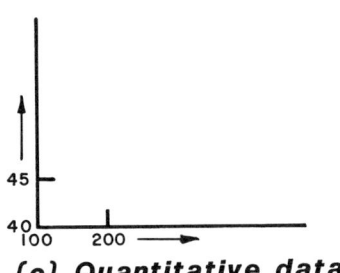
(c) Quantitative data

Fig. 7.5 Axes scale.

(greater than 40°) indicate that the data pictured is significant while flat curves (less than 10°) indicate insignificant change, Fig. 7.7 (b).

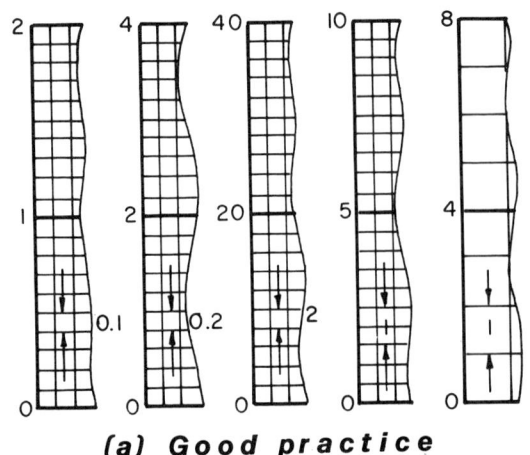

(a) Good practice

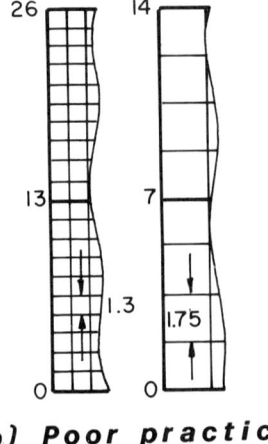

(b) Poor practice

Fig. 7.6 Scale unit designation.

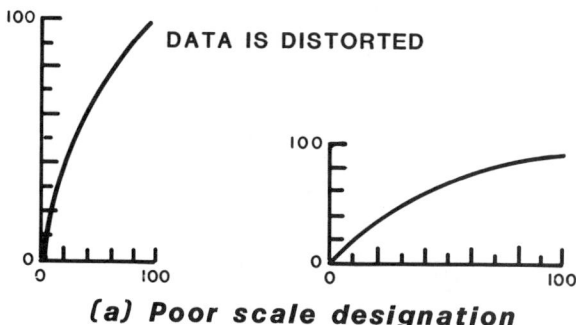

(a) *Poor scale designation*

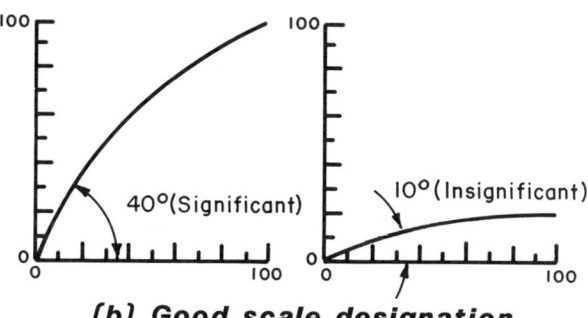

(b) *Good scale designation*

Fig. 7.7 Scale designation and its possible effects.

In general, scale units should be used that will give a grid system that will not create false impressions for the data plotted, Fig. 7.8. Graphs can be designed that are misleading and should always be studied carefully before jumping to any conclusions.

4. <u>Plotted points</u>. The plotted points on a graph representing empirical or experimental observations should be indicated on the final graph. Symbols for marking plotted points, when more than one curve is shown on a single graph, are illustrated in Fig. 7.9. The open point symbols are recommended in the order numbered. Normally, the filled-in symbols are used when three of more curves are drawn. The curve should never pass through the point symbols.

Data based on mathematical expressions should not have the plotted points indicated on the final graph.

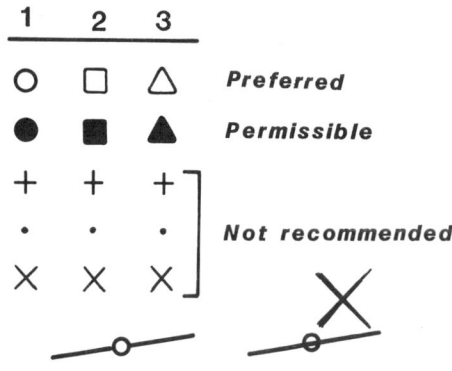

Fig. 7.9 Plotted point symbols.

GRID SYSTEM–WIDER THAN HIGH

- *For data extending over a long time period*
- *For a series requiring frequent observed points*
- *When actual rate of growth is slow*

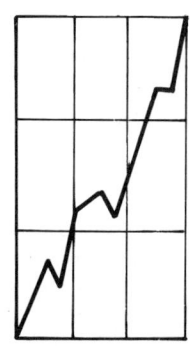

GRID SYSTEM–HIGHER THAN WIDE

- *For data over a short period of time*
- *When it is desirable to accentuate minor fluctuations*
- *For separating overlapping curves*
- *For showing series which differ greatly in magnitude*

(From Human Engineering Guide for Equipment Designers)

Fig. 7.8 Preferred grid systems.

5. _Drawing the curve._ The type of data that is plotted dictates the nature of the curve to be drawn. Continuous data (physical phenomena) is represented by a smooth curve that is drawn through the average of the plotted points, Fig. 7.10 (a). The average line, both straight and curved, should hit as many of the plotted points as possible. To help draw a smooth curve (fairing), it is helpful to first sketch lightly the curve and then to connect each point to the next with a straight line both above and below the curve, Fig. 7.10 (b). The final curve is then drawn in such a way that the area formed between the curve and the boundaries of the data points on one side of the curve is approximately equal to the area on the other side. A curve may also be balanced by trying to have an equal number of points above and below the line with the summation of ordinate distances above and below the line equal to zero, Fig. 7.10 (b).

Discontinuous or discrete data where there is no direct relationship between the variables, like time and rainfall, should show the curve connected by straight lines between the plotted points, Fig. 7.10 (c).

Curves are frequently extrapolated or extended beyond the plotted points for analyzing and predicting future trends in the data, Fig. 7.10 (a). A dashed line is recommended for this purpose.

The curve should be drawn using a solid heavy line that is in proportion to the graph size. A finer line should be used to draw the curve if the data represented is mathematical and it is necessary to make computations. When more than one curve is drawn, the lines should stand out prominently and be clearly differentiated from one another. Confusion or difficult comparisons may be reduced by using color, different lines (solid, dashed, dotted, etc.), different point symbols (circles, squares, and triangles), and by labels placed close to each curve, Fig. 7.11. The number of curves for a single graph should be limited to four to avoid an appearance of clutter.

6. _Lettering._ The effectiveness of a graph is largely dependent upon the neatness of the lettering. Poor

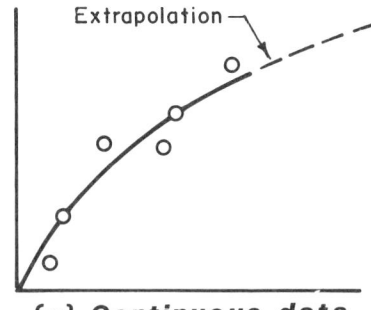
(a) Continuous data

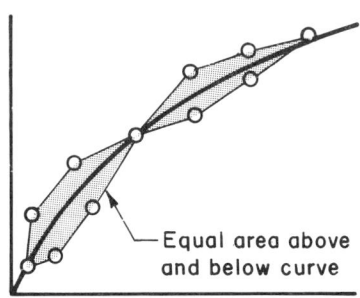

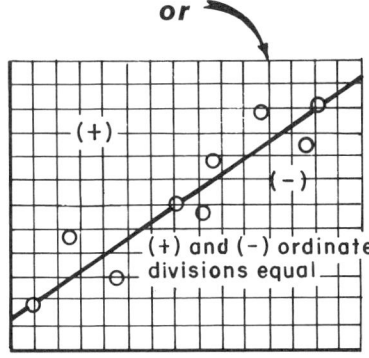
(b) Balancing the curve

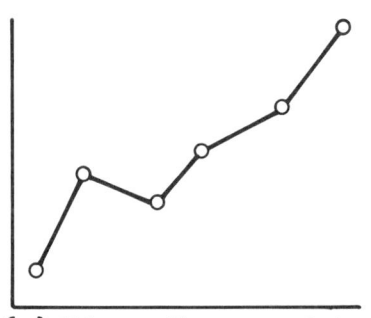
(c) Discontinuous data
Fig. 7.10 Curve fitting.

lettering may seriously reduce the overall impression of an otherwise well designed graph and should be considered carefully in terms of legibility, artistic appeal, time available, and ease of drawing. The lettering may be executed by mechanical methods (template or set type), typing (carbon paper placed on

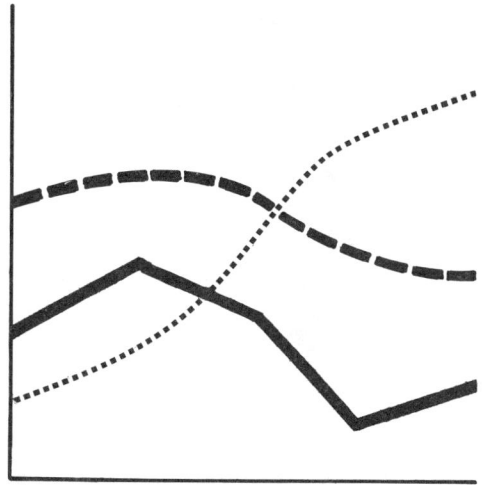

Easy comparisons. The use of appliqué materials (see Chapter 2) in different line patterns and colors are helpful for preparing graphical information.

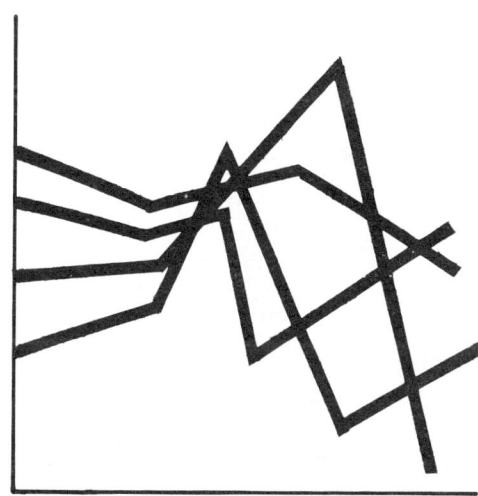

Difficult comparisons. Confusion occurs when several lines fall close to one another or cross each other.

Fig. 7.11 Curve representation.

the back side of the drawing media like tracing paper will allow the image of the typed character to be deposited on both sides of the paper producing a dense black character that will reproduce with excellent results), or freehand. Allowance should be made for lettering reduction on any graph intended for reproduction.

(a) Scale identification. The numbers identifying the scale units on both axes should read from the bottom of the graph, Fig. 7.12 (a). Scale units should be designated only at major intervals and are normally spaced 12.7 to 25.4 mm (½" to 1") apart. Values having three or more digits should have the zeros omitted, Fig. 7.12 (b). The actual value may be identified by including this information in the scale caption, Fig. 7.12 (c), or through the use of powers of 10, Fig. 7.12 (d). Powers of 10 should not be used in the scale caption, since it may not be clear whether the scale number is to be multiplied by the power of 10 or already has been, Fig. 7.12 (e).

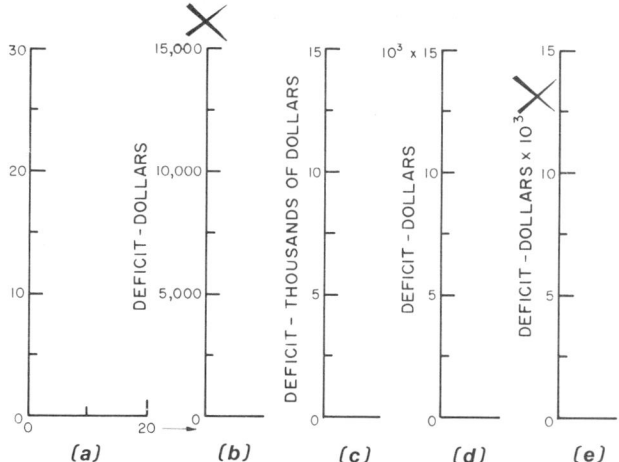

Fig. 7.12 Numerical scale identification.

Non-numerical scale values such as days, weeks, months, and years are designated using standard abbreviations, Fig. 7.13. Depending on the nature of the variable, it may be necessary to place the lettering vertically or slightly inclined rather than horizontally.

(b) Scale caption. The scale captions should be arranged along the axes so as to be read from the bottom and right side of the graph, Fig. 7.14 (a). Each scale caption should include the name of the variable, symbol used for the variable, if any, and the units of measure. Lettering arranged vertically is extremely difficult to read and should be avoided, Fig. 7.14 (b).

(c) Curve identification. When more than one curve is drawn on a graph, individual identification may be obtained by labels placed close to each curve or through the use of a legend appearing within the grid system, Fig. 7.15.

(d) Title. A short, well-thought-out title that completely identifies the

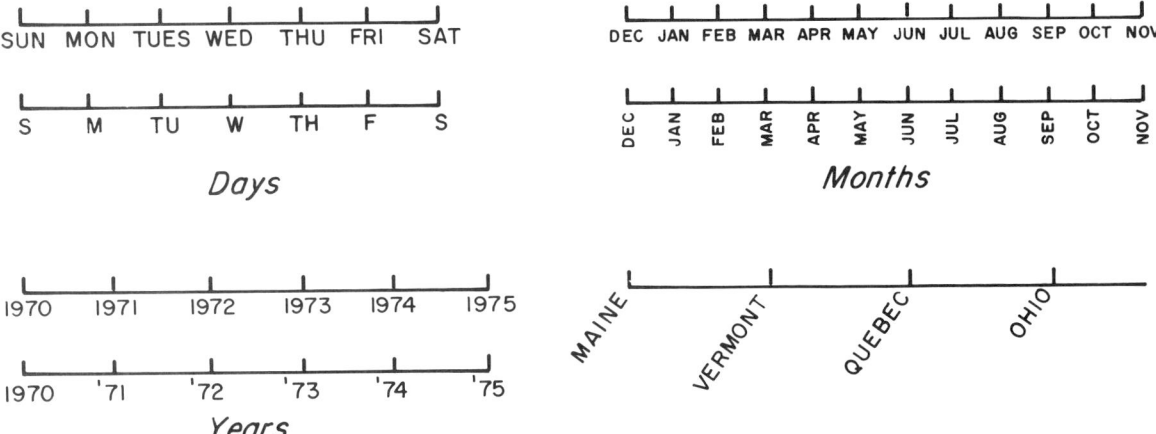

Fig. 7.13 Non-numerical scale identification.

data, including source references, should be placed above the graph, below the graph, or in an open area of the graph. A subtitle may be used to augment the title for clarity. Abbreviations should not be used in the title and only standard abbreviations used elsewhere on the graph. The title for experimental data should also include the date and the name of the experimenter. For titles lettered on the graph, a rectangular box may be drawn around the title if desired. The placement of title information is illustrated in many of the figures of this chapter.

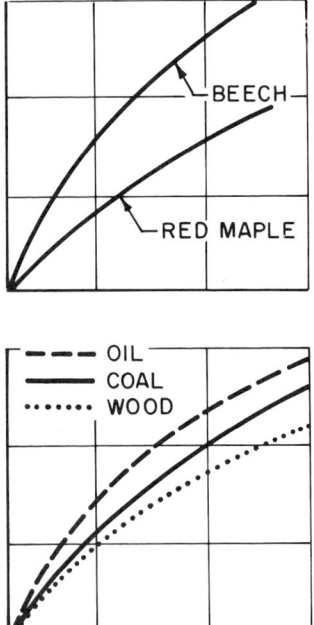

Fig. 7.15 Curve identification.

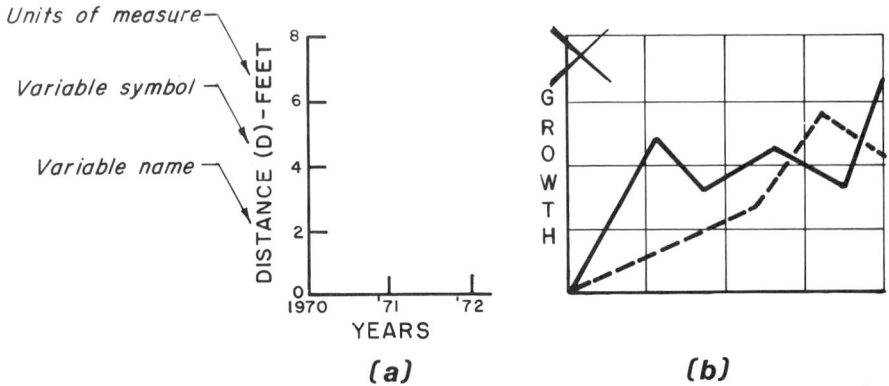

Fig. 7.14 Scale captions.

7.5 SEMILOGARITHMIC GRAPHS

The grid on semilogarithmic paper has equally spaced lines along one axis and logarithmic-spaced lines in the other direction, Fig. 7.16 (a). The major scale divisions numbered 1, 2, 3, ...10, referred to as one cycle, are determined by finding the logarithm to the base 10 of numbers 1 through 10. A unit distance to represent this cycle may be laid out on blank paper using any scale. Usually, commercially prepared semilogarithmic grid paper is used. The paper may have any number of cycles for increasing the range of values to be plotted. A full cycle always begins with the power of 10 and ends with the next highest power of 10. The groups of numbers, 0.1 to 1, 1 to 10, and 10 to 100, illustrated in Fig. 7.16 (b) each represent one cycle.

The terms ratio graph and rate of change graph are frequently used to describe graphs plotted on semilogarithmic paper. Graphs of this type are used where it is desirable to show the rate of change of a variable instead of the quantitative change. The slope of the tangent to the curve at any point gives the rate of increase or decrease of the data at that point. A graph plotted on rectangular coordinate paper, where the curve is not a straight line, may give a false impression of the trend of a curve and should be plotted on semilogarithmic paper to reveal if the true rate of change is constant, decreasing, or increasing, Fig. 7.17 (a). It is not uncommon to find that the same data that produces a curve on a rectangular coordinate graph will produce a straight line curve on a semilogarithmic graph. The curve will be a straight line whenever the increments of the dependent variable increase in geometric progression while the increments of the independent variable increase arithmetically.

Semilogarithmic graphs are also helpful in the derivation of empirical equations. The data plotted on rectangular coordinate paper, Fig. 7.17 (b), indicates an initial rapid decrease in the intensity of radiation that suggests an exponential type of equation. This assumption can be tested by rectifying the data using semilogarithmic paper to replot the data, Fig. 7.17 (c). The curve now plots as a straight line resulting in an equation on this type of graph paper of the form $Y = b(10)^{mx}$ or $Y = bm^x$, where b is the Y-intercept of the curve and m is the slope of the curve.

7.6 LOGARITHMIC GRAPHS

The grid lines on logarithmic paper for both axes are ruled in divisions proportional to the logarithms of numbers 1 through 10. Commercial paper is available with one or more cycles in each direction. The advantages of using paper with multiple cycles is that large numbers may be shown by short distances.

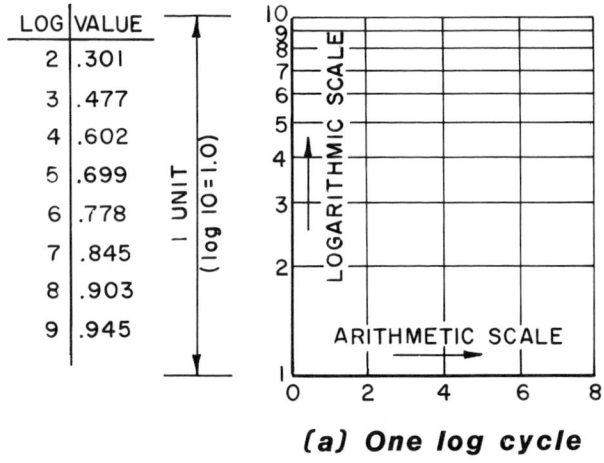

(a) One log cycle

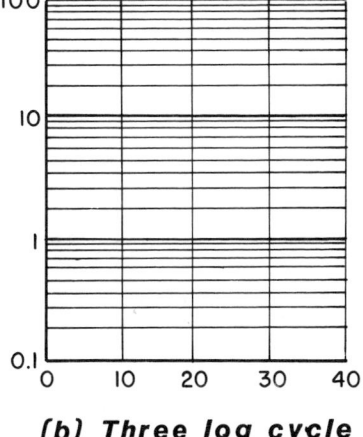

(b) Three log cycle

Fig. 7.16 Semilogarithmic graph paper.

Logarithmic paper is not used for presenting factual kinds of data or where it is desirable to emphasize changes in absolute values. Its primary use is for plotting equations involving multiplication, division, powers and roots where the curve appears as a straight line. Curves of this type are easy to draw since only two points need to be calculated and plotted to draw the curve. A typical logarithmic graph is shown in Fig. 7.18.

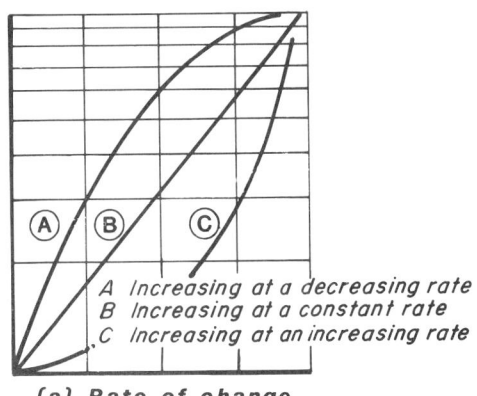

(a) Rate of change

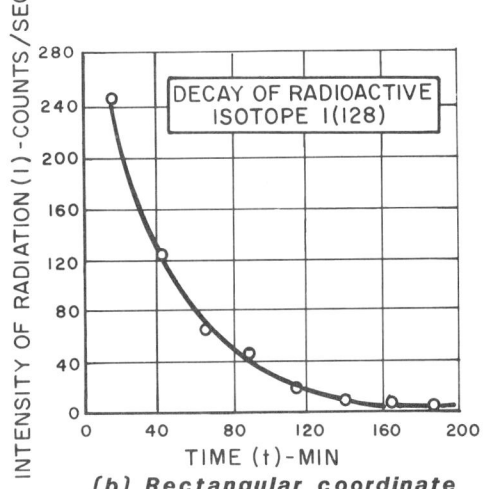

(b) Rectangular coordinate

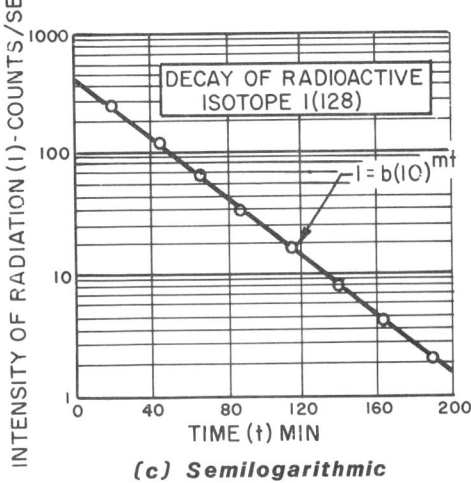

(c) Semilogarithmic

Fig. 7.17 Rectangular coordinate and semilogarithmic graph comparisons.

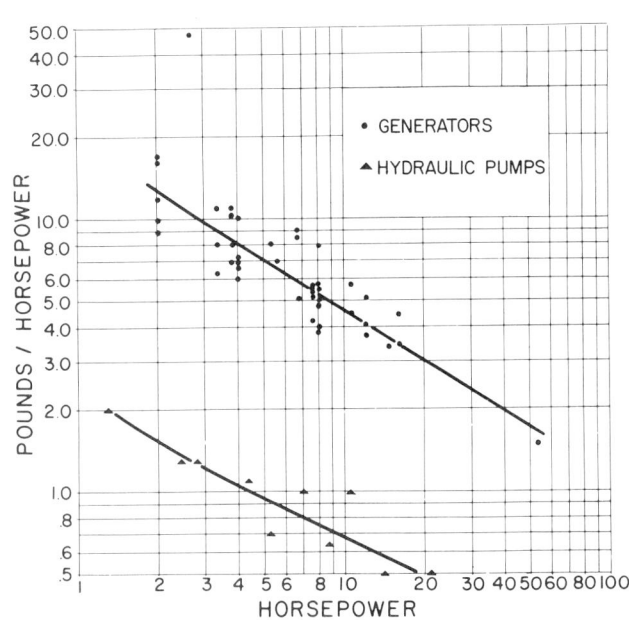

Fig. 7.18 Logarithmic graph for specific weight vs. horsepower of electric generators and hydraulic pumps.

7.7 BAR CHARTS

Illustrations appearing in publications of a popular nature or those intended to be used for visual communications commonly use bar or column charts since they are easy to make and understand. They are useful in comparing data for a given item at different times. Other attributes include an effective means for emphasizing total magnitudes in a single series of data, showing a few components of totals, and showing a range of values or deviations from a normal or standard. Numerical values are represented by vertical or horizontal bars that should always be drawn from a common zero base line in order that the lengths of the bars represent values accurately, Fig. 7.19.

Many variations of bar charts are possible. They may be additive, Fig. 7.20 (a), or comparative, Fig. 7.20 (b). The histogram or connected-column chart, Fig. 7.20 (c), is particularly useful for plotting statistical information that involves frequency distributions. The bars are drawn touching each other and should be equal in width so that the reader is not distracted from the differences in bar height. If the plotted values at the center of each bar are connected by straight lines, a frequency polygon results and if the line is smoothed out, a frequency curve results. Other possible variations include the deviation chart, Fig. 7.20 (d), and the combination bar chart showing a line curve, usually a trend line, and column comparisons, Fig. 7.20 (e). The use of overlays should be considered for combination line-column charts since they are helpful with charts of this type in conveying one point at a time.

Occasionally bar charts are drawn as if they were three-dimensional, Fig. 7.21. This is effective if the chart does not give a distorted picture of the quantities involved.

All charts require careful planning to make them easy to comprehend. The bars should be drawn so as to make it possible for the data to be translated emphatically. A variety of colors and patterns should be used to distinguish each column and to separate segments within each column. Pressure-sensitive tapes are helpful for this

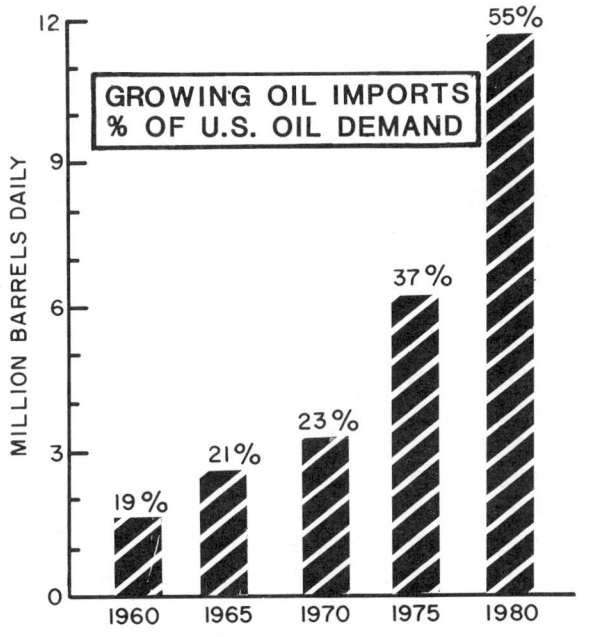
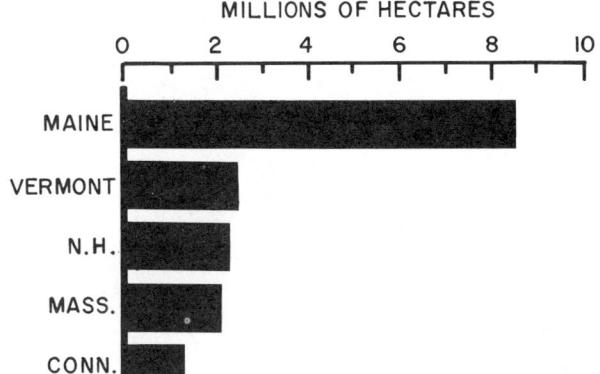

Fig. 7.19 Vertical and horizontal bar chart.

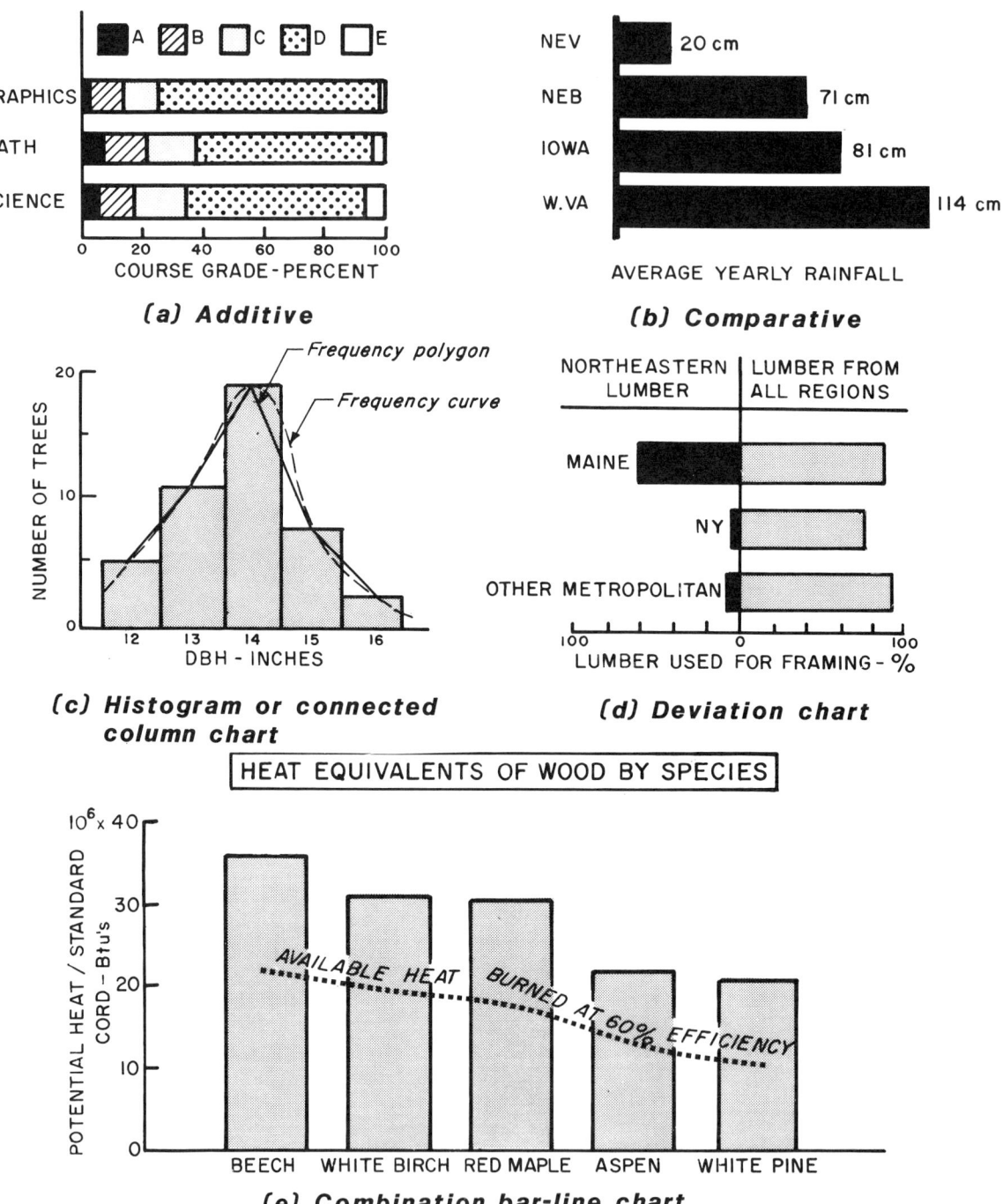

Fig. 7.20 Variations for drawing bar charts.

purpose. Optical illusions may occur if crosshatching and shading patterns, or colors are incompatible, Fig. 7.22 (a). Where more than one pattern or color is used, the stronger one should be placed on the bottom. If one color or pattern is bolder than the others, that particular item in the chart will stand out, Fig. 7.22 (b). Bars which vary in width as well as length are disproportionate and all widths should be equal. Spacing between the bars will depend on the nature of the chart and should be uniform.

It is poor practice to leave spaces in bars that break up their continuity in any way, Fig. 7.22 (c). False impressions usually result because correct visual comparison is not possible.

The use of grid lines across a chart is confusing and may hinder rather than help the reading of a chart, Fig. 7.22 (d). Where the magnitude of each bar is given, the grid axis may be omitted entirely. Labeling of the bars should appear outside base lines or borders. Lettering at the end of the bars should be avoided, if possible, since it makes them appear longer, causing distortion.

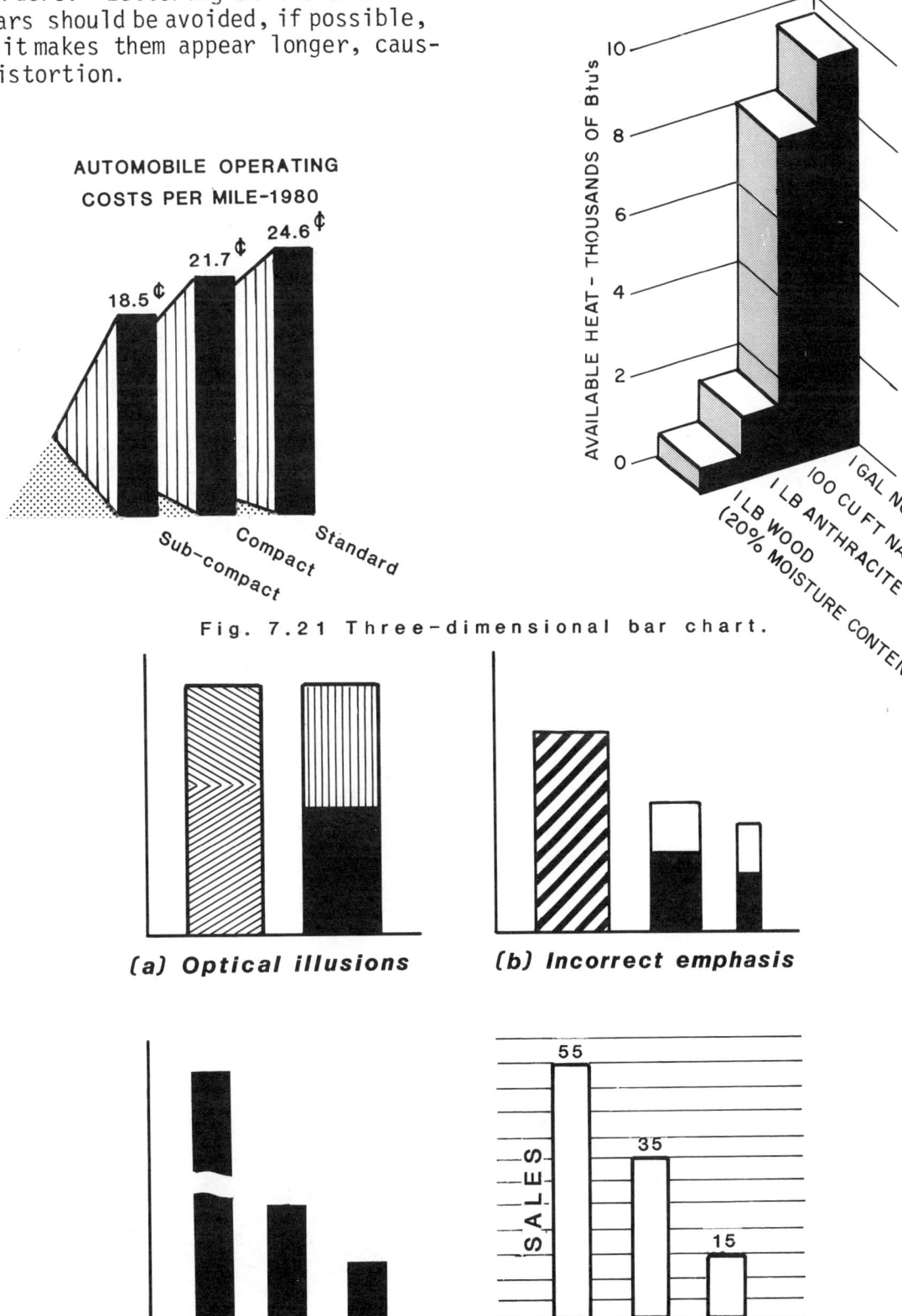

Fig. 7.21 Three-dimensional bar chart.

Fig. 7.22 Common charting errors to avoid.

7.8 AREA CHARTS

Area charts are used to show changes in the component parts of a total. The point of area charts should be obvious immediately. Their main advantage is their simple, direct, "non-statistical" form of presentation. An area of some shape is used to represent the total, expressed as percentages, and it must be carefully ascertained that the spaces enclosed within the total area accurately represent the proportional parts of the data in the area.

The pie chart, sometimes called a circle graph, is the most popular form of area chart, Fig. 7.23 (a). The area of the circle represents 100 percent, and the slices out of the circle represent percentages of the total. Shading of the sectors, for emphasis, should be light and contrasts limited. A wide variety of patterns and color tints in pressure-sensitive films offers many compatible combinations. Overloading is a common fault in the planning and drawing of pie charts. Lettering is likely to become crowded and confusion will result from the use of many patterns, shadings, or varieties of color, Fig. 7.23 (b). To simplify interpretation, limit the number of sectors to as few as practicable. Lettering may be placed either within or outside the sectors.

Other effective variations for drawing pie charts are illustrated in Fig. 7.24.

A single bar representing 100 percent of the total quantity may be subdivided and used instead of a circle, Fig. 7.25. The techniques for drawing, shading, and lettering the segments are the same as for pie charts.

RECOMMENDED BALANCED DAILY DIET

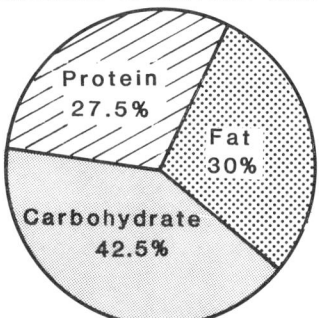

(a) Good relationship

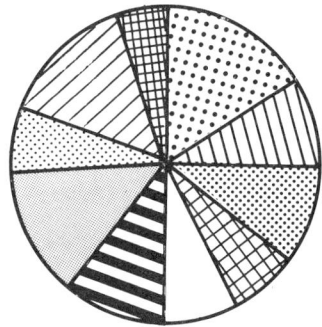

(b) Avoid complex relationship

Fig. 7.23 Pie chart.

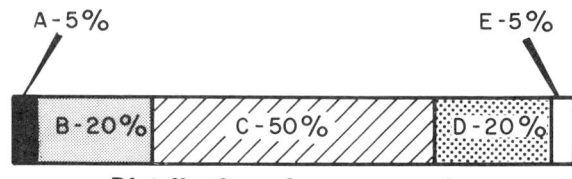

Distribution of course grades

Fig. 7.25 Percentage bar chart.

Segmented form

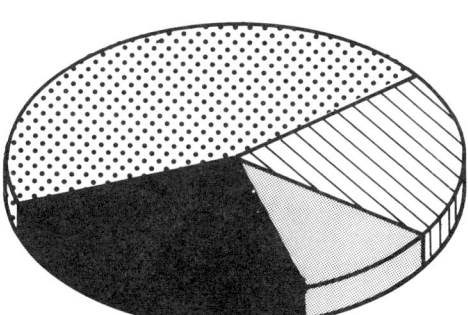

Pictorial form

Fig. 7.24 Pie chart variations.

A dramatic variation of the basic curve chart is the surface chart which represents data by filling in the areas to which the data applies, Fig. 7.26. Components of totals can be easily shown when it is desired to focus attention on the distance between curves. Various types of shading film may be used to create the areas between the lines. To avoid a top-heavy appearance, the darkest area should be at the bottom and the lighter tones at the top.

7.9 PICTORIAL CHARTS

Pictorial charts, sometimes referred to as pictographs or picture graphs, are used for showing simple comparison of relationships symbolically. Statistical facts in newspapers and magazines are often presented this way to attract immediate attention and interest. They should not be used where it is necessary to convey to the reader detailed accurate information. An unlimited variety of charts may be drawn using various symbols, as eye-catchers, to represent subject and quantity, Fig. 7.27.

Fig. 7.26 Surface chart.

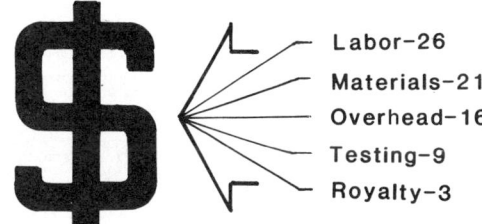

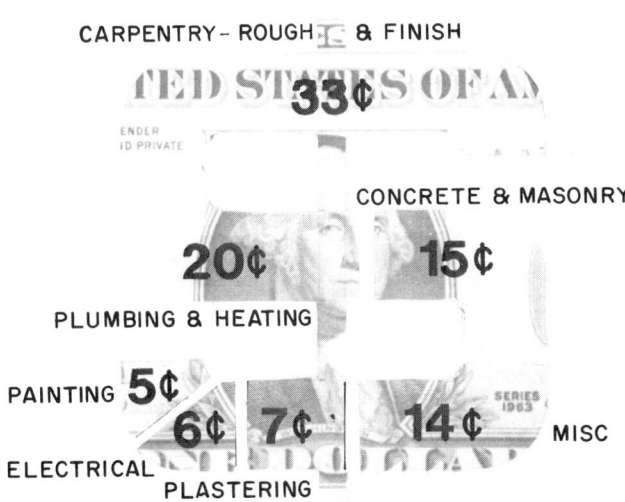

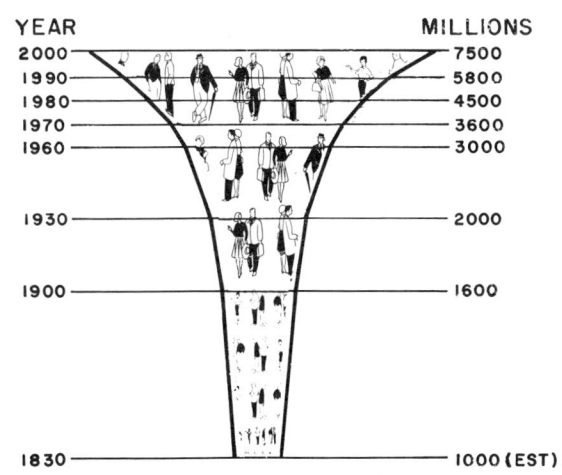

Fig. 7.27 Pictorial charts.

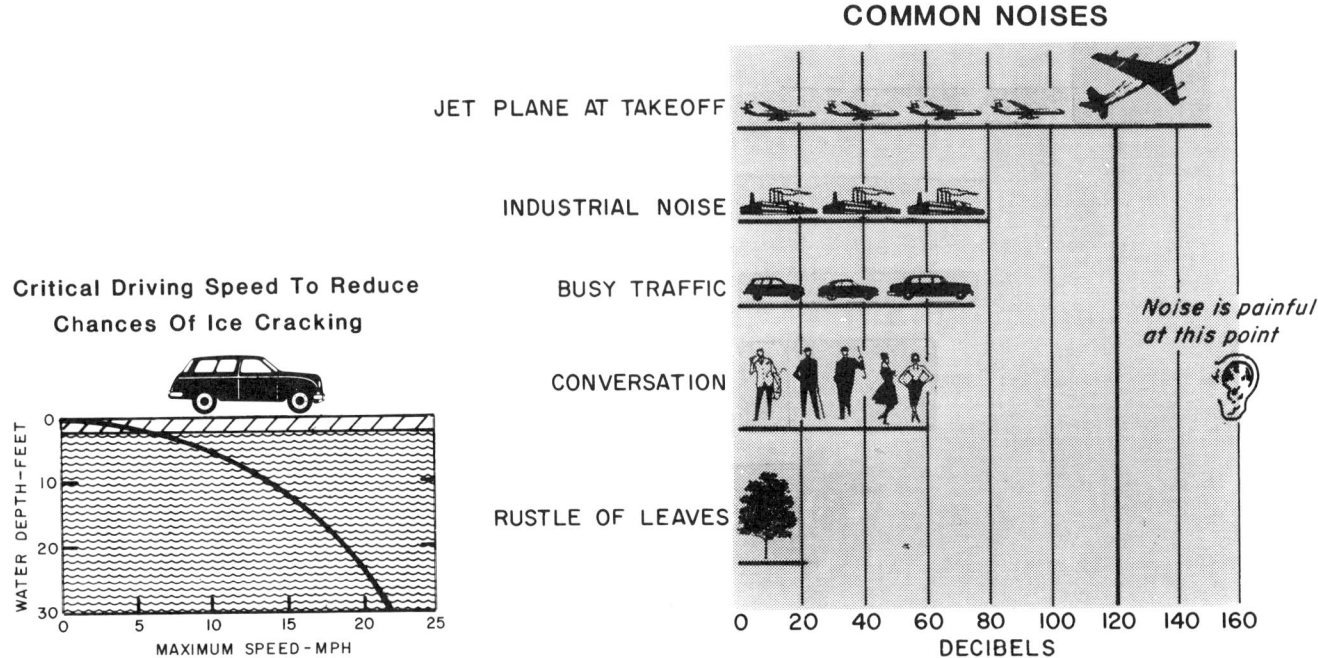

Fig. 7.27 Pictorial charts (continued).

7.10 ORGANIZATION AND FLOW DIAGRAMS

Schematic type drawings are used to portray both organization and flow diagrams. Organization diagrams, Fig. 7.28 (a), are used to show the distribution of functions and personnel while flow diagrams, Fig. 7.28 (b), are used to indicate the various stages of a process. Data is enclosed in rectangles, squares, circles, or other geometric shapes and connected by lines. Schematic symbols may be used to represent or enclose the data to make the drawing interesting and easier to read.

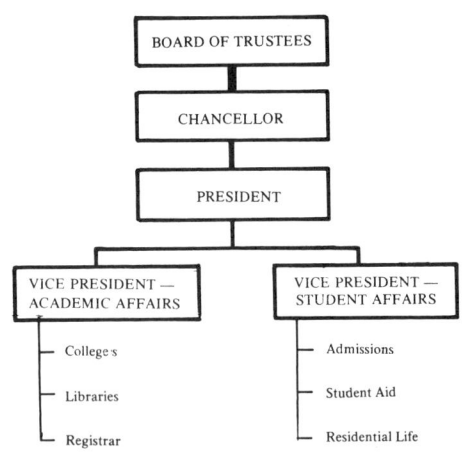

(a) Organization diagram

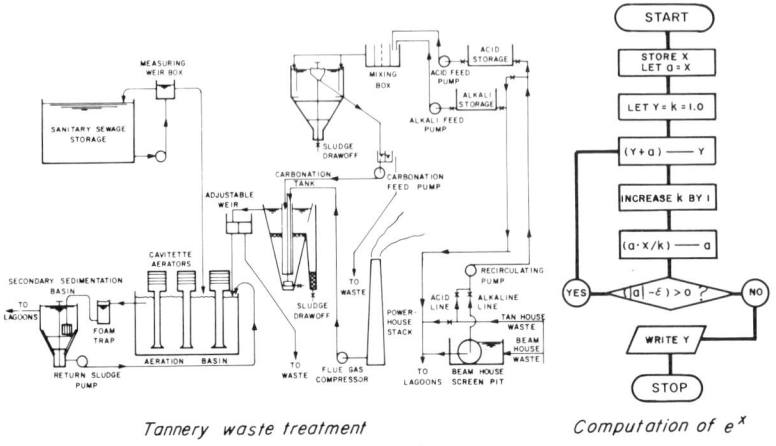

Tannery waste treatment   *Computation of e^x*

(b) Flow diagrams

Fig. 7.28 Organization and flow diagrams *(Flow diagram courtesy Camp Dresser & McKee Inc).*

ORTHOGRAPHIC PROJECTION

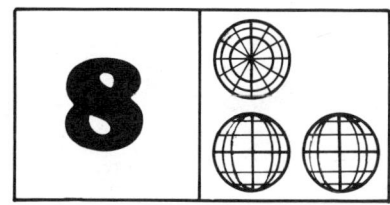

8.1 INTRODUCTION

Photographs and pictorial drawings are realistic in appearance and easy to visualize, since they show objects as viewed by the observer. Since the exact shape of individual parts of an object is needed by industry, photographs and pictorial drawings are usually not satisfactory for describing an object except for relatively simple parts. The true representation of individual parts of objects requires the use of orthographic projection to define clearly an object on a two dimensional surface... a single, flat plane. The word *orthographic* is derived from two Greek words: "orthos" meaning at right angles and "graphikos," to write or describe by drawing lines. A single orthographic view shows only two-dimensions and will not usually reveal all the necessary information about the object. For most objects a series of related orthographic views must be drawn carefully to scale along with the necessary dimensions, notes, symbols, and instructions for manufacturing. This perpendicular arrangement of views for showing three dimensions without distortion is called *multiview projection*.

In the area of cartography, the study of orthographic projection is helpful for understanding how the earth's surface may be drawn as a single view in a flat plane. A number of specialized map projection systems have been developed to meet different needs. Orthographic projection should be understood to allow individuals to work effectively with map projections or to provide the foundation for additional work.

8.2 PROJECTION THEORY

A variety of projections is used to define and represent objects. Each of these has advantages and disadvantages which should be understood in selecting the best type of projection to use for drawing a given object. All forms or projection for drawing an object have four components, Fig. 8.1: (1) the station point or observer's eye,

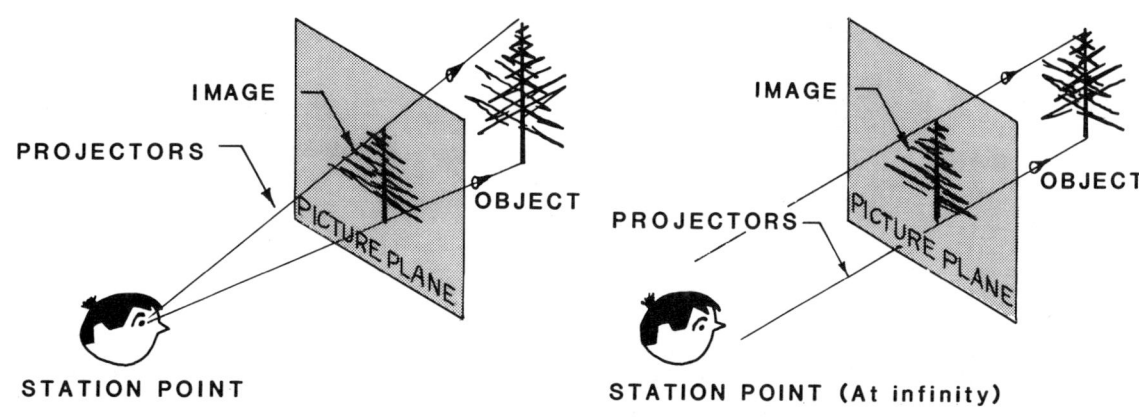

(a) Perspective *(b) Parallel projection*

Fig. 8.1 Fundamental types of projections and components.

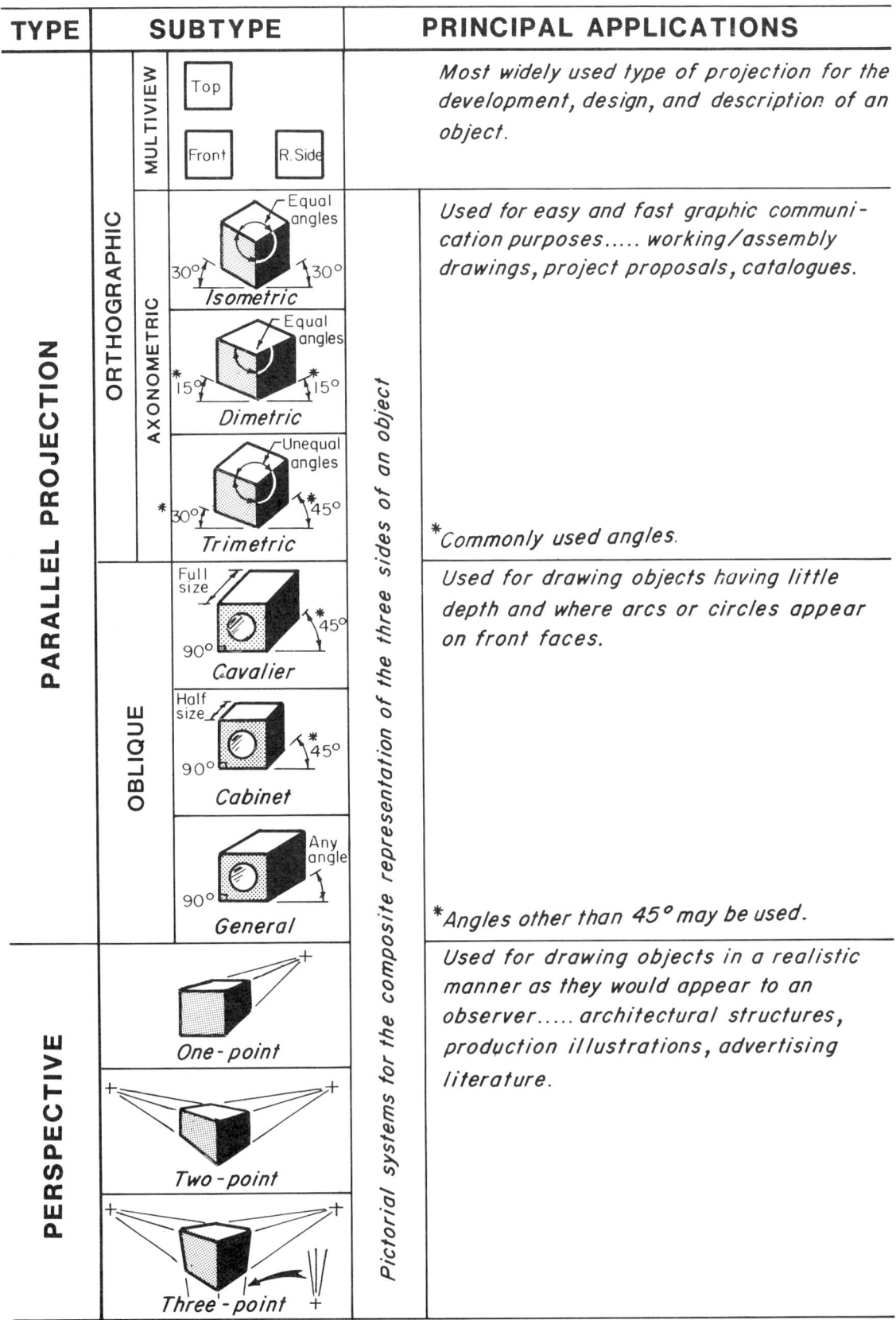

Fig. 8.2 Types of projections.

(2) the object, (3) the picture plane or plane of projection, and (4) the projectors or visual rays or lines of sight. The two fundamental types of projection are *perspective* and *parallel projection*. In perspective, the projectors converge to a single point, the station point, that is a finite distance from the object. In parallel projection the station point is considered to be at infinity with the projectors parallel to each other. Other types of projection may be developed from these two main types of projection and are discussed in subsequent chapters. It is recommended, however, that the relationship of the different types of projections to each other, Fig. 8.2, be studied before proceeding to orthographic projection.

8.3 PRINCIPLES OF MULTIVIEW ORTHOGRAPHIC PROJECTION

Multiview projection is the means for showing three basic dimensions of an object in a clear and concise manner. If the viewer is considered to be an infinite distance from the object and the projectors are parallel to each other and perpendicular to the picture plane, then this is considered *orthographic projection*. The two ways for obtaining the views are the glass box and direct method.

Glass box method. An imaginary glass box with the object in the center is commonly used to illustrate how the orthographic multiviews of an object are drawn, Fig. 8.3. The individual orthographic views are obtained by projecting perpendicular lines from points on the object toward one plane of the box. Each of the points on the object forms a point on the plane of the box. Lines connecting all these points form the principal views: top, front, right side, rear, and bottom. The top, front, and right side views are used most frequently since the remaining views duplicate information and are seldom necessary. The surfaces of the box on which the three conventional views are projected are referred to as the *horizontal, frontal*, and *profile* planes of projection. If the box is unfolded, Fig. 8.4 (a), and laid out flat, like a sheet of paper, the views will be arranged as illustrated in Fig. 8.4 (b). The top,

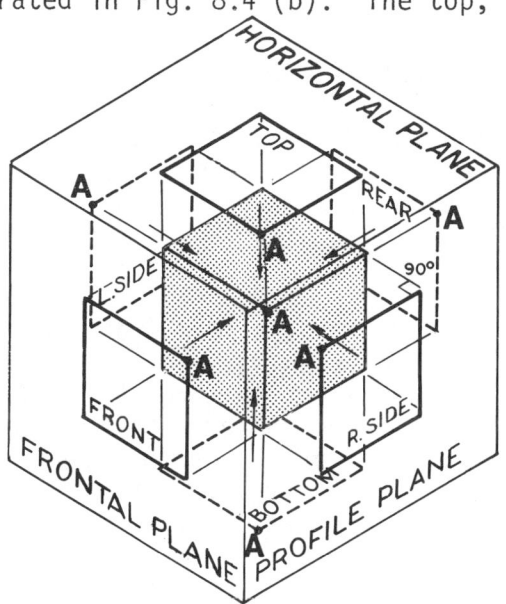

Fig. 8.3 The glass box.

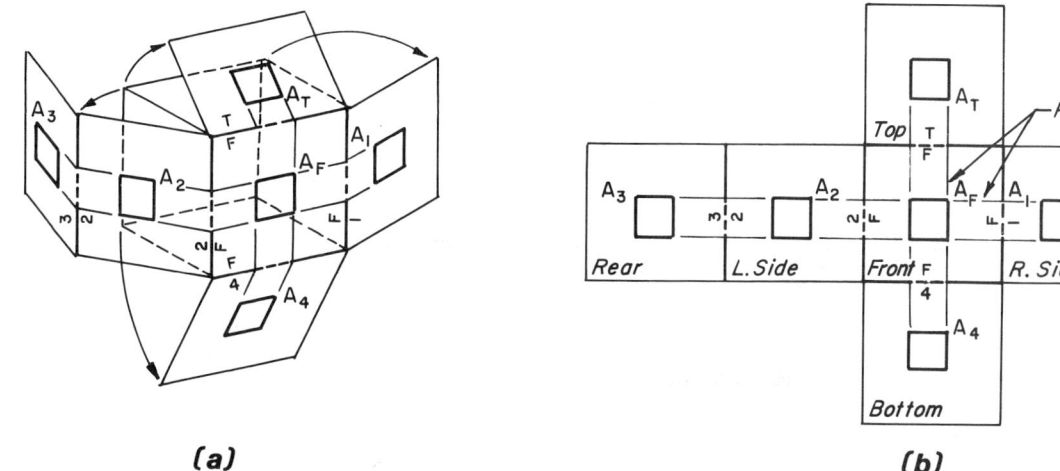

Fig. 8.4 Orthographic arrangement of views.

front, and bottom views will all line up vertically while the left side, front, right side, and rear views will line up horizontally. Occasionally the side view will be shown as lining up horizontally with the top view (see section 8.6). This arrangement of views facilitates the location of the image of points on the object. Projection lines drawn from the images of the points in any one view will project to the corresponding images of the same points in any adjacent view. The ability to locate the image of a point in several related views will now make it possible for the images of any object to be drawn.

In Fig. 8.4 the line of intersection between two image planes or the line one image plane is folded on to bring it into the plane of the other image plane is termed a *folding* or *reference line*. The folding lines may be drawn as thin lines using dashes, one long and two short, and are always perpendicular to the projection lines between adjacent views. The line marked T-F represents the edge view of the frontal plane as seen from the top and the horizontal plane as seen from the front, line F-1 represents the edge view of the profile plane as seen from the front, etc. The notation for labeling the folding lines and different views is the capital letter T for the top view, the capital letter F for the front view, with all other views numbered 1, 2, 3, etc. in the order in which they are drawn. A capital letter with a small subscript number which is the same as the view number is used to identify points in each view.

The use of folding lines is helpful in solving difficult problems since they provide a convenient place from which to make measurements. In drawing simple objects which require only two or three views, the folding lines are not generally shown.

Direct method. The direct method for drawing the views of an object requires the observer to move around the object to view the side required, Fig. 8.5. Only those points, lines, and planes that are visible from any one position are drawn. In some instances hidden features must be indicated to completely describe the object. The use of hidden lines is discussed in subsequent sections.

Third angle projection. The arrangement of views as illustrated in Fig. 8.4 is referred to as *third angle projection*. Since the frontal, horizontal, and profile planes of projection intersect at 90° with each other, four quadrants or angles may be obtained, Fig. 8.6 (a). In Fig. 8.6 (b), the object is placed in the third quadrant resulting in the arrangement of the three conventional views; top, front, and right side, as illustrated. This is the standard projection system used throughout the United States and Canada.

It is possible to place the object in any of the quadrants. In Europe, drawings are made with the object

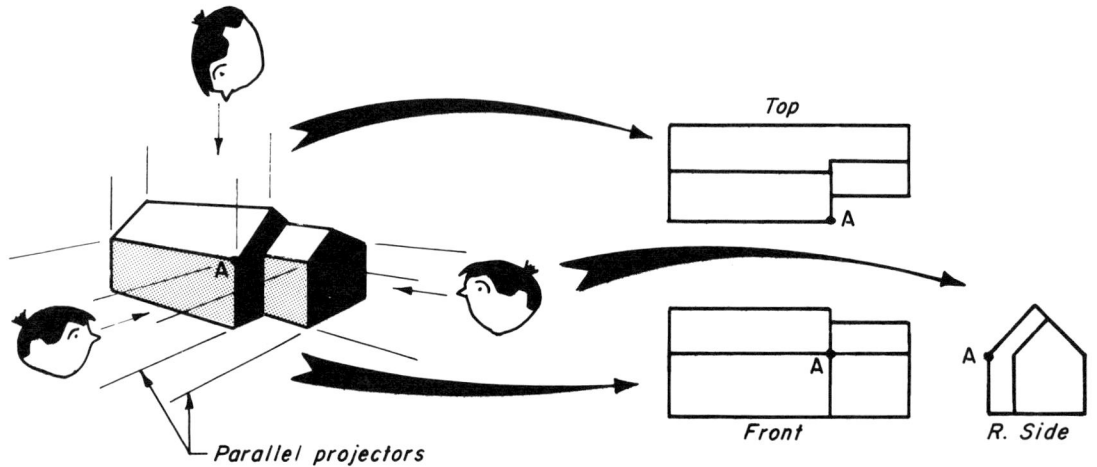

Fig. 8.5 Direct method to obtain views.

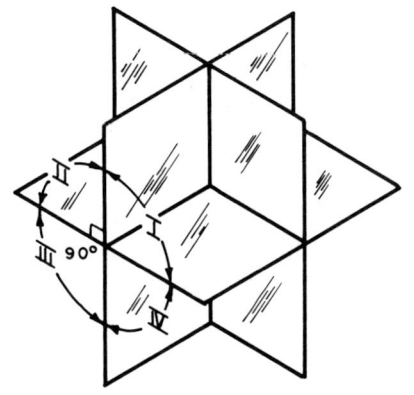

(a) Principal planes and quadrants

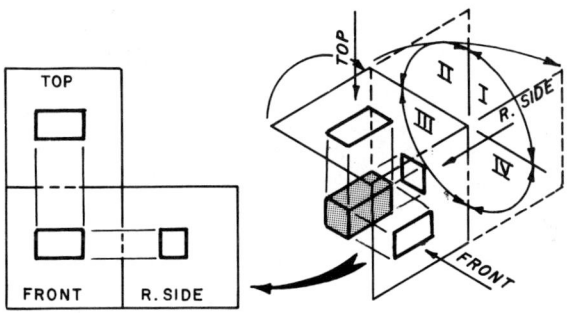

(b) Third angle projection

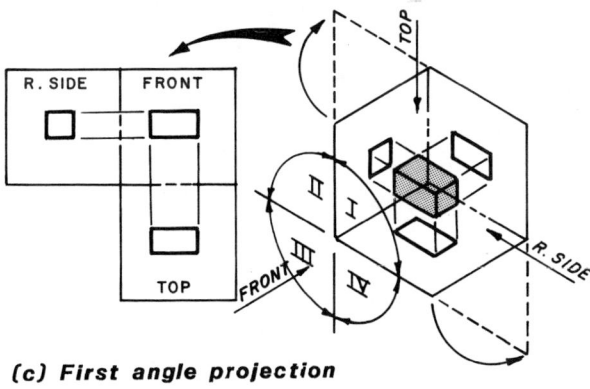

(c) First angle projection

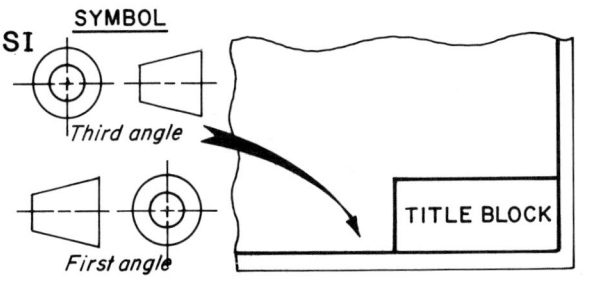

(d) ISO projection symbol

Fig. 8.6 Third and first angle projection.

located in the first quadrant. Fig. 8.6 (c) shows an object in the first quadrant together with the arrangement of the views.

The type of projection system is identified by the symbol recommended by the International Standards Organization (ISO), Fig. 8.6 (d). It is recommended that the symbol be located in the lower right-hand corner of the drawing, adjacent to the title block on all drawings made for international distribution. The large letters SI are used to indicate that the measurements are metric.

8.4 DIMENSIONS

The three principal dimensions are *height, width,* and *depth*. Since an object is composed of points, lines, and planes, the dimensions of each should be recognized, Fig. 8.7. A point has no dimensions, a line has one dimension, that is generally referred to as length, and a plane has two dimensions. In Fig. 8.7 (c) the width dimension is common to both of the planes. The remaining dimension, height or depth, is determined by the orientation of the plane. Height is used unless the plane recedes from the observer.

An orthographic view of an object shows only *two* of the three principal dimensions without distortion or foreshortening, Fig. 8.7 (d). Adjacent views have one dimension in common and the views are arranged to allow this common dimension to be projected between two adjacent views.

The principal dimensions are measured or transferred in a horizontal or vertical direction only in all three views. Errors will result if measurements are taken along inclined lines. It is emphasized that the width dimension is horizontal in the top and front views, the height dimension is vertical in the front and right side views, and the depth dimension is vertical in the top view and horizontal in the right side view.

Transferring dimensions. In drawing the views of an object, projection lines should be drawn from one view to another, and retained as light construc-

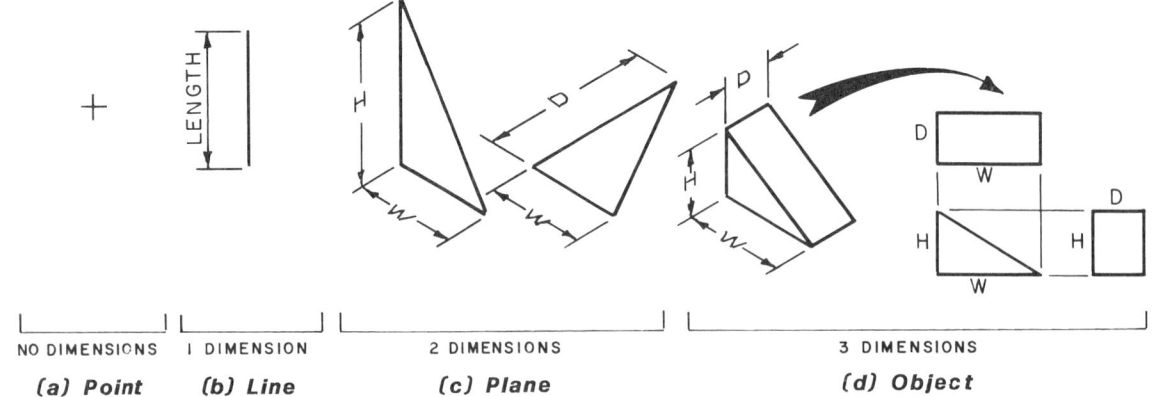

Fig. 8.7 Dimensions.

tion, to maintain the vertical and horizontal alignment of the views. The width dimension may be projected from the front to the top view, height projected from the front to the right side view, and vice versa. It is recommended that only the dividers be used to transfer the depth dimension from the top to the right side view and vice versa, Fig. 8.8. This method for transferring distances is fast, accurate, and will result in a better understanding of the dimensions and their relationship to each other. The use of a 45° mitering line should be avoided.

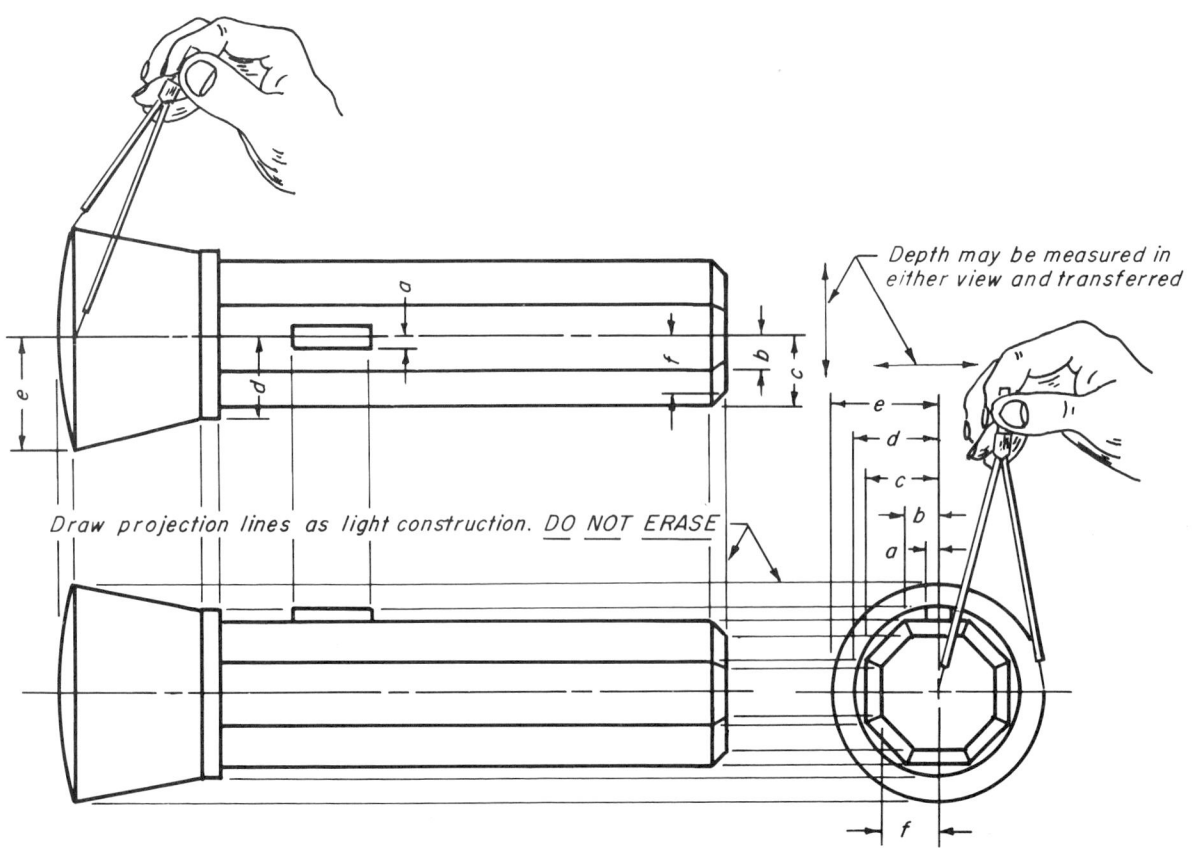

Fig. 8.8 Transferring depth dimensions using dividers.

8.5 PRIMARY CONSIDERATIONS FOR ORTHOGRAPHIC DRAWING

The following considerations should be analyzed before drawing the orthographic views of an object.

<u>Object orientation</u>. In drawing any object, it is possible to orient the object in a variety of ways for drawing the views. Several good practices for the proper orientation of an object are:

1. The object should be kept in its natural position.

2. The principal shape or the most characteristic contour of the object should appear in the front view, regardless of the natural front of the object.

3. An attempt should be made to have the longest dimension appear in the front view as the width dimension.

4. Orient the object so that a minimum of hidden lines will appear in the views.

*Front and right side views

<u>Selection of views</u>. In describing an object, only those views necessary to completely and clearly define the object should be drawn. Views that confuse the reader or repeat information should be avoided. Unnecessary views require additional work to draw and this is not an efficient use of time.

It is sometimes possible to draw simple objects, e.g., flat shapes, spherical, and cylindrical pieces, using a single view. Additional information concerning the description of the object is given in the form of notes, dimensions, or lettered symbols. In Fig. 8.9 (a) a note is used to reveal the thickness of the object. The diameter of the cylindrical piece in Fig. 8.9 (b) may be indicated by lettering the abbreviation DIA for diameter after the dimension or by using a note.

For any object where the features are not clear in one view, additional views will be required. Two views will describe objects that are symmetrical, Fig. 8.10 (a). A careful analysis of other shapes is necessary to determine if two views will clearly describe the object. The right side view in Fig. 8.10 (b) should be omitted since no additional information is given that is not clearly represented in the front and top views. The two views used to represent an object must always be related or adjacent views, e.g., front and top or front and right side. This relationship will always allow one of the three principal dimensions to be projected be-

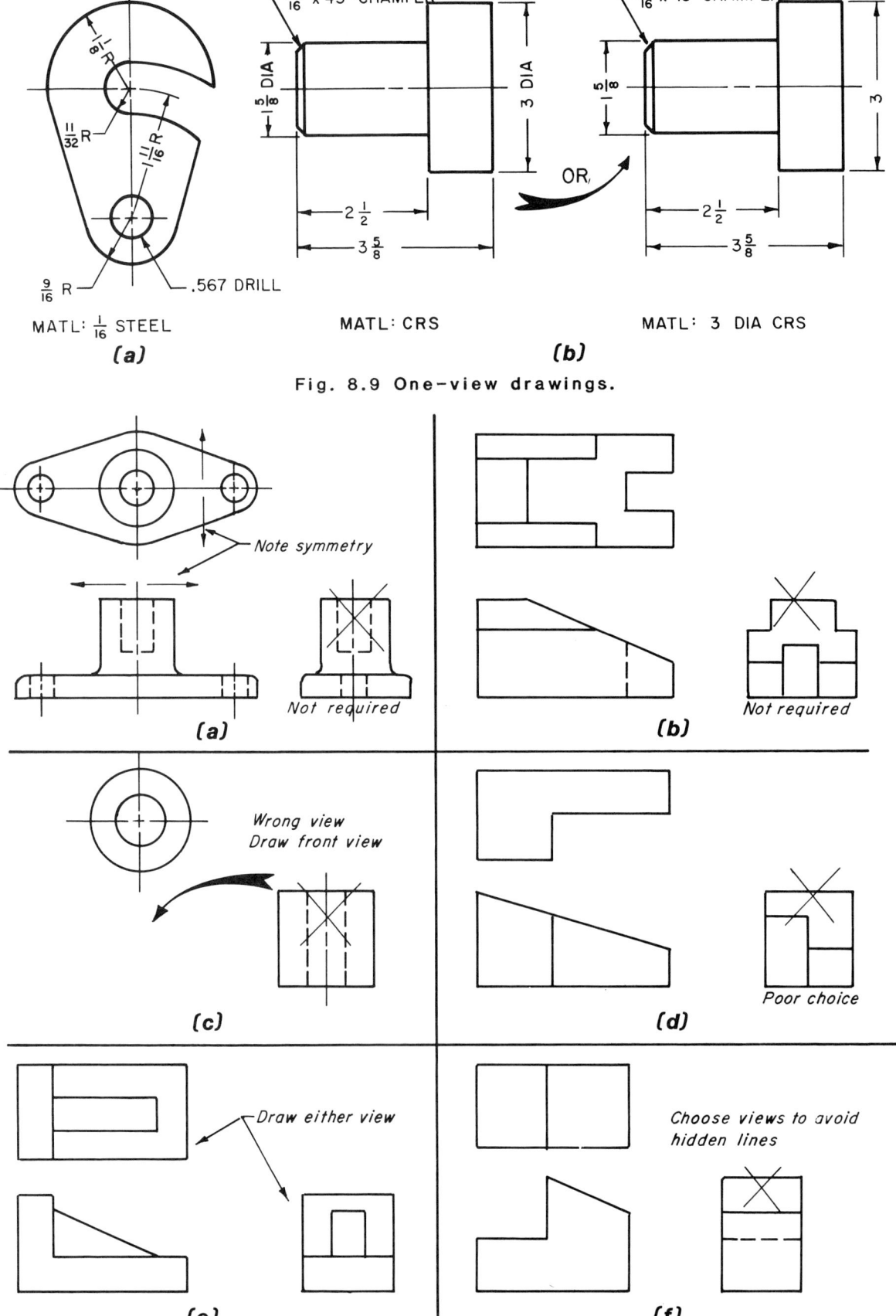

Fig. 8.9 One-view drawings.

Fig. 8.10 Two-view drawings.

tween the views. The top and right side views should never be drawn, for this reason, to describe an object, Fig. 8.10 (c). The two views that best describe the shape of the object should be drawn. In Fig. 8.10 (d) the right side view would be a poor choice to use in representing the object since the top view is more descriptive. In some instances the top and right side views will be identical or equally descriptive of the object and either view may be drawn, Fig. 8.10 (e). Other factors to be considered in selecting the best view to draw are: select the view that will have the fewest number of hidden lines, Fig. 8.10 (f), and select the view that best fits the paper to achieve a pleasing balance.

The complexity of many objects will require that three views be used for complete shape description. Irregular objects, Fig. 8.11, frequently fall in this category. The hole, the curved outline, the inclined surfaces, and the U-shaped appearance in the right side view would not be clear if any one of these views was omitted. At times an object may also require more than the three standard views to show the shape and features clearly.

<u>Spacing the views.</u> A finished drawing should be reasonably positioned on the drawing sheet with sufficient space for notes and title information.

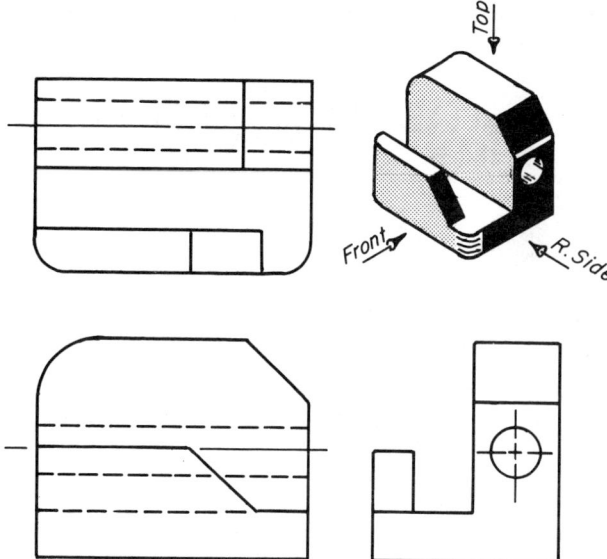

Fig. 8.11 Three-view drawing.

The procedure outlined in Fig. 8.12 will allow the views to be balanced to obtain a pleasing over-all effect. Sufficient space between the views should be provided without creating a crowded appearance and yet far enough apart for the views to be distinguishable and easily read. As a rule the distance between views should allow for easy placement of dimensions between the views. Where possible, it is suggested that approximately 35 mm (1½") be left between the views. The space between the front and top and the front and right side views does not have to be the same.

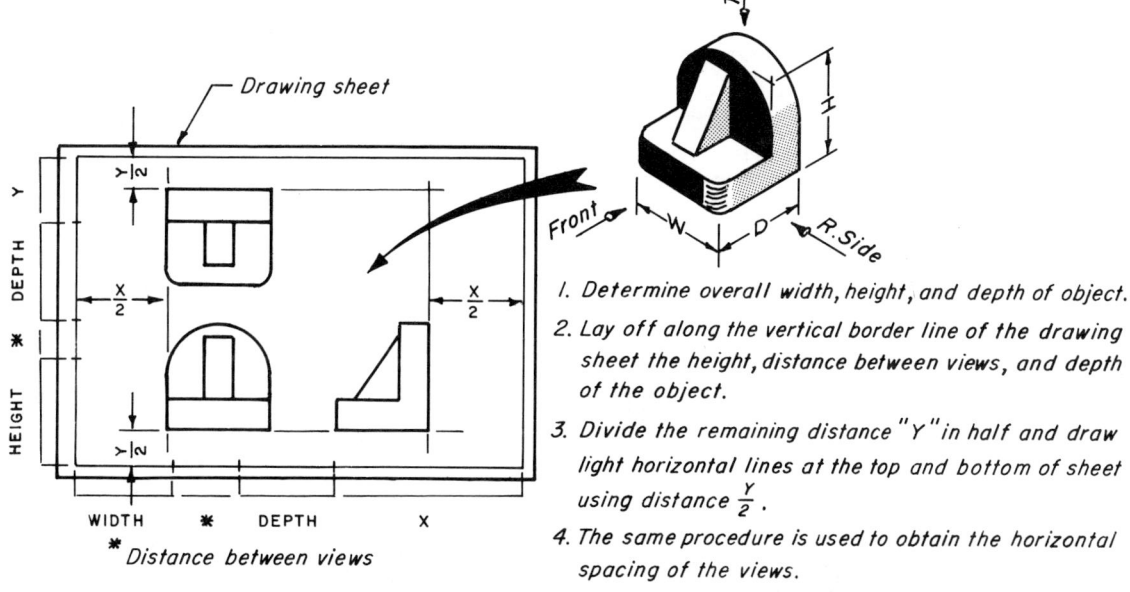

1. Determine overall width, height, and depth of object.
2. Lay off along the vertical border line of the drawing sheet the height, distance between views, and depth of the object.
3. Divide the remaining distance "Y" in half and draw light horizontal lines at the top and bottom of sheet using distance $\frac{Y}{2}$.
4. The same procedure is used to obtain the horizontal spacing of the views.
5. Draw views within the rectangular area.

Fig 8.12 Spacing the views on paper.

8.6 ALTERNATE POSITION-RIGHT SIDE VIEW

An alternate position for drawing the right side view is obtained when the profile plane is hinged to the horizontal plane rather than to the frontal plane of projection, Fig. 8.13. This arrangement is sometimes convenient to save space or to fit the paper better. The right side view in this position is turned on its side and the same basic priniciples of projection presented still apply.

8.7 CENTER LINES

Center lines are thin lines with alternate long and short dashes as shown in chapter 2, Fig. 2.16. These lines are useful since they are frequently the first lines to be constructed when laying out a drawing involving cylindrical forms and circles, Fig. 8.14 (a). The center lines establish the common axes of symmetry in the views and are of great assistance in making measurements. Center lines are also used for arcs, except for small arcs, and may be extended for dimensioning circular features such as holes, Fig. 8.14 (b).

The intersection of two mutually perpendicular center lines is used to

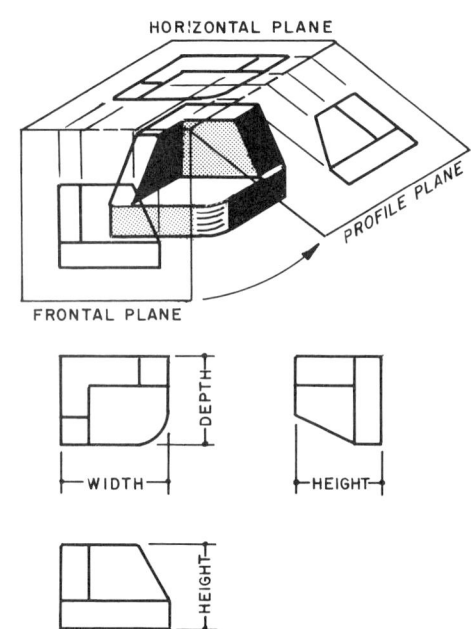

Fig. 8.13 Alternate position-right side view.

locate the center of circles and arcs. The 3 mm (1/8") short dashes should cross at the intersection of center lines and should not be used to end a center line. In drawing the long dashes, 19 mm (3/4") to 38 mm (1½"), it is recommended that they be extended about 6 mm (¼") beyond the outline of the feature to which they apply. The center lines should not be extended to connect adjacent views. Since center lines for small holes are difficult to draw, they

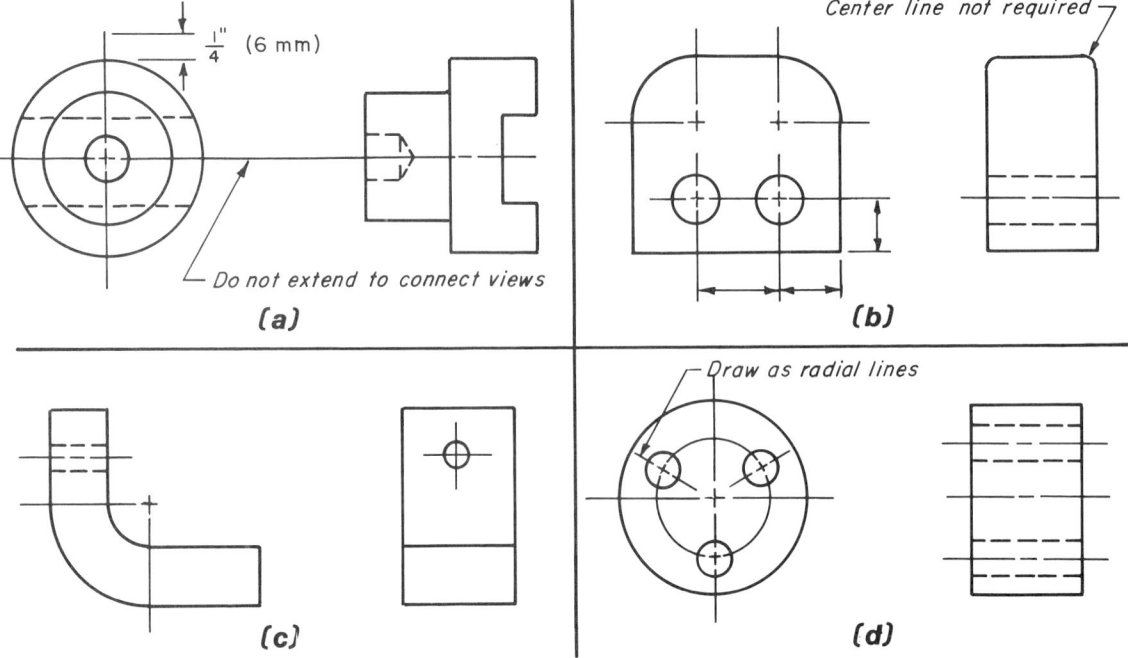

Fig. 8.14 Use of center lines.

may be drawn as solid lines, Fig. 8.14 (c). In some instances center lines are circular, instead of straight, as in the case of the center line used to locate the holes shown in, Fig. 8.14 (d). A circular center line drawn in this manner is frequently referred to as a *bolt circle*. The straight center lines for the small holes should be drawn as radial lines.

8.8 HIDDEN LINES

Hidden lines (see chapter 2, Fig. 2.16) are drawn with short 3 mm (1/8") dashes and are used to show features of an object that are not visible in the direction from which a view is taken. The individual dashes should be separated by a space of about 0.8 mm (1/32") to maintain line continuity and to make the drawing easier to read. A drawing that has many hidden lines may confuse the reader rather than clarify the representation. In some instances it may be desirable to omit hidden lines if they are not needed to completely describe the object.

The illustration in Fig. 8.15 shows nine examples of incorrect practice in the drawing of hidden line intersections. As a general rule, the hidden lines should not cross other kinds of lines. Study the drawing carefully to become familiar with the correct technique required for drawing hidden lines.

8.9 PRECEDENCE OF LINES

Several different kinds of lines may coincide on a drawing. Since it is impossible to show more than one kind of line, a decision must be made on which line to draw. An order of precedence has been established for this purpose and is illustrated in Fig. 8.16.

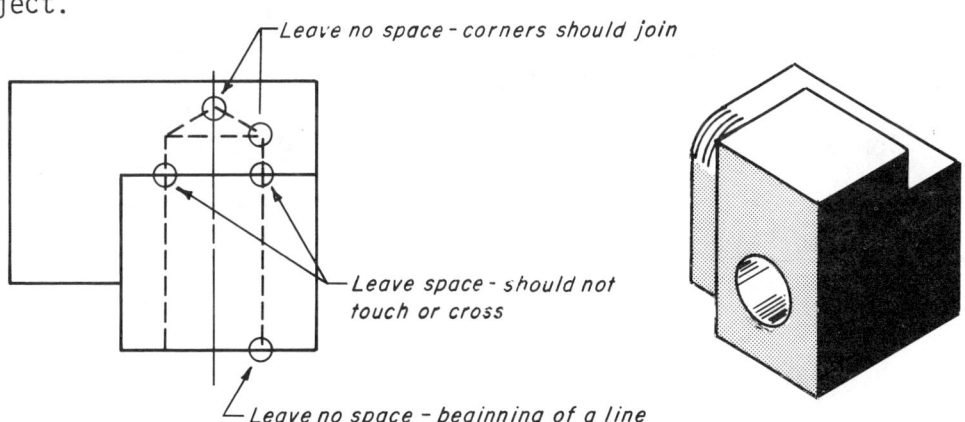

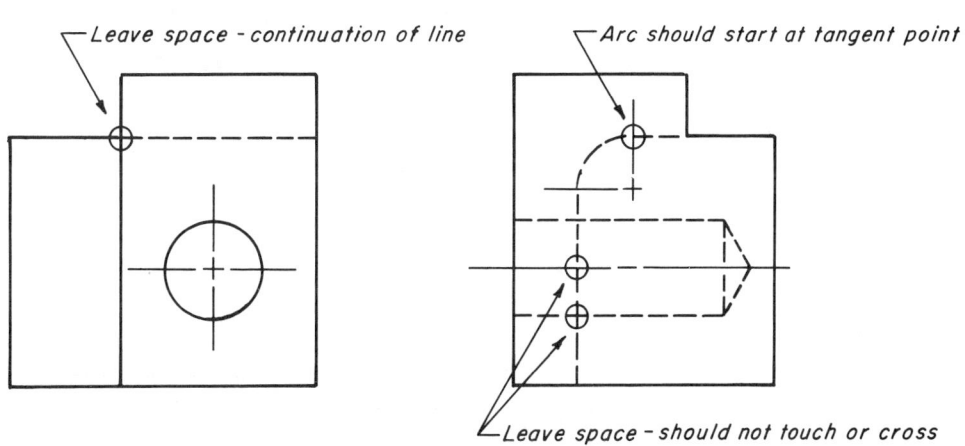

Fig. 8.15 Use of hidden lines.

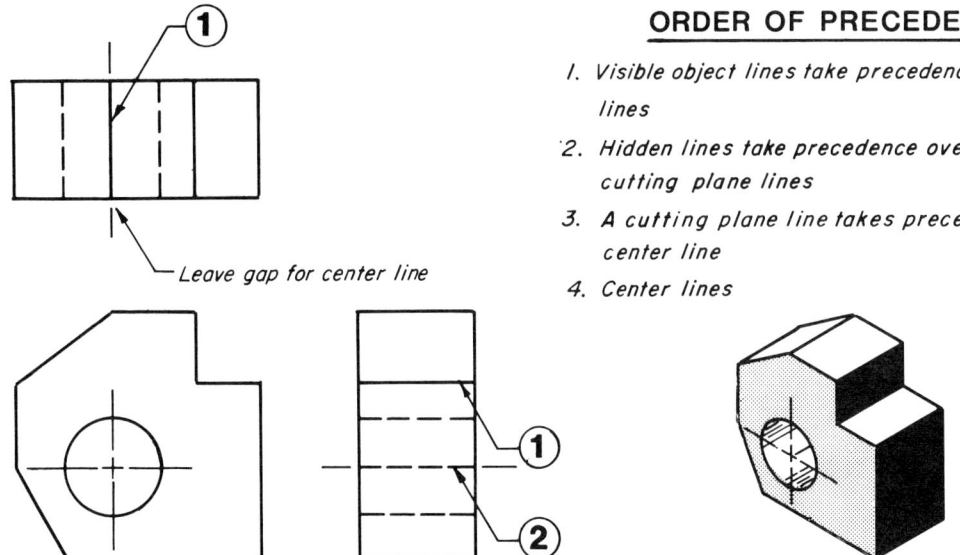

ORDER OF PRECEDENCE

1. Visible object lines take precedence over all other lines
2. Hidden lines take precedence over center and cutting plane lines
3. A cutting plane line takes precedence over a center line
4. Center lines

Fig. 8.16 Precedence of lines.

8.10 HALF VIEWS AND PARTIAL VIEWS

The available space for drawing is sometimes limited and it may be easier to draw the object using a *half view* or *partial view*. If the object is symmetrical, one half of the object may be omitted, Fig. 8.17. Since the front view shows the most characteristic shape of the object, it is not a good idea to draw this view as a half view. The half drawn should always be the front half of the top or right side view if the front view is an exterior view.

The use of a partial view can save time since the entire drawing of complicated or irregular pieces is frequently unnecessary. In Fig. 8.18 a number of hidden lines in the front view would be required to identify the object if only the front and top views were drawn. The object is easier to visualize by drawing a partial right side view. A break line is used to indicate that the view is incomplete. It is also permissible to draw the partial view at a larger scale for clarification or dimensioning provided a note is used to indicate this change.

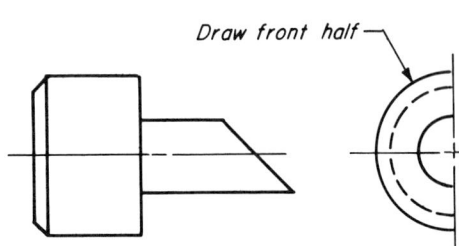

Fig. 8.17 Half views.

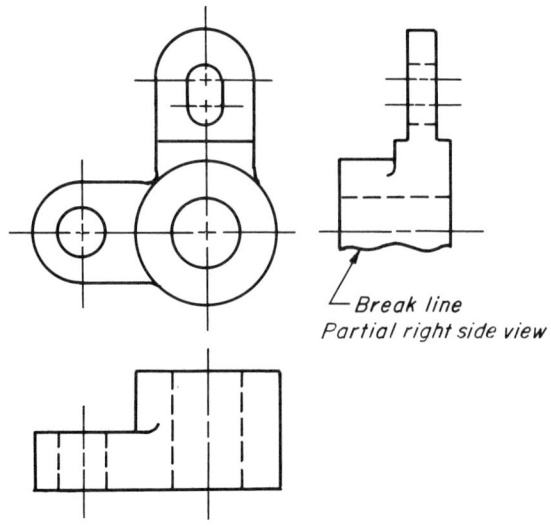

Fig. 8.18 Partial views.

8.11 MULTIVIEW PROJECTION

Multiview projection requires the ability to work with points, lines, and planes. The views of any object may be drawn or the graphic solution of many problems can be solved logically by working with these basic elements.

Points. The location of a point is measured in a perpendicular direction from the three planes of projection of the glass box, Fig. 8.19 (a). In Fig. 8.19 (b) the planes of projection of the glass box are unfolded revealing clearly how the distances x (width), y (height), and z (depth) are measured from the respective planes of the glass box to locate the images of point A. In order to become familiar with the principles of orthographic projection, the multiview drawing of point A, Fig. 5.19 (c), should be studied carefully. In the front view, distances are measured right or left and up or down. In the top view, distances right or left and front or back are measured. The right side view will show distances up or down and front or back. Folding lines may be located wherever convenient as long as the z (depth) measurement is the same distance from the folding line in the top and right side views.

Lines. A *line* may be defined as a path or locus of a moving point. If the line is straight, its location may be determined by two points. Additional points will be necessary to establish the location of curved lines. Fig. 8.20 illustrates several types of lines and their meaning. For the line types, Fig. 8.20 (a) through (e), it is suggested that a pencil be used to represent the line as an aid in understanding or visualizing its position. The top view will show which end of the line is in front while the front view will show which end of the line is higher. A line may also appear true length, foreshortened or as a point in adjacent views. It is very important to understand that the viewer must look at a line perpendicularly to see its true length. When the line is parallel to a plane of projection, it will appear true length in the view projected onto the plane of projection to which it is parallel.

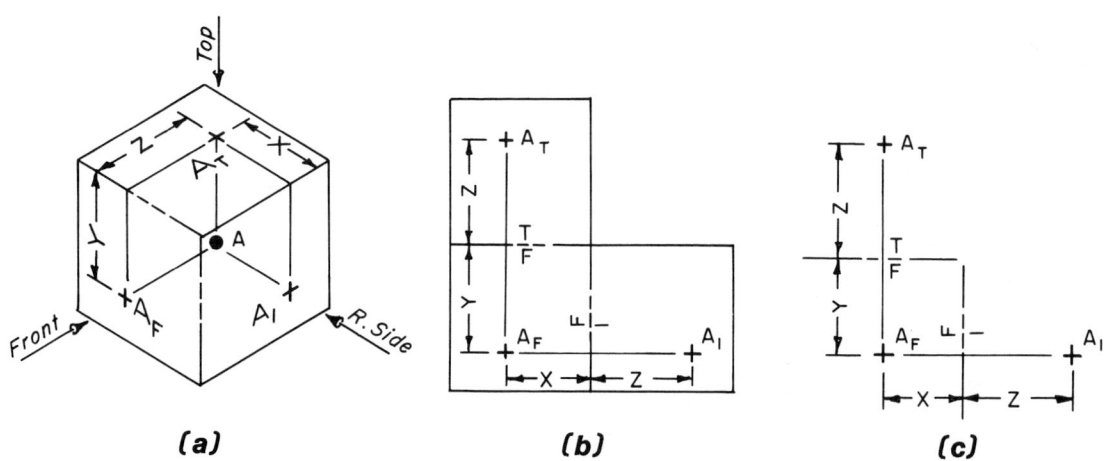

Fig. 8.19 Multiview projection of a point.

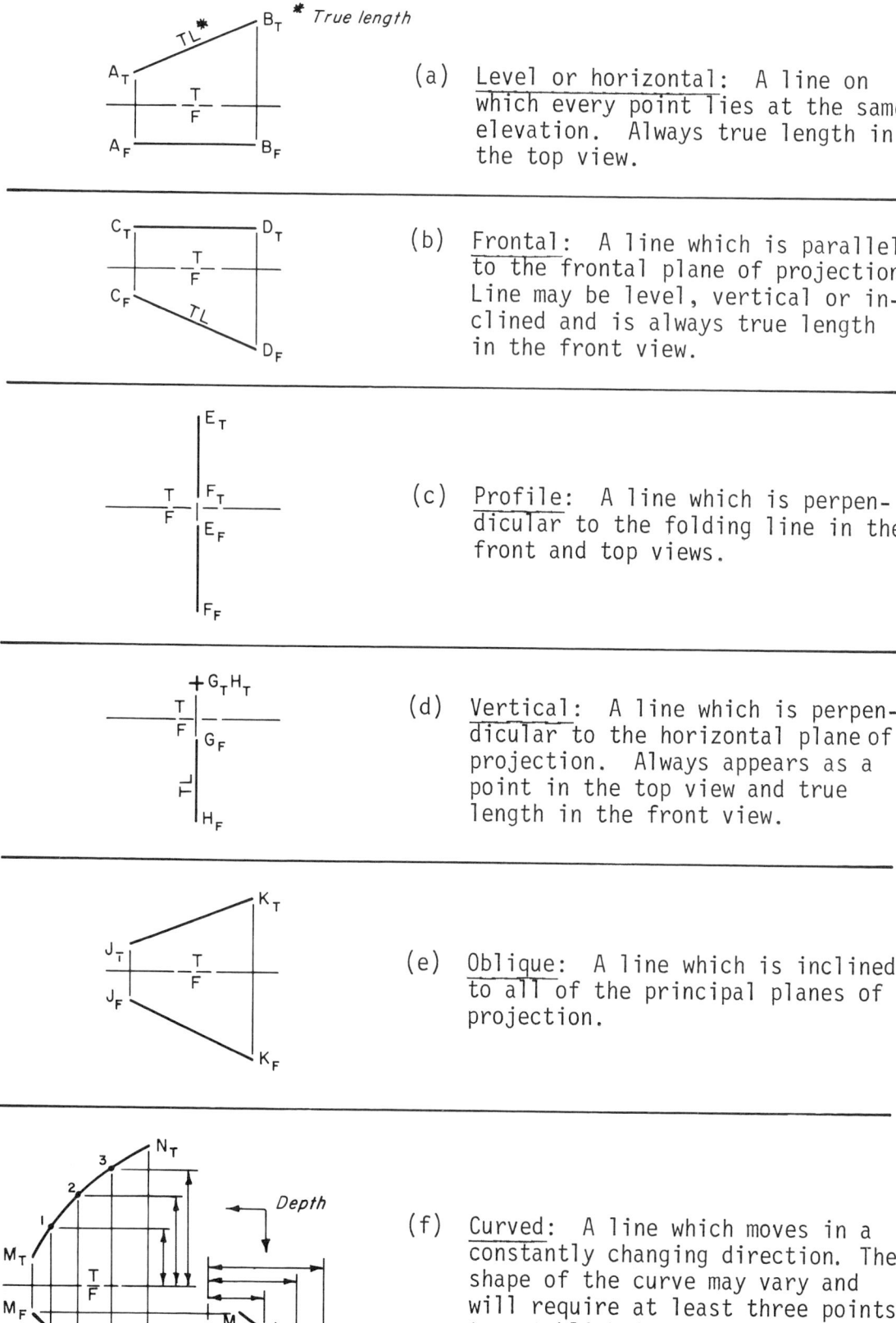

(a) <u>Level or horizontal</u>: A line on which every point lies at the same elevation. Always true length in the top view.

(b) <u>Frontal</u>: A line which is parallel to the frontal plane of projection. Line may be level, vertical or inclined and is always true length in the front view.

(c) <u>Profile</u>: A line which is perpendicular to the folding line in the front and top views.

(d) <u>Vertical</u>: A line which is perpendicular to the horizontal plane of projection. Always appears as a point in the top view and true length in the front view.

(e) <u>Oblique</u>: A line which is inclined to all of the principal planes of projection.

(f) <u>Curved</u>: A line which moves in a constantly changing direction. The shape of the curve may vary and will require at least three points to establish its contour.

Fig. 8.20 Types of lines.

Planes. A *plane* is a flat surface on which any two of its points may be connected by a straight line lying wholly on the surface; or every point in that line is on the surface. Four different ways to represent a plane are illustrated in Fig. 8.21 and they may be classified into several types similar to lines, Fig. 8.22. It is essential to be able to recognize and visualize planes in various positions in multiview projection. A plane will retain the same general shape in every view unless it projects as an edge or line. The plane may also appear true size or foreshortened. If the plane is parallel to a plane of projection, it will appear true size in the view projected onto the plane of projection to which it is parallel. An edge view will occur when the plane is perpendicular to the plane of projection and foreshortened when it is inclined to the plane of projection.

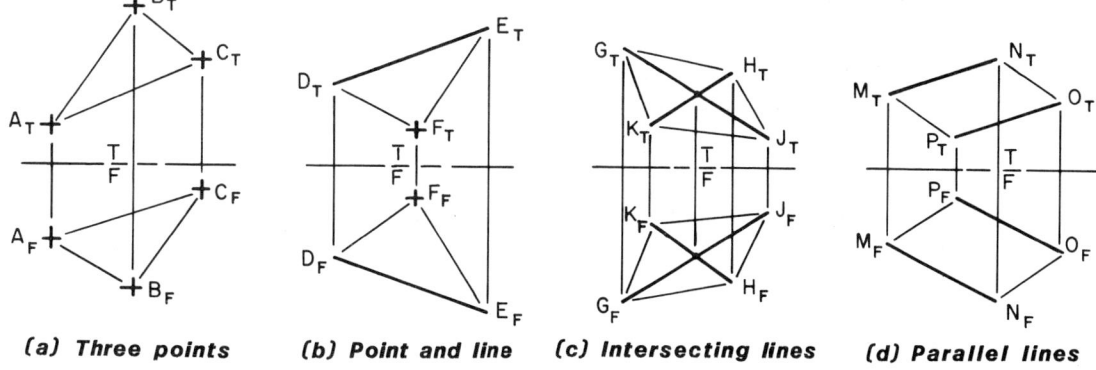

Fig. 8.21 Methods for representing planes.

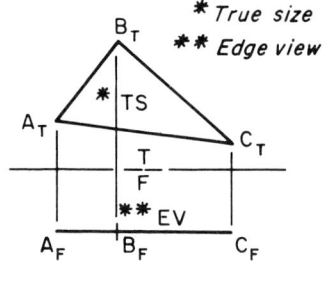

(a) Level of horizontal: A plane in which any of its points lie at the same elevation. Always appears as an edge view in the front view and true size in the top view.

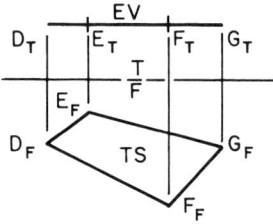

(b) Frontal: A plane which is parallel to the frontal plane of projection. Always appears as an edge view in the top view and true size in the front view.

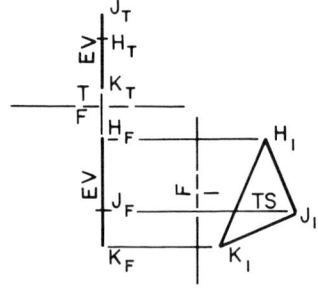

(c) Profile: A plane which is perpendicular to the folding line in the front and top views.

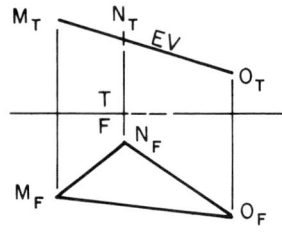

(d) Vertical: A plane which is perpendicular to the horizontal plane of projection.

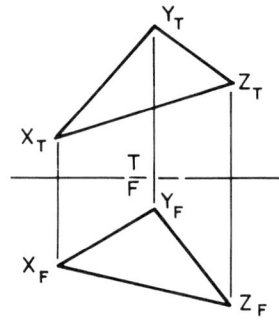

(e) Oblique: A plane which is inclined to all of the principal planes of projection.

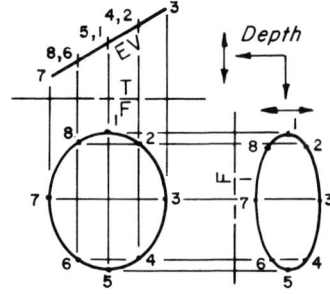

(f) Single-curved surface. A plane generated by a straight line moving so that it is always in contact with some curved line.

Fig. 8.22 Types of planes.

143

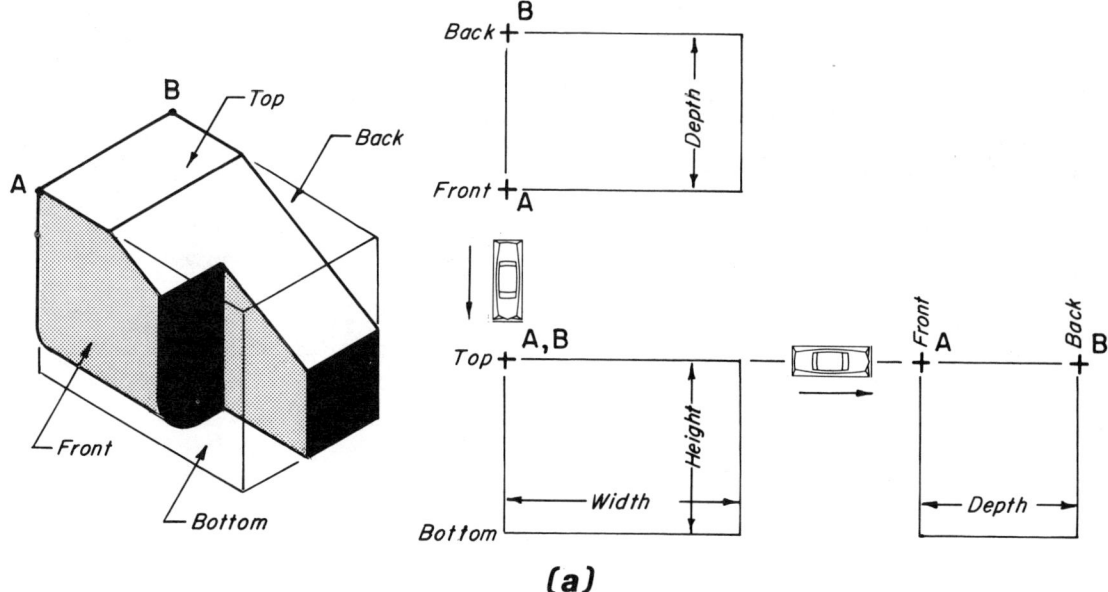

Solid objects. Three dimensional objects are made up of points, lines, and planes. The ability to work with these elements will allow any object to be drawn. Folding lines may be used for drawing the views but are frequently omitted. The use of folding lines and the notational system, discussed previously, is often necessary for the drawing of complicated problems.

The views of the object in Fig. 8.23 may be drawn if the principles of orthographic projection are understood. The drawing is started by first blocking in the height, width, and depth in the three views, Fig. 8.23 (a). It is reiterated that several adjacent views are necessary to describe most three dimensional objects. The front view will be required to define and show what features of the object are at the bottom and top. The top and right side views are needed to give descriptive information concerning features that are at the front and back of the object. Point A on the pictorial is located at the top front left corner of the object. The top view shows point A located at the left corner and the front view shows point A located at the left top corner. These two views now fix the location of point A. Before progressing further, it may be helpful to point out that multiview projection is frequently and unnecessarily made difficult by many individuals. It can be compared to driving an automoile. Since

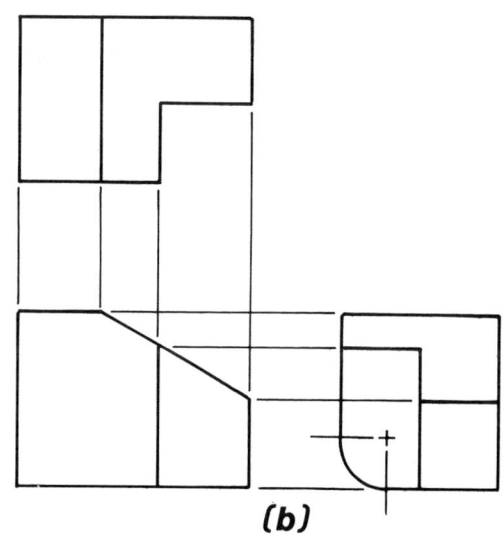

Fig. 8.23 Multiview drawing.

all points must line up between related views, the projection lines may be considered the roadway. To locate point A in the right side view the automobile starts at point A on FRONT STREET in the top view and moves vertically to point A in the front view, turning the corner and traveling horizontally until it reaches FRONT STREET in the right side view. Point B is on BACK STREET in the top view, and is located by traveling vertically to point B in the front view, turning the corner and traveling horizontally to BACK STREET in the right side view. The views of the object may now be completed by locating the remaining points delineating its shape, Fig. 8.23 (b).

In drawing an object all required views should be drawn simultaneously. Features should be drawn first in the view that is easiest to work with and then constructed in the remaining views by projecting back and forth between the views.

An object having curved features must have a sufficient number of points to ensure a smooth curve, Fig. 8.24. The top view is drawn by locating a series of points 1,2,3......11 on the curved surface in the right side view. The points are projected first to the inclined line representing the surface as an edge in the front view and then to the top view. The distance (depth) from the center line to any point on the curved surface must be the same in both the right side and top views.

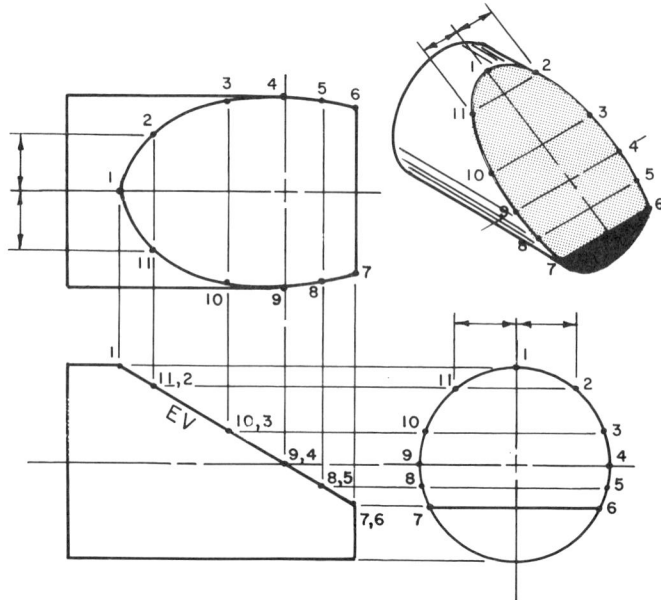

Fig. 8.24 Plotting curved features.

8.12 READING THE VIEWS

Reading the views is essentially the reverse procedure used for constructing a multiview drawing given the object. No single view will provide all of the information necessary to visualize the entire shape. The individual details must be sorted out in each view and then combined to form a composite picture of its form. It is like piecing together the individual parts of a puzzle. After a number of pieces in the puzzle have been fitted together, the image of the picture begins to take

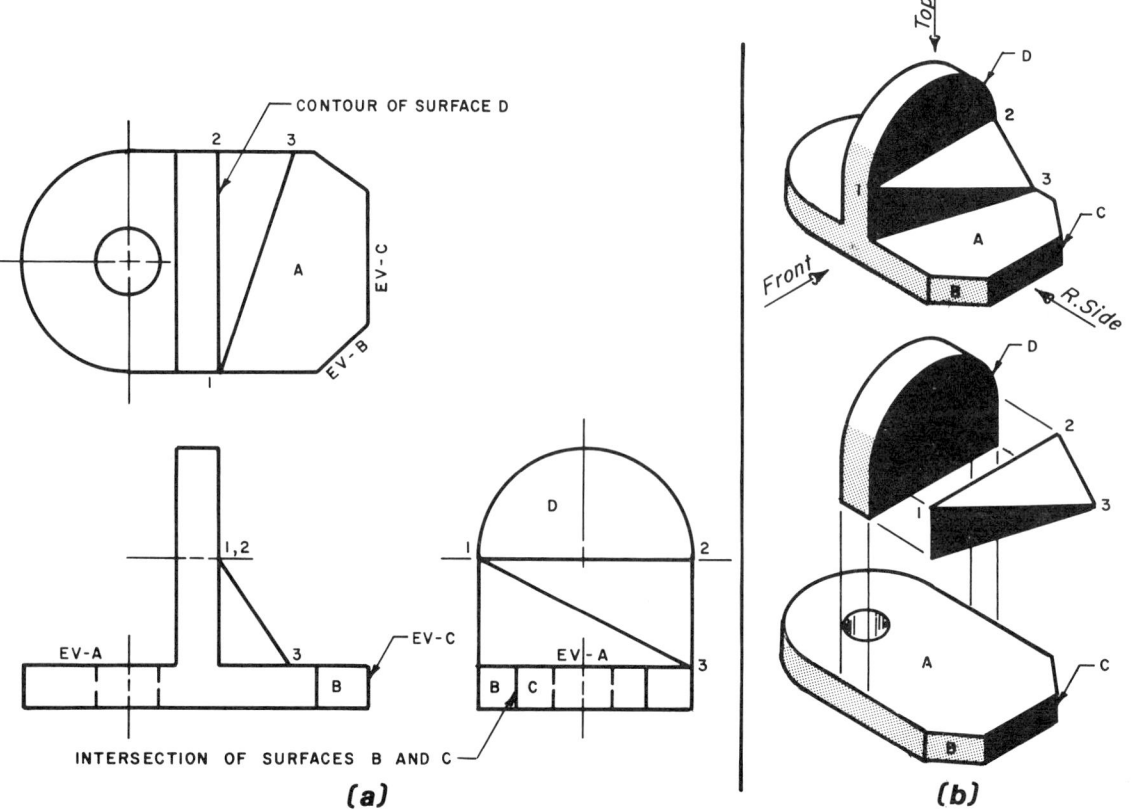

Fig. 8.25 Reading a multiview drawing.

shape and the remaining parts fall easily into place. The points, lines, and planes may be labeled, if necessary, to help identify the component parts in reading the views, Fig. 8.25 (a). A pictorial sketch of these component parts will also be helpful in developing the final picture, Fig. 8.25 (b).

A line on any one view of a drawing may represent (1) the edgeview of a plane surface, (2) the intersection of two plane surfaces, or (3) the contour view of a curved surface, Fig. 8.25 (a). All the views must be studied carefully to identify the features completely. Surface A appears as an edge in the front and right side views. No two adjacent areas can lie in the same plane and the short vertical line in the right side view represents a change in planes between surfaces B, and C. In the top view a line represents the contour view of the semi-circular surface as seen in the right side view.

A plane must always retain the same shape in all views except when it occurs as an edge. Surface C appears as a true size rectangular plane in the right side view. Both the top and front views show the surface as an edge in a vertical plane. The plane labeled 1, 2, 3 is an inclined surface occurring as an edge in the front view and is a foreshortened triangular plane in both the top and right side views.

8.13 CONVENTIONAL REPRESENTATION

A number of conventional practices have been developed for representing information on drawings. It is sometimes easier to convey information using symbols or to violate a principle of projection to save time and to simplify a drawing. The use of established conventional procedures should be understood if they are to be used to clarify certain presentations.

Representation of holes. Machined holes of different kinds are one of the most common features occurring on objects, Fig. 8.26. The hole size is always specified by the *diameter* and never the radius. The description of holes using notes is described in chapter 9. A hole that goes through a piece is a through

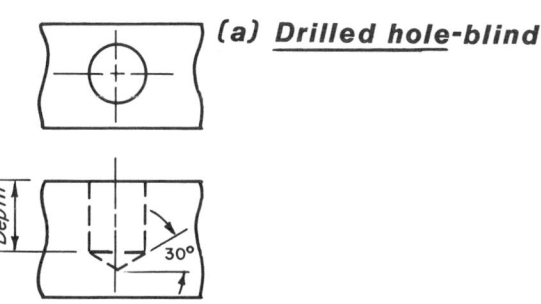
(a) Drilled hole-blind

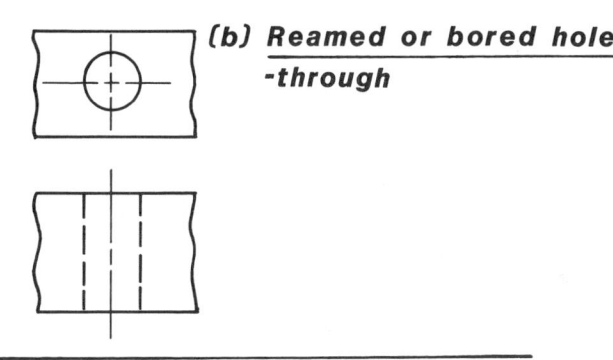
(b) Reamed or bored hole -through

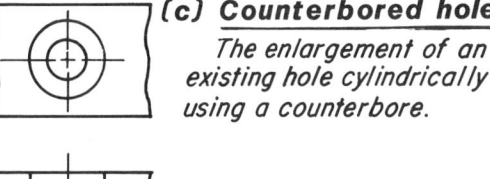
(c) Counterbored hole
The enlargement of an existing hole cylindrically using a counterbore.

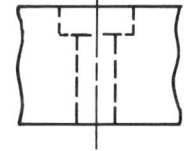

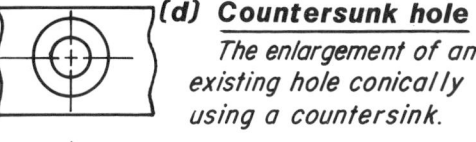
(d) Countersunk hole
The enlargement of an existing hole conically using a countersink.

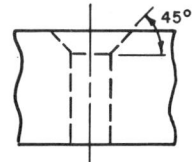

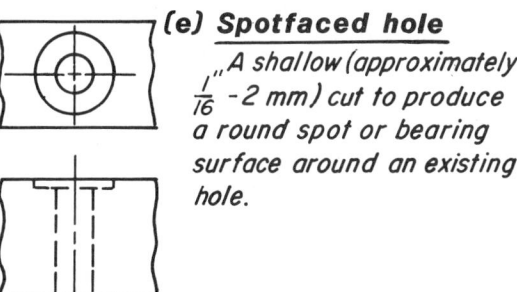
(e) Spotfaced hole
A shallow (approximately $\frac{1}{16}''$ - 2 mm) cut to produce a round spot or bearing surface around an existing hole.

Fig. 8.26 Reprsentation of holes.

hole and a hole that does not go through a piece is termed a blind hole. By convention the drill point Fig. 8.26 (a), is drawn using an angle of 30°. The sides of the countersunk hole, Fig. 8.26 (d), are drawn at an angle of 45°.

Fillets, rounds, and runouts. The design or methods used to produce many objects requires the use of small radii at the intersections formed by surfaces of an object. A rounded interior corner is termed a *fillet* and a rounded exterior corner is called a *round*, Fig. 8.27. Fillets and rounds are used to improve the appearance, strength, and comfort handling characteristics of a part. If the object is formed by casting, fillets and rounds are usually formed intentionally by the pattern maker since sharp corners are difficult to obtain on the final cast part. On a drawing of a casting, fillets and rounds are shown except where a surface is to be machined. A corner formed by two intersecting surfaces, one or both of which are to be machined, should be drawn sharp. In representing the views of an object, projection lines should be drawn from the intersections of surfaces, as if the fillets and rounds were non-existent.

A *runout* is used to represent rounds and fillets where a plane and curved surface intersect. In Fig. 8.28 (a) the arc of the runout ends at the point of tangency between the surfaces. The cross section of ribs, webs, etc. will determine the direction in which the arcs of the runout turn. If the cross section is square or rectangular the arcs turn outward, Fig. 8.28 (b).

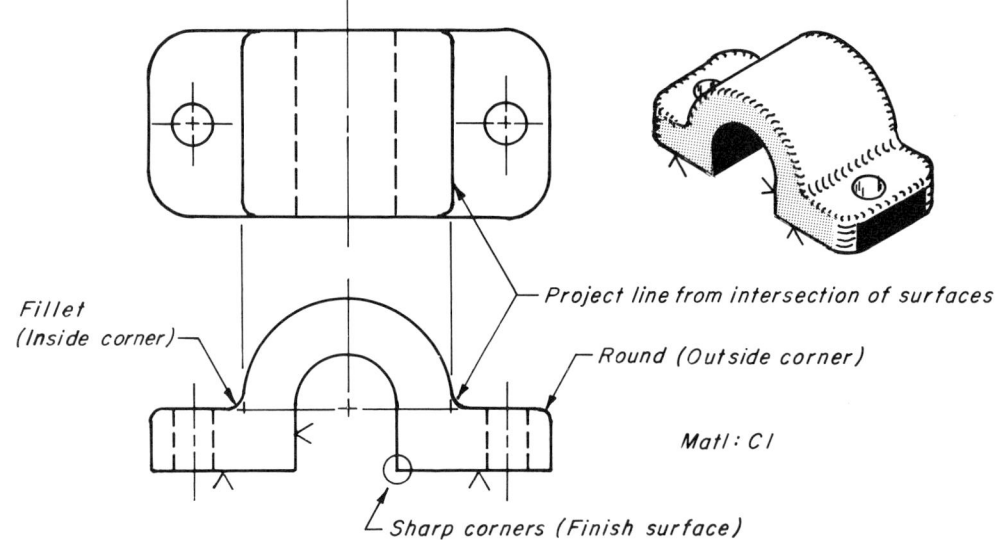

Fig. 8.27 Fillets and rounds.

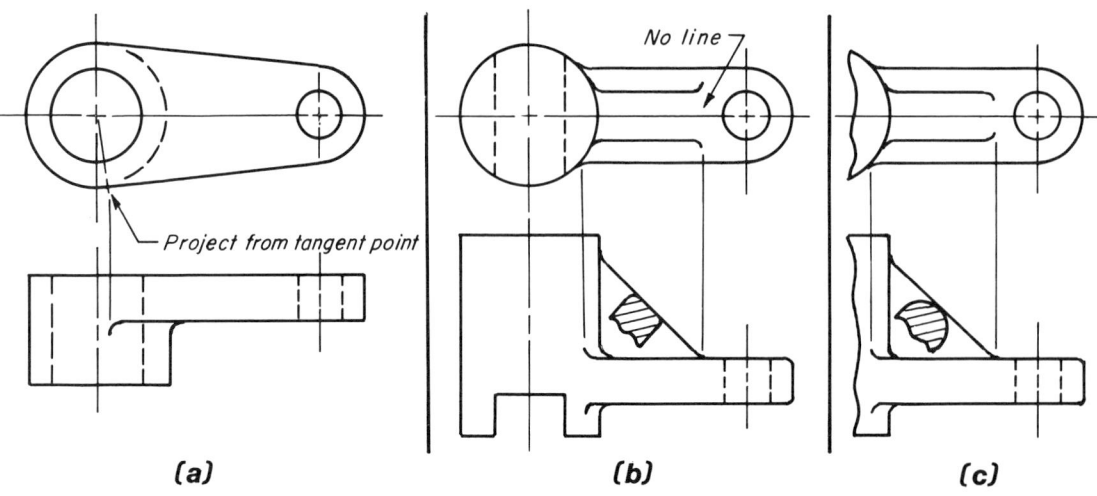

Fig 8.28 Runouts.

The arcs turn inward, Fig. 8.28 (c), where the cross section is oval or round. The arcs for fillets, rounds, and runouts are usually drawn freehand.

Tangent surfaces. Lines are usually omitted when drawing the views of an object where a curved surface is tangent to a plane surface. Where surfaces blend together smoothly, the addition of lines will generally not make the drawing any easier to read. Some discretion will have to be used, however, in deciding whether to show the tangencies. The tangency in the top view, Fig. 8.29 (a) should be shown because the curved surface is tangent to a vertical plane. No lines should be drawn to show the tangencies in Fig. 8.29 (b) and (c). In the front view, Fig. 8.29 (c), a small round is located at the point where the inclined and horizontal surfaces intersect. Lines are drawn to show the point of intersection where small arcs such as fillets and rounds occur.

Breaks. A *break line* may be used to save space for elongated objects of uniform cross section, Fig. 8.30 (a). The actual length of the object is always dimensioned. An end view may be drawn to reveal the shape of the cross section or the cross section may be shown at the point of the break. By breaking out a piece, the object itself may be drawn to a larger scale. Break lines are frequently drawn freehand; however, it is better to draw large round sections with a compass or template, Fig. 8.30 (b).

In some instances long parts having a series of identical features may be shortened and represented using ditto lines, Fig. 8.30 (c). The dashed lines indicate that the feature repeats itself.

8.14 SECTIONAL VIEWS

The interior of many objects is complex and the use of hidden lines to describ these features makes it difficult to readily comprehend the drawing. In situations of this kind, one or more sectional views may be drawn to clarify the interior parts of an object. A *section* is defined as an imaginary cut made through the object with all of the material removed on one side of the cut so as to reveal the interior, Fig. 8.31. A view may now be drawn using visible lines to clearly delineate the interior. Hidden lines are not generally shown in sectional views unless required for clarity.

Cutting plane line. A *cutting plane line* is used on the drawing to indicate the edge view of the imaginary cutting plane which passes through the object. The line is drawn as thick as an object line and may be drawn with alternate long and short dashes or a series of dashed lines, Fig. 8.32 (a). Arrow heads are placed at the ends of the cutting plane line if required to show the direction which the section is viewed. A reference letter is usually placed near each arrowhead to identify the sec-

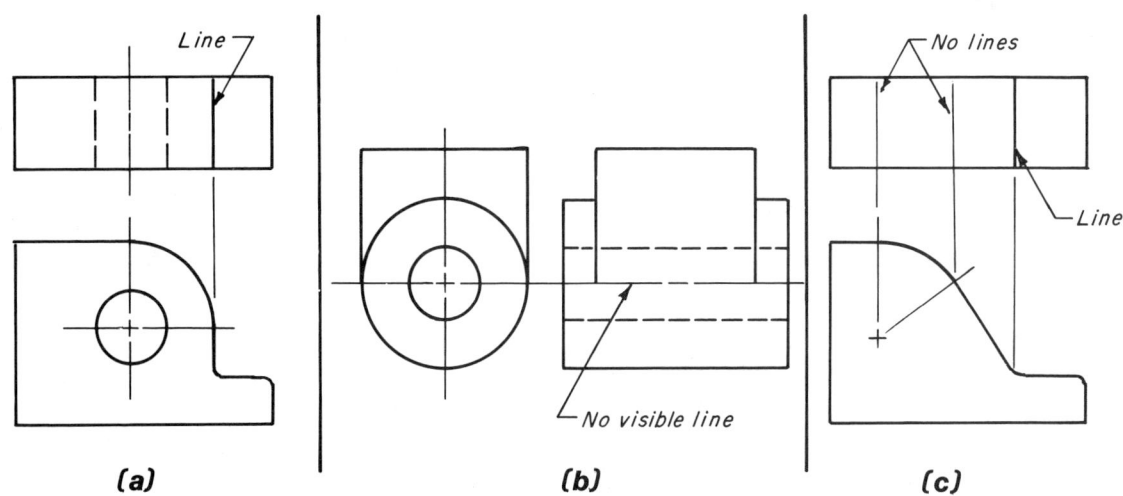

Fig. 8.29 Tangencies.

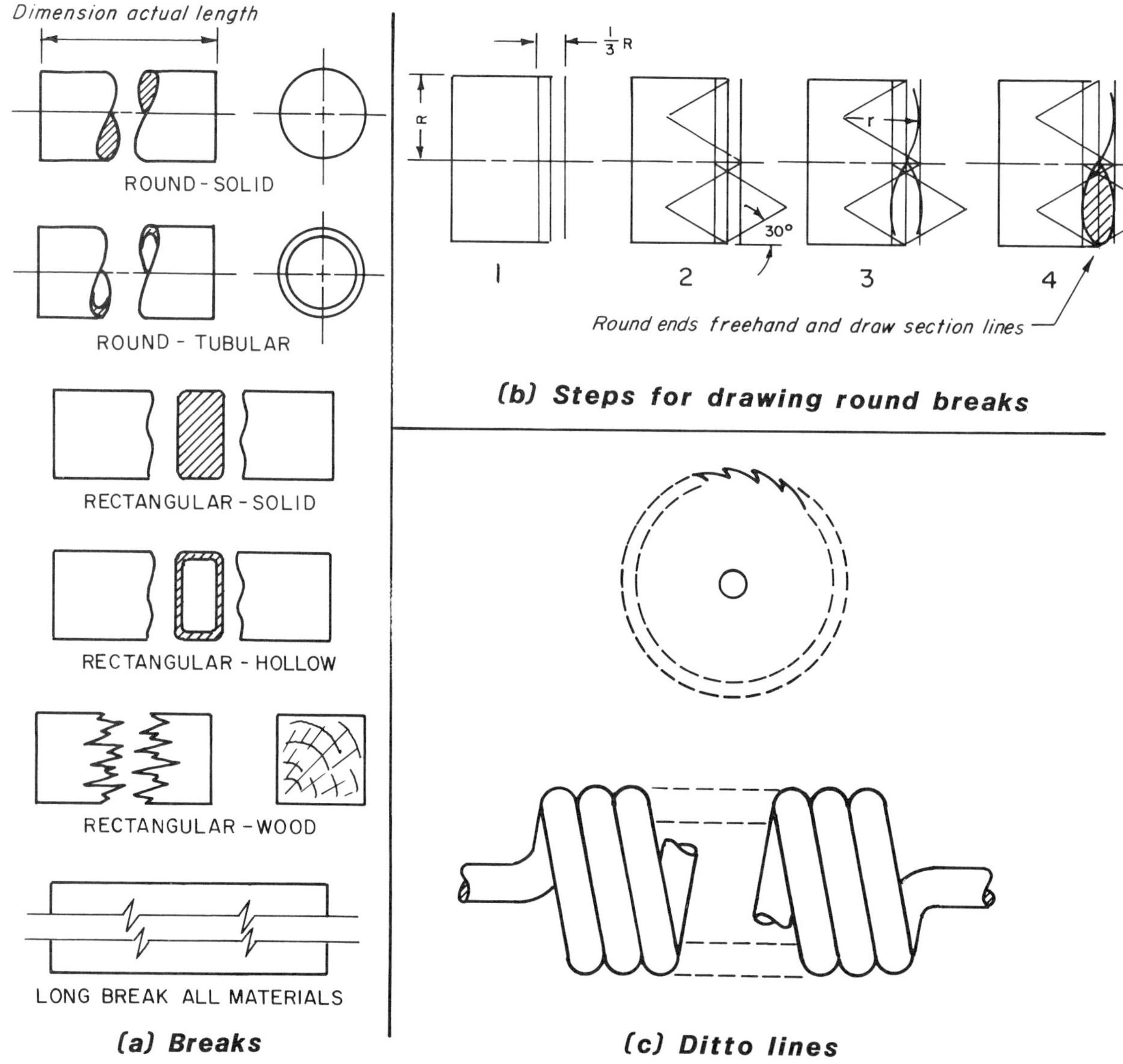

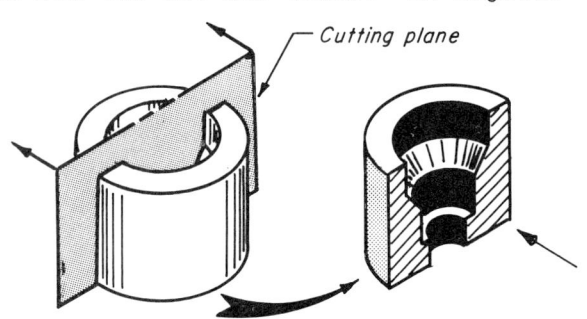

Fig. 8.30 Conventional breaks and ditto lines.

tion. Several sections may appear on a drawing and each section must be related to the proper cutting plane by lettering each, e.g., "Section A-A, Section B-B," etc., Fig. 8.32 (b). Cutting plane lines are used only when needed to show where and how the cut was taken. On objects that are symmetrical the location of the cutting plane is obvious and it is not necessary to indicate its representation on the drawing, Fig. 8.32 (c).

Section lining. A sectional view is easier to interpret if *section lines*, sometimes called crosshatch lines, are drawn on the solid parts of the object cut by the cutting plane. Section line symbols for several materials are shown in Fig. 8.33. Symbols for section lines are usually used on sectioned assembly drawings where the component parts are made of different materials. For single parts it is recommended that the all-purpose cast iron symbol be used with a note to indicate the actual material.

Section lines are used to create a "shading effect" and should be drawn

Fig. 8.31 Theory of sectional views.

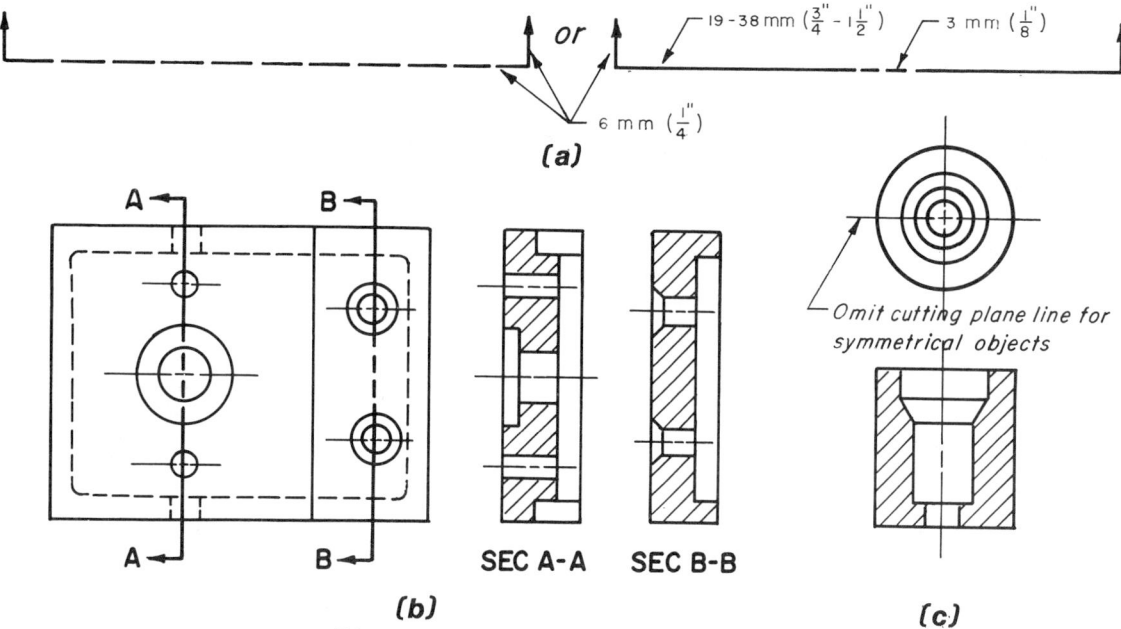

Fig. 8.32 Cutting plane line.

as thin lines at any convenient angle, e.g., 30°, 45° or 60°, Fig. 8.34 (a). Spacing will depend on the material represented and the size of the area. A uniform spacing of approximately 3 mm (3/32") works well for most average size drawings. For cut areas of any one single part, the direction of the section lines should remain constant, Fig. 8.34 (b). On assembly drawings the adjacent parts should have the section lines indicated in different directions or angles, Fig. 8.34 (c). It is a conventional practice for clarity not to section some kinds of standard parts, e.g., bolts, nuts, ribs, webs, screws, shafts, rods, pins, gear teeth, etc.

The section lines should be drawn in a direction so they will not be parallel or perpendicular to the major outlines of the object, Fig. 8.34 (d). Care should also be exercised not to draw section lines across other kinds of information. Large areas of machine parts as well as map features may be section-lined around the perimeter only, Fig. 8.34 (e). This technique saves time and will frequently improve the appearance of the drawing.

<u>Types of sections.</u> The use of sectional views is a conventional practice to help clarify a drawing. No one type of section will satisfy all conditions and several types of sections have been devised to meet various needs. The lo-

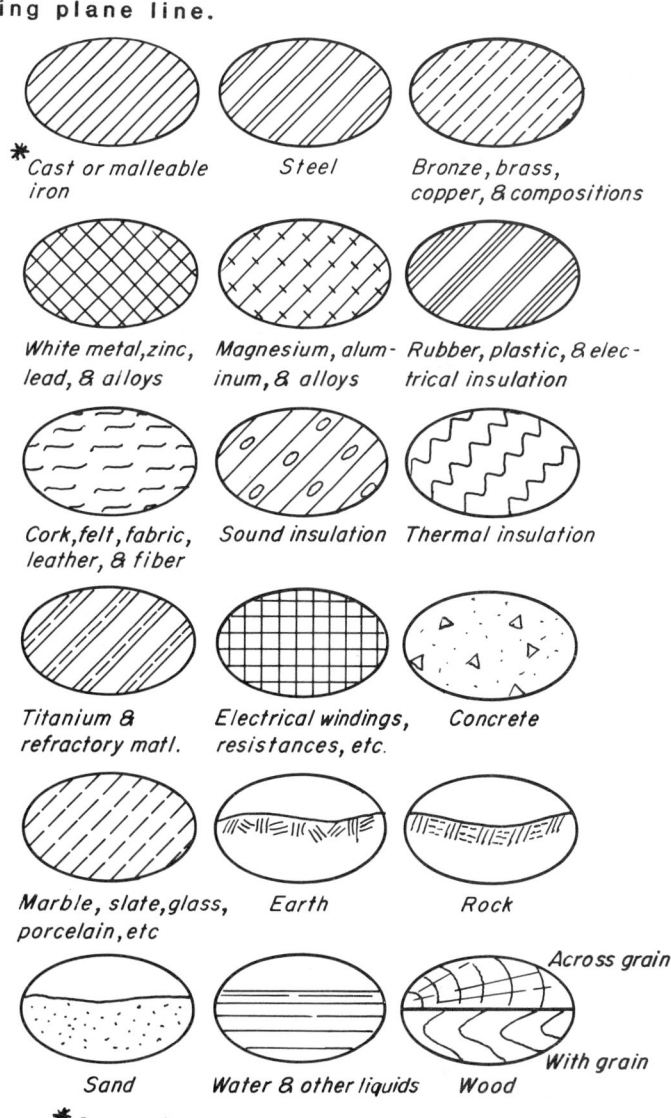

*General purpose symbol for all materials

Fig. 8.33 Symbols for section lining.

cation and extent of the cutting plane passing through the object determines the type of section. Several common types of sectional views are described in Fig. 8.35.

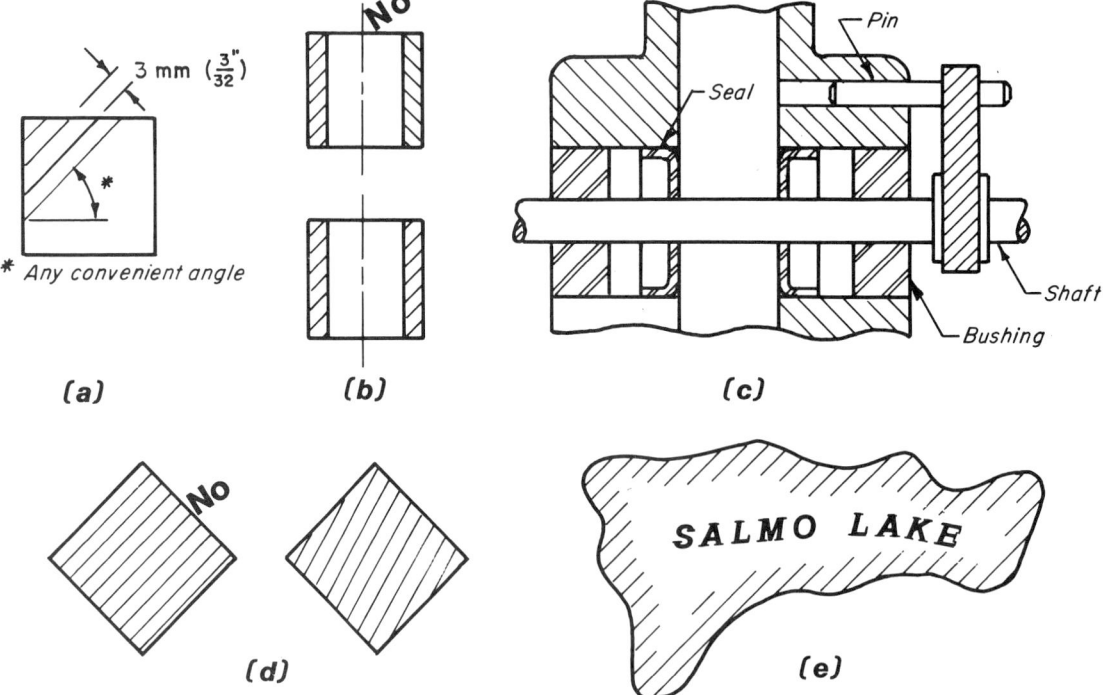

Fig. 8.34 Section lining.

TYPE	*EXAMPLE	CUTTING PLANE	APPLICATION
FULL	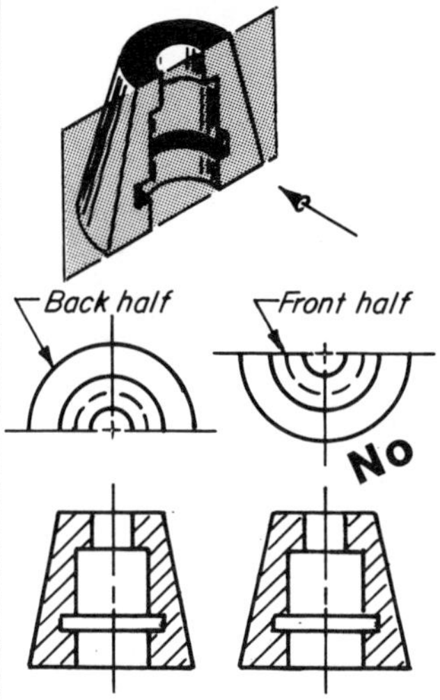	The cutting plane passes entirely through the object removing one half of the object.	Full sections are used when it is necessary to show the complete interior. A partial view may be used when sectioning symmetrical objects. The back half should be drawn since the front half is imagined to be removed for viewing the interior features.
OFFSET	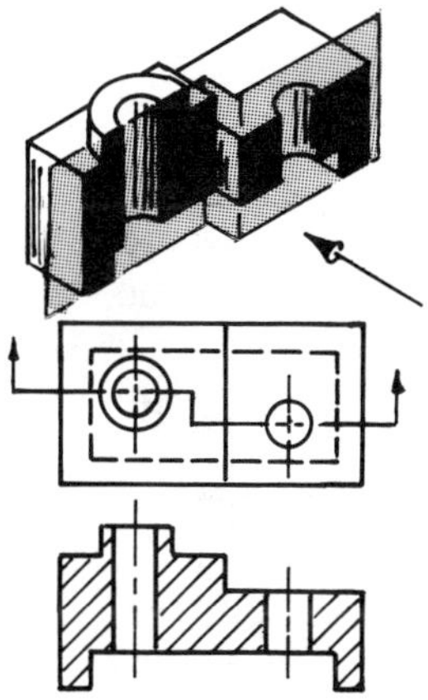	The cutting plane line is offset (staggered) to cut through all necessary features that are to be sectioned. The location of the cutting plane line should be indicated.	An offset section is used to show the complete interior of an object, similar to a full section, for features that are not in a single plane. The change of direction of the cutting plane is not indicated in the sectional view.
HALF	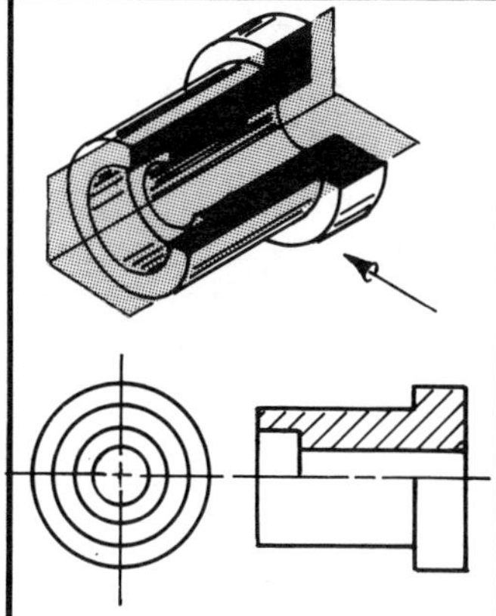	The cutting plane passes halfway through the object, removing only the front quarter of the object.	Half sections are used when the object is symmetrical and when it is desirable to show both the interior and exterior in one view. A center line is used to separate the interior and exterior portions.
BROKEN-OUT	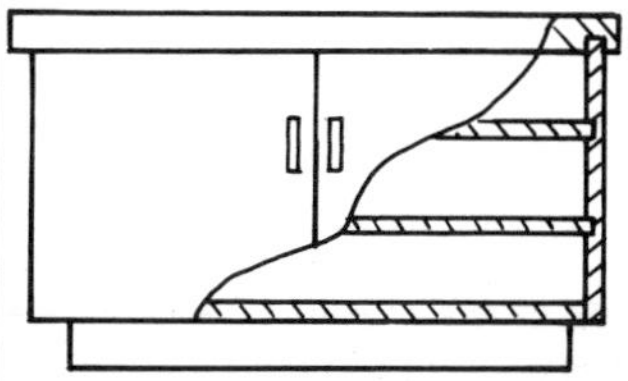	No cutting plane is shown. An irregular break line is used to indicate the limits of the sectioned area.	Broken-out sections are used to show important interior features without drawing an extra view.

Fig. 8.35 Types of sections.

TYPE	*EXAMPLE	CUTTING PLANE	APPLICATION
REVOLVED	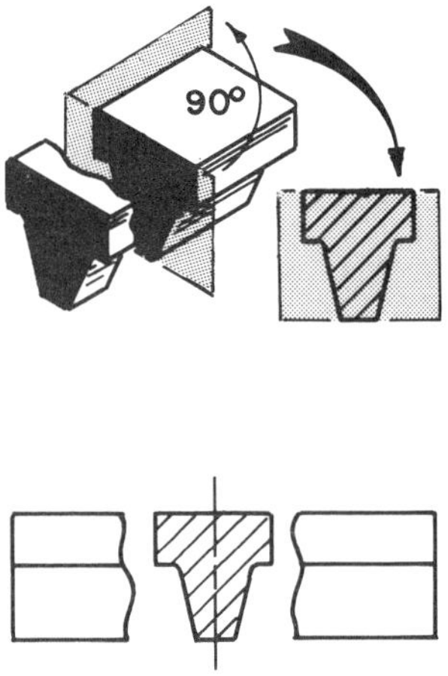	The cutting plane passes entirely through the object to show the cross section of the object rotated 90° with the resulting section superimposed on another view. A center line is used to locate the axis of revolution.	A revolved section is used to show the shape of the cross section of an elongated object. Frequently a break is made in the object to provide an open space for drawing the cross section.
REMOVED	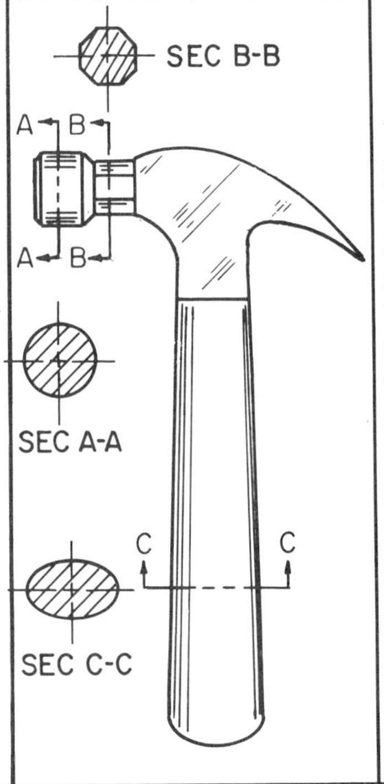	The cutting plane passes entirely through the object to show the cross section of the object rotated 90°. The resulting section is not superimposed on the view but drawn at some other location. The location of the cutting plane line should be indicated on the object. A center line is used to locate the axis of revolution on the section.	A removed section is similar to a revolved section and is used to obtain additional clarity by removing sections to another area on the drawing. The sections may be drawn to a larger scale to show detail and to facilitate dimensioning. All sections must be identified by labelling the sections.
PHANTOM	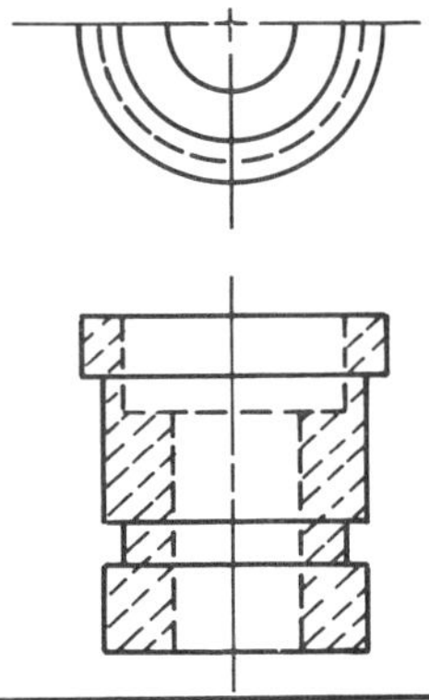	No material in front of the cutting plane is removed. Section lines are superimposed on the exterior view using dashed lines to represent the hidden interior cut parts.	Phantom sections are used to emphasize the interior and to also preserve the exterior construction of an object.

*Avoid hidden lines unless they are needed for the description of the object.

Fig. 8.35 Types of sections (continued).

<u>Projection theroy exceptions</u>. It is conventional practice to violate certain procedures of true projection when drawing sectional views. Parts such as ribs, spokes, and holes are drawn as if they were rotated into the cutting plane, Fig. 8.36, and should not be sectioned lined. This procedure preserves symmetry and helps to avoid misleading implications.

8.15 AUXILIARY VIEWS

An object may have lines or planes that cannot be seen true length or true size in one or more of the regular views. Auxiliary views are frequently required to show the true size of a line or plane or to describe the object more clearly. Any view other than the six principal views, top, front, right side, left side, rear, and bottom may be classified as an *auxiliary* view.

The same theory for drawing the principal views for points, lines, planes, and solid objects applies to the drawing of auxiliary views. In Fig. 8.37 (a), point A could be viewed from an infinite number of directions as indicated by the sight arrows. Each of these views would be classified as an auxiliary view. The correct measurements to take in order to draw any new auxiliary view may be summarized in the following rule.

In all views that are related to a common view (or project from it) any point of the object must be the same distance away from the folding line if used to transfer measurements.

Since the top view and the auxiliary view of point A are both related to the front view of point A, the depth distance from the folding line must be the same.

The lines of sight must be at right angles to a line for any view to show its true length. Line BC, Fig. 8.37 (b), represents a ski-tow cable between two towers. The true length may be found by drawing an auxiliary view either by viewing the line at right angles from the top or front. Height measurements for drawing line B_1C_1 are taken from the front view since both the front and auxiliary views are related to the top view.

A plane must always appear as an edge before a view can be drawn showing it in its true size. View 1, Fig. 8.37 (c), shows plane EFG as an edge. Since the lines of sight are at right angles to the plane, the auxiliary view will show all lines true length and all angles true size. Width measurements to draw the auxiliary view must be taken from the front view.

Successive auxiliary views may sometimes be required to solve problems. In Fig. 8.38 lines MN and ON (views T and F) represent existing storm sewers.

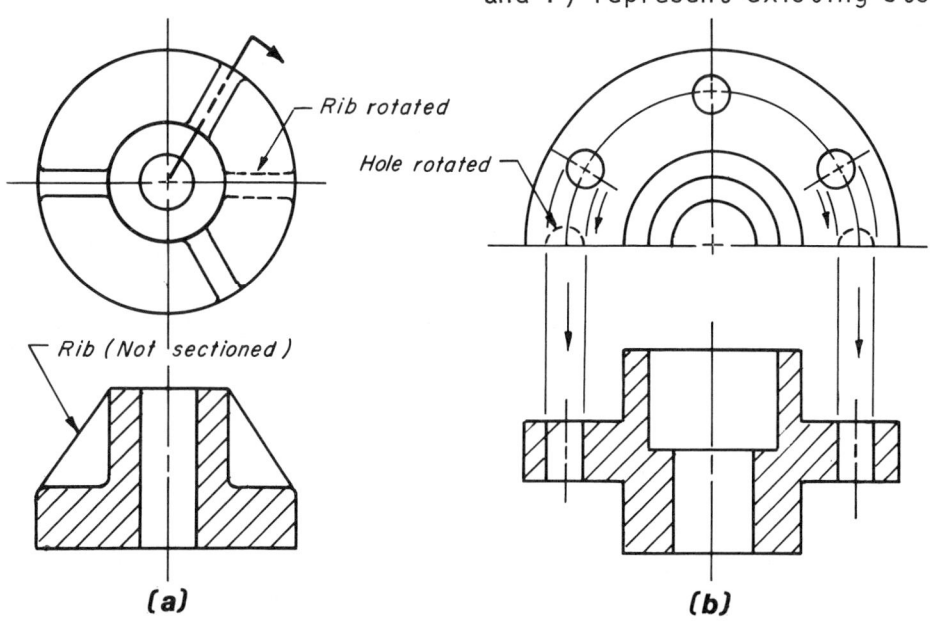

Fig. 8.36 Rotation of ribs and holes.

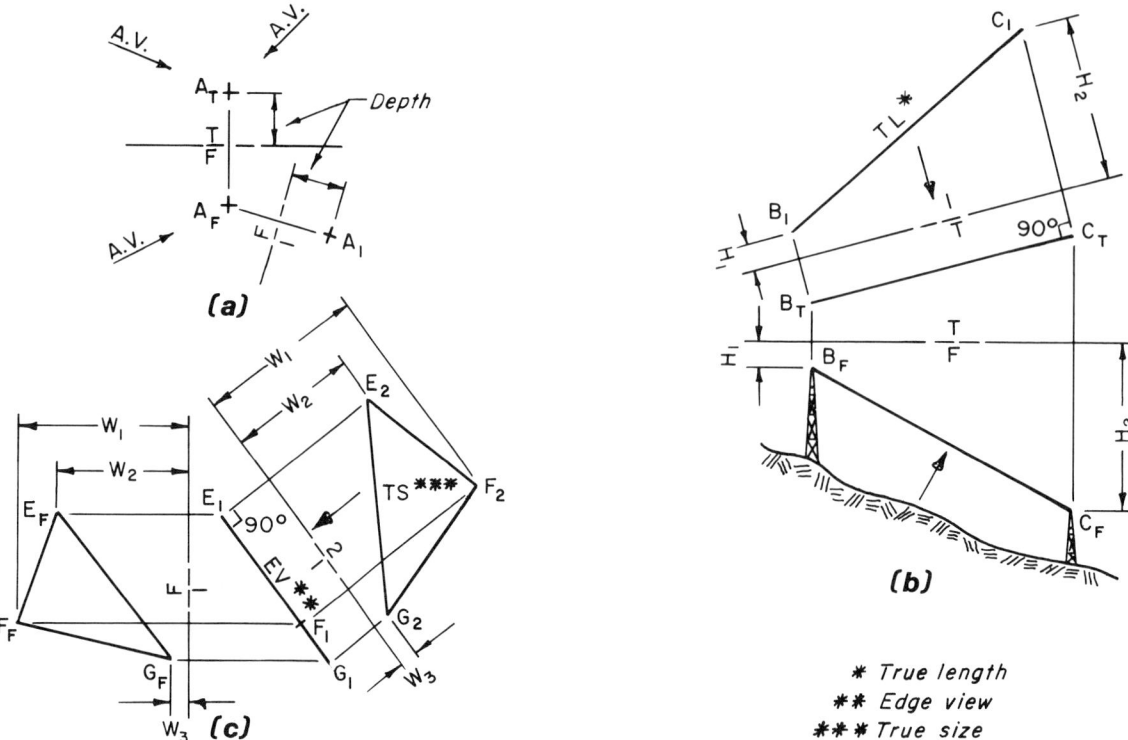

Fig. 8.37 Auxiliary views of a point, line, and plane.

* True length
** Edge view
*** True size

A new storm sewer is to be constructed from a catch-basin at point P to MN or ON, whichever is the shorter line. A careful analysis of the problem reveals that a true size view is required of the plane formed by lines MN and ON to determine the shortest connection. Before a true size view can be drawn of plane MNO, a view must first be drawn which shows the plane as an edge. In order to obtain this edge view of the plane, some line on the plane will have to show in its true length. A level line MS may be drawn in the front view. This line will appear true length in the top view. The edge view, View 1, of the plane is found by drawing a view where the lines of sight are parallel to the true length line. This line will always appear as a point in the edge view of the plane. View 2 is now drawn, looking at right angles to the plane in View 1, to obtain the true size. Point P should also be drawn in

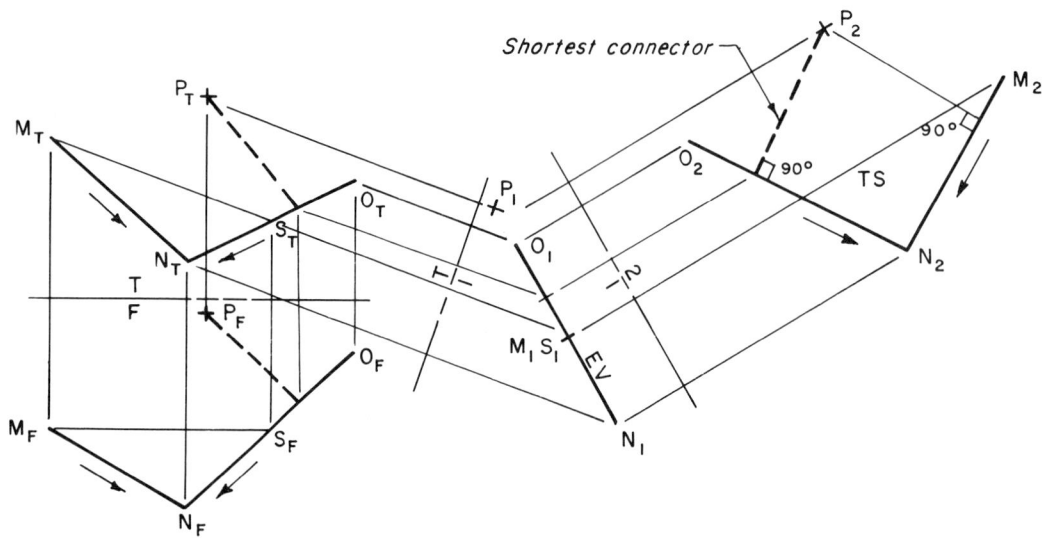

Fig. 8.38 Successive auxiliary views.

all views. A line drawn from point P at right angles to lines MN and ON will reveal that the shortest connection is from point P to ON. The location of this connection is projected and shown in all views.

An object may have one or more surfaces that are inclined away from either a horizontal or vertical plane, Fig. 8.39. The regular views will not show the true shape of the surfaces A, B, and C since the lines of sight are not 90° to these surfaces in the front, top, and right side views. An auxiliary view is constructed for each surface by projecting from the view that shows the inclined surface as an edge. Folding lines may be omitted when drawing simple objects, but measurements must still be taken from a related view.

Auxiliary views are commonly drawn as partial views. Fig. 8.39 shows auxiliary views in which only the inclined surfaces are included. A projection showing the entire object in these three views would add very little to its shape description.

Curved features. An object having curved features must have a sufficient number of points to ensure a smooth curve, Fig. 8.40. The auxiliary view is drawn by locating a series of points, 1, 2, 3. . . 11 on the curved surface in the right side view. The points are projected first to the inclined line representing the surface as an edge in the front view and then to the auxiliary view. Depth measurement for any point from the center line in the right side view is the same as its distance from the center line in the auxiliary view.

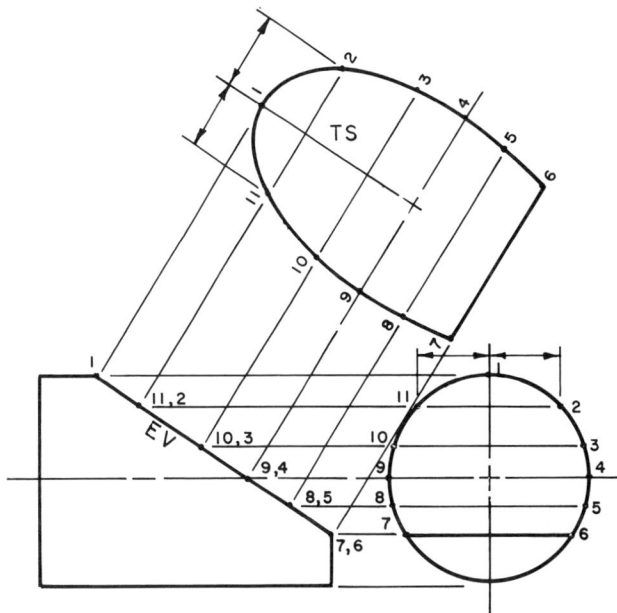

Fig. 8.40 Plotting curved features.

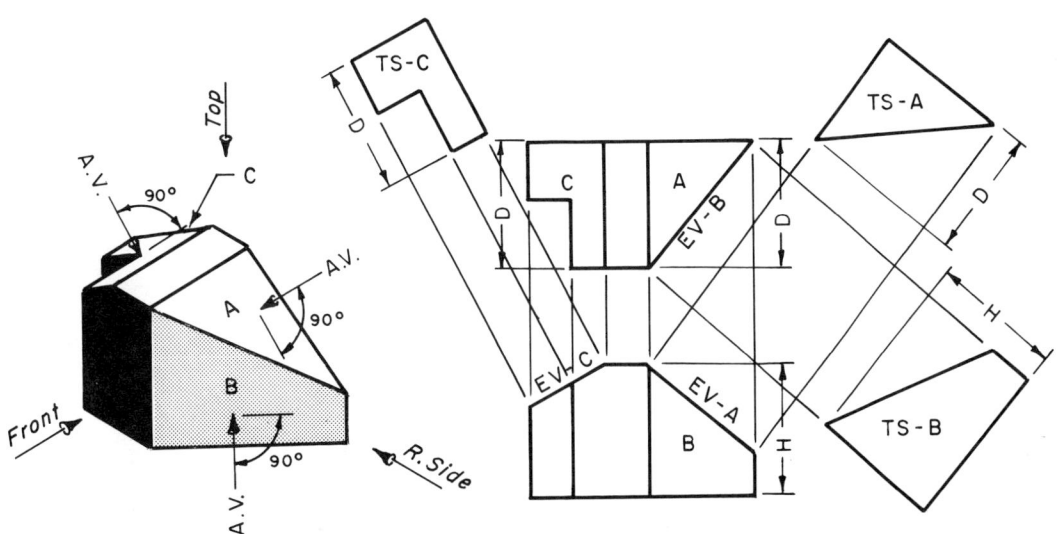

Fig. 8.39 Auxiliary views.

BASIC DIMENSIONING

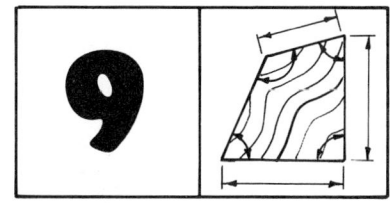

9.1 INTRODUCTION

The individual parts of objects may be represented and defined clearly by drawing a series of related views. Drawings that are to be used in the production or construction of an object, however, can only be considered half finished if additional information in the form of dimensions, notes, and symbols has been omitted. This information must be thoughtfully placed on the drawing to include all of the descriptive information required to manufacture the object. An understanding of the following fundamentals is essential in determining the best method to use for dimension selection and placement.

9.2 TOOLS OF DIMENSIONING

Good dimensioning requires the use of certain techniques that include items like different kinds of lines, lettering directions, notes, and symbols. These items form the basis for dimensioning and are discussed below. All dimensions illustrated in the figures are in millimetres (mm) except where noted.

<u>Lines</u>. The lines used in dimensioning are dimension lines, extension lines, and leaders that are all drawn as solid thin lines. *Dimension lines* are always drawn parallel to the direction measured and are terminated at each end with an arrowhead, Fig. 9.1 (a).

It is preferable to place the dimensions from 6 mm (1/4") to 12 mm (1/2") outside the views with additional dimension lines spaced from 6 mm (1/4") to 9 mm (3/8") apart. A break in the dimension lines, near the center, is provided for the dimension figure. Dimension lines should be carefully placed and drawn so they avoid crossing each other. For small features where space is restricted, it may be necessary for clarity to place the dimensions as illustrated in Fig. 9.1 (b).

An *extension line*, as the name implies, extends from some point on the object slightly beyond the dimension line, Fig. 9.2 A space of about 0.8 mm (1/32") should be left between the ob-

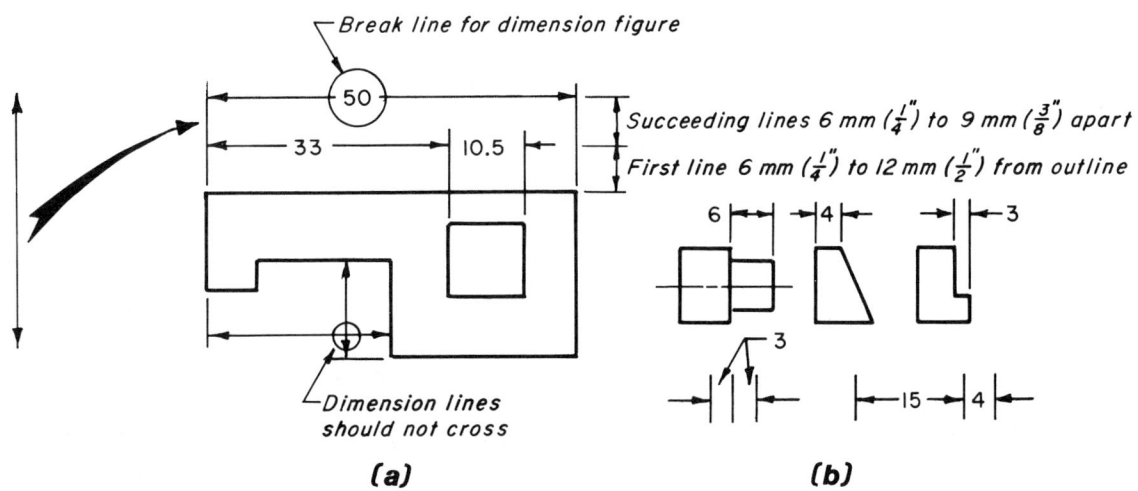

Fig. 9.1 Dimension lines.

ject and the beginning of the line and it should extend about 3 mm (1/8") beyond the last dimension line. Dimension

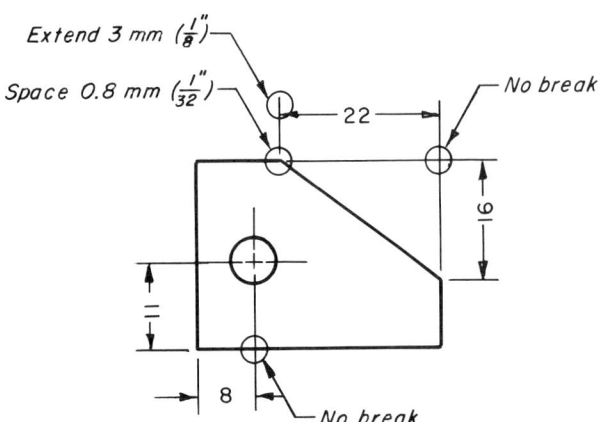

Fig. 9.2 Extension lines.

and extension lines should not cross unless necessary. It is permissible for extension lines to cross each other or other lines on the object without breaks. To help in locating holes, center lines may be used as extension lines by continuing the line.

Leaders are straight lines drawn at a convenient angle (30°, 45°, or 60°) that lead, from the beginning or end of a lettered note describing a feature, to the feature on the object to which it applies, Fig. 9.3 (a). The horizontal shoulder on the leader is drawn about 6 mm (1/4") long with an arrowhead at the other end pointing to the feature. Leaders that point to circular features should be radial and drawn long enough just to clear the object outline. If possible, it is desirable for easy reading to avoid crossing leaders and to keep the leader angles the same throughout a drawing. Several common errors to avoid in drawing leaders are illustrated in Fig. 9.3 (b).

Arrowheads. An *arrowhead* is used on dimension lines and leaders. The size of the arrowhead may vary but should be the same size for any one drawing. Arrowheads are drawn freehand and should have good proportions. For most work it is recommended that arrowheads be drawn 3 mm (1/8") long, Fig. 9.4. The sides are drawn with a short single curved stroke and then filled in solid. If the arrowhead is kept extremely narrow, it will generally look more professional.

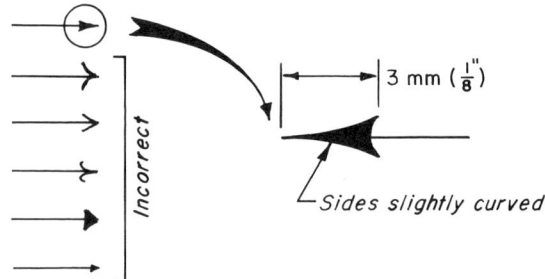

Fig. 9.4 Arrowheads.

Dimension figures and units. All lettering should be carefully executed and should be in conformity with the rules of good practice discussed in chapter 6 on lettering. The direction of the dimension figures may be arranged using either the aligned or unidirectional system. In the *aligned* system, Fig. 9.5 (a), all vertical dimensions

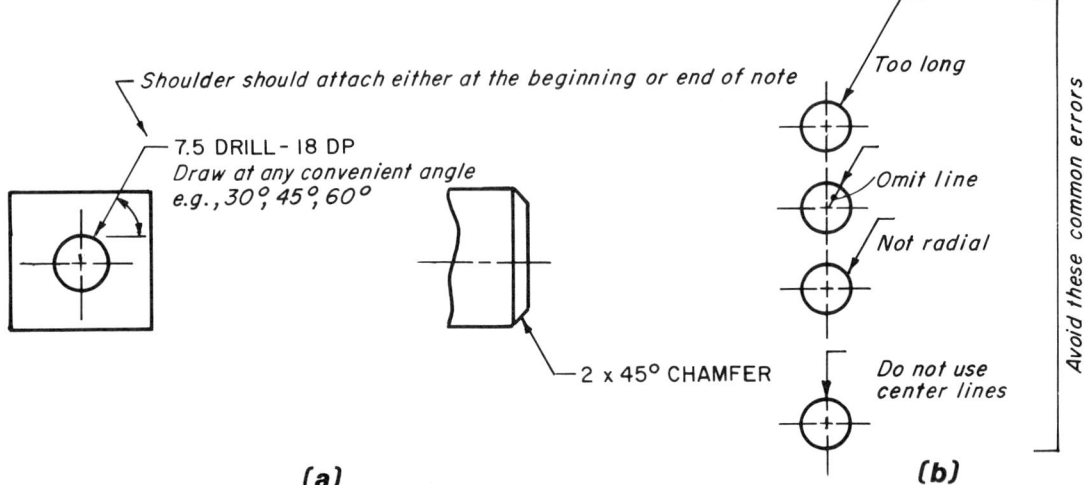

Fig. 9.3 Leaders.

are placed to read from the right side of the drawing and all horizontal dimensions read from the bottom of the drawing. The dimension figures in the *unidirectional* system, Fig. 9.5 (b) all read horizontally or from the bottom of the drawing.

The dimension values may be lettered using metric units or English units (common fractions or decimals), Fig. 9.6. Dimensions for all product type and engineering drawings in the metric system are *millimetres* or decimal parts of a millimetre. Centimetres or other graduations should not be used. In the English system, dimension values are given in *inches*. Dimensions that exceed 72 inches should be given in feet and inches, 7'-0", 7'-0½", 7'-1", 7'-2½", using the accent marks to avoid misinterpretation. It is common practice to omit the units for dimensions expressed in millimetres or inches. The number 1 is often expressed with a unit so that it is not confused with a dash.

Common fractions, 1 3/4, 1/2, 1/4, 1/8, etc., will continue to be used for some time on drawings in the United States. It is difficult to work with an accuracy closer than ± 1/64 using this system. Where greater accuracy is required, the decimal-fraction system is used to express dimensions that must be closer than 1/64, .1875, 1.250, 2.375, etc.

The trend in dimensioning is to use the complete decimal system. Values for dimensioning may be expressed with extreme accuracy. Two-place decimals are used for dimensions for tolerance limits of ± .01 or more. The second digit in a two-place decimal should preferable be an even number, .02, .04, .06, so that when halved in obtaining a radius from a diameter, two-place decimals will still result. Three or more decimals are used for tolerance limits less than ± .010.

It will be necessary for some time for individuals to work with both the English and metric systems. In convert-

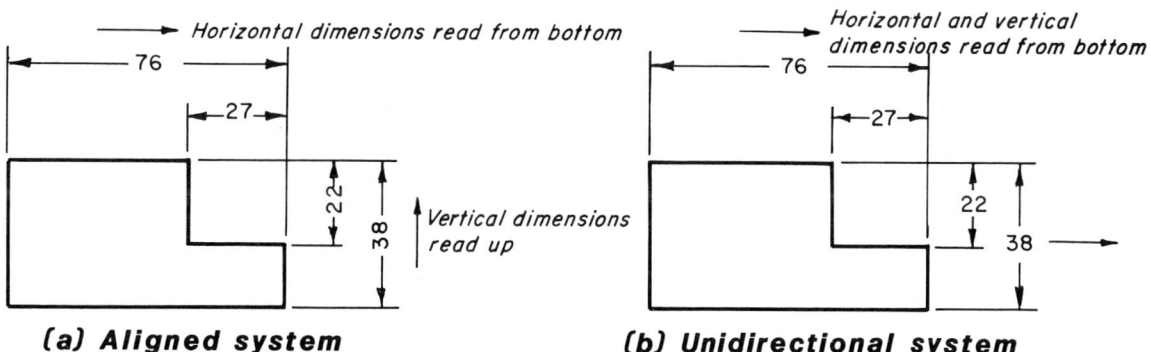

Fig. 9.5 Direction of dimension numerals.

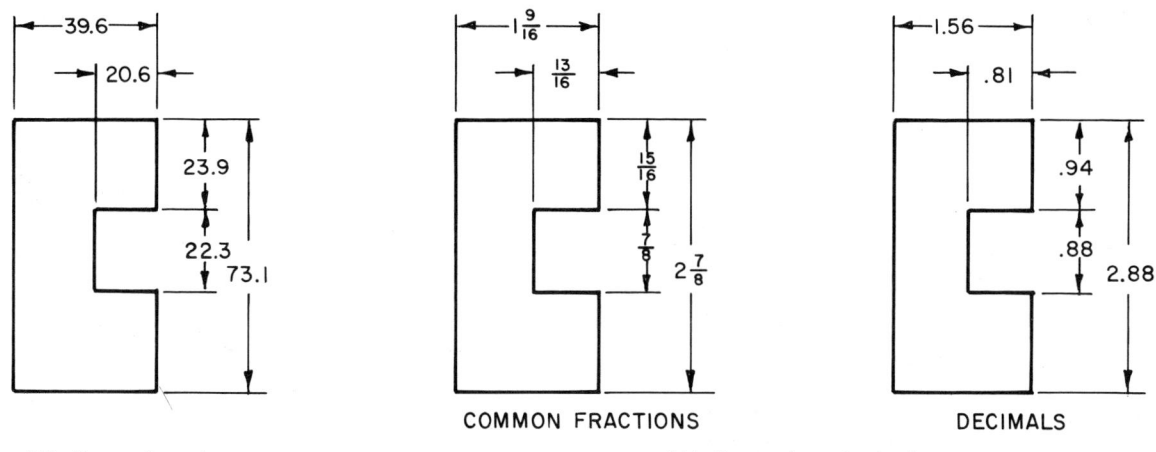

Fig. 9.6 Metric and English dimensioning.

ing units from one system to the other, the rounding of any conversion should be consistent with the expected accuracy of the original quantity. As a general rule, the following practices may be followed:

- Converting inches to millimetres-carry one more decimal place for millimetre dimension than inch.

    1.386 in. = 35.20 mm not 35.204 mm

- Converting millimetres to inches-carry one more decimal place for inch dimension than millimetre, except the inch dimension should carry a minimum of three decimal places.

    32.2 mm = 1.268 in. not 1.27 in.

Several additional rules to follow when a number is to be rounded to fewer decimal places are:

- When the first digit discarded is less than 5, the last digit retained must not be changed.

    3. 463 25 → 3. 463 → 3.46

- When the first digit discarded is greater than 5, or if it is a 5 followed by at least one digit other than 0, the last figure retained must be increased by one unit.

    8. 376 52 → 8. 377 → 8.38

- When the first digit discarded is exactly 5, followed only by zeros, the last digit retained must be rounded upward if it is an odd number, but no adjustment made if it is an even number.

    4. 365 → 4.36
    4. 355 → 4.36

- Dimensions are rounded directly to the nearest value having the desired number of decimal places.

    28.46 → 28 (correct since .46 is less than one-half)

- Rounding must not be done in successive steps to less places.

    1. 28.46 → 28.5 (Correct)
    2. 28.5 → 29 (Incorrect)

Notes. In addition to the dimensions on the views, drawings require statements in the form of notes to supplement this information. Notes may be classified as being general or specific. *General notes* apply to the entire drawing such as MATERIAL, CI, FILLETS AND ROUNDS 3R, FINISH ALL OVER, ALL DIMENSIONS ± .01. *Specific notes* apply to individual features on the drawing, e.g., 12 DRILL-3 HOLES, and leaders are usually used with the note to describe the feature.

All notes should be lettered horizontally and are generally capitalized. The general notes should be grouped together and placed in uncluttered areas of the drawing. For easy readability 3 mm (1/8") high lettering is recommended for all notes and dimensions.

Symbols. In dimensioning a drawing, standard symbols are used to indicate finished surfaces, types of welds, surface quality, etc. Many parts made by casting or forging require that surfaces be machined or finished. Finished surfaces are necessary for mating parts to function properly, to improve handling, and to enhance appearance. It is important that these surfaces be marked so that extra material will be left on the work piece for this to be accomplished. Machined holes, parts made to a specified tolerance and parts made from rolled stock do not need to be marked.

A 60° "V" symbol 3 mm (1/8" high is used to represent the surface to be finished, Fig. 9.7. The symbol should be drawn, on the outside of the view

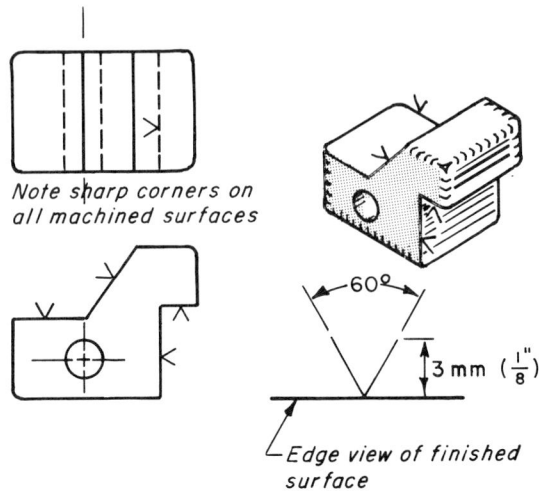

Fig. 9.7 Finish marks.

where possible, using the 30° - 60° triangle with the point of the "V" touching the surface to be machined. Only the line representing the edge view of all finished surfaces is marked including visible and hidden edges in all views. The marks are omitted if a part is to be machined on all surfaces and a note FINISH ALL OVER (FAO) is used.

9.3 GENERAL DIMENSIONING APPROACHES

An object regardless of its complicated features may be considered an assembly of a number of separate parts. Each of these parts has its shape represented on the views of the drawing. A knowledge of how a part is to be made and used is helpful in selecting the best dimensions to show the size, shape, and location of all features completely. Only those dimensions that are pertinent for manufacturing or for the proper operation of the part should be given. All parts must be dimensioned completely so that it is not necessary for the drawing to be scaled or mathematical computations made to obtain information for the production of the object. Dimensions selected may be classified as either size dimensions or location dimensions.

Size dimensions. The shape description of an object should be studied carefully before any dimensions are placed on the views of the drawing. Different shapes like cylinders, holes, and prisms should be identified on the two most characteristic views. Since three dimensions are necessary for the complete size description of a part, two of the dimensions should be placed on the most characteristic view and the third dimension on the next most characteristic view, Fig. 9.8. This procedure known as *size dimensioning* should be repeated for each of the parts of the drawing.

Location dimensions. After the size dimensions have been placed on the views of the various parts, another set of dimensions called *location dimensions* must be placed on the views of the drawing, Fig. 9.8. All parts and dimensions

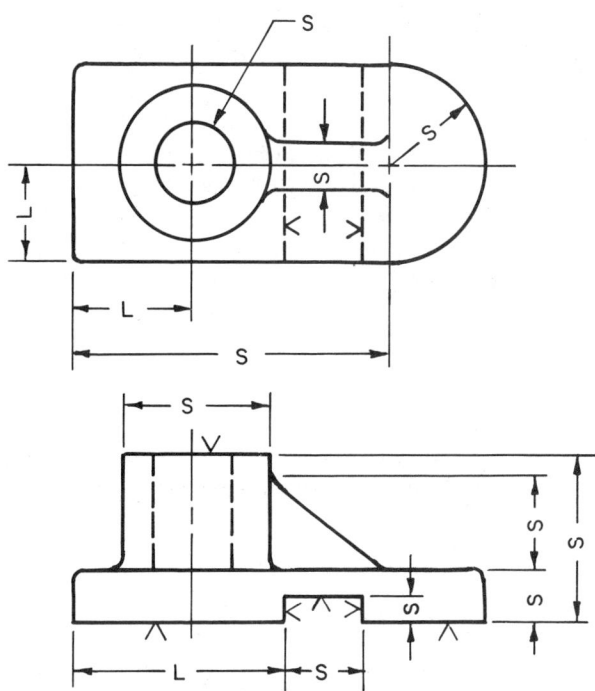

Fig. 9.8 Size and location dimensions.

between related parts must be located. The most characteristic view should carry the location dimensions for each part as was done in size dimensioning.

9.4 RULES FOR DIMENSIONING

Dimensioning for the beginner can best be accomplished by remembering to place dimensions on drawings in the clearest possible manner. The dimensioning process should be carried out on all views of the object simultaneously. It is undesirable to try to dimension one view completely before starting to dimension another view. Attempts at such practice usually lead to poorly placed dimensions and omissions. Examples of objects that have been correctly dimensioned should be studied to help determine the best way to dimension similar parts for other objects. A few standard practices have been set forth in the form of rules and should continually be referred to when dimensioning. Several of these rules are in the form of guiding principles to which there are occasional exceptions while others are absolute practices which should never be violated.

Placement of dimensions with respect to the views of an object.

1. Place the dimensions *outside* the views unless clarity will result from placing them on the views, Fig. 9.9. The use of long extension lines should be avoided and may require a dimension to be placed inside the view.

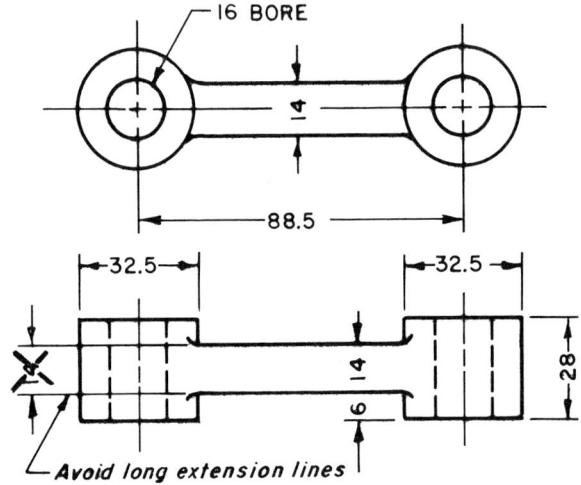

Fig. 9.9 Dimensioning inside a view.

2. Place the dimensions *between* the views where possible, since adjacent views have one dimension in common, Fig. 9.10. Dimensions should be placed next to the feature on the most characteristic view and it will not always be possible to keep all dimensions between the views. As a general rule most of the dimensions will probably be clustered around the front view since it is selected as being the most characteristic view to draw.

3. Place dimensions to *visible* lines, Fig. 9.10. Avoid dimensioning to hidden lines if possible.

4. Dimension the *overall* height, width, and depth of the object unless the end is rounded, Fig. 9.11.

5. Duplication of dimensions or repetitive information must be avoided, Fig. 9.12.

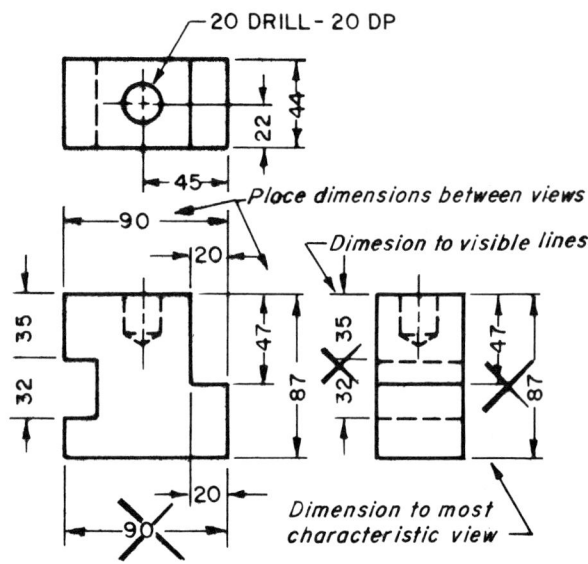

Fig. 9.10 Dimensioning between views.

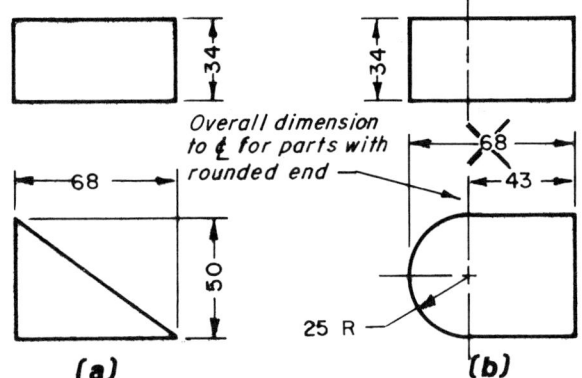

Fig. 9.11 Overall dimensions.

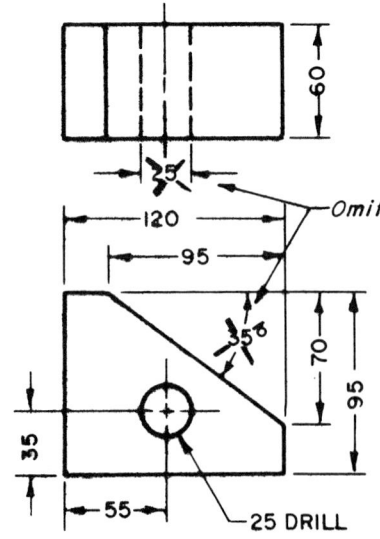

Fig. 9.12 Repetitive dimensions.

6. In a chain or one long line of dimensions, omit one dimension unless it is noted "REF" for reference, Fig. 9.13 (a). Note that the dimension lines should not be staggered, Fig. 9.13 (b).

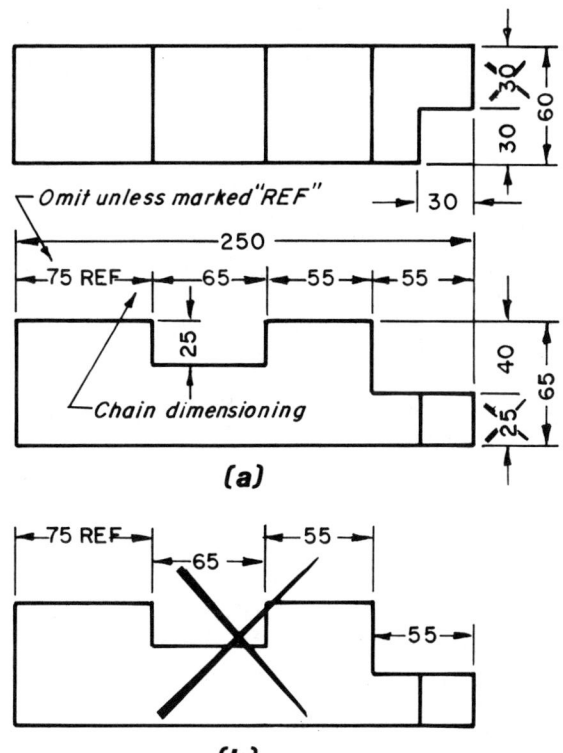

Fig. 9.13 Unnecessary dimensions.

7. The longer or *overall* dimensions should be placed outside the shorter or *intermediate* dimensions, Fig. 9.14, to prevent dimension and extension lines from crossing. Intermediate and overall dimensions should be kept together where possible.

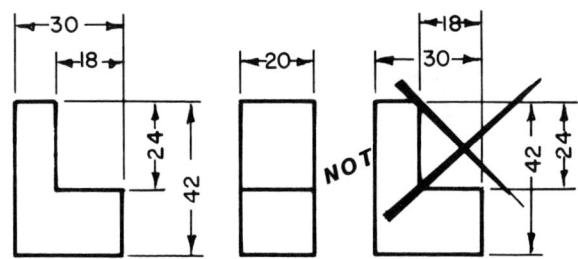

Fig. 9.14 Overall and intermediate dimensions.

8. Section lines should be broken when placing dimensions in a section area, Fig. 9.15.

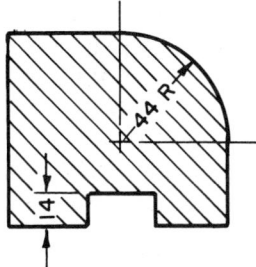

Fig. 9.15 Dimensioning sectional areas.

9. Dimensions that will not be used make the drawing difficult to read and must be omitted, Fig. 9.16.

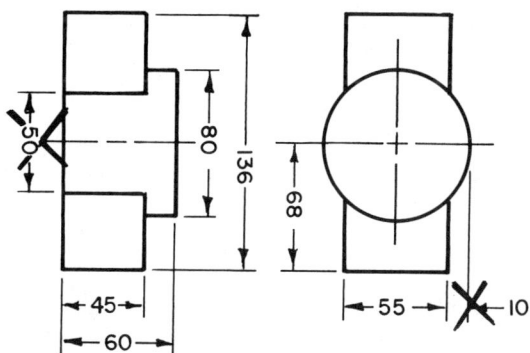

Fig. 9.16 Needless dimensions.

10. Location dimensions should be specified to a finished surface when possible, Fig. 9.17.

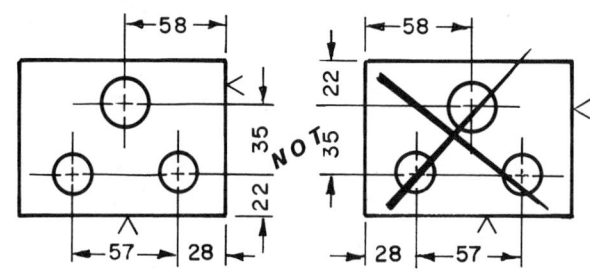

Fig. 9.17 Dimensions to a finished surface.

Placement of dimensions in relation to each other:

11. A series of parallel dimensions should have the figures staggered to avoid a crowded appearance and to make the numerals easier to read, Fig. 9.18.

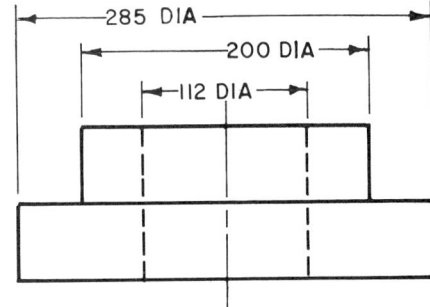

Fig. 9.18 Staggering dimensions.

12. A dimension line should not be the continuation of some line on the drawing, Fig. 9.19 (a).

    Dimension lines and numerals should not be placed on or along centerlines, Fig. 9.19 (b).

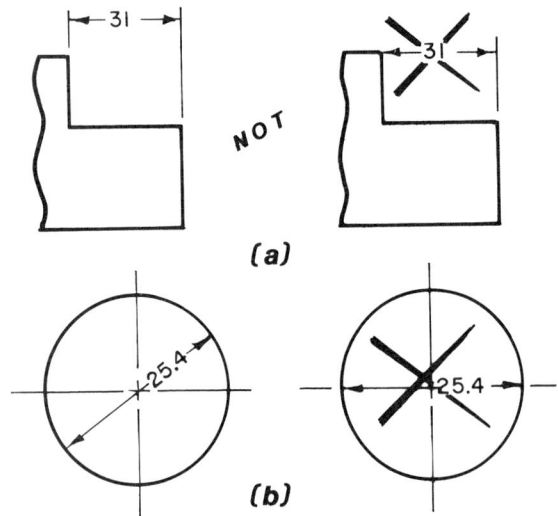

Fig. 9.19 Dimension line placement.

Placement of dimensions for circles. Circular features should be identified as representing holes or cylinders before dimensioning. The *diameter* and never the radius should be specified. The abbreviation "DIA" should not accompany the dimension numeral unless the object can be described by one view, Fig. 9.18.

13. Circular features representing holes generally are dimensioned on the circular view using a leader with a note describing the hole, Fig. 9.20. The centers of the holes should be located on the view of the object where the holes show as circles. Never give a location dimension to the edge of a circular part.

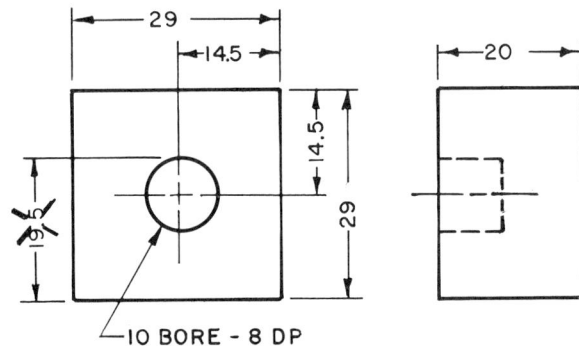

Fig. 9.20 Dimensioning holes.

14. Specific notes for holes should contain only the information that applies to that feature (Fig. 9.20, 9.21, and 9.22) and given in the following sequence:
    - Diameter of hole
    - Operation to produce hole
    - Depth of hole
    - Number of holes
    - Spacing of holes

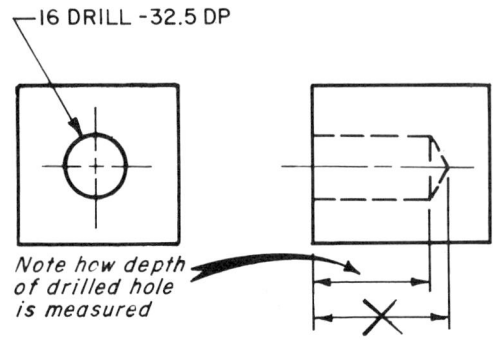

Fig. 9.21 Notes for holes.

15. A series of equally spaced holes in a circular pattern are located by giving the diameter of the center line circle, commonly referred to as a bolt circle, Fig. 9.22 (a). Holes that are not equally spaced should be located by dimensioning the angles between successive holes in degrees, Fig. 9.22 (b). For greater accuracy holes may be dimensioned by coordinates using two datum lines to locate all holes, Fig. 9.22 (c).

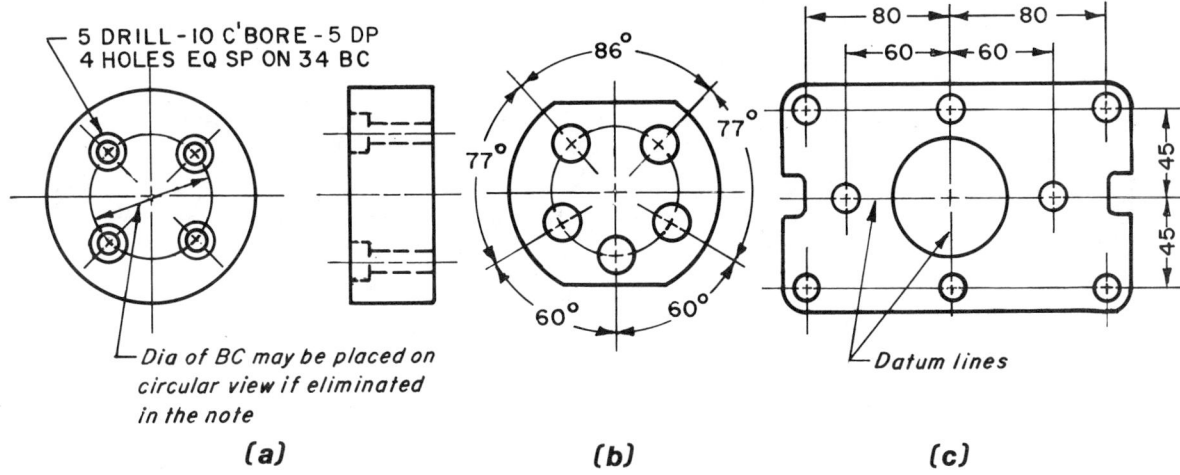

(a)   (b)   (c)

Fig. 9.22 Location of holes.

16. Circular features representing cylinders are dimensioned on the rectangular view, Fig. 9.23.

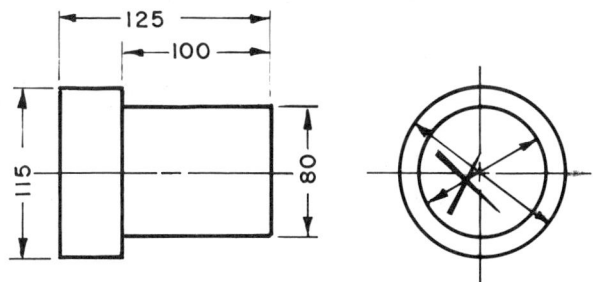

Fig. 9.23 Dimensioning cylinders.

17. Noncircular curves are dimensioned by coordinates from two datum lines, Fig. 9.24.

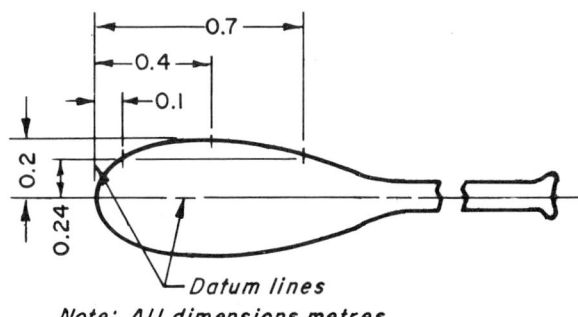

Note: All dimensions metres

Fig. 9.24 Dimensioning noncircular curves.

Placement of dimensions for arcs.

18. Circular arcs are dimensioned by the *radius* with the abbreviation "R" following the dimension numeral, Fig. 9.25. Small arcs termed fillets and rounds should be dimensioned using a note or one typical dimension for each type.

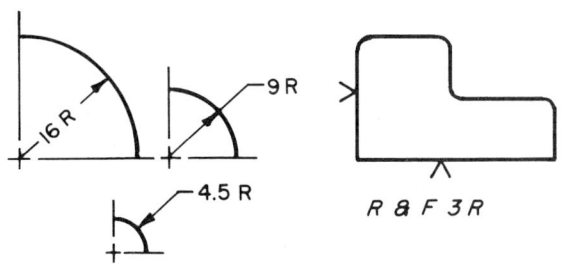

Fig. 9.25 Dimensioning arcs.

Placement of dimensions for angles.

19. Angles may be dimensioned by degrees or by coodinates, Fig. 9.26 (a). The vertex of the angle is used as the center to draw the circular dimension line. Dimension figures showing the number of degrees in the arc should be placed to read from the bottom of the drawing, except for arcs greater than 90°, Fig. 9.26 (b).

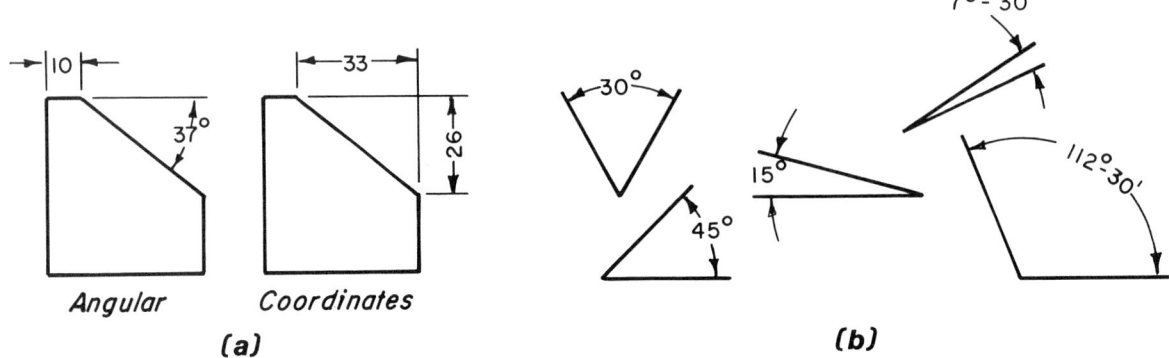

Fig. 9.26 Dimensioning angles.

Dimensioning other features.

20. A chamfer is dimensioned by specifying the angle and length, Fig. 9.27.

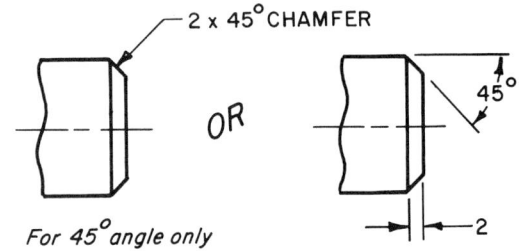

Fig. 9.27 Dimensioning chamfers.

22. A keyway is dimensioned as illustrated in Fig. 9.29.

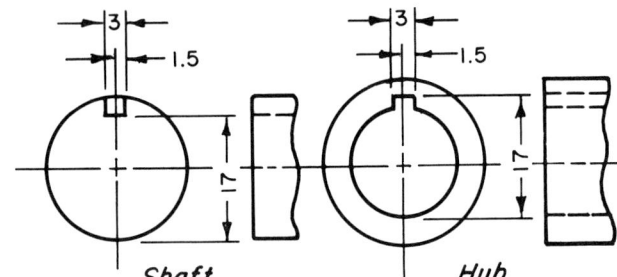

Fig. 9.29 Dimensioning keyways.

21. A slot is dimensioned by specifying the diameter of the cutting tool and center to center distance between both ends or the radius of the slot and its length, Fig. 9.28. Location dimensions are given to the center lines or to one end of the slot and its longitudinal center line.

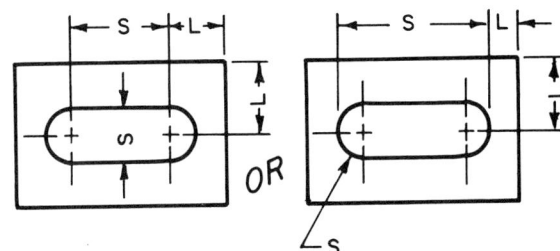

Fig. 9.28 Dimensioning slots.

9.5 PICTORIAL DIMENSIONING

A pictorial drawing is dimensioned following the same general procedures used for dimensioning a multiview drawing with the exception of the following specific rules:

1. All extension and dimension lines should be parallel to the pictorial plane to which they apply.

2. Arrowheads should lie in the pictorial plane they dimension.

3. Dimensions should lie in a visible pictorial plane where possible.

4. Guide lines and dimension figures are drawn parallel to the pictorial axes. All lettering is vertical and great care must be exercised to make the lettering appear pictorial.

5. All specific notes are lettered horizontally using a leader.

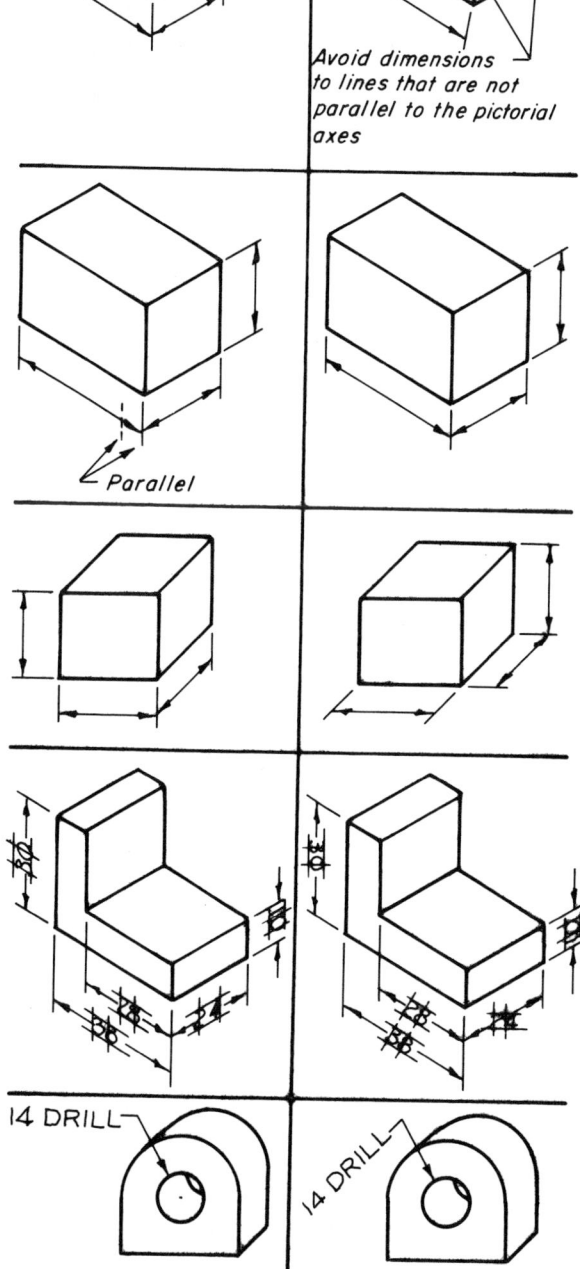

Fig. 9.30 illustrates pictorial dimensioning for an isometric and oblique drawing.

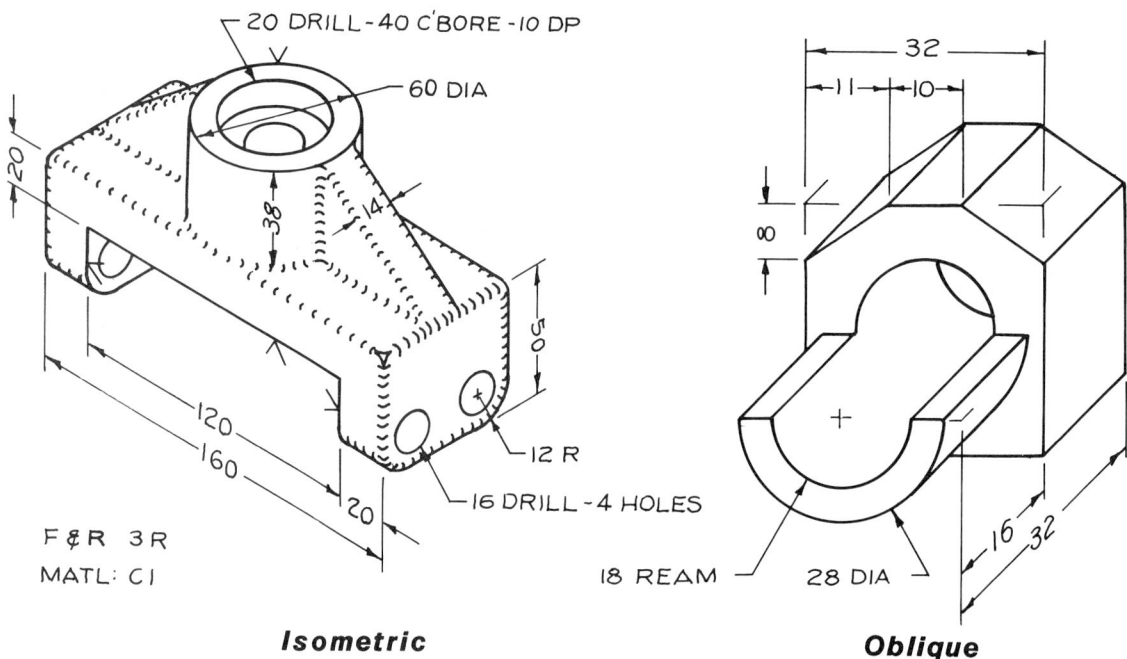

Isometric **Oblique**

Fig. 9.30 Pictorial dimensioning.

PICTORIAL SYSTEMS

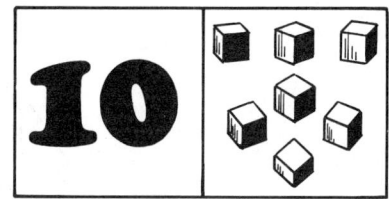

10.1 INTRODUCTION

Any drawing that has the appearance of a three dimensional object may be considered a *pictorial*. In multi-view projection a view can show only *two* of the *three* principal dimensions. Since a thorough understanding of orthographic projection theory is required to read drawings of this kind, pictorials are commonly used where easy and fast graphic communication is necessary. Many working/assembly drawings, project proposals, and catalogues present information in pictorial form. Visualization is easier since all three dimensions, height, width, and depth, for the composite representation of the object are viewed at one time, Fig. 10.1.

10.2 TYPES OF PICTORIALS

Pictorials may be classified into three primary types, *axonometric*, *oblique*, and *perspective*. The difference between these types of pictorials as a group can be compared in chapter 8, Fig. 8.2 and are discussed in greater detail in the following sections. The kind of pictorial to draw will depend primarily on the shape of the object and how the drawing is to be used.

10.3 AXONOMETRIC PROJECTION

Axonometric projection is a form of orthographic projection where the object is rotated and tipped forward to reveal three faces of the object in a single plane, Fig. 10.2. The projectors are always parallel to each other and perpendicular to the picture plane. An infinite number of axono-

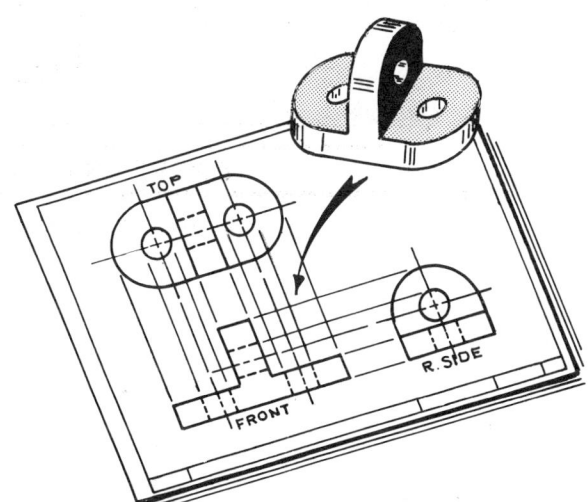

Fig. 10.1 Pictorials enhance understanding and visualization.

metric views may be drawn depending upon the angle through which the object is rotated and tipped. Any change in the angle α and θ will alter the proportions of the axonometric view, e.g., size of angles and lengths of edges. Different axonometric views are therefore divided into three subtypes, *isometric*, *dimetric*, and *trimetric projection*.

10.4 ISOMETRIC PROJECTION

In *isometric projection* the object is rotated 45° about a vertical axis, Fig. 10.3 (b), and tilted forward until the front edges of the cube make an angle of 35°-16' with the picture plane, Fig. 10.3 (c). The isometric projection of the object is viewed in the front view. A disadvantage of this type of projection is that all lines are foreshortened and do not show in their true lengths. The three front or leading edges are called *isometric axes* and make an angle of 120° with each other. Any lines that are drawn par-

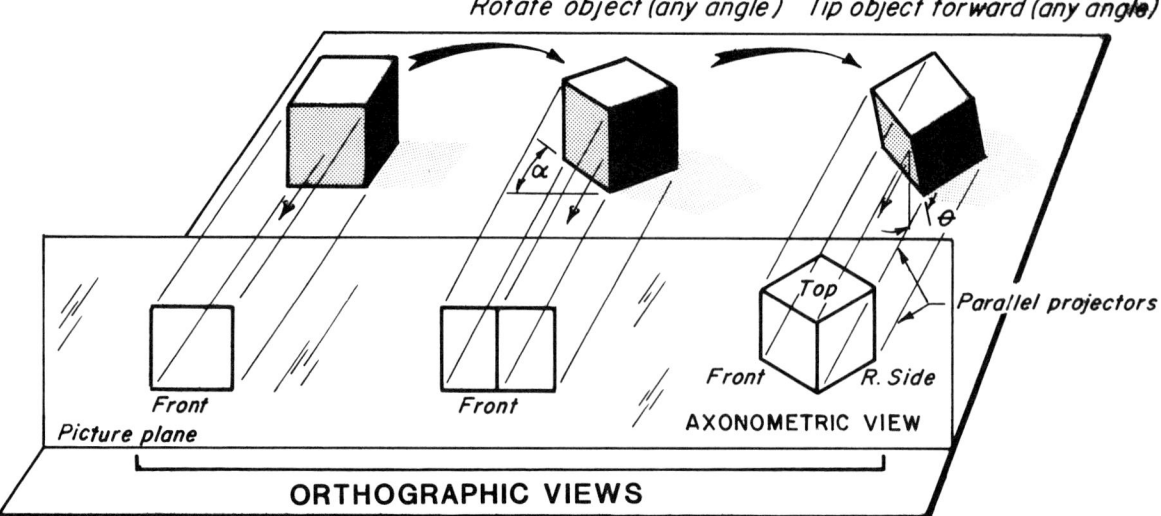

Fig. 10.2 Theory of axonometric projection.

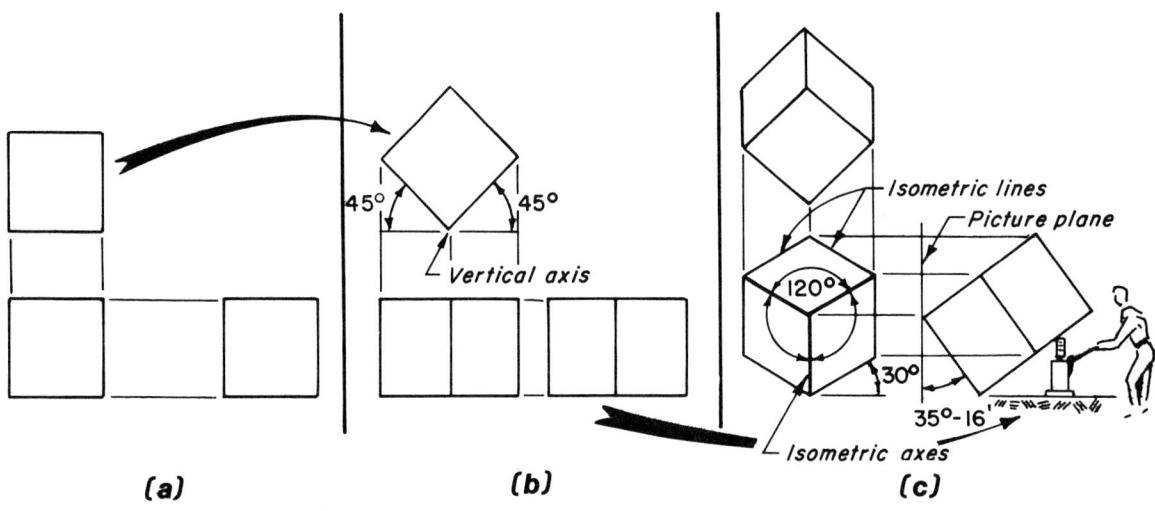

Fig. 10.3 Isometric projection.

allel to the isometric axes are called *isometric lines*. The receding isometric lines make an agle of 30° with the horizontal.

10.5 ISOMETRIC DRAWING

In isometric projection the pictorial view is obtained by actual projection from orthographic views of the object. Since pictorials of this kind are time consuming to draw and actual lengths of lines are difficult to measure, isometric drawings are preferred. Pictorially an isometric projection and an isometric drawing will look the same, Fig. 10.4. The only difference is one of size, with the isometric drawing approximately 22.5% larger than the isometric projection.

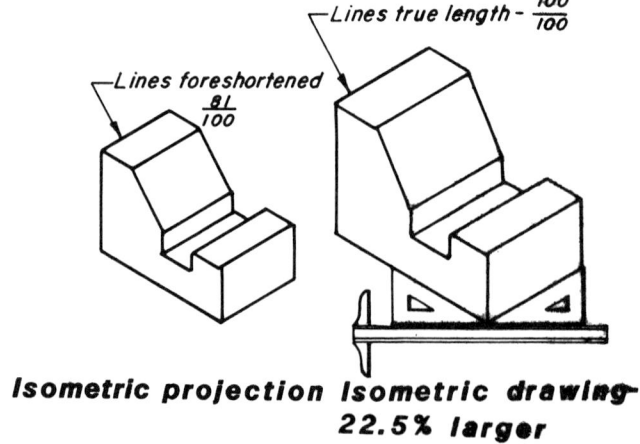

Fig. 10.4 Comparison of isometric projection and isometric drawing.

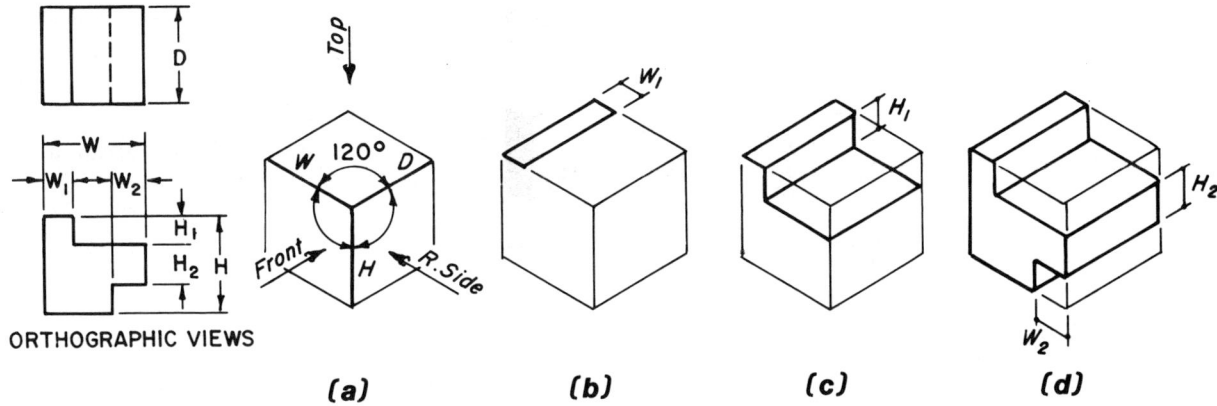

Fig. 10.5 Constructing an isometric drawing.

Constructing an isometric drawing. The object should always be oriented to allow the important features to be viewed. Normally the hidden lines in the pictorial are not drawn unless needed for clarity. The steps in constructing an isometric drawing are:

(1) Draw the isometric axes at 120° with each other, using the T-square and 30°-60° triangle, Fig. 10.5 (a), and lay off the dimensions height, width, and depth along the axes. A box equal to these dimensions may now be drawn to enclose the object.

(2) Draw parts or shapes of the object by setting off measurements only on isometric lines of the drawing as illustrated in Fig. 10.5 (b) thru (d). The object is completed by drawing lines, through measured points, parallel to the axes. All parallel lines on the object must appear parallel to each other in the pictorial.

It may be desirable or necessary to view the object in some other direction to show all features clearly, Fig. 10.6. The isometric axes may be drawn in any position as long as the 120° angles between the axes are maintained. An object that is oriented with the long dimension placed on one of the receding axes may appear distorted. Rotating the object with the long axes drawn in the horizontal position will help reduce the effect of distortion.

All drawings should be approximately centered in the working space to create a balanced pleasing effect. The isometric drawing of an object may be centered within a specific area as illustrated in Fig. 10.7.

10.6 ISOMETRIC ANGLES

An angle in isometric drawing will normally not appear true size unless it is in a plane parallel to the picture plane. The orthographic views of the object are used to find the end points of a line forming an angle. The

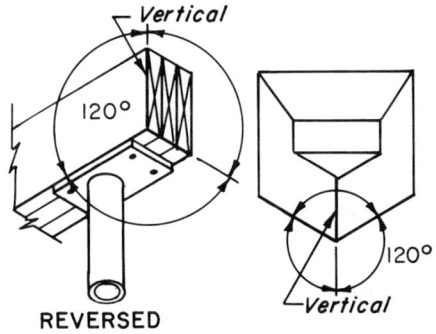
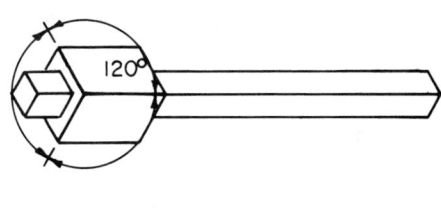

Fig. 10.6 Position of isometric axes.

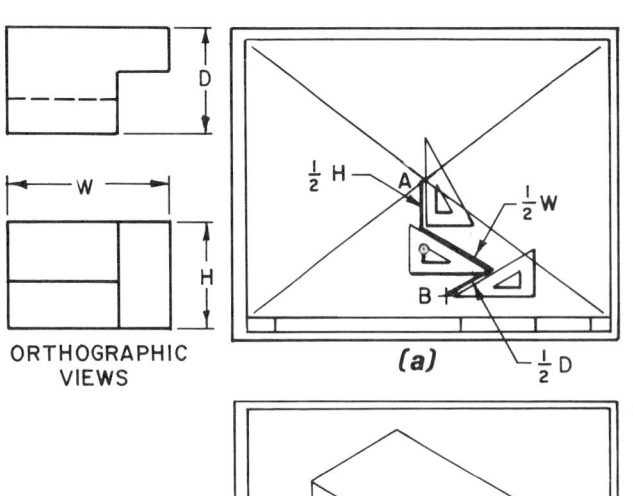

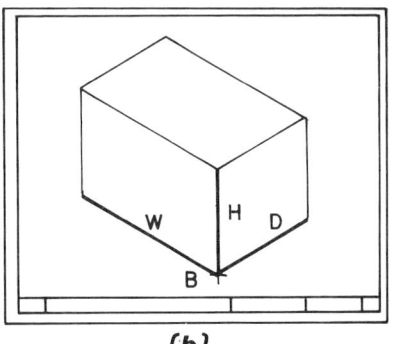

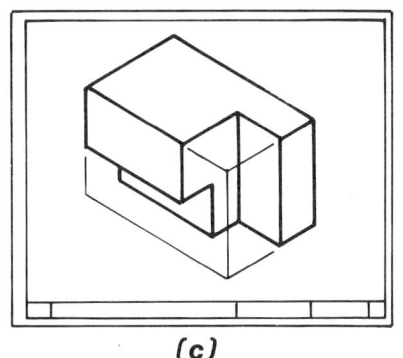

1. Locate center of working space, point A.
2. Draw isometric lines illustrated and lay off one-half the height, width, and depth in order along the lines.
3. The terminal point B is the starting point on the paper.

4. Draw edges of box through point B.
5. Complete box to enclose the object by laying off total height, width, and depth along the isometric lines.

6. Draw object using procedure previously described.

Fig. 10.7 Centering an isometric drawing.

scale of the orthographic drawing and the pictorial must be the same to lay off these distances. In Fig. 10.8 (a) the angle is dimensioned by coordinates. To locate points A and B in the pictorial, lay off the horizontal distance X and the vertical distance Y along isometric lines of the object. All lines that are not parallel to the isometric axes are called *non-isometric lines*.

The angles in Fig. 10.8 (b) are specified in degrees and again must be located by coordinates since the true size of angles normally cannot be measured on an isometric drawing. It is necessary to find the coordinate value X, X', and Y to lay off along isometric lines in the pictorial to draw the 45° and 60° angles.

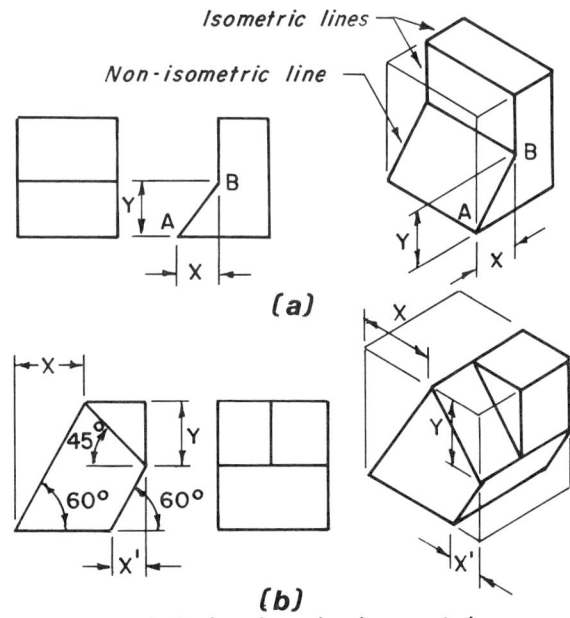

Fig. 10.8 Angles in isometric.

10.7 ISOMETRIC CIRCLES AND ARCS

A circle or an arc will appear elliptical in isometric drawing. The four-center method is used to draw an approximate ellipse in an isometric plane, Fig. 10.9 (a), that is satisfactory for most work. This method is generally used since it is easy to construct and requires only the 30° - 60° triangle, T-square, and compass. The following procedure may be used for drawing both circles and arcs, Fig. 10.9 (b).

(1) Draw an isometric square (rhombus) with sides equal to the diameter of the cirle. The rhombus must always be oriented and drawn in the same isometric plane containing the circular feature. The orientation for drawing the circular hole in the top, front, and right side surface of the object is illustrated in Fig. 10.8 (a) and (b).

(2) Erect perpendicular bisectors to each side from opposite corners of the rhombus.

(3) The two points where the perpendicular bisectors intersect are used as centers to draw the ends of the ellipse using radius r.

(4) The opposite corners of the rhombus used to erect the perpendicular bisectors are now used as centers to draw the sides of the ellipse using radius r_1.

Since the centers of circular features are necessary for their construction and location, the alternate four-center method to draw an ellipse is preferred by many individuals. The construction of the ellipse is very similar to the four-center method and is constructed as follows, Fig. 10.10.

(1) Draw isometric center lines and draw a circle equal to the diameter of the circular feature.

(2) From the points (A, B, C, and D) where the circle intersects the center lines, erect perpendicular bisectors to the two center lines.

(3) The two points where the perpendicular bisectors intersect are used as centers to draw the ends of the ellipse using radius r.

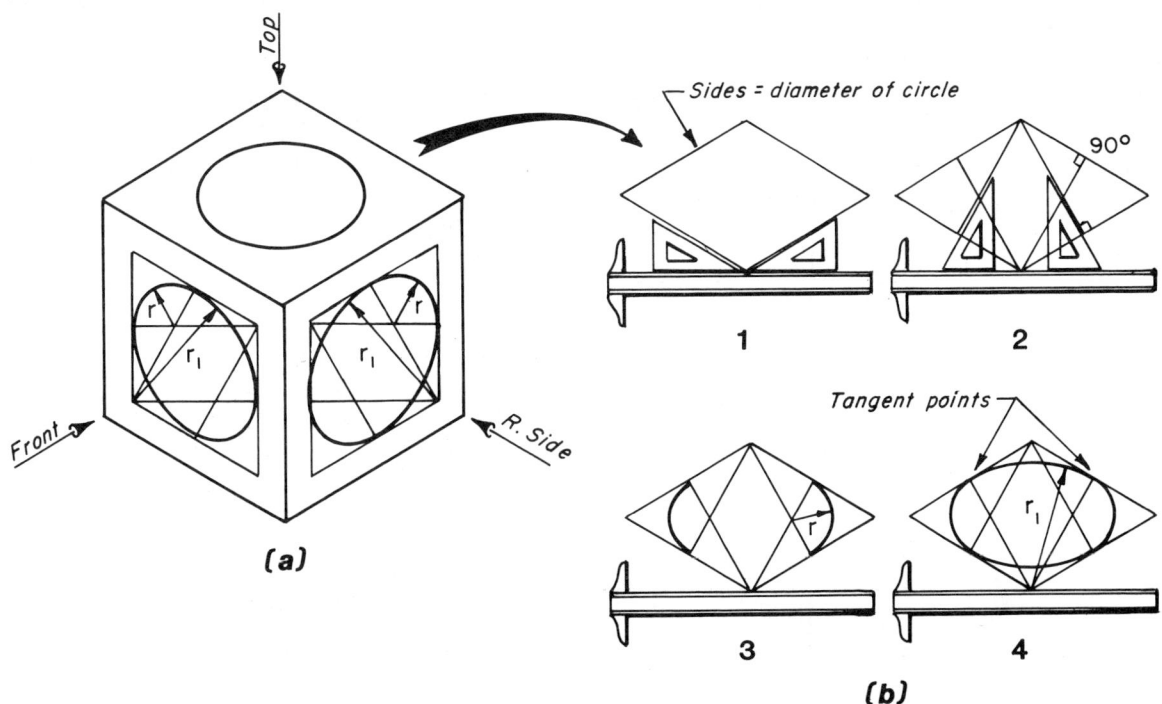

Fig. 10.9 Four-center isometric ellipse construction.

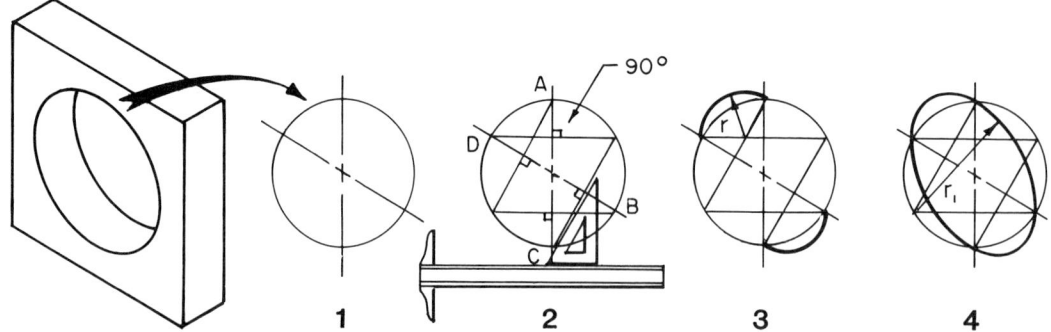

Fig 10.10 Alternate four-center isometric ellipse construction.

(4) The two points on the edge of the circle where the perpendicular bisectors intersect are used as centers to draw the sides of the ellipse using radius r_1.

A curved feature may show in more than one parallel plane in an isometric drawing requiring construction in each plane containing the curved feature to draw the ellipse. Considerable time can be saved by finding the centers to draw the ellipse in one plane and moving the centers to draw the same ellipse in a parallel plane, Fig. 10.11. The arc in the top surface at the right end of the object is first constructed to locate the centers a and b for drawing the ellipse. When drawing an arc, only as much of the four-center method is drawn as necessary to find the centers for drawing the elliptical segment required. Moving the centers a and b vertically, the distance Y representing the height of the right end, locates the centers a' and b' for drawing the ellipse on the lower surface.

The same procedure is used to draw the arc at the left end of the object using the four-center method to find the centers c, d, e, and f for the ellipse. Points c and d are now moved back along an isometric line the distance Z, representing the depth of the curve in the back plane, to locate points c' and d' for drawing the visible portion of the arc.

The construction of an isometric ellipse is time consuming and difficult to draw for small circles and arcs. Where considerable isometric drawing is contemplated, an isometric ellipse template is recommended. Isometric centerlines are drawn and the isometric centerlines on the template, represented by four marks on the periphery of each el-

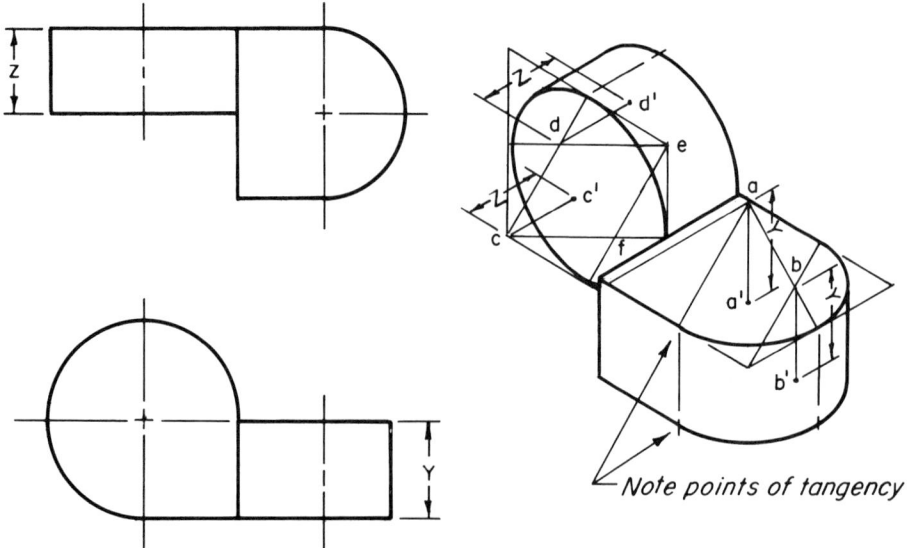

Fig. 10.11 Four-center isometric ellipse.

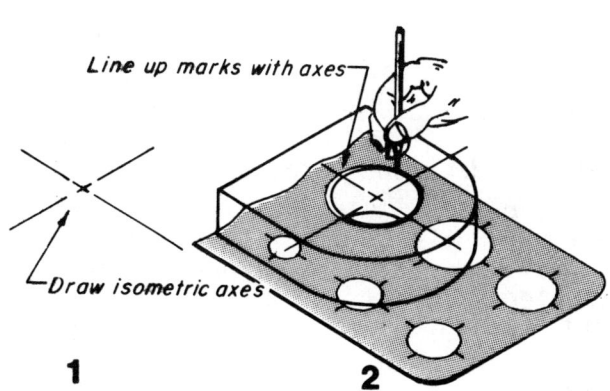

Fig. 10.12 Use of isometric template.

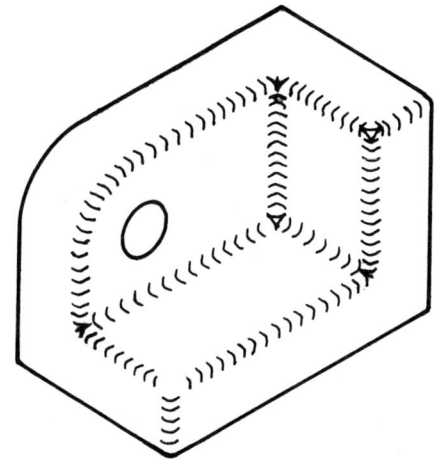

Fig. 10.13 Pictorial fillets and rounds.

lipse, are lined up with the centerlines to draw the part of the isometric circle that is required, Fig. 10.12. The pictorial drawing of small arcs representing fillets and rounds are usually drawn freehand as represented in Fig. 10.13.

10.8 NON-ISOMETRIC CIRCLES AND ARCS

A circle or an arc frequently occurs in a non-isometric plane and may be plotted by the coordinate method. The orthographic views, Fig. 10.14, are first drawn and a series of points are located on the curved surface in both views. All coordinate measurements for each point on the orthographic views are transferred with the dividers to corresponding isometric lines in the pictorial. The procedure for finding the coordinate measurements X, Y, and Z for point 6 on the front inclined surface and point 6' on the back surface are illustrated.

10.9 IRREGULAR CURVES

In isometric drawing irregular curves are drawn using coordinates, Fig. 10.15. Points are randomly located on the curve in the orthographic views and their coordinate values are transferred to the framework of the pictorial. The final curve is drawn with the irregular curve.

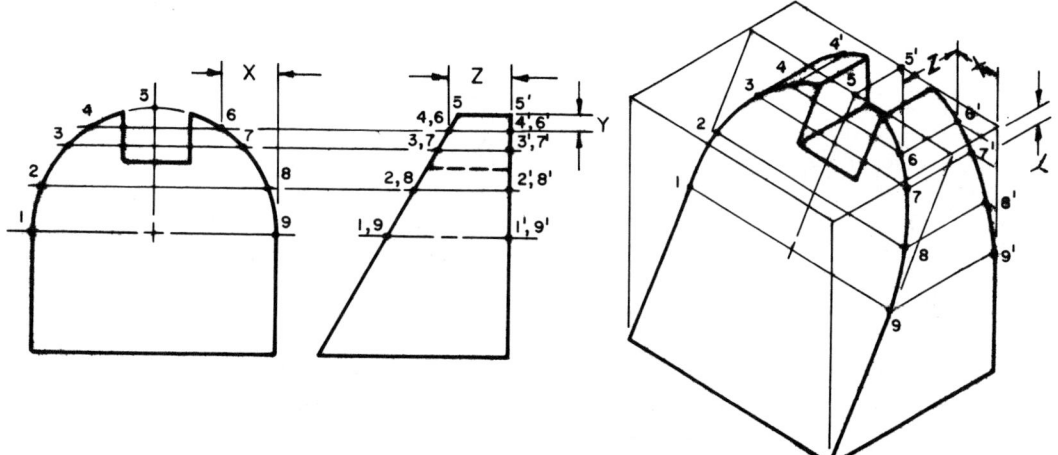

Fig. 10.14 Ellipse construction in a non-isometric plane.

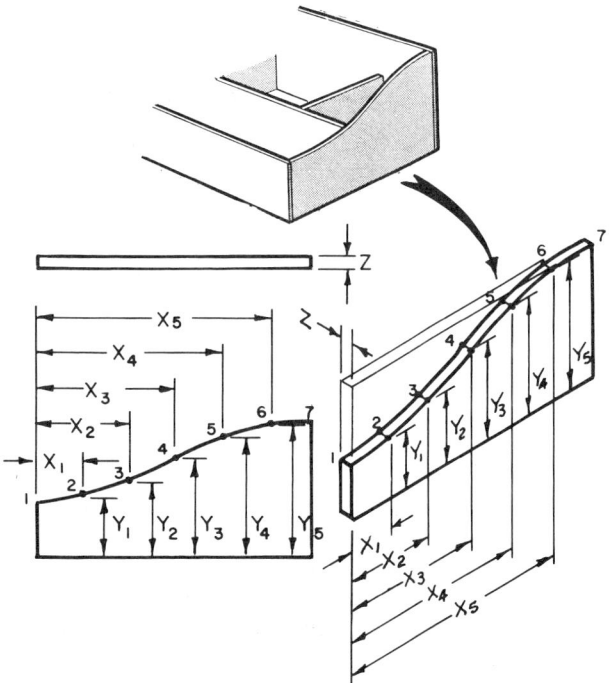

Fig. 10.15 Plotting irregular curves by coordinates in isometric.

10.10 DIMETRIC AND TRIMETRIC PROJECTION

A dimetric or trimetric projection of an object is more difficult to draw than an isometric and both are seldom used. Each type of projection should be recognized, however, and a generalized discussion is presented for comparison purposes.

In *dimetric projection* the object is rotated 45° about a vertical axis and tilted forward. The object may be tilted into any number of positions resulting in two of its axes making equal angles with the picture plane, Fig. 10.16 (a). A dimetric drawing may be constructed similar to the procedure described for isometric drawing, Fig. 10.16 (b). An angle of 15° is commonly used to draw the receding axes. Measurements are laid off along the axes using two scales (full size and 3/4 size) since only two axes are foreshortened equally.

In *trimetric projection* the object is rotated and tilted so that all three axes form unequal angles with the picture plane, Fig. 10.17 (a). The three approximate scale ratios described in Fig. 10.17 (b) may be used to construct a trimetric drawing where the receding axes are drawn at 30° and 45°. Other angles and scale ratios for drawing the axes have also been worked out for both dimetric and trimetric drawing.

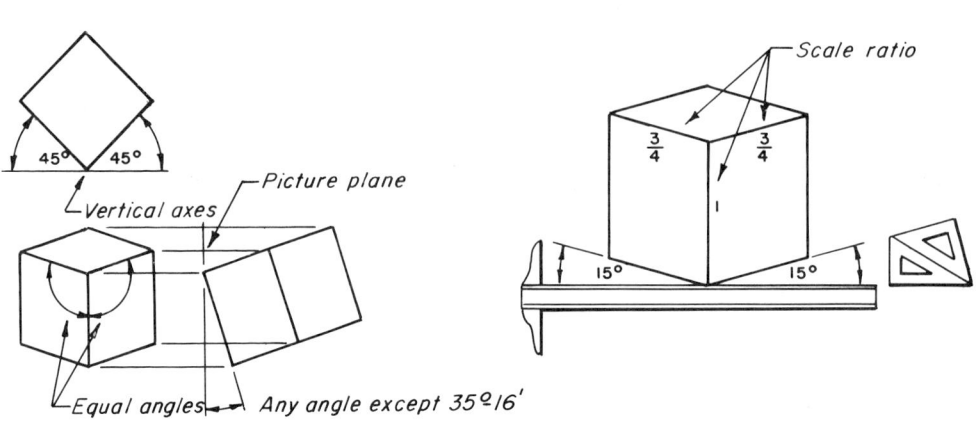

(a) Dimetric projection   **(b) Dimetric drawing**

Fig. 10.16 Dimetric projection and dimetric drawing.

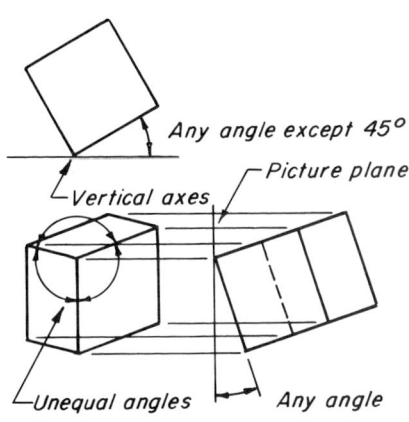

(c Trimetric projection

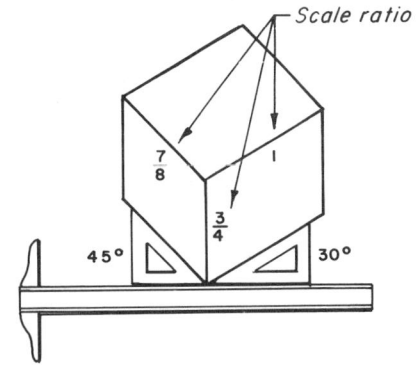

(d) Trimetric drawing

Fig. 10.17 Trimetric projection and trimetric drawing.

10.11 OBLIQUE PROJECTION

In *oblique projection* the projectors remain parallel to each other but make an angle other than 90° with the picture plane, Fig. 10.18. The object may be oriented in any position, but one surface of the object is generally placed parallel to the picture plane. All features in this plane will appear true size and shape as in orthographic projection. This is the main advantage over isometric projection. Since the projectors are not 90° to the picture plane, the top and side receding surfaces are also seen to create the pictorial.

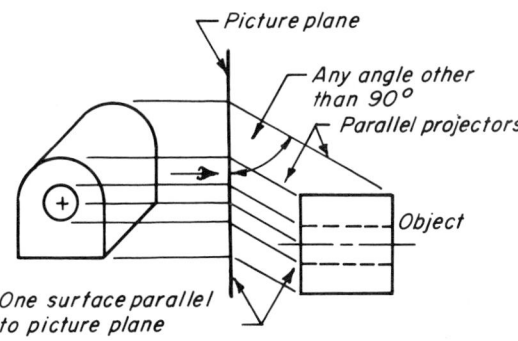

Fig. 10.18 Oblique projection.

Receding lines on the pictorial may project different lengths depending on the angle the projectors make with the picture plane. When the projectors make an angle of 45° with the picture plane, the receding lines will project in their true length, Fig. 10.19 (a). If the projectors make a greater angle than 45° with the picture plane, the lines will be shorter, Fig. 10.19 (b); if the angle is smaller, the lines will be longer, Fig. 10.19 (c).

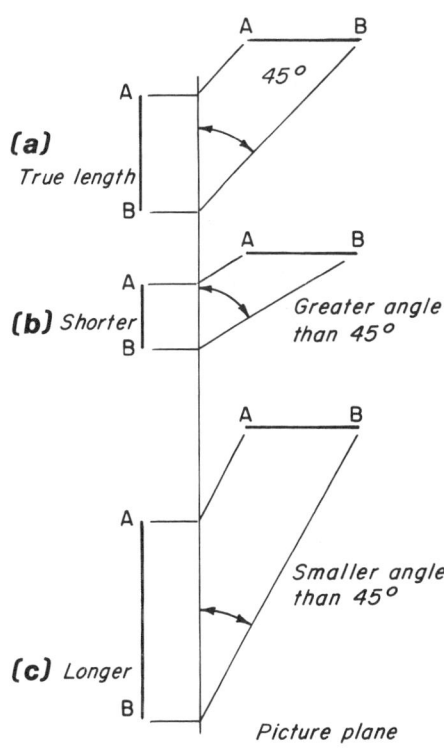

Fig. 10.19 Oblique projection of a line.

10.12 OBLIQUE DRAWING

In *oblique drawing* two of the axes are perpendicular to each other and the third or receding axis is drawn at some convenient angle such as 30°, 45°, or 60° with the horizontal, Fig.

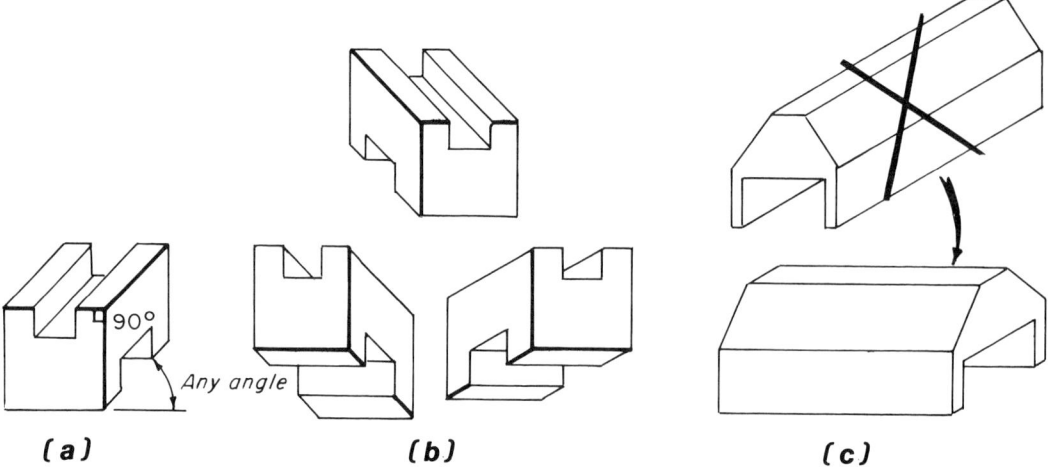

Fig. 10.20 Position of oblique axes.

10.20 (a). The pictorial effect of the drawing is altered by varying the angle of the receding axis. Smaller angles will produce less distortion for objects of great depth and allow the side of the object to be emphasized. Other pictorial effects may be created by varying the position or angle of the receding axes to show different surfaces or features of the object, Fig. 10.20 (b).

To construct an oblique drawing, the object is oriented so that the front contains the circular or irregular contour. Where possible, the longest dimension of the object should also be placed parallel to the picture plane to lessen the effect of distortion, Fig. 10.20 (c). The three axes are drawn and the dimensions are laid off similar to the procedure described for isometric drawing.

Types of oblique drawing. Oblique drawing may be classified into three subtypes, *cavalier*, *cabinet*, and *general*. When the projectors are 45° to the picture plane, depth dimensions will project true length on the picture plane regardless of the angle the receding axis makes with the horizontal. In *cavalier drawing* the receding axis is commonly drawn at 45° with all distances along the receding lines laid off true length, Fig. 10.21 (a). When the measurements for height and width are drawn at the same scale and the receding distances are laid off one-half their true length, the drawing is known as *cabinet*, Fig. 10.21 (b). Cabinet drawings have the advantage of reducing the distortion for objects having considerable depth. Any oblique drawing made by using a scale other than full or one-half for laying off the receding measurements is called *general*, Fig. 10.21 (c).

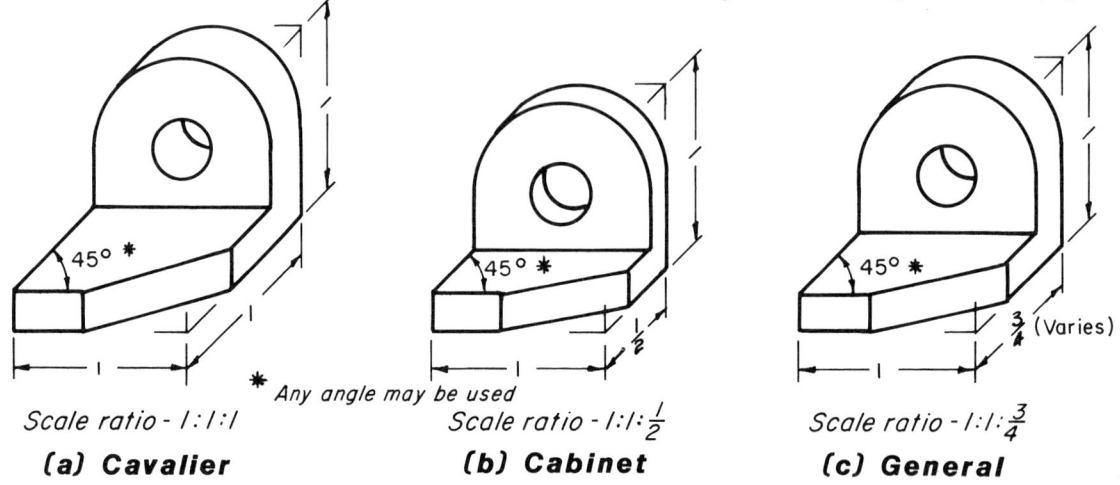

Fig. 10.21 Types of oblique drawings.

10.13 OBLIQUE ANGLES, ARCS, AND IRREGULAR CURVES

All measurements for oblique drawings must be made only along the oblique axes similar to the procedures described for isometric drawing. For angles and irregular curves the coordinate method is used, Fig. 10.22.

Circles and arcs that occur in a receding plane may be drawn using the four-center method for all cavalier drawing. The method of construction is similar as for the ellipse in isometric. Different receding axes may change the shape of the rhombus and the center for drawing the large arcs usually falls outside the rhombus, Fig. 10.23.

The ellipse is constructed by coordinates when a different scale is used to lay off distances along the receding axes or the circular feature is on an inclined plane, Fig. 10.24. A sufficient number of points should be located on the curved feature to plot its location. All points are located using the procedure described for the points illustrated.

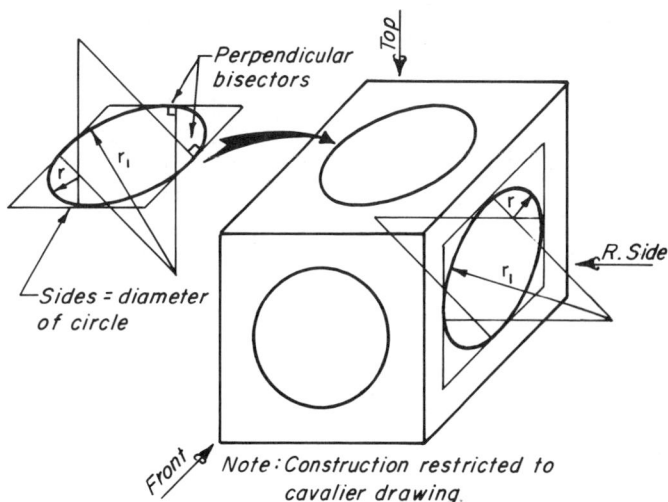

Fig. 10.23 Four-center oblique ellipse construction.

10.14 PICTORIAL SECTIONS

Pictorial drawings are used to reveal clearly the exterior features of an object. Hidden lines are seldom used and interior features of complicated objects are best shown by sectional views. The cutting planes should be isometric or oblique planes and the sectioning principles presented for orthographic sections should be followed. Fig. 10.25 shows several examples of pictorial sectioning for isometric and oblique drawing.

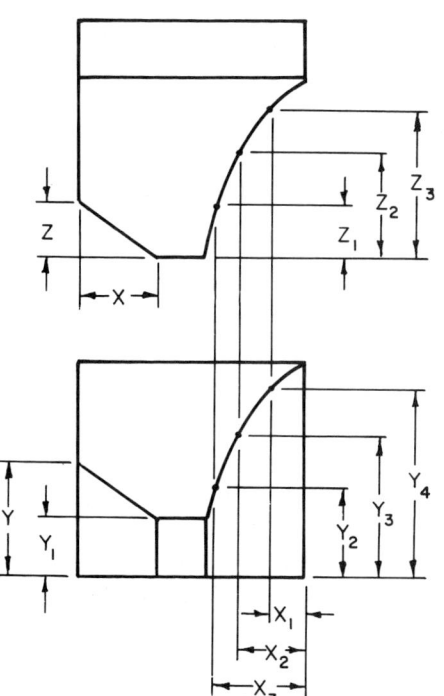
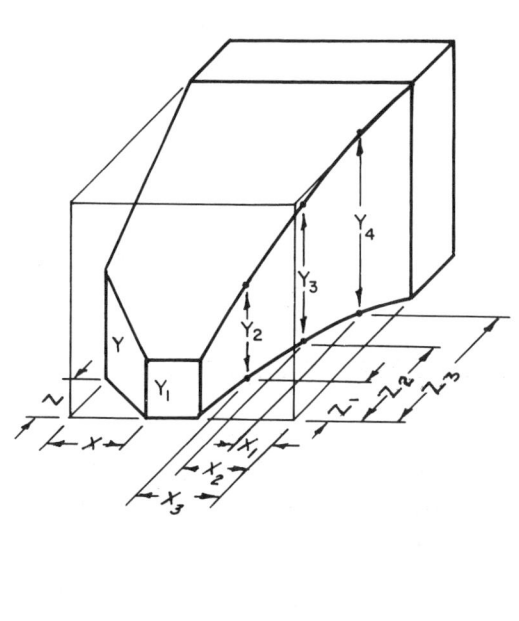

Fig. 10.22 Plotting angles and irregular curves by coordinates.

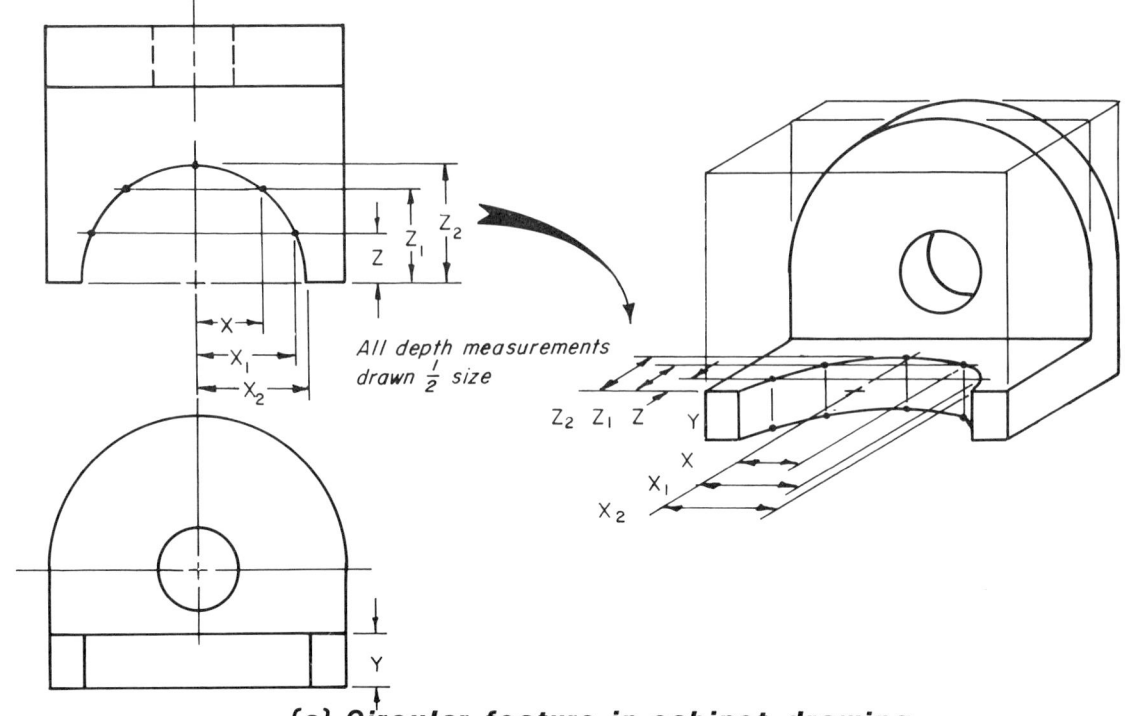

(a) Circular feature in cabinet drawing

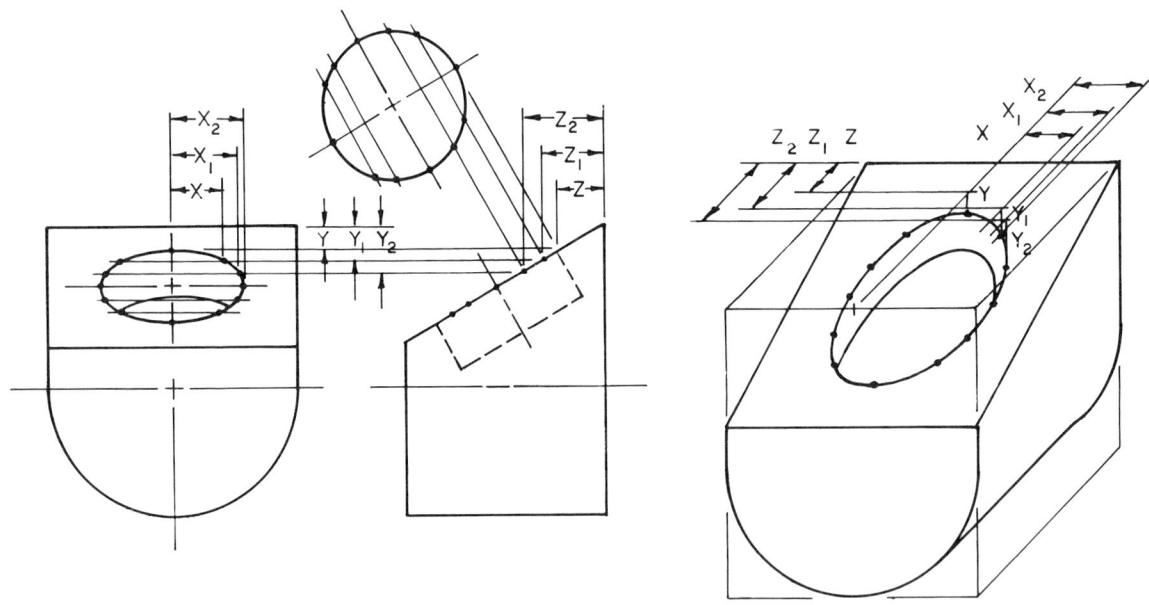

(b) Circular feature on an inclined plane

Fig. 10.24 Construction of curved outlines using coordinates.

10.15 PICTORIAL ASSEMBLIES

Pictorial assembly drawings are used to convey information concerning the relationship of objects composed of several parts. Drawings of this type are very useful for individuals who are not skilled in reading multi-view drawings. Excellent examples of pictorial assemblies are usually found in sales literature, construction details, and product instructions to be used by the consumer. They may be drawn to show the composite placement of individual parts, Fig. 10.26 (a), or in an exploded form showing the parts arranged in the proper order for assembly, Fig. 10.26 (b).

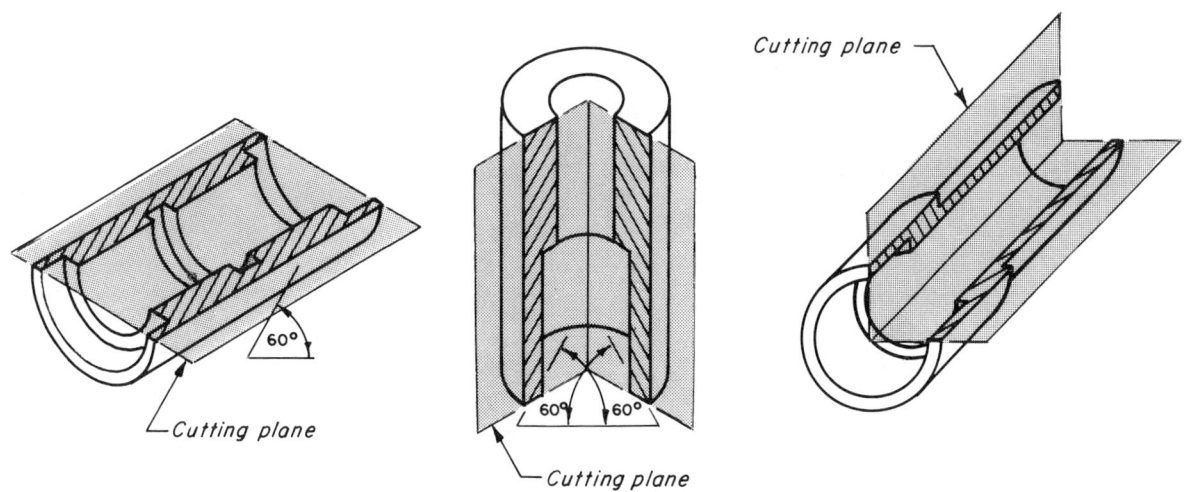

(a) Isometric full section (b) Isometric half section (c) Oblique half section

Fig. 10.25 Pictorial sectioning.

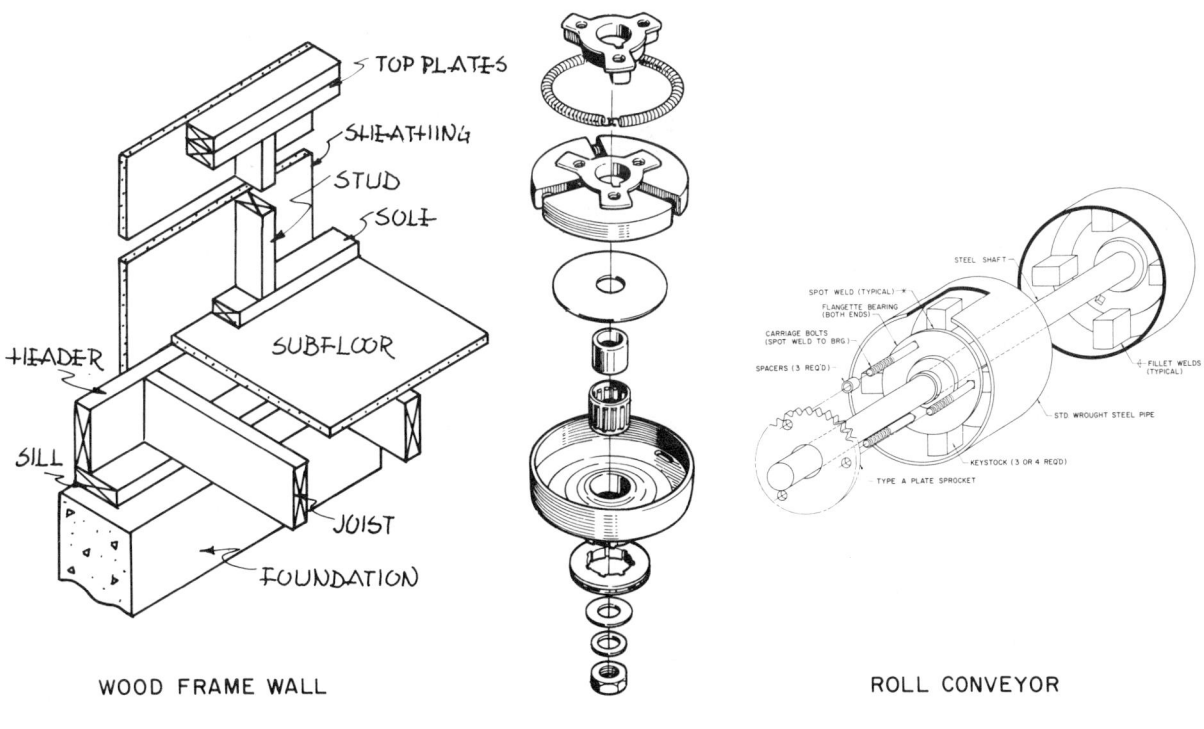

                                Isometric   *Oblique*

(a) Isometric assembly   (b) Exploded assembly

Fig. 10.26 Pictorial assembly drawing.

10.16 PERSPECTIVE

The human eye views everything in perspective when the lens forms an image of the object on the retina. When viewing an object, the projectors converge to a single point, the station point, that is a fixed distance from the object, Fig. 8.1. Photographs of objects are excellent examples of perspective since the camera lens records the image on film as seen by the observer. A drawing in perspective is considered the most effective way to represent the three-dimensional features of an object in a natural realistic way. The true size and shape of features are seldom seen, however, and is a disadvantage of this pictorial type.

Perspective drawings are used widely by architects in preparing preliminary planning sketches, Fig. 10.27, and presentation drawings, Fig. 10.28. Other uses include production and advertising illustrations prepared by professional artists, Fig. 10.29.

Fig. 10.27 Preliminary architectural sketch *(Courtesy Architectural Record)*.

10.17 PERSPECTIVE NOMENCLATURE

Certain terms must be understood before constructing a perspective drawing. In perspective the image is projected onto the picture plane. The points where the projectors pierce the picture plane are connected to form the outline of the perspective drawing. The relationship of the component parts and nomenclature that follows is illustrated in Fig. 10.30.

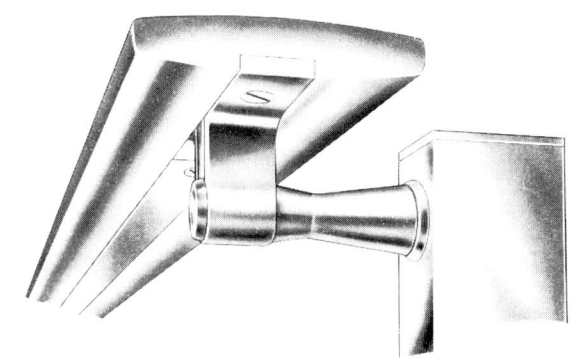

Fig. 10.29 Advertising illustration.

Fig. 10.28 Architectural presentation drawing *(student design)*.

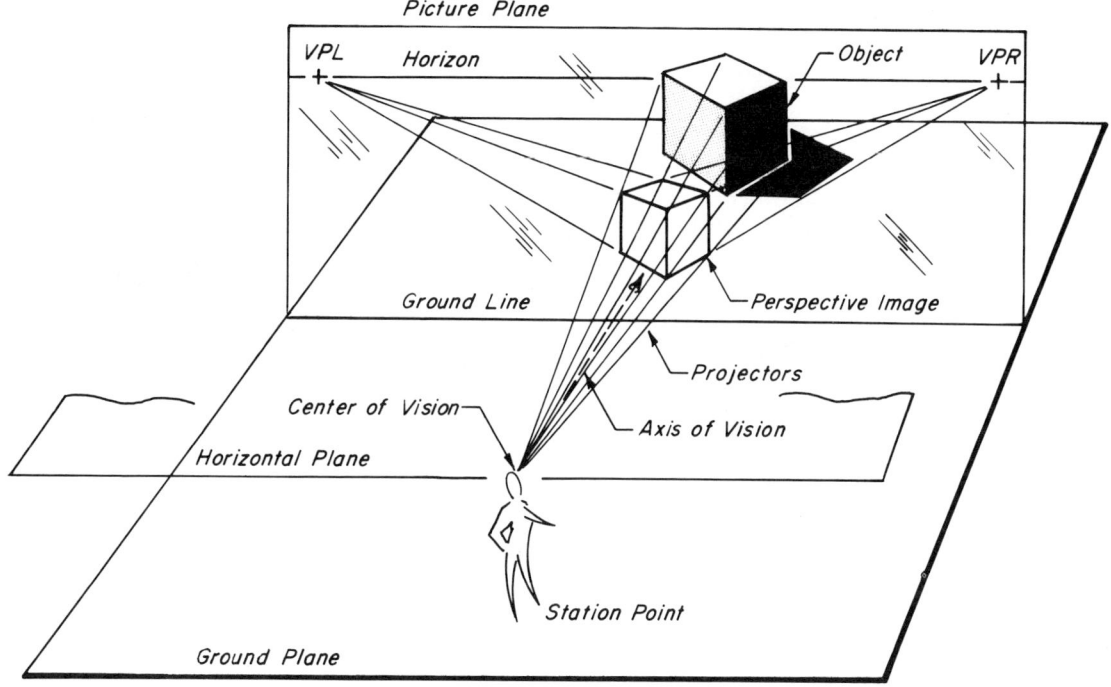

Fig. 10.30 Perspective nomenclature.

Station point: Point at which observer views object.

Center of vision: The line at the eye of the observer lying in a horizontal plane perpendicular to the picture plane.

Picture plane: A vertical plane on which the perspective is drawn.

Ground plane: A horizontal plane representing the surface of the ground.

Ground line: The line of intersection of the ground plane with the picture plane.

Horizontal plane: A horizontal plane through the center of vision.

Horizon: The line of intersection of the horizontal plane with the picture plane.

Vanishing point: The point where all parallel horizontal lines which are not parallel to the picture plane vanish.

10.18 PERSPECTIVE CONSIDERATIONS

The drawing of a perspective requires consideration and understanding of how to locate the different perspective components to obtain a pleasing pictorial.

Picture plane. The picture plane is usually placed between the object and the station point. Its location will affect the size of the perspective. Any lines touching the picture plane will project true length and will touch the ground line on the perspective, Fig. 10.31 (a). One corner of the object is normally placed on the picture plane so that vertical measurements may be made along the line.

Lines on the object behind the picture plane will be foreshortened and will appear above the ground line on the perspective, Fig. 10.31 (b). Lines in front of the picture plane will be longer than true length and will extend below the ground line on the perspective, Fig. 10.31 (c).

Plan angle. The orientation of the object in reference to the picture plane will affect the pictorial qualities that each side of the object will exhibit, Fig. 10.32. One side of the object is usually placed at a 30° or 45° angle with the picture plane. An angle of 30° for the most important

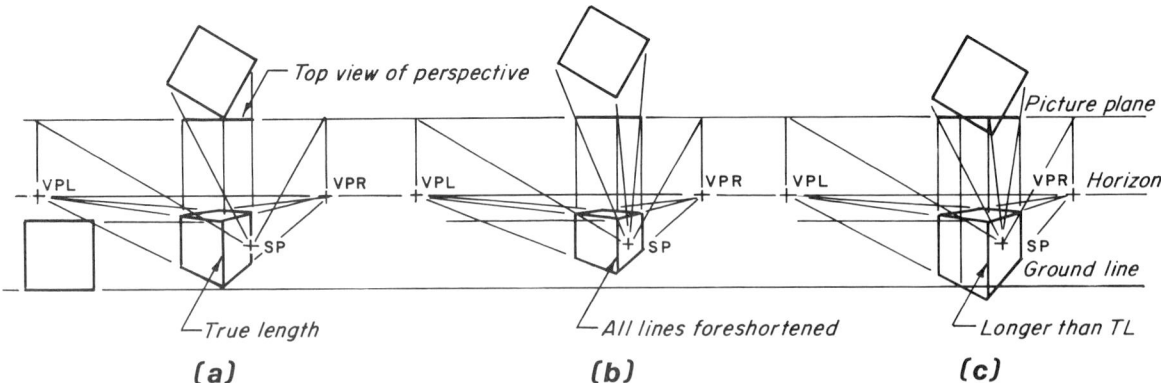

Fig. 10.31 Location of picture plane.

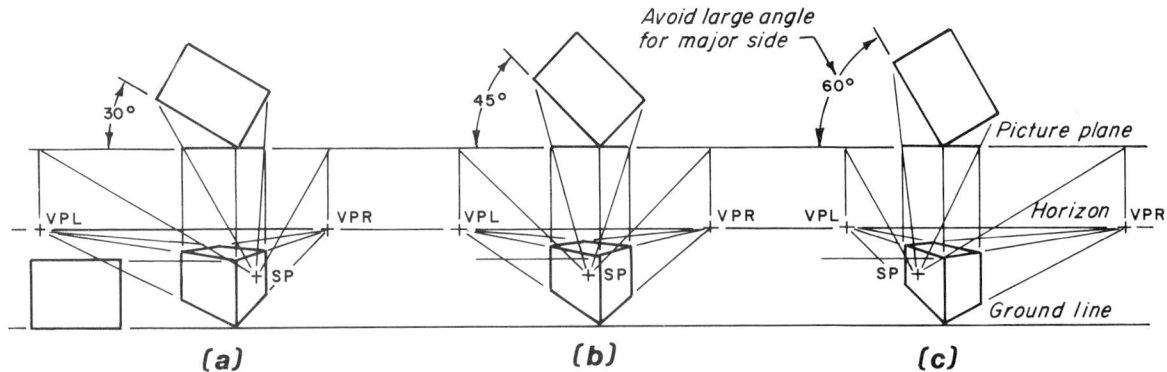

Fig. 10.32 Plan angle of object.

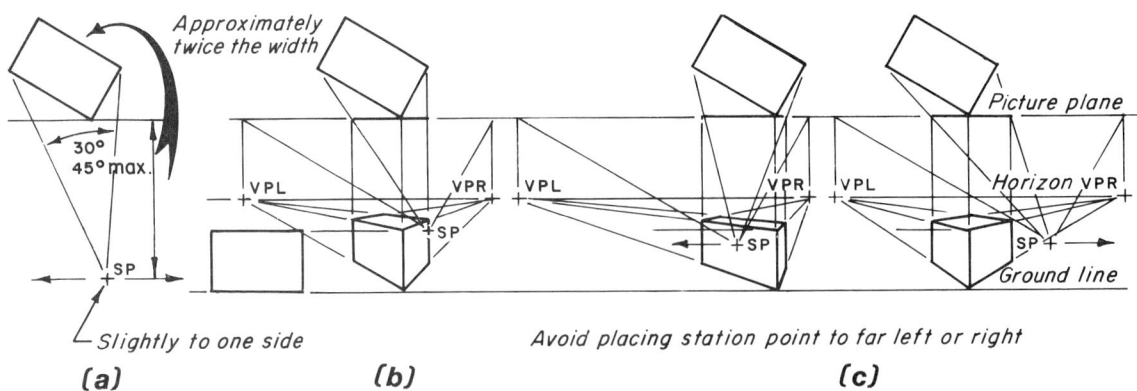

Fig. 10.33 Location of station point.

side of the object works very well and allows the major features to be emphasized. A 45° angle will produce equal emphasis to each of the sides.

Station point. The location of the station point determines the appearance of the final perspective. A location should be selected to produce a pleasing well proportioned drawing. In general, the station point should be placed far enough from the plan view to produce a 30° angle of vision with respect to the extremities of the plan, Fig. 10.33 (a) and (b). Locating the station point a little to one side of the exact center of the object will help to emphasize major features. Placing the station point too far to the right or left will produce considerable distortion, Fig. 10.33 (c). Another rule that is sometimes used is to place the station point twice as far from the picture plane as the width of the object.

The size of the perspective is also affected by the location of the station point. As the distance between the station point and picture plane increases, the perspective will become larger. If the station point is placed too close to the picture plane, the angle of the receding lines in the perspective will be too large, resulting again in distortion.

Horizon. The horizon represents the level of the observer's eye and is generally placed above the ground line at a height that will produce an interesting perspective, Fig. 10.34 (a). For buildings, the horizon is commonly placed above the ground at a height representing the level of a man's eye, about 182 cm (6'-0). Locating the horizon above the elevation view of the object will result in a bird's-eye view, Fig. 10.34 (b). If the horizon is placed below the elevation view, the perspective will appear as if the observer is looking up at the object, Fig. 10.34 (c).

10.19 TYPES OF PERSPECTIVE

There are three general subtypes into which perspectives may be divided: *one-point* or *parallel*, *two-point* or *angular*, and *three-point perspectives*, Fig. 10.35

In *one-point perspective* one surface of the object is parallel to the picture plane and may be drawn in its true size and shape as in orthographic projection. The receding lines all converge at one point on the horizon. This type of perspective has many architectural applications including interior room layouts and street scenes.

The object in *two-point perspective* is rotated so that the principal faces are at an angle with the picture plane. All vertical lines are parallel to each other and the picture plane. The horizontal lines converge at two vanishing points located on the horizon. This is the most popular form of perspective and is used extensively by architects for showing the exterior of buildings and by illustrators for drawing all kinds of products.

A *three-point perspective* is obtained when the object is rotated so that the vertical edges are not parallel to the picture plane. All horizontal lines converge at two vanishing points and all vertical lines will converge at a third vanishing point. Three-point perspective is difficult to draw and is seldom used due to the extreme distortion which appears at the top and bottom of objects.

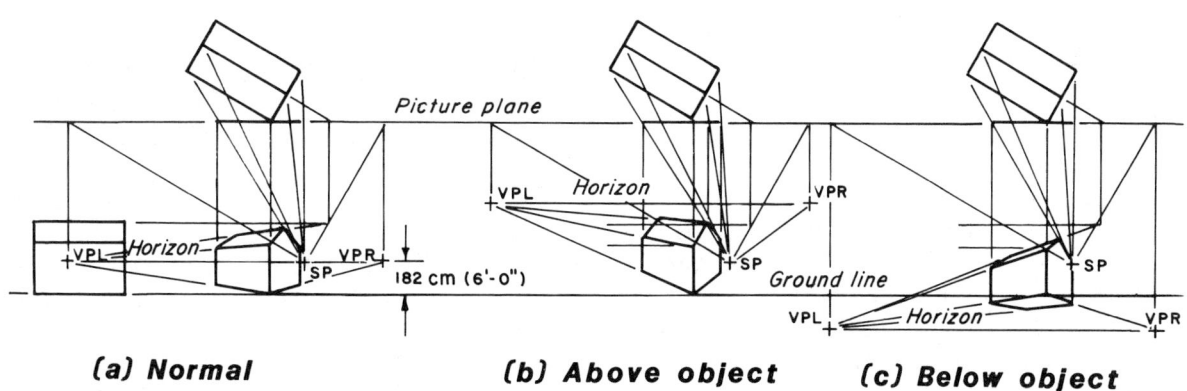

(a) Normal   *(b) Above object*   *(c) Below object*

Fig. 10.34 Location of horizon.

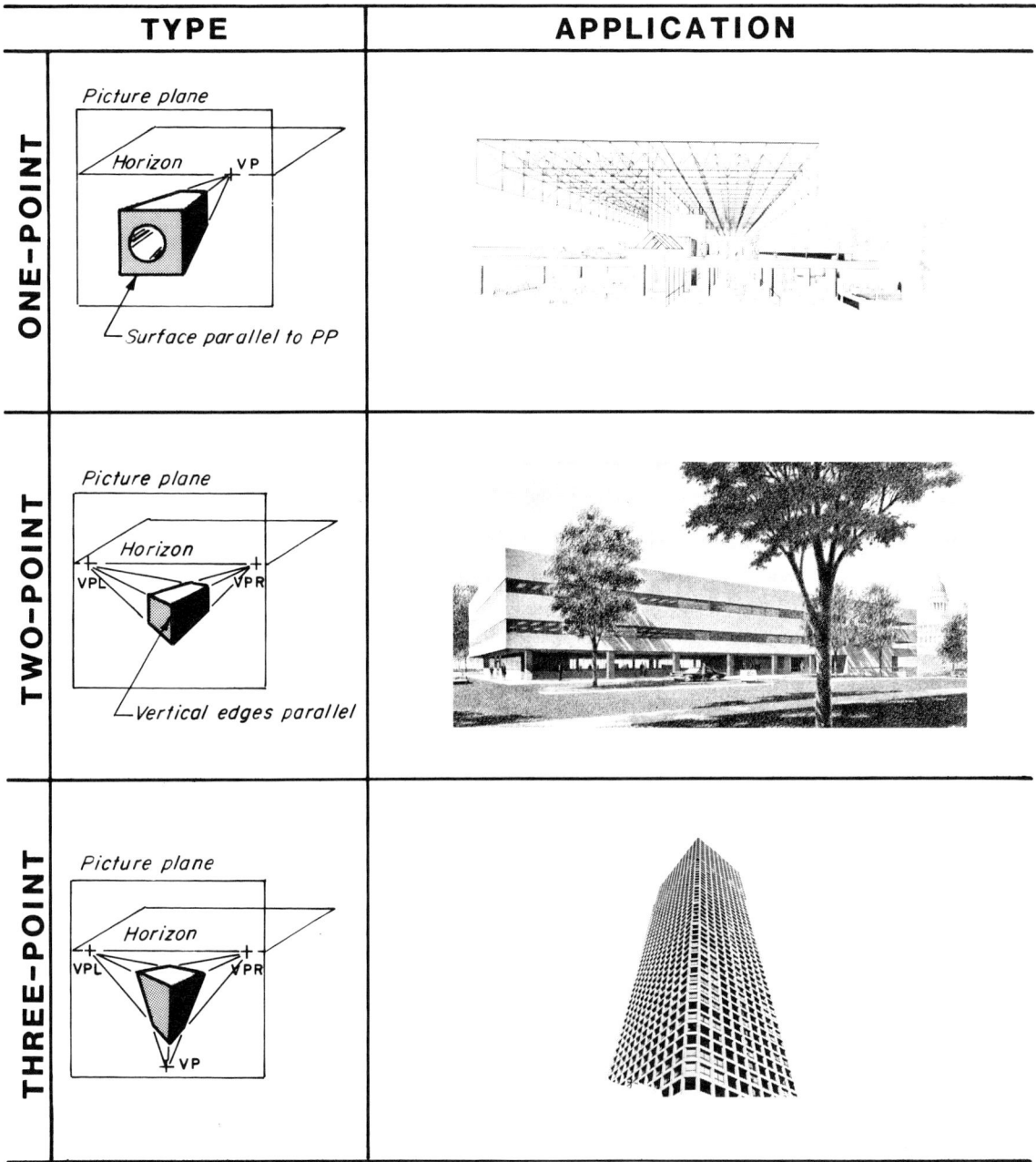

Fig. 10.35 Types of perspective (Pictures courtesy Architectural Record).

10.20 CONSTRUCTION OF A PERSPECTIVE

All of the basic concepts and variables previously discussed must be understood thoroughly before proceeding to set-up and draw a perspective.

Perspective of a line. The construction of a horizontal line in perspective, Fig. 10.36, is basic to drawing the perspective of any object.

After the location of the perspective components is decided, the piercing point of line AB in the picture plane is found, Fig. 10.36 (a), by extending the line until it pierces the picture plane in the top view. The front view of the piercing point is found by projecting vertically to the height or elevation of line AB.

The vanishing point of line AB is found by drawing a line from the station point parallel to line AB until it pierces the picture plane, Fig. 10.36 (b). Since the vanishing point must always lie on the horizon, this

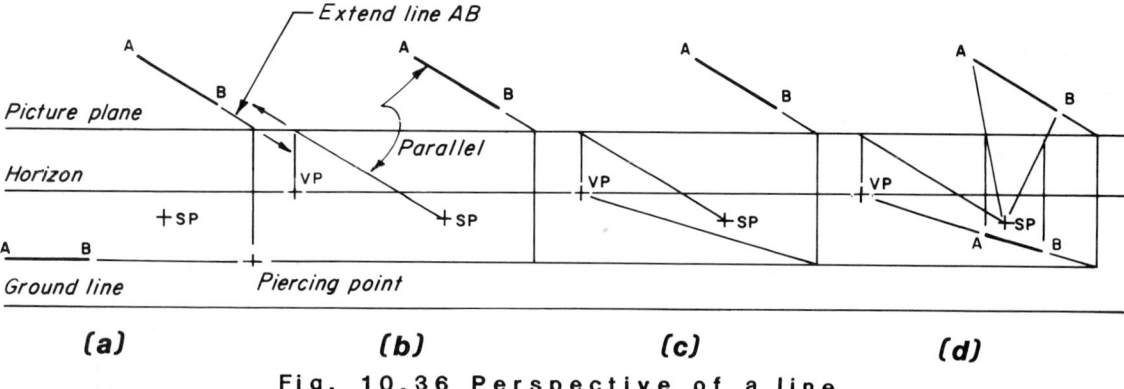

Fig. 10.36 Perspective of a line.

point is projected vertically to the horizon line.

In Fig. 10.36 (c) a straight line is drawn from the piercing point to the vanishing point. This line represents a perspective line of infinite length on which the perspective of the line segment AB is found.

The final perspective of line AB is found by locating the end points of the line, Fig. 10.36 (d). A visual ray is drawn from the station point to points A and B in the top view to locate the piercing points in the picture plane. Both points are projected vertically to draw the perspective view.

One-point perspective. In drawing a one-point perspective the top and front views of the object are drawn, Fig. 10.37. The front view is placed to one side to allow space for drawing the perspective view. One surface of the object is commonly placed on the picture plane so it will appear true size in the perspective. All other lines which do not lie on the picture plane are found by drawing visual rays from the station point to the end points of the line. Vertical projection lines are drawn from the piercing points of these lines in the picture plane to obtain the final perspective.

The true height of lines that are not in the picture plane such as line AB are found using a measuring line. Since the corner A'B' is on the picture plane, this true height may be laid off. A perspective line is drawn from A' and B' to the vanishing point and the top view of AB in the picture plane is projected vertically to locate AB in the perspective.

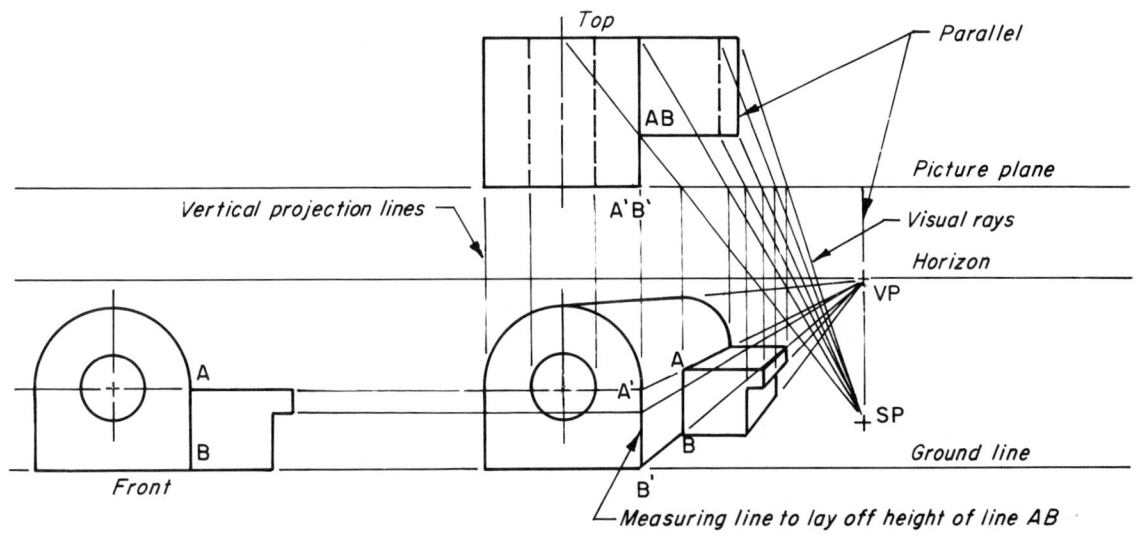

Fig. 10.37 Construction of a one-point perspective.

Two-point perspective. The construction in Fig. 10.38 illustrates how an object may be drawn in two-point perspective. The two vanishing points are formed by drawing a line from the station point parallel to the sides of the top view. Vertical projection lines are dropped from the piercing points of these lines in the picture plane to the horizon to locate the left (VPL) and right (VPR) vanishing points. To draw the perspective it is necessary to find the perspective of some line, e.g., line AB as previously explained. The perspective of all horizontal lines representing the width and depth of the object must recede to one of the vanishing points. One corner of the object is located in the picture plane and will appear true length in the perspective. This true length line may be used as a measuring line to project vertical measurements from the front view. Heights are then projected around the object to locate their true positions. All horizontal measurements are established on the picture plane and then projected vertically to the perspective.

The procedure for drawing the perspective of all remaining lines is repeated as previously explained.

10.21 BLOCK DIAGRAMS

Landforms may be drawn in pictorial form and are referred to as *block diagrams*, Fig. 10.39. Relief models of this type showing a block cut out of the earth's crust are used to illustrate mining geology problems, land use, land types, and other terrain relationships. The three-dimensional aspects of the drawing may be improved by shading since this helps to emphasize slopes, small details, and depth perception. Color may be used to supplement shading and is excellent for calling attention to relationships such as hydrology, vegetation, and geological sections. Other information that should be on the final drawing of the block diagram includes a title, a scale, the amount of vertical exaggeration, and an orientation diagram. A location diagram or small inset map is help-

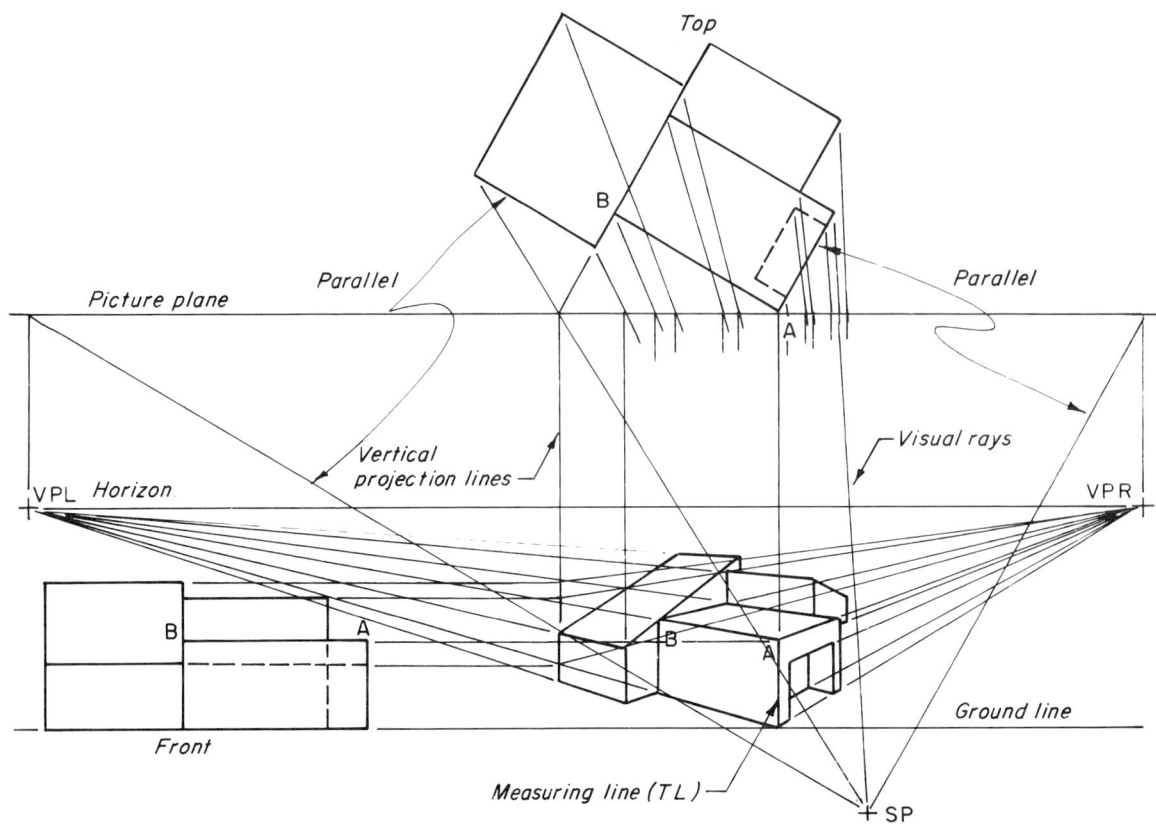

Fig. 10.38 Construction of a two-point perspective.

Fig. 10.39 Block diagram.

ful to establish where the block diagram is in relation to a larger area. Block diagrams are relatively easy to draw either in isometric or oblique using layers or sections.

Layer method. A block diagram in isometric may be drawn in layers from a contour map. The area to be drawn should be enclosed in a rectangle and subdivided into squares, Fig. 10.40 (a). The same plan area is drawn in isometric square by square, Fig. 10.40 (b). A careful analysis will reveal the best way to orient the drawing. In general, the highest land should be in the background. The contour range on the map is from 430 to 510. Using a horizontal scale of 1:25 000 for horizontal measurements and a vertical scale of 1:3000 to layoff height, an isometric box is drawn on tracing paper to enclose the area, Fig. 10.40 (c). A vertical scale for laying off contour heights that will provide a good impression of relief should be used. Vertical elevations are labeled and light horizontal lines are drawn around the box for each elevation. The tracing of the isometric box is placed over the isometric plan view of the map and moved vertically until the highest contour is in the same plane as indicated by the vertical scale. The contour is traced and the box is moved vertically again one contour interval at a time until all contours have been traced. Where a lower contour cuts a higher contour, the line is not continued. The edges of the block are formed by joining the ends of the contours. Roads, streams, lakes, and other details are drawn to fit the relief, Fig. 10.40 (d).

Profile method. The area to be represented is enclosed in a rectangle and subdivided into squares, Fig. 10.41 (a). All terrain should be analyzed to determine the best way to orient the features to be represented on the block diagram. The same grid represented on the map is drawn in oblique with vertical lines erected at each grid corner, Fig. 10.41 (b). A suitable vertical scale that will portray the relief is chosen to lay off the elevation of the grid corners on the surface of the ground along the vertical lines. The vertical scale may be laid off on a strip of paper and moved from line to line to quickly lay off the vertical distances. Interpolation will be necessary since most map contours do not fall on the grid corners. After all grid elevations have been plotted, a series of profiles are drawn through the points, Fig. 10.41 (c). Depending on the relief, some profiles may be hidden and should be omitted. The profiles are used as a guide to shape and shade in the map features, Fig. 10.41 (d).

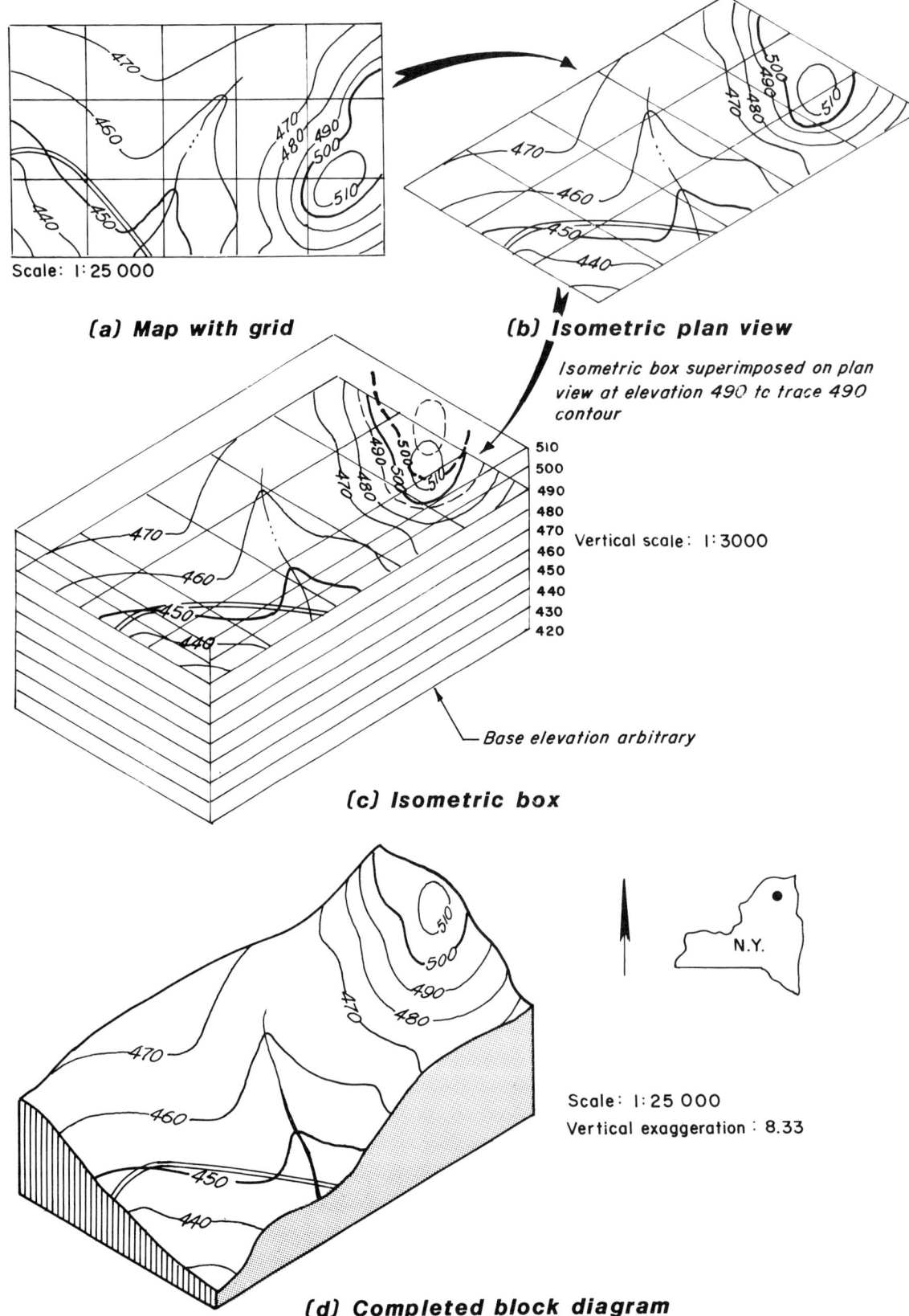

Fig. 10.40 Isometric block diagram using layer method.

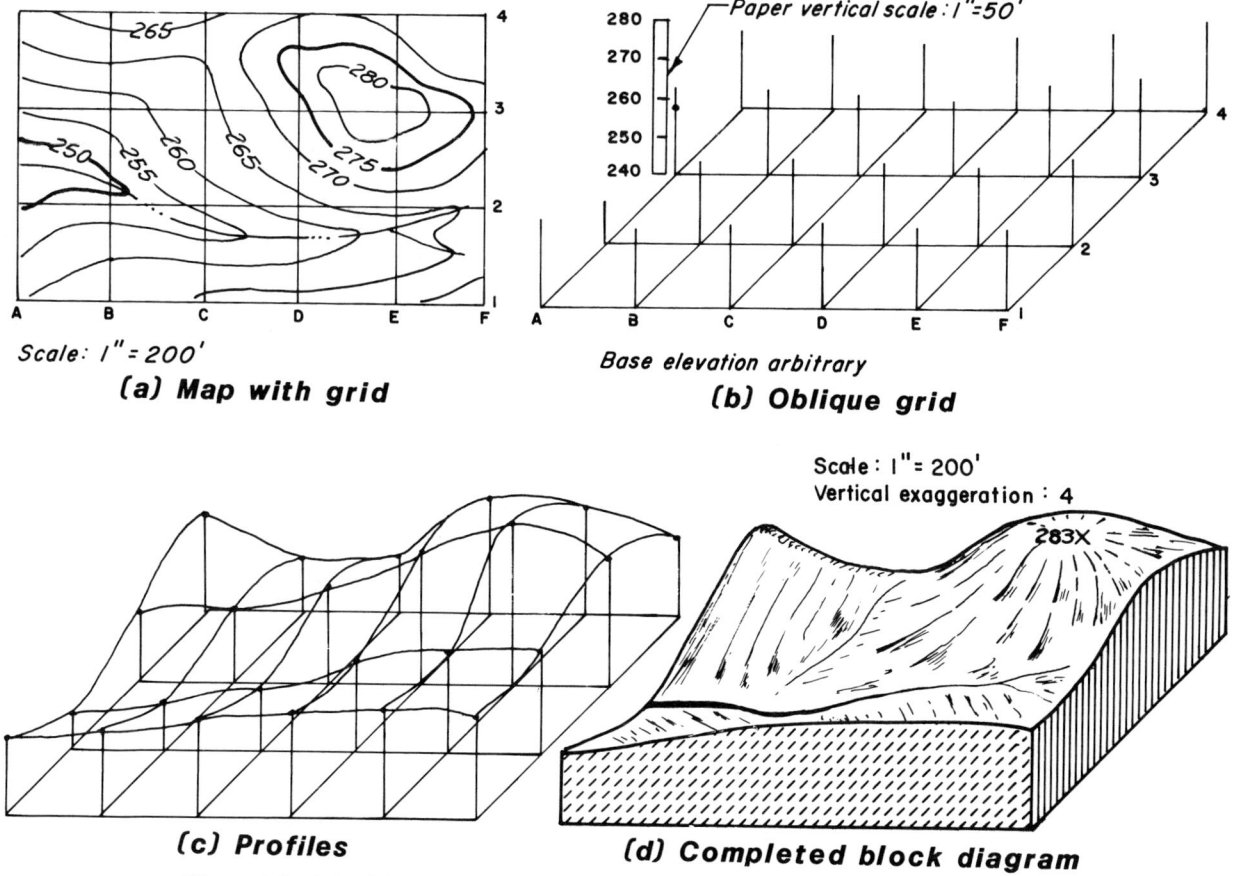

Fig. 10.41 Oblique block diagram using profile method.

SKETCHING

11.1 INTRODUCTION

The ability to make clear, concise sketches is important for individuals to record their ideas on paper. For many purposes a sketch will be satisfactory with a real advantage realized in time saved and cost loss. A sketch is helpful to plan and organize mental thoughts in graphic form, and at the same time enable one to see how additional improvements or alternate solutions may be made. Design problems requiring several solutions should be in the form of preliminary sketches to aid analysis before a final design and drawing is prepared. Sketches in the form of diagrams, exploded views, perspectives, isometrics, and other forms of graphic representation are widely used to communicate ideas. In cartographic work it is often necessary to be able to make sketches as part of the field work in gathering and recording information. These serve as valuable graphic records to further clarify the tabulated numerical data and explanatory notes for boundary outlines, relative locations, topographic features, etc. The sketch more than anything else will help to convey correct impressions that will allow anyone to interpret the data. Several examples of sketches are shown in Fig. 11.1 to indicate their uses.

The principles and concepts of graphic representation have been presented and discussed in the preceding chapters. Although there are no standards for sketching, the information discussed in the following sections should conform to these principles. A knowledge of this information is necessary for both instrument drawing and sketching.

11.2 MATERIALS

To make a freehand sketch, all that is required is paper, pencil, and eraser. A good quality paper with a surface texture that will allow a dense black line to be drawn should be used. Tracing paper is popular for sketching since modifications may be made by placing the transparent paper over previous sketches to eliminate repetitious construction. Commercial paper with light background grid lines of different kinds is of considerable help to sketch an object to an approximate scale. It is important to always maintain the proportions of the object being sketched. Where the drawing is to be photo-reproduced, it is suggested that the sketch be drawn at least twice as large as it will appear in final form. This will sharpen all linework when reduced.

Pencil leads for sketching are normally softer than those used for instrument drawing. An HB lead or a common No. 2 pencil may be used.

A sketch is usually thought of as being freehand so as to save valuable time. Any device or technique that saves time should be considered, however. Due to the difficulty of drawing long lines, circles, and other kinds of repetitive symbols, some instruments may be worthwhile to use. It may be easier and faster for many individuals to use a straight-edge, compass, or some type of symbol template when sketching.

11.3 SKETCHING TECHNIQUES

The aim of any drawing is to keep it neat. By following a few basic rules a neat sketch may be achieved with practice.

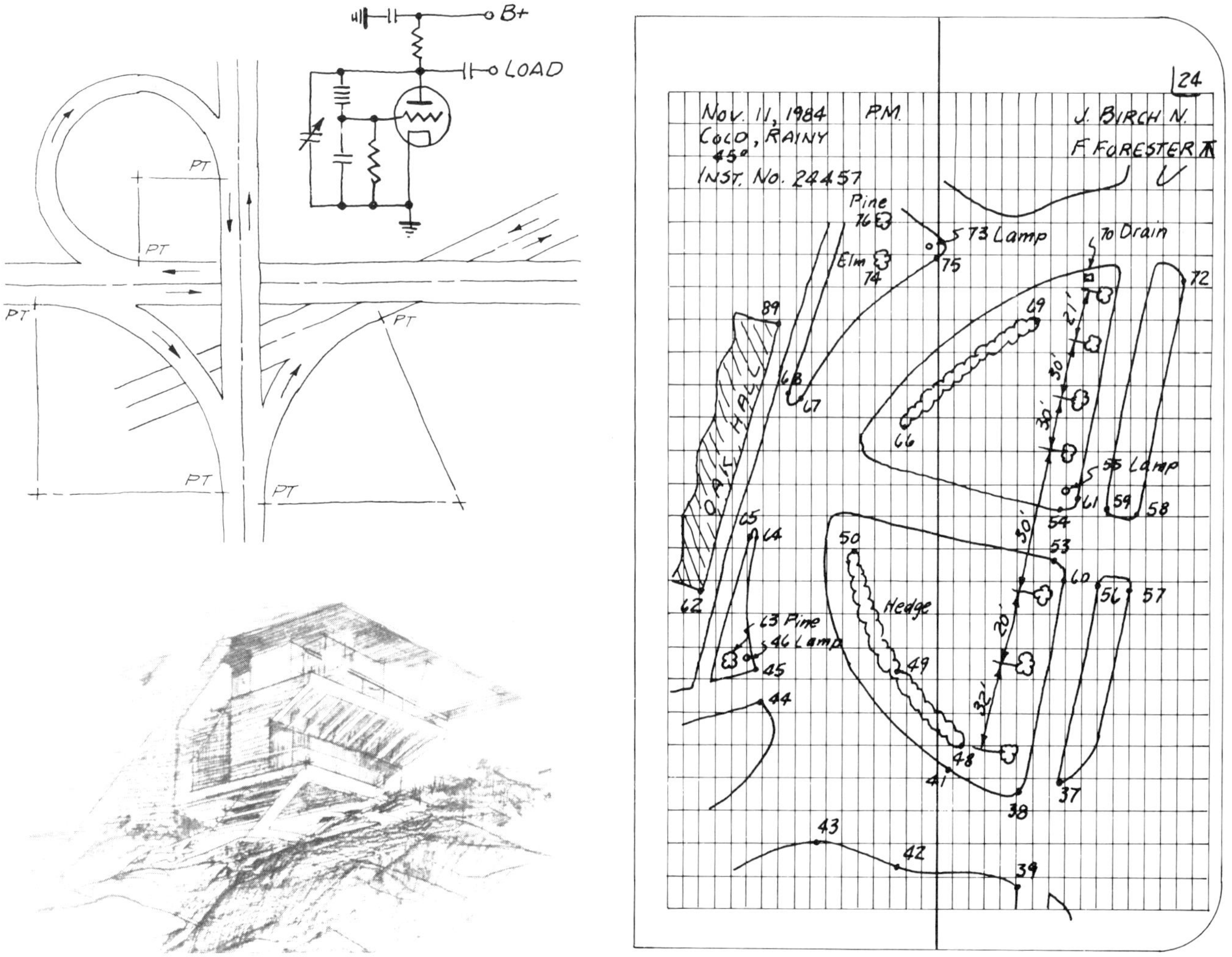

Fig. 11.1 Types of sketches.

Straight lines. The lines in sketching will not be as smooth and uniform as those drawn with instruments. Straight lines are commonly drawn with a series of short strokes, instead of trying to draw each line with one stroke. By using short strokes, better control is possible in maintaining the direction of the line. One way to sketch lines is to locate each end with a small dot and then to draw the line between the two points, Fig. 11.2. Others may find it easier to place the pencil on one of the two points, look to the other point, and then to draw a single smooth line. It is sometimes helpful to rotate the paper for drawing vertical or slanted lines.

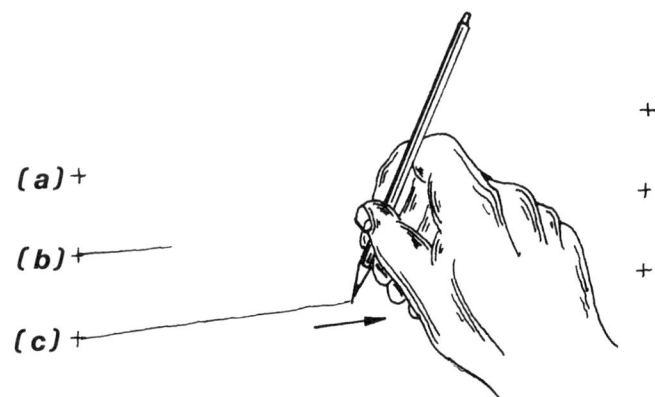

Fig. 11.2 Sketching lines.

Curved lines. Sketches frequently include many circles or arcs. For large circles or arcs the progressive steps shown in Fig. 11.3 should be followed. The centerlines are sketched and a square whose sides equal the diameter of the circle is drawn. Diagonal lines are drawn across the square with the radius of the circle estimated and lightly marked on the diagonals. The circle or arc is then sketched through these points and the tangent points along the sides of the square. The completed circle should be well rounded and not flattened or elliptical in appearance.

In sketching an ellipse, Fig. 11.4, the centerlines are drawn and the major and minor diameters marked. A rectangle is sketched through the two diameters. The small arcs form the ends of the ellipse and the larger or flatter arcs the sides. It is important to maintain symmetry about the centerlines when the ellipse is completed.

The ellipse in isometric, Fig. 11.5, is sketched by drawing centerlines with an enclosing parallelogram or rhombus whose sides equal the diameter of the circle. Where the corner angles of the rhombus are acute, the ends of the ellipse are sketched using a small radius arc. The sides of the ellipse are relatively flat requiring a larger radius arc since the two remaining corner angles of the rhombus are obtuse.

Lettering. Lettering is required on sketches to communicate information that is not contained on the drawing. Good, clear, precise lettering is necessary for that professional look. A

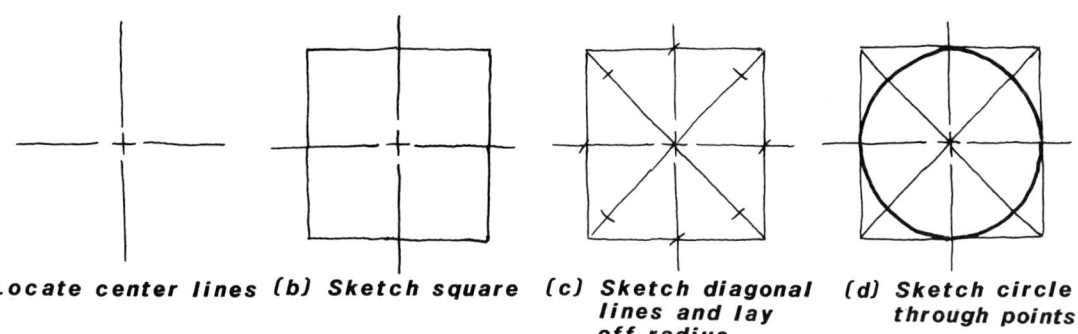

(a) Locate center lines (b) Sketch square (c) Sketch diagonal lines and lay off radius (d) Sketch circle through points

Fig. 11.3 Sketching circles or arcs.

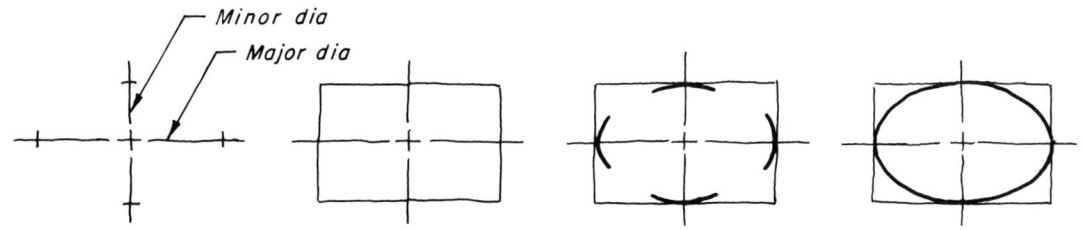

(a) Locate center lines, major, and minor diameters (b) Sketch rectangle (c) Sketch end and side arcs (d) Complete ellipse

Fig. 11.4 Sketching the ellipse.

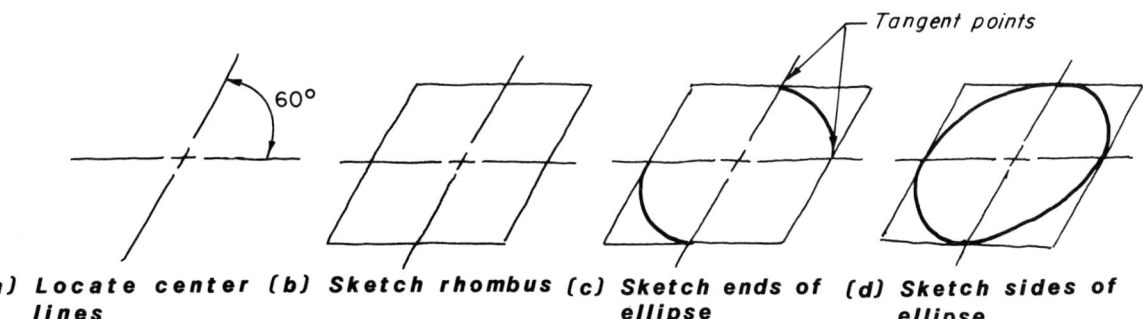

(a) Locate center lines (b) Sketch rhombus (c) Sketch ends of ellipse (d) Sketch sides of ellipse

Fig. 11.5 Sketching an isometric ellipse.

single stroke Gothic style of lettering should be used. Personal preference will dictate whether to use vertical or inclined letters. Guide lines are not required, but care should be taken to keep the letters of uniform size. The important idea is to remember the shape and form of the letters. Letters should be about as wide as they are high and closely spaced together in words.

Proportion. The completed sketch should portray the object as it actually looks. Without good proportions the sketch may be misleading and of little value. Good sketching is again easily accomplished by following a few simple basic rules or principles. The parts of an object should first be studied to see how the different shapes relate to each other. The overall dimensions of the object should be blocked in next, Fig. 11.6 (a), rather than starting at one corner and then attempting to draw detail by detail. Major outlines of the object are now sketched, Fig. 11.6 (b), and centerlines established for circular parts, Fig. 11.6 (c). The sketch is completed by darkening in the details, Fig. 11.6 (d). All of the component parts should maintain their identity. Lines that are perpendicular or parallel must remain this way.

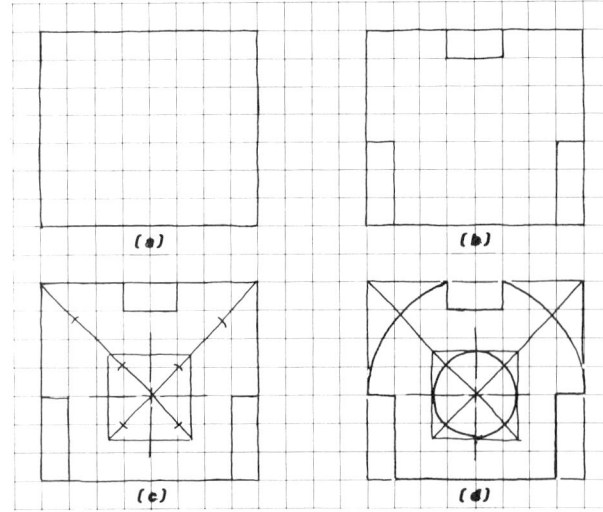

Fig. 11.6 Blocking in a sketch.

It is advisable to sketch lightly at first. Depending upon the object, and as construction lines are drawn to develop the drawing, it will soon become clear which lines are to be darkened in to make the object stand out. There is no need to erase the construction lines.

The technique of blocking in a drawing may be used for preparing multiview and pictorial sketches. Examples of these are shown in Figures 11.7 through 11.9.

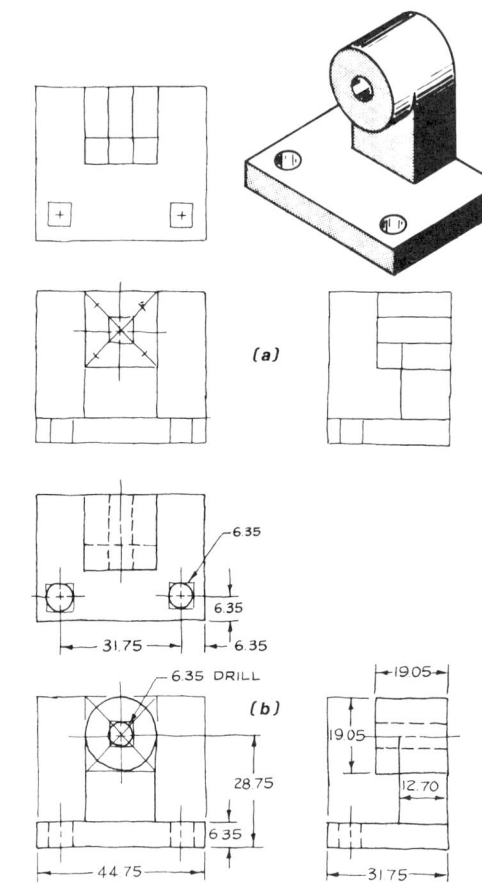

Fig. 11.7 Multiview sketching.

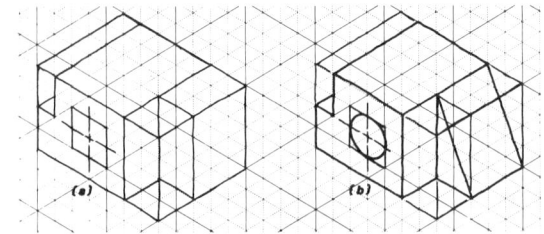

Fig. 11.8 Isometric sketching.

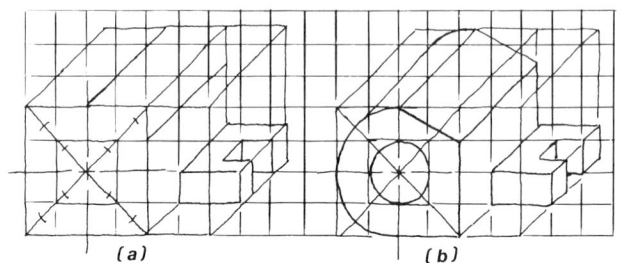

Fig. 11.9 Oblique sketching.

Map sketches. Sketches are useful when maps are not available or existing maps are not adequate. The purpose, the scale, the accuracy required, and the amount of detail to be shown on the map will influence how a map is to be sketched. Additional equipment such as an alidade, sketching board, compass, and protractor are helpful tools for sketching in the field.

A few suggestions to keep in mind are: neatness and legibility, orienting the map with north at the top, if possible, and sketching (using conventional symbols) of the large, or major features first followed by the detail.

All features should be lettered clearly along with any explanatory notes that are pertinent. A lot of valuable information may be conveyed by lettering items carefully like elevations, feature names, etc., Fig. 11.10.

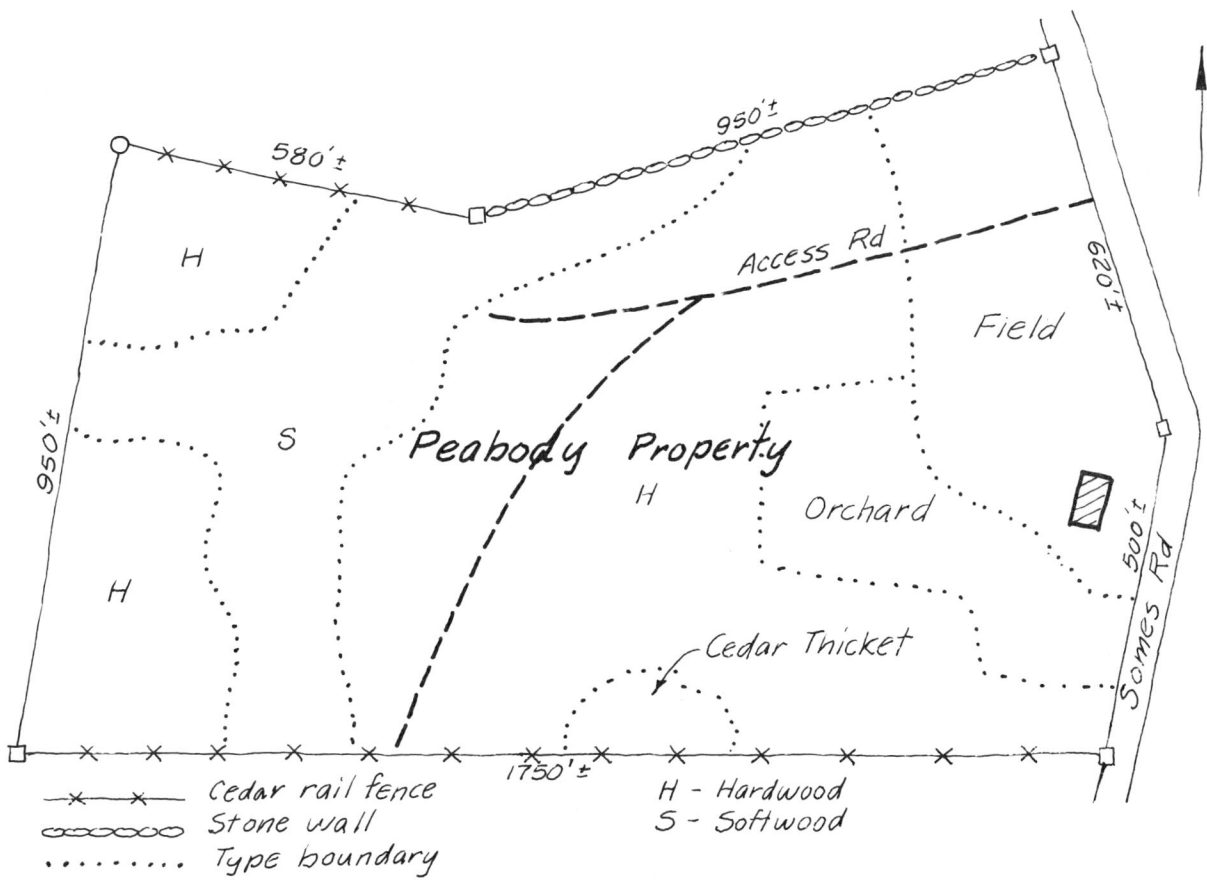

Fig. 11.10 Map sketch.

CONVENTIONS AND SYMBOLS

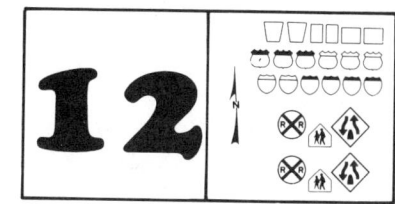

12.1 INTRODUCTION

All forms of drawings require the use of graphic symbols to communicate information quickly, accurately, and consistently. Architects, engineers, and cartographers have all devised symbols to express the kinds of information commonly involved in these disciplines. Map drawing requires that one learn a new language and the study of *semiotics*, the science of signs and symbols, is of major concern to cartographers. Disciplines like architecture and engineering use well developed and adopted standard symobls. While this is true to a certain extent in mapping, it is common to find different agencies adopting and using many kinds of graphic symbols. Confusion is often the result of this practice. Although many conventional signs and symbols are used, the development of a group of universally accepted mapping symbols would be helpful. If all symbols were standard and if everyone using a map knew the symbols, there would be no need to define them on each map constructed. Since this is not the case, every map contains a legend showing the symbols used and what they represent. Obvious features which are self-explanatory may be omitted from the legend when well known symbols are employed.

Most mapping agencies adopt their own symbols of representing various symbols or adhere to standards set forth in work orders. Several excellent examples of agencies conventionalizing symbols are the Defense Mapping Agency and the United State Geological Survey. Some common symbols that may be valuable for reference are shown in Fig. 12.1. In addition to these symbols, the legend of published maps should be checked carefully for other signs and symbols as well as the colors used in producing maps.

Since it would be a difficult task to develop a universal set of map symbols to meet all kinds of map requirements, a few guidelines should be adhered to in designing symbols suitable for good visual communication.

12.2 TYPES OF SYMBOLS

The types of symbols used on maps may be classified as being *point*, *line*, and *areal* in shape or form, Fig. 12.2. The actual appearance of the symbol on a map determines its form. Point symbols are most commonly portrayed by using geometric shapes. Linear features are represented by lines of different types, solid, dashed, dotted, etc. Other symbols are pictographic or physiographical in form.

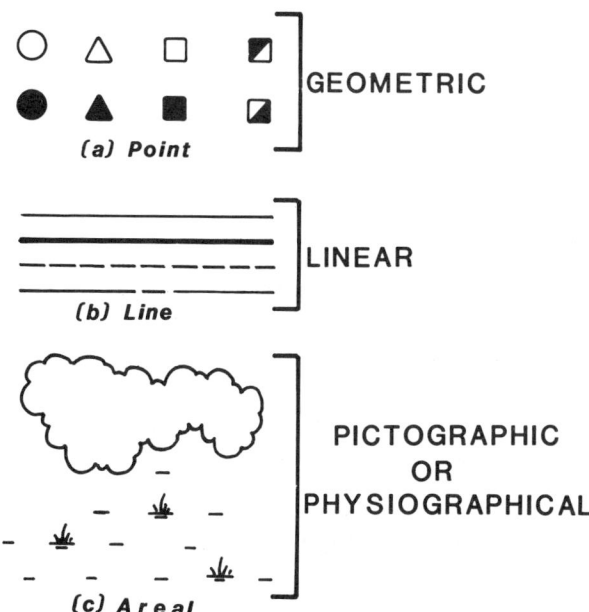

Fig 12.2 Types of symbols.

Hard surface, heavy-duty road	════════
Hard surface, medium-duty road	▬ ▬ ▬ ▬
Improved light-duty road	════════
Unimproved dirt road	=========
Trail	- - - - - -
Railroad: single track	—+—+—+—
Railroad: multiple track	—+++—+++—
Bridge	—+ +—
Drawbridge	—+ o +—
Tunnel	—+)====(+—
Footbridge	- - + + - -
Overpass—Underpass	† ‖ ‖ †
Power transmission line	•—•—•—•
Telephone line, landmark line (labeled as to type)	TELEPHONE
Dam with lock	
Canal with lock	
Large dam	
Small dam: masonry — earth	
Buildings (dwelling, place of employment, etc.)	▪ ▦ ▨
School—Church—Cemeteries	⊡ † [Cem]
Buildings (barn, warehouse, etc.)	▫ ▨ ▨
Tanks; oil, water, etc. (labeled only if water)	• • ⊙ Water Tank
Wells other than water (labeled as to type)	o Oil o Gas
U.S. mineral or location monument — Prospect	▲ X
Quarry — Gravel pit	✕ ✕
Shaft—Tunnel entrance	▪ Y
Campsite — Picnic area	⅄ ⋏
Located or landmark object—Windmill	o ⚘
Exposed wreck	⌒⌒⌒
Rock or coral reef	⌒⌒⌒
Foreshore flat	
Rock: bare or awash	* ⊙
Horizontal control station	△
Vertical control station	×672
Road fork — Section corner with elevation	⨯429 +58
Checked spot elevation	× 5970
Unchecked spot elevation	× 5970

Boundary: national	— — —
State	— — —
county, parish, municipio	— — —
civil township, precinct, town, barrio	— — —
incorporated city, village, town, hamlet	— — —
reservation, National or State	—·—·—
small park, cemetery, airport, etc.	- - - -
land grant	—··—··—
Township or range line, U.S. land survey	───────
Section line, U.S. land survey	───────
Township line, not U.S. land survey	·········
Section line, not U.S. land survey	
Fence line or field line	- - - -
Section corner: found—indicated	+ +
Boundary monument: land grant—other	▫ ▫
Index contour	⌒	Intermediate contour	⌒
Supplementary cont.	⌐	Depression contours	⊙
Cut — Fill		Levee	
Mine dump		Large wash	
Dune area		Tailings pond	
Sand area		Distorted surface	
Tailings		Gravel beach	
Glacier		Intermittent streams	
Perennial streams		Aqueduct tunnel	→====→
Water well—Spring	o ⌒	Falls	
Rapids		Intermittent lake	
Channel	======	Small wash	
Sounding—Depth curve	10	Marsh (swamp)	
Dry lake		Inundation area	- - - -
Woodland		Mangrove	
Submerged marsh		Scrub	
Orchard		Wooded marsh	
Vineyard		Bldg. omission area	

Fig. 12.1 Map symbols of the U.S Geological Survey (Variations may be found on older maps).

12.3 SYMBOL CONSIDERATIONS

Symbols may be varied or emphasized by several different methods that include *line thickness*, *size*, *shape* or *form*, *pattern*, and *color*, Fig. 12.3.

Line thickness and color. Lineweights used for all symbols should always be of sufficient thickness to insure legibility when reduced. For large or major features the use of bold lines should be considered. Maps drawn using different line weights or color add interest and make the drawing easier to read. The discussion of line thickness and color has been taken up in chapters 2 and 5.

Size and shape. Good judgement must be used in determining the size and shape to draw symbols. For maps drawn at a scale of 1:250 (1" = 20') or larger, it is common practice to draw features to scale. On maps drawn at a scale of 1:5000 (1" = 400') or smaller, symbols of different kinds are used to represent the map features. In general, the size of symbols should be kept in proportion to the drawing or varied with the scale of the map on which they are drawn. The size may be varied to indicate importance or to distinguish between large and small for quantitative map information.

The shape of symbols used on maps should always be distinct to avoid confusion and are basically *geometric*, *linear*, or *pictographic* in form. Pictographic symbols resemble the feature they are to depict, although they are frequently generalized. A pictograph kind of symbol that looks like the feature and reveals its content is ideal. The form of these symbols take on the appearance of objects as they would appear from above or the side, Fig. 12.4 (a). Individuals may easily relate and recognize this form of symbol through association. Symbols that are abstract or made up of multiple elements are more difficult to recognize and should be avoided.

Another very common pictograph symbol is functional in form, Fig. 12.4 (b). Symbols of this type may be quickly associated with a particular task or purpose.

Geometric shaped symbols like circles, triangles, and squares are more commonly used to represent features due to their ease of drawing and the ability to represent features at any map scale. Since many geometric shapes are similar, with only slight differences, they must be drawn carefully so as to be identified quickly. A very small circle cannot be easily distinguished from a very small square and will have to be drawn larger, Fig. 12.5.

Other related or similar symbols that are positioned close together may be differentiated easier than when spread apart, Fig. 12.6. Two adjacent lines having a slight difference in thickness are more easily perceived than when they are far apart. The same criteria applies to variations in tints and shades.

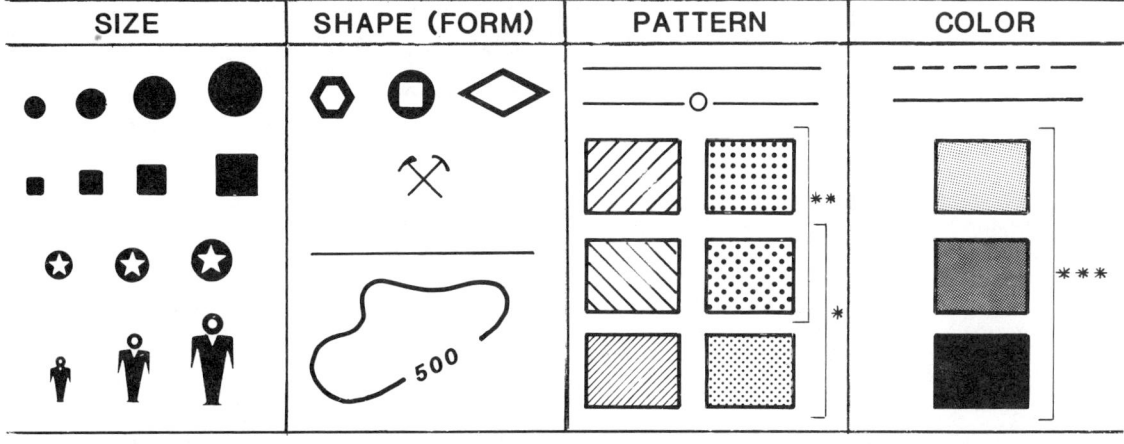

* *Note texture*
** *Note orientation*
*** *Note color value*

Fig. 12.3 Symbol variations.

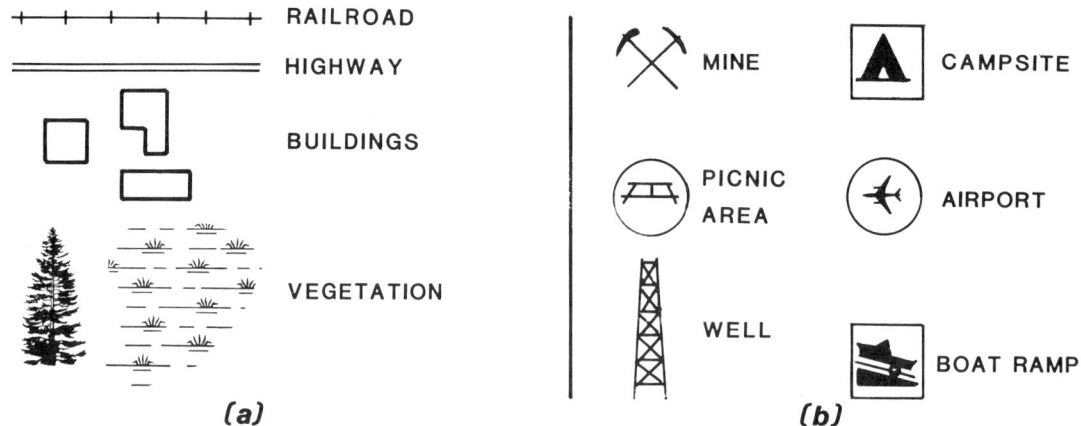

Fig. 12.4 Pictographic type symbols.

Fig. 12.5 Symbol size.

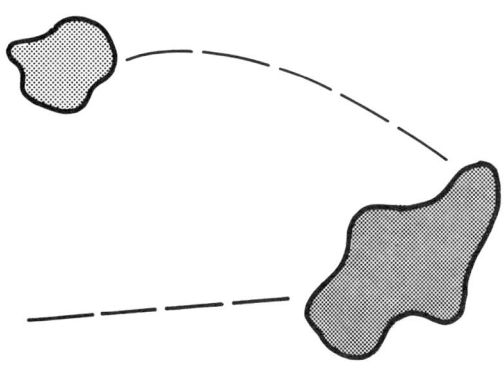

(a) Symbols difficult to distinguish

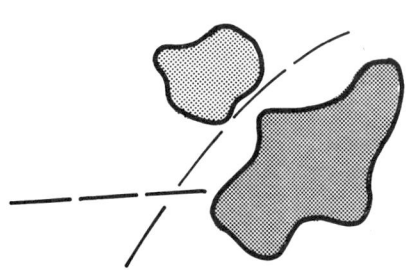

(b) Symbols easier to distinguish

Fig. 12.6 Symbol placement and perception.

If they are placed far apart on the map, it may be difficult to determine whether they are the same or different.

Pattern. The symbols used on a map should above all other considerations be clear, easily understood, and visually appealing. Shading of the symbols will add dominance and allow emphasis to be given to different features or to clarify a confusing situation, Fig. 12.7. Bold kinds of symbols should be used for major or important features and those with less shading for minor features.

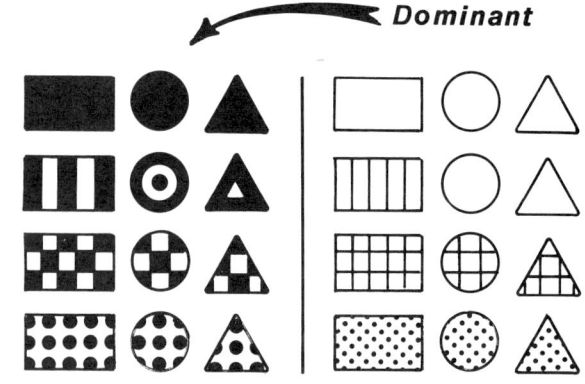

Fig. 12.7 Shading techniques for symbols.

Appliqué shading materials are widely used to fill in an area or symbol. The designs used must be selected carefully and should not clash. Symbols that vary in density like dot patterns or line patterns of different thickness and direction are very distinctive, Fig. 12.8. Patterns that are adjacent to each other should be considerably different from each other, in line thickness and density, to avoid creating an irritating appearance.

Fig. 12.8 Distinctive shading mediums.

Areas having definite boundaries may be shaded using a solid tint or color. Compact shading materials may also be used with the boundary line omitted, Fig. 12.9 (a). An area having no well defined boundary, like a marsh, should be shaded using open type symbols, Fig. 12.9 (b).

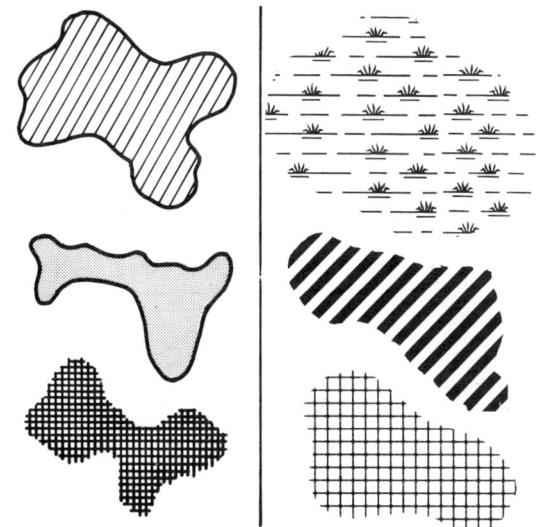

(a) Boundary delineated (b) Boundary not delineated

Fig. 12.9 Boundary shading.

12.4 THEMATIC MAPS

Thematic maps of various kinds are used to show the geographical distribution of a wide variety of statistical data. A good outline map is generally made with boundary and division lines of the geographical units. Other detail such as transportation and major water features may also be helpful. The objectives of the map will determine the amount of detail that will be necessary or desirable. In any event, sufficient base data must be presented to aid the user in obtaining the relationship presented. Topics related to the distribution of information like meterological statistics, populations, and vegetation may be indicated on the outline map by several methods that include choropleths, isopleths, or the use of symbols and shading patterns to indicate both qualitative and quantitative data. A properly designed map using any one of these methods should again be clear, interesting, and instantly understood.

Choropleths. Statistical maps of all types use *choropleths* to depict average values per unit of area such as population density and land cultivation - percent. Various topics are delineated by boundary lines to distinguish density of distribution by using different line - dot patterns, color or shading, Fig. 12.10. Appliqué shading patterns or colors that are both harmonious and distinctive are excellent materials for use on maps of this kind. The number of patterns used should normally be limited to 8-10 to insure that different areas may be readily identified. The density of shading should be increased for areas of increasing density values. Pattern symbols of different densities or lines of different thickness, direction, and spacing will achieve this result. Care

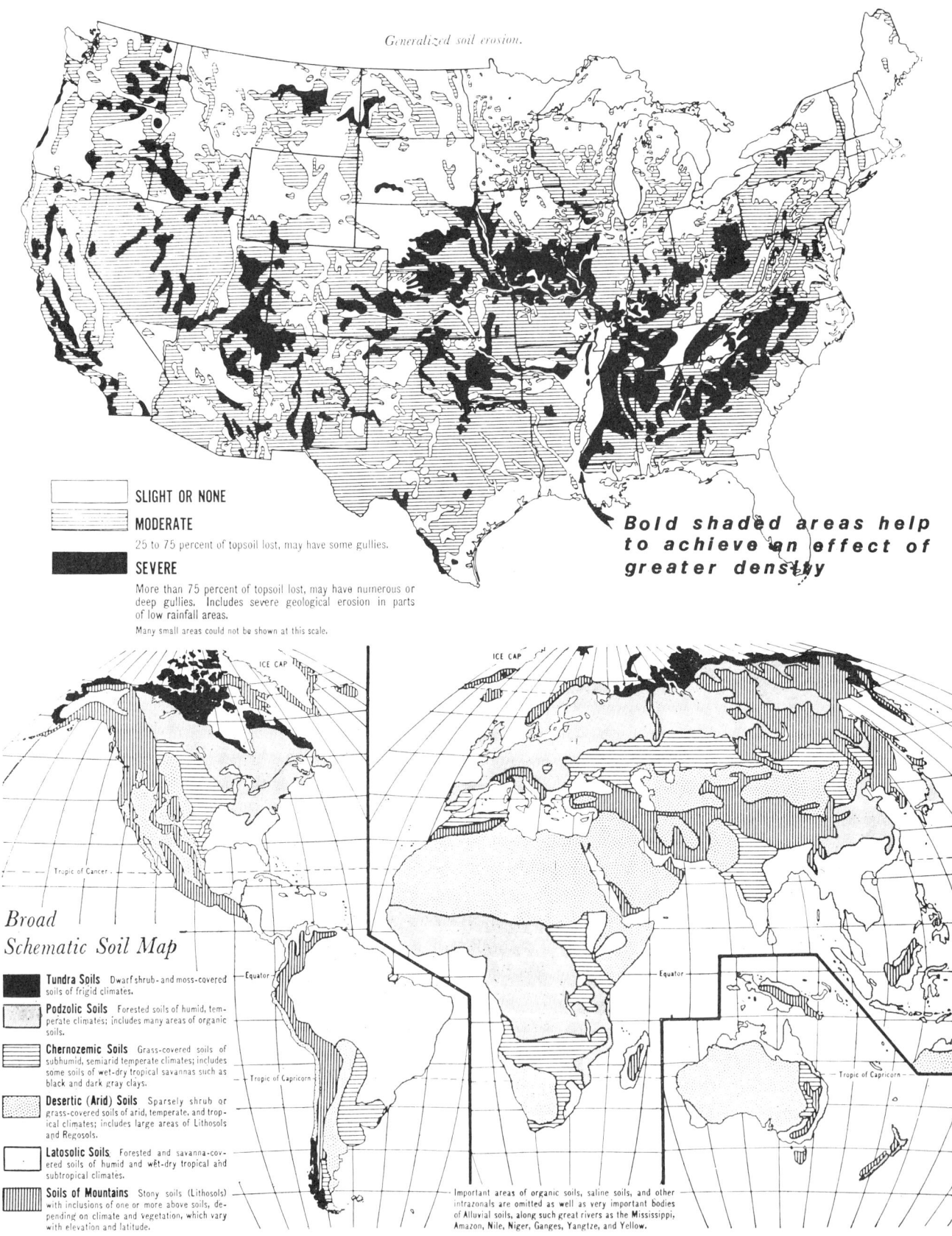

Fig. 12.10 Choropleth maps *(Courtesy Yearbook of Agriculture).*

Fig. 12.10 Choropleth maps (continued).

should be exercised to use symbols that do not obliterate necessary underlying map detail. It may be necessary to also leave an open window around lettering for added clarity. A solid black or white area must be used with great care or avoided entirely since this creates the effect of "emptiness and plenty." Choropleth maps are sometimes misleading since adjacent areas are delineated by boundary lines that indicate an abrupt change when in reality the change is a gradual one.

Isopleths. An isopleth map represents data in the form of lines that connect points of equal value. Numerical values, rather than discrete variables, for continuous distributions such as contours and climatic data may be shown using *isopleths*; also referred to as *isarithms* or *isolines*. Exact information will be shown only along the lines. The interval between the lines

Qualitative data. For qualitative information like vegetation, soil,

will not reveal what takes place and should be carefully interpreted. In drawing an isopleth map the principles of drawing contours discussed in chapter 17 apply. Several examples of isopleths are illustrated in Fig. 12.11.

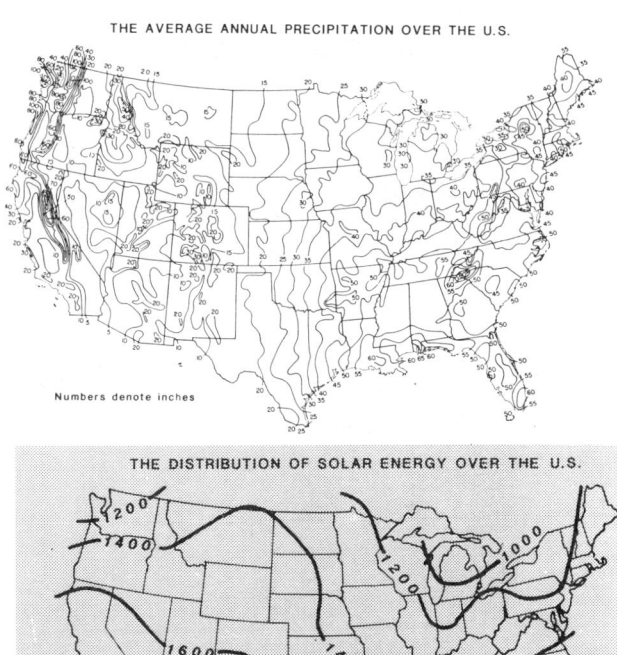

Fig. 12.11 Isopleth maps.

and geology it is not necessary to distinguish between features according to size and importance. The features may be drawn on the map using symbols, letters, or shadings and color to locate an item. Identical patterns may be oriented on the map in different ways or different textures of the same pattern may be used to further differentiate between features. The color (hue) used adds another dimension to the drawing and will make it easier to distinguish features.

Quantitative data. It is often necessary for symbols to provide information that is quantitative. The value or magnitude of an item may be required in addition to knowing where it is located. The size of the symbol may represent its level of importance in locating a feature or the area represented by the symbol may allow value assess-

ments to be made.
  Quantities that occur as a series of points may be drawn on the map by using clear, easy to read symbols such as squares, triangles, or other geometric shape figures, Fig. 12.12. A small circle is easy and quick to draw and may be

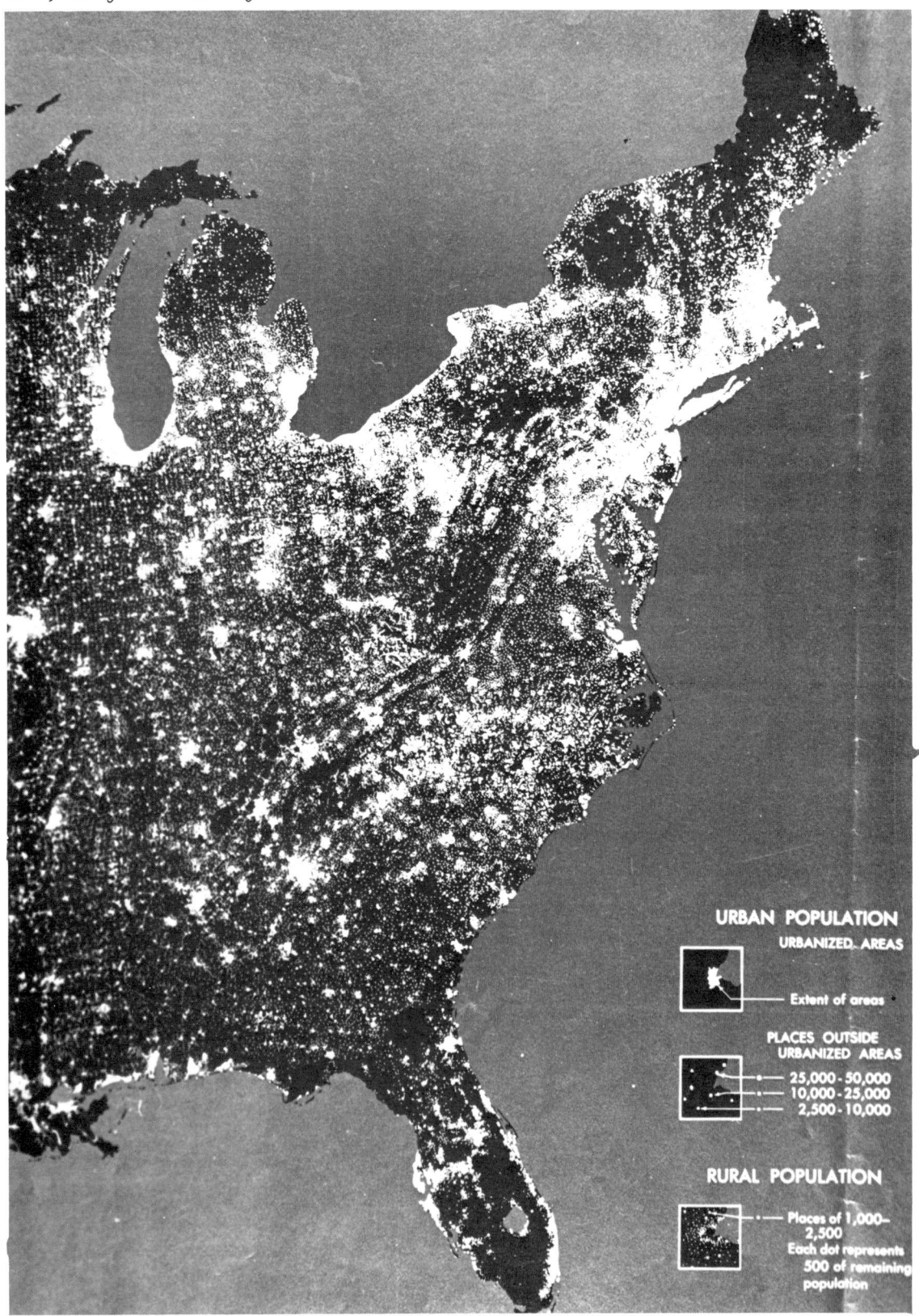

Fig. 12.12 Portion of 1970 population distribution map, night-time view *(Courtesy Bureau of the Census)*.

used to represent almost any quantity. Maps of this kind are commonly referred to as dot maps, Fig. 12.13. The dots are repeated to form a clear distribution picture of the data. Dot maps may be difficult to draw and must be planned to answer questions concerning their size, units, and location. The size of the dots should be drawn to provide some background space to allow counting. A reservoir type inking pen works well for drawing uniform dots of any size. The pen should be held in a vertical position to insure that all dots are the same size and round.

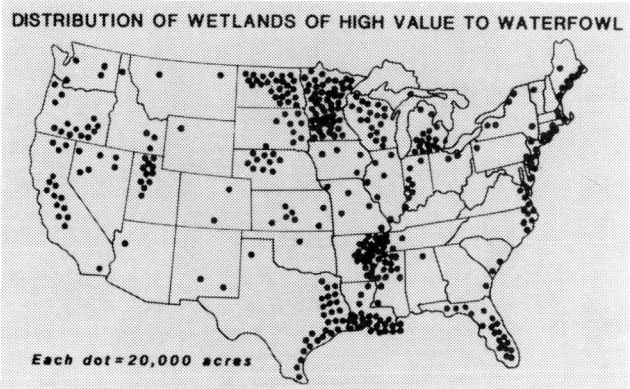

Fig. 12.13 Dot type maps.

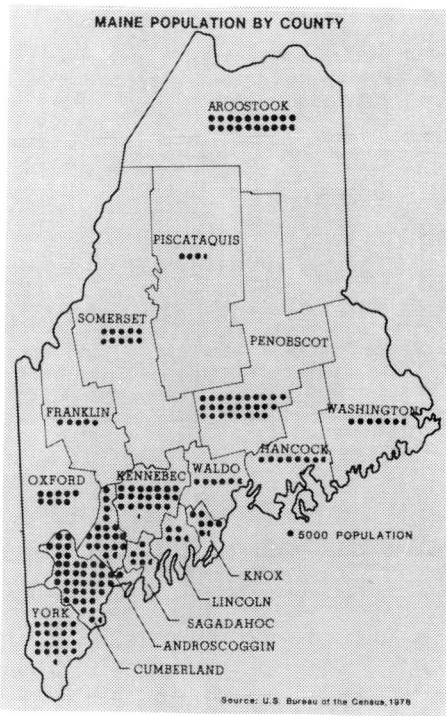

Fig. 12.14 Repeated symbols.

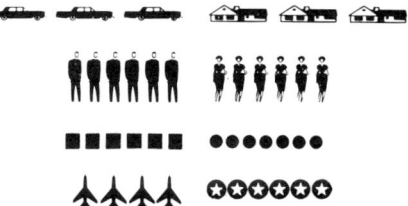

Fig. 12.15 Repetitive symbol shapes.

The number of dots to draw will depend on the unit value assigned to one dot. Units should be assigned to each dot that will allow a sufficient number of dots to be drawn to create a pattern showing relationships.

The location of the dots indicates density changes and do not necessarily give precise locational information. Dots in many instances should not be evenly spaced to avoid misrepresentation in distribution. How the data is collected, dot size, and number of dots to be drawn will all affect their location.

Fig. 12.14 illustrates the use of repeated symbols for showing quantitative information. Maps of this kind are easily understood since visual comparisons or distributions are readily perceived. Appliqué repetitive symbols available in tape or sheet form are recommended since they are easy and fast to use, Fig. 12.15. Pictographic shapes rather than geometric shapes are frequently used to enhance visual appeal. For geometric shapes such as circles, squares, and triangles, a template will work well. The drop bow compass may also be used to draw large numbers of circles very fast. A reservoir type inking pen is used to fill in the circles. All symbols must have the units indicated and the total quantity will equal the sum of all symbols. A value should be assigned to the symbol to allow a sufficient number to be drawn for creating an effective distribution pattern. Quantities that are less than that of the assigned unit are shown by using a partially completed symbol. Too many symbols will require a large space that may not be available and their numbers make counting difficult. The problem of too many symbols may be solved by using a larger symbol having a different unit value, Fig. 12.16.

Proportional circles, squares, bars, spheres, cubes, or other geometric shapes may be used to convey quantita-

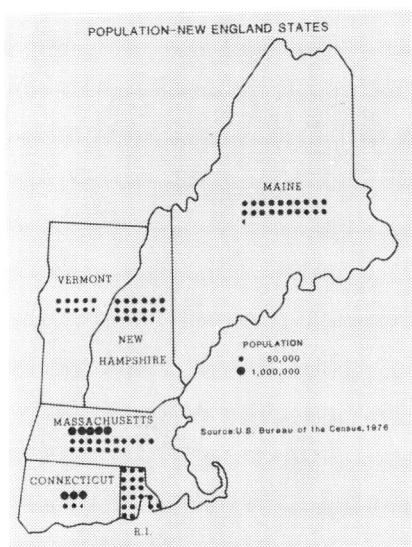

Fig. 12.16 Repeated symbols using different unit values.

tive information, Fig. 12.17. These may be shaded in various ways to improve the appearance of the symbols. The size of the symbols used to make comparisons must be drawn carefully to avoid a tendency to distort data and exaggerate differences between amounts compared. For example two circles, cubes, or spheres used to represent 1000 hunters and 500 hunters respectively will not show the data accurately if the size of the symbol is exactly half of the larger symbol.

The circle is the most commonly used graduated symbol. Since the area of a circle is πr^2 and π is a constant, the size of circles may be determined by taking the square roots of the data. The circles may be drawn using the calculated values to lay off the radius of the circle proportional to the square roots.

A scale should be selected to give a good effect or impression with circles neither too small or too large. The effect created by drawing circles at two different scales is illustrated in Fig. 12.18.

Another method for determining the size of the circle is to prepare a linear scale from which the radii of the circles may be obtained. The methods illustrated in Fig. 12.19 (a) and (b) may be used. Squares may be used instead of circles by finding the square roots of the data and drawing the sides to scale, Fig. 12.19 (c).

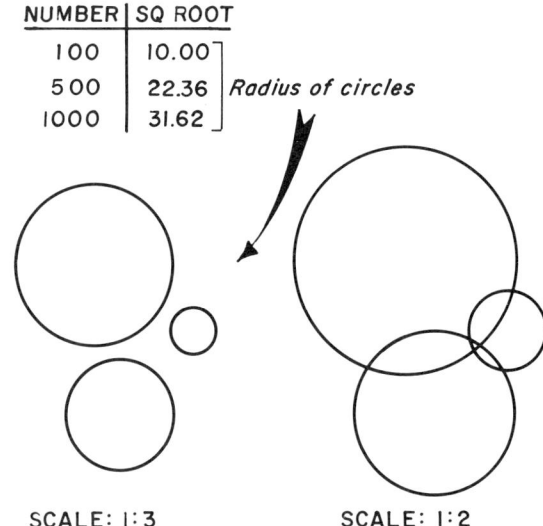

Fig. 12.18 Proportional circles.

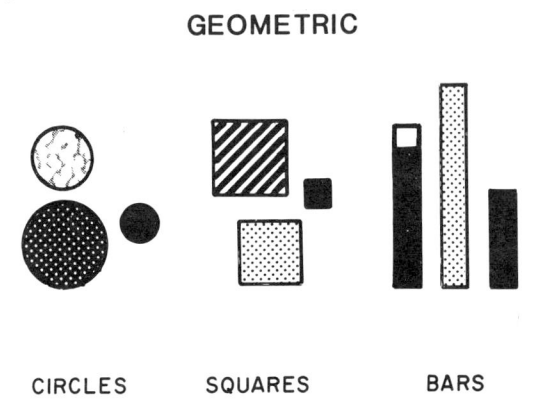

(a) Quantities proportional to areas

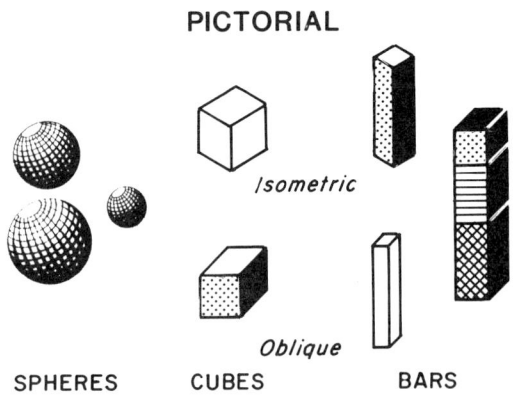

(b) Quantities proportional to volume

Fig. 12.17 Proportional type symbols.

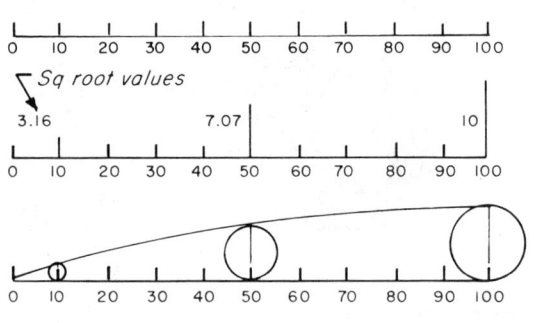

* 1. Draw a line of conveient length and lay off equal major units involved.

2. Calculate the square root of several numbers, e.g., 10, 50, and 100. Erect a perpendicular at calculated points and lay off values using any convenient scale.

3. Draw a smooth curve through the ends of the perpendicular lines. Measure diameter of circle for any value required.

(a)

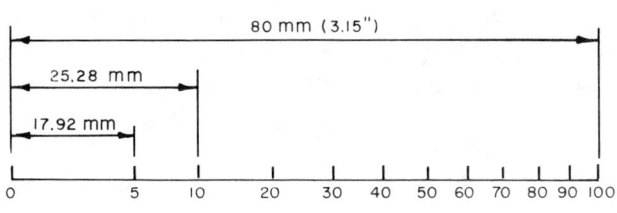

* 1. Draw a line of any pre-determined length, e.g., 80 mm (3.15"), and lay off units from the zero point proportional to the square roots of the numbers involved. If the maximum value is 100, the length of the line is proportional to its square root value of 10, (see Table 1).

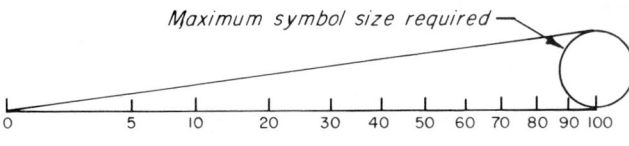

2. Draw maximum symbol size required at the end of the scale. From the zero end draw a line tangent to the circle.

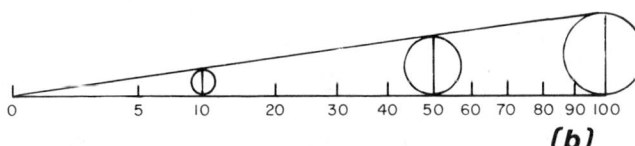

3. A perpendicular line drawn from any point on the horizontal scale to the inclined line will give the diameter of the circle required.

(b)

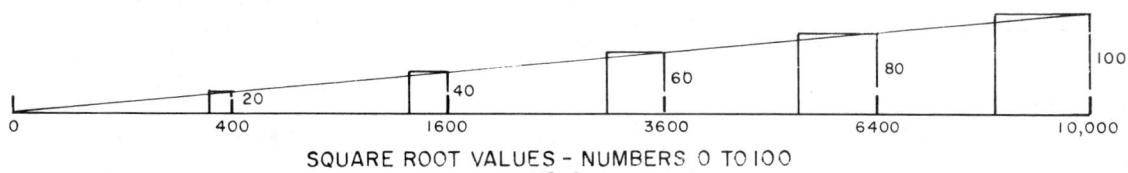

SQUARE ROOT VALUES - NUMBERS 0 TO 100

(c)

* Any scale may be used. The use of a large scale will help to determine values between major units

Fig. 12.19 Procedure for finding the size of proportional circles and squares.

Table 1. Values for scale measurement, Fig. 12.19 (b)

Number	Sq. Root	Measurements (mm)
5	2.24	17.92
10	3.16	25.28
20	4.47	35.76
30	5.48	43.84
40	6.33	50.64
50	7.07	56.56
60	7.75	62.00
70	8.37	66.96
80	8.94	71.52
90	9.49	75.92
100	10.00	80.00

$\dfrac{10}{80} = \dfrac{2.24}{x} = 17.92$

A proportional symbol map involving the use of circles is shown in Fig. 12.20. A key or legend should always be used to identify symbols and values. An alternate method for representing the legend is shown at the bottom of the map. The symbols are drawn as a series of graduated nested circles.

In congested areas where symbols overlap, the circles may be drawn with the smaller circles appearing to be on top, Fig. 12.21 (a). The smaller circles should be drawn first and filled in using a reservoir type inking pen. In Fig. 12.21 (b) the circles actually overlap and are either left entirely

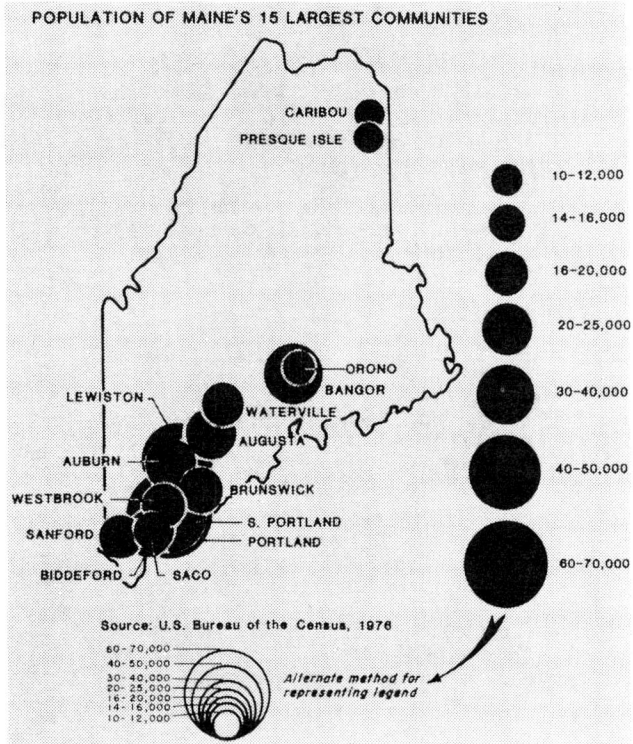

Fig. 12.20 Proportional circle map.

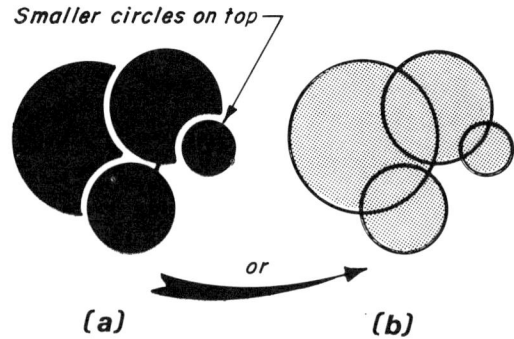

Fig. 12.21 Overlapping symbols.

open or lightly shaded using appliqué materials.

Where a three dimensional effect is desired, pictorials in the form of spheres and cubes are used. The use of circles or squares frequently presents difficulty in showing both the low and high values for a large range of data. A sphere or cube representing volume, on the other hand, will allow a larger range of data to be represented. It is difficult, however, to relate the volume of spheres and cubes of varying size. The volume of a sphere two times larger than another, for example, will be eight times greater ($2^3 = 8$). Cube root values must be used to determine the size to draw spheres and cubes. The same general procedure outlined in Fig. 12.19 (b) and (c) is used to find the size of spheres or cubes, Fig. 12.22. The cubes unlike

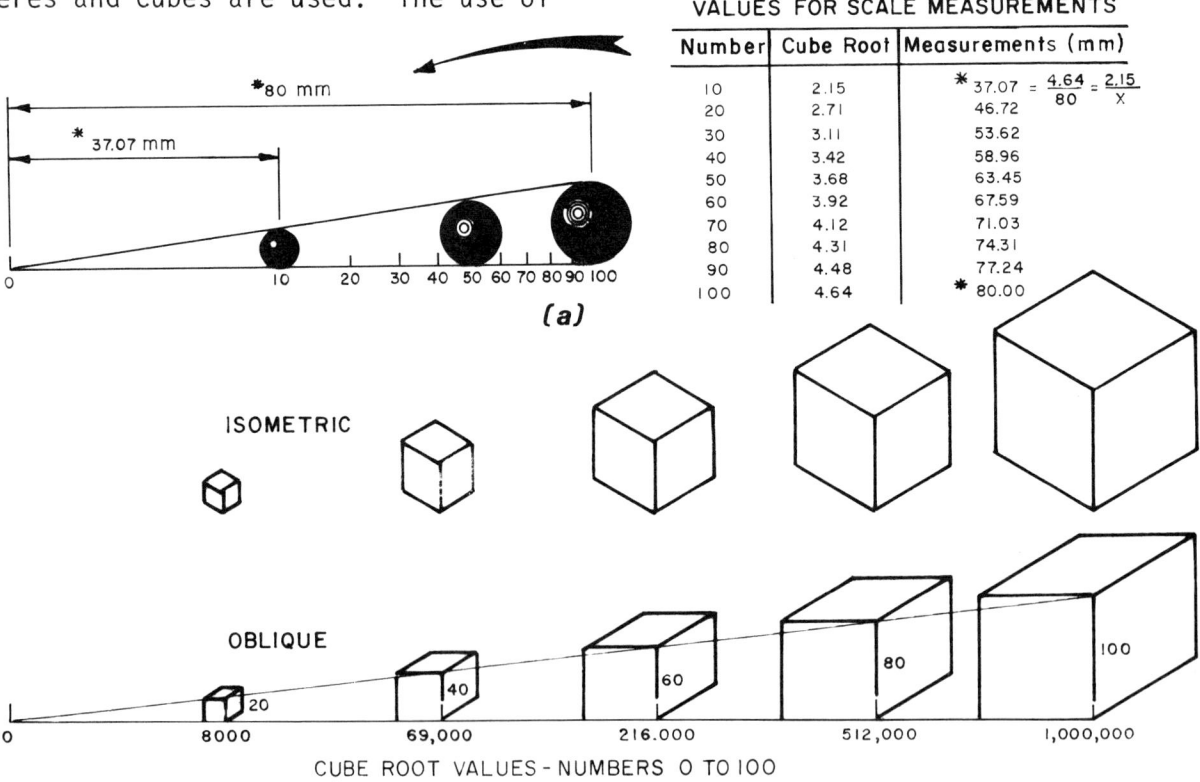

Fig. 12.22 Proportional spheres and cubes.

the spheres are easily drawn in pictorial form, either in oblique or isometric.

Spheres are usually represented as a globe with meridians and parallels, Fig. 12.23 (a), or as a shaded ball, Fig. 12.23 (b). The use of appliqué symbols in the form of globes is the easiest way for representing a sphere. Another method for shading a sphere is to draw a series of concentric circles with a brilliant point left near the top, Fig. 12.23 (c). The brilliant point may be approximated. Geometrically it is the point where the bisector of the angle between a light ray to the center and the visual ray from the center to the eye pierces the external surface, Fig. 12.23 (d). This point may also be used for drawing a series of concentric circles. The thicker lines will occur where the parallel rays of light are tangent to the sphere along the shade line.

12.5 MAP CHARTS

The use of proportional bars and circles, divided or undivided, on maps may be used to create interesting effects, Fig. 12.24. Where it is desirable to allow the reader to make simple comparisons or to emphasize location, this technique lends itself to a wide range of possibilities. The reader is referred to chapter 7 for additional information and ideas for drawing bar or circular charts. The base map should be uncomplicated with sufficient open space to clearly draw the bars or circles. The primary problem with maps of this type is that the user may find it difficult to evaluate a number of variables or to ascertain total other than local distributions for the whole map.

12.6 SYMBOL DETAIL AND GENERALIZATION

The method and amount of detail to include on a map will depend primarily on compilation costs, the map scale, purpose, and how the map is to be used. In compiling a map it is best to work at the largest possible scale since map data derived from small scale maps has been generalized. Symbols used to represent the features should be drawn mechanically except those that have elements that can only be drawn freehand, e.g., timberline, flowage, etc.. templates and appliqué materials having map symbols of many types are easy and fast to use. For repeated symbols the use of templates and appliqué materials may save valuable drawing time resulting in an economic savings.

Generalization is the process in determining the amount of detail and the way it is to be drawn on a map. This is a common problem for the mapmaker as the scale of the map becomes smaller. The cartographer must analyze and select the best way to indicate features symbolically by simplifying and eliminating small details. Line weights, size of symbols,

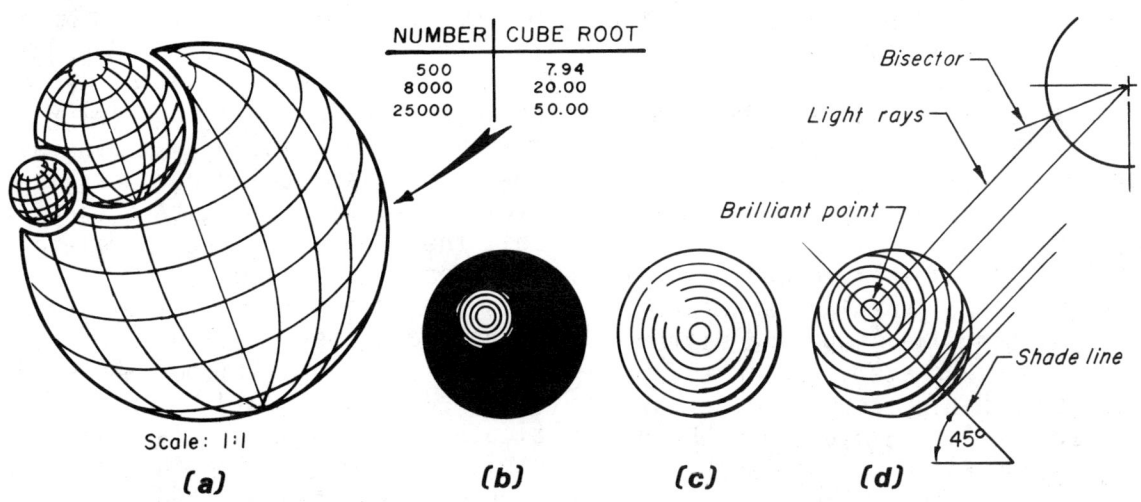

Fig. 12.23 Methods for representing a sphere.

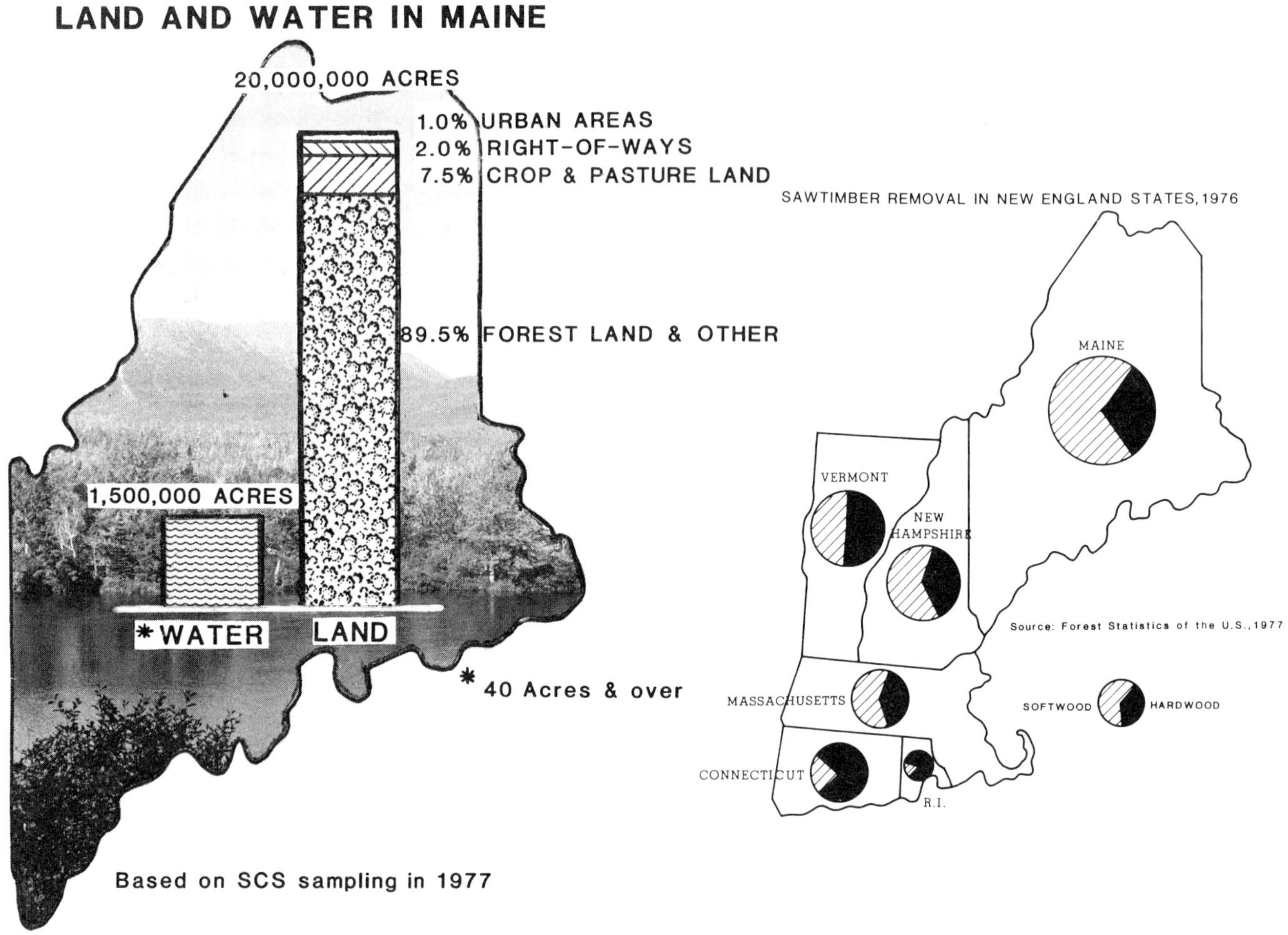

LAND AND WATER IN MAINE

20,000,000 ACRES
- 1.0% URBAN AREAS
- 2.0% RIGHT-OF-WAYS
- 7.5% CROP & PASTURE LAND
- 89.5% FOREST LAND & OTHER

1,500,000 ACRES

*WATER   LAND

*40 Acres & over

Based on SCS sampling in 1977

SAWTIMBER REMOVAL IN NEW ENGLAND STATES, 1976

MAINE, VERMONT, NEW HAMPSHIRE, MASSACHUSETTS, CONNECTICUT, R.I.

SOFTWOOD   HARDWOOD

Source: Forest Statistics of the U.S., 1977

Fig. 12.24 Bar and circular maps.

and lettering will be of major importance since legibility must always be maintained. The most important single factor when deciding on which features to draw and how they are to be represented will be the intended purpose of the map. The individual that uses good judgment and possesses a sense of proportion will find the task of generalizing map features easier.

Fig. 12.25 illustrates a map drawn at two different scales. Many map users will not want to know every little twist and turn in the lines representing map items. Since it is not necessary to include all feature elements or irregularity, some lines are smoothed out. The distinctive shape and pattern of all features, however, should always be preserved. Beginners should be careful of over simplification since the representation of known shapes may become totally unrecognizable.

The size and importance of features must be considered as the scale of the map becomes smaller. Decisions will have to be made on which features to show without overloading the map which will reduce its usefulness. After selecting the detail to be drawn, it may become necessary to further simplify the features, Fig. 12.26. A major river or town, e.g., would be shown on a small scale map while a minor river or small town might be omitted. In generalizing features by smoothing out detail, sufficient information must be shown to enable the map user to still relate to the features represented.

The combination of similar features is another generaliztion technique, Fig. 12.27. Individual timber types on a large scale map could be combined to indicate a more general classification on small scale maps. The individual buildings in a built-up section could be

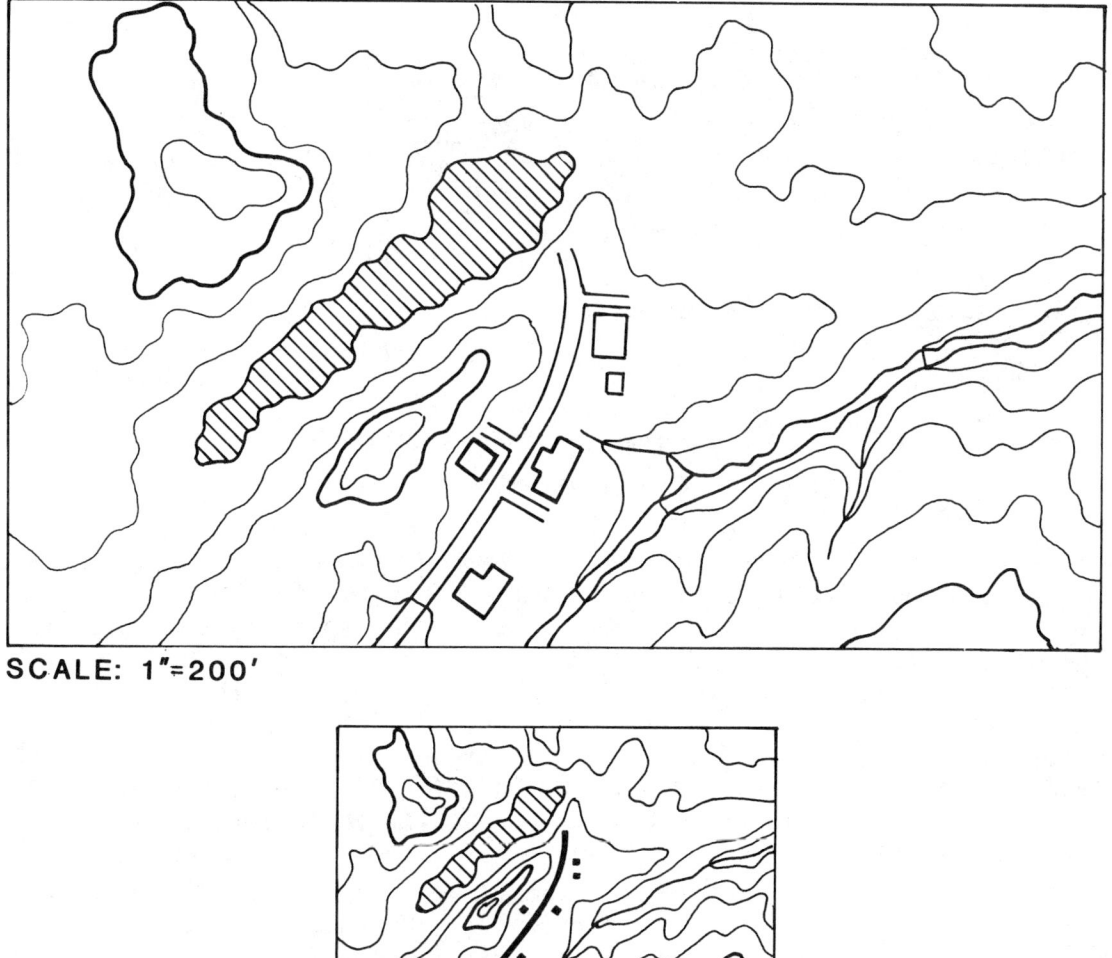
Fig. 12.25 A comparison of two maps of the same area drawn at two different scales.

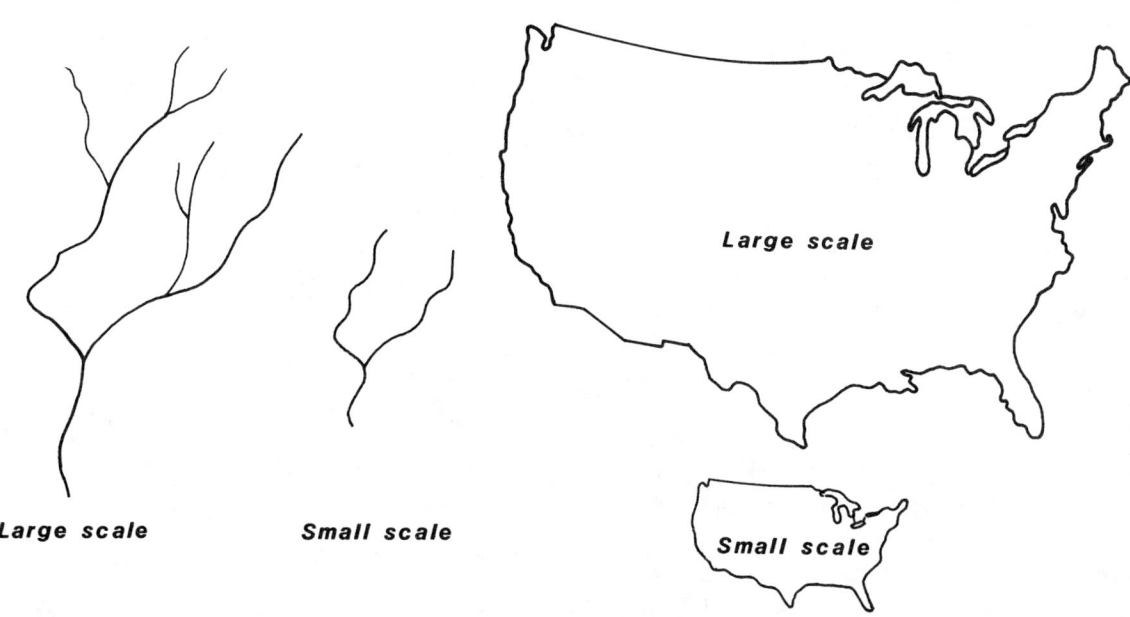
Fig. 12.26 Selection and simplification of map features.

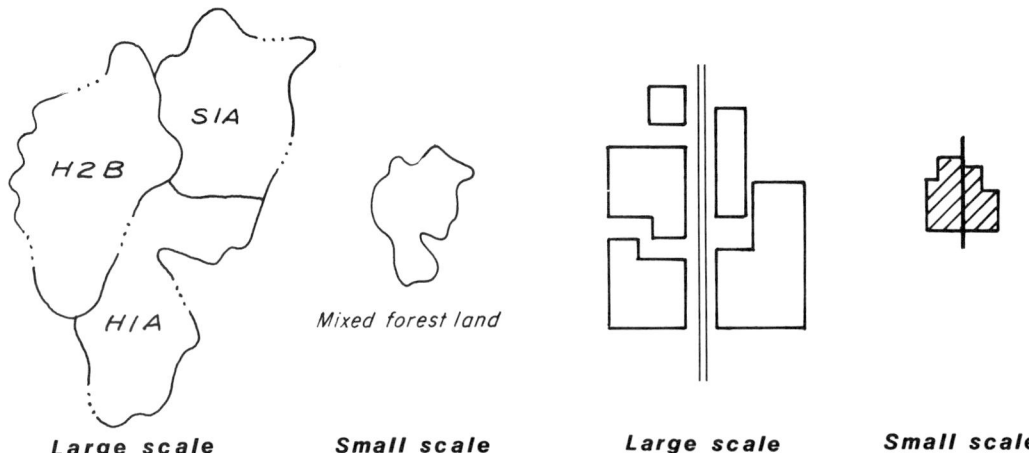

Fig. 12.27 Combination of map features.

shown on a large scale map but would have to be combined on a small scale map.

The generalization of features requires that all items represented be clear. Several features may parallel each other and it may be necessary to shift the position of the less important feature as the map scale is reduced. The correct relative position of the feature should be maintained, however.

It is very difficult to set forth a set of rules that will pertain to all situations since the amount of detail that may be drawn on a map is endless. Generalization all too frequently has been carried out subjectively by cartographers and is often not very consistent. A general discussion of symbol detail and generalization for different kinds of information follows.

Transportation routes. Roads of different kinds and railroads are among the most important features to be shown on many types of maps. The transportation routes to omit is frequently a difficult question and must be left to the judgment of the mapper. A good transportation network on the map is helpful in compiling and showing other features and will give the map user a better idea of the area mapped. The scale and purpose of the map, in general, will determine the best way to represent different classes of roads and how complete a transportation network should be drawn. Main routes should normally be included with minor routes eliminated except for undeveloped areas where all traffic arteries may be of utmost importance. It is desirable to draw roads true size when the map scale and road width allow for clear delineation, Fig. 12.28. For small scale maps roads are symbolized according to class. In addition to using various symbols it is common to use a color fill for all or parts of the road for further classification.

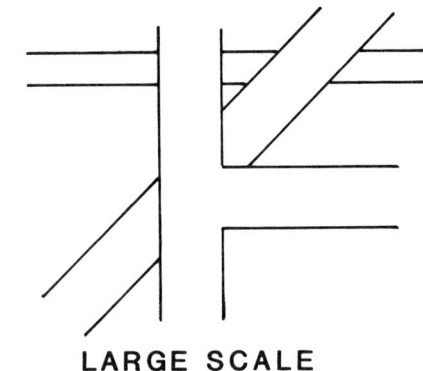

LARGE SCALE

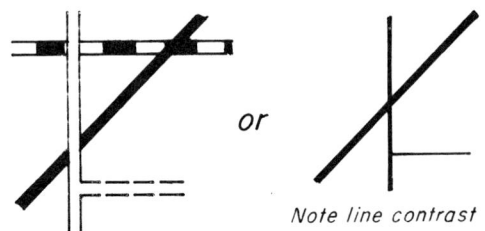

SMALL SCALE

Fig. 12.28 Large and small scale road representation.

Buildings and urban areas. Buildings are represented either by conventional symbols or, if large enough, by actual outline showing the general building configuration plotted to scale, Fig. 12.29. The building symbol may be drawn solid, open or cross hatched. All structures should be correctly oriented in respect to other map features. This may result in buildings touching or overlapping other symbols representing physical features. For conditions of this kind one of the features should be moved slightly to improve clarity. The building symbol is moved no more than necessary, at right angles to linear symbols like a road or railroad, Fig. 12.30 (a). Irregular shaped linear features, such as a trail or stream, are displaced slightly with the buildings maintaining their true position, Fig. 12.30 (b). Features that are not at ground level like powerlines and non-physical features such as boundaries are drawn across buildings and roads, Fig. 12.30 (c). In general the symbols used to meet the purpose or objectives of the map should retain their correct positions. The displaced symbols should be those that are not as important in meeting the purposes for which the map is intended.

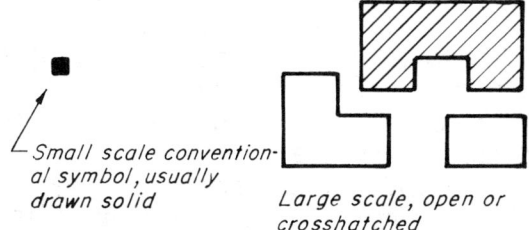

Fig. 12.29 Buildings.

Within areas, defined as urban or built-up, it may be impossible due to scale limitations to show all structures with the exception of prominent landmark buildings. The extent of the urban area may be shown using a color tint or appliqué shading materials, Fig. 12.31. It is common to omit the area tint for features like water, parks, and large open areas.

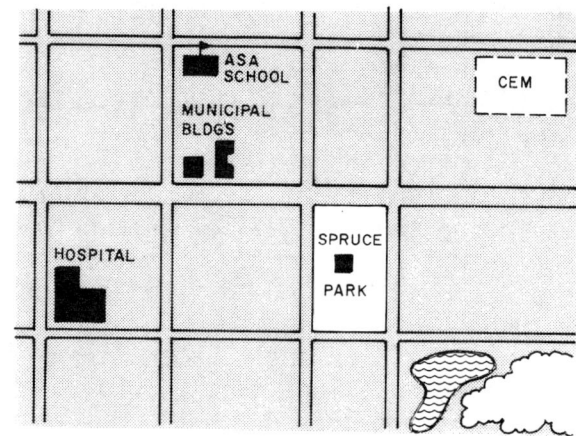

Fig. 12.31 Urban areas.

Boundaries. Boundary lines are used to delineate jurisdictional authority for the various levels of government. The boundary lines which may be required on the map are international, national, state, county, and primary divisions of counties and townships. Where two or more types of boundaries coincide, the order of precedence for drawing is the order listed. All boundary lines should be shown with appropriate standard symbols.

Hydrographic features. Water features are relatively permanent and are one of the more important elements on

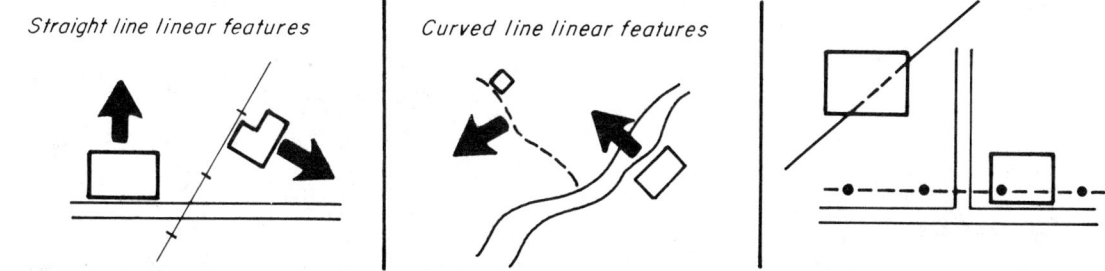

(a) Buildings displaced (b) Features other than buildings displaced (c) No displacement

Fig. 12.30 Symbol displacement for features that touch or overlap.

maps. Again, the inclusion of these features will depend on their significance. Point, line, and areal symbols may be used to distinguish features, Fig. 12.32.

Streams should be drawn double lined if large enough at map scale to justify showing both shore lines. For small streams a finer line may be drawn to represent the head waters with the line gradually becoming heavier on the lower course.

The shore line of a body of water should correspond to the normal water-level mark and be shown with a solid line or other appropriate symbol. At one time fine lines, termed water lines, within the shore lines were drawn to emphasize water features. This technique is a holdover from the days of copper plate engraving when it was not possible to print solid colors. Water lining is very time consuming and is now obsolete. It is now more common to show the outline of the water and to then crosshatch, apply an appliqué pattern or color the area within. Water surface elevations are commonly lettered within the outlines of lakes and ponds.

Water features are normally generalized on small scale maps. Lakes and ponds will require considerable smoothing out as the scale decreases. Large streams should be shown with minor tributaries eliminated. Small bends will also have to be smoothed out. Again, the characteristic shape of the features should be maintained.

Wet areas of low land like a marsh and swamp are represented using the same symbol, Fig. 12.33 (a). A marsh is characterized by low lying herbaceous growth while tree-like growth is associated with a swamp. The marsh symbol is drawn in blue or black with a blue color printed over the area. For a swamp the symbol is drawn in green or black with a green color over the area. A line should not be drawn around the marsh or swamp symbol since it is difficult to delineate the feature boundaries. Areas that are periodically inundated with water are termed flowages and may be represented as illustrated in Fig. 12.33 (b).

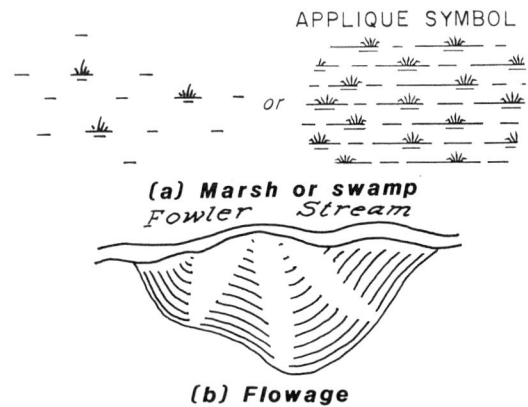

Fig. 12.33 Wet area symbols.

Surface cover. Areas of vegetation may easily be mapped by indicating the area over which it extends and the type of cover represented. Due to the variety of vegetation that exists, it is common practice to group different types and generalize their outlines. For topographic maps woodlands are shown with crops omitted since they are not of great importance. A woodland boundary or other vegetative type, e.g., may be indicated by a suitable line symbol, appliqué material, or color tint overprint, Fig. 12.34. A clearing should be left around features such as roads, water, and power transmission lines.

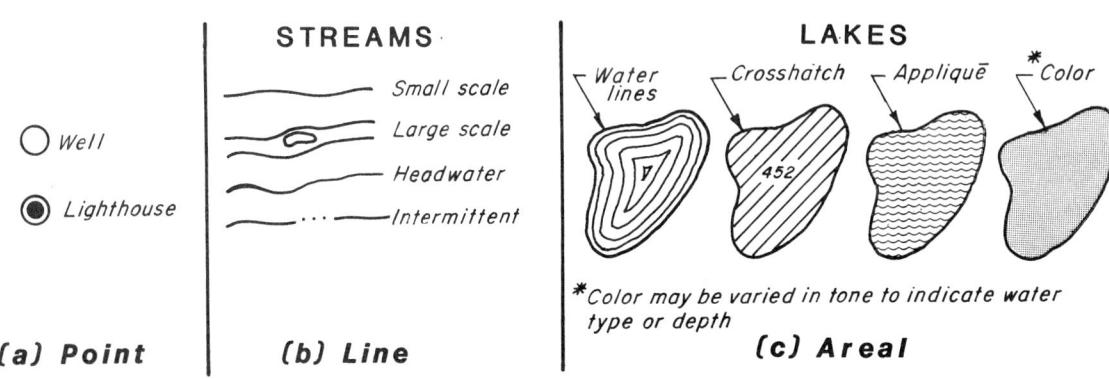

Fig. 12.32 Hydrographic symbol variations.

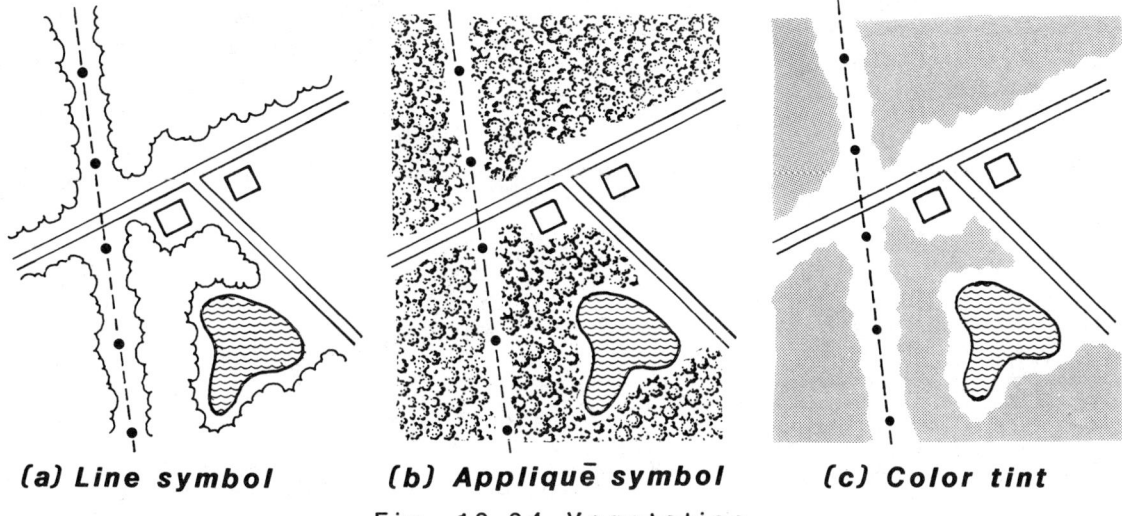

(a) Line symbol **(b) Appliqué symbol** **(c) Color tint**
Fig. 12.34 Vegetation.

Woodland areas on topographic maps may be divided into many types. Areas exceeding approximately 61 x 61 metres (200 x 200 feet) are generally shown and the outline is broken for clearings 30 metres (100 feet) or more in width. The woodland areas are classified and a symbol pattern or key letter is used to indicate the type. A dashed outline symbol may be used to show the boundaries between the various types. Submarginal land having only a scattered growth of trees or brush is often difficult to classify. As a general rule, if less than 10 percent is covered, it is classified as nonforest.

In forest stand mapping, a variety of classification systems has been devised to provide information needed in forest land management. Most stand mapping is done from aerial photography using the following variables: species or species group, tree height, stand density, site, and crown diameter. A classification code is generally devised using a letter or number for each of the above variables that is used with the stands being identified by using a combination of letters and numbers. For example, H2A may indicate hardwood land of medium growth and fully stocked. Codes may be developed to meet most conditions and will depend upon the objectives of the mapping project and the forest region involved. Fig. 12.35 illustrates a portion of a stand map.

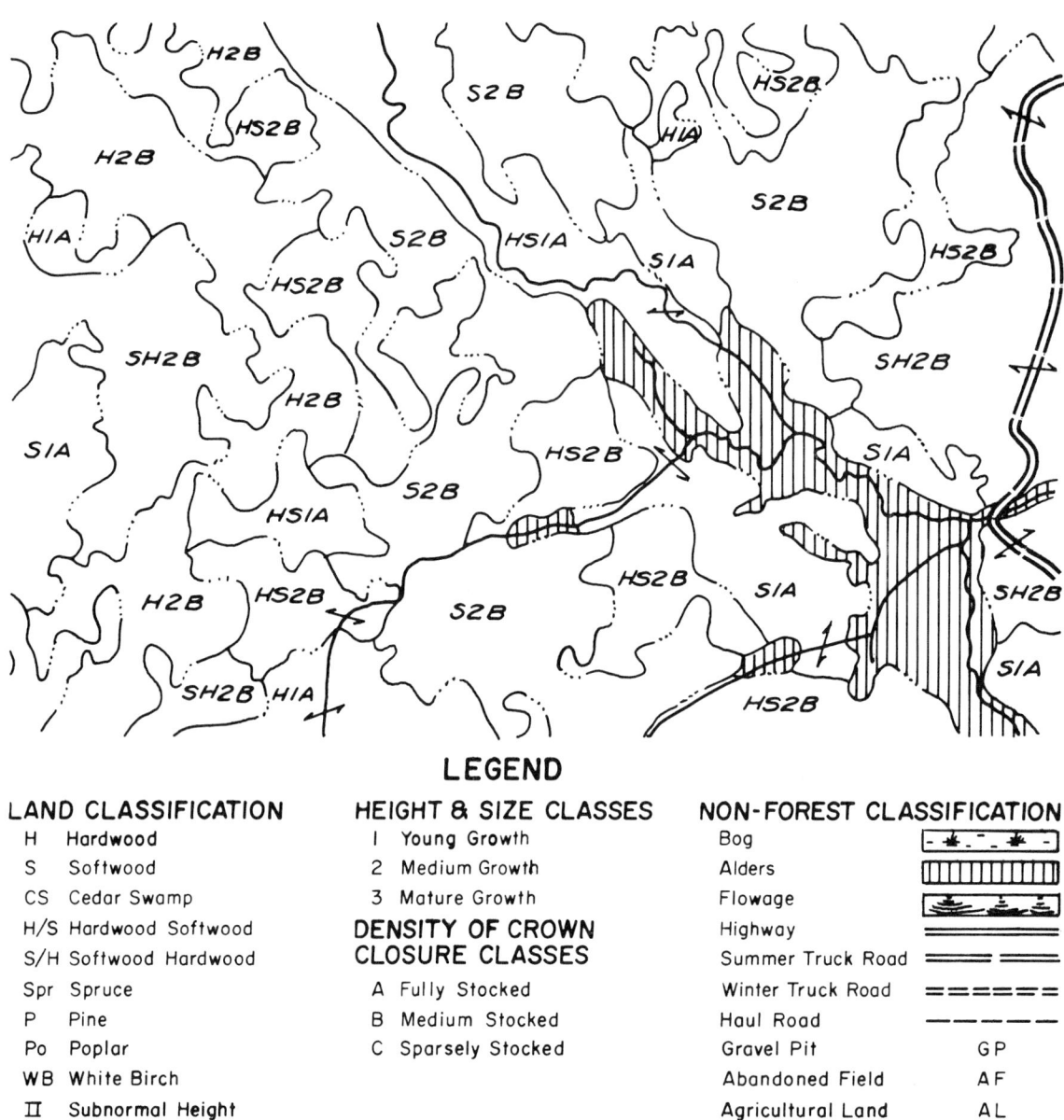

Fig. 12.35 Forest stand mapping.

MAP PROJECTIONS

13.1 INTRODUCTION

The earth's surface can only be truly represented on a globe, Fig. 13.1. Several limitations of globes, however, are scale and convenience. It is obvious that a globe would have to be extremely large to show the detailed information represented on flat large scale maps. Since the globe represents the spherical earth in reduced form, only general kinds of information are portrayed. The globes greatest utility is for showing land mass location, directions, and distributions of information such as vegetation, political data, and general relief.

Fig. 13.1 Typical globe *(Courtesy The George F. Cram Company, Inc).*

The curvature of the earth on small area, large scale maps such as plot plans, type maps, and engineering plans, can be ignored. The map may be constructed to show the relative location of objects without measurable distortion. When a map is made of a large area and to a small scale, map projections are employed. A *map projection* is defined as the systematic drawing of lines representing parallels and meridians (graticule) of the earth's surface onto a flat or plane surface. It is a well-known fact that the surface of the earth, like that of a ball, cannot be developed (to lay out in a flat plane like a piece of paper) without some distortion. The closest method for representing the surface of the earth on a flat plane would be to draw globe gores (lune-shaped segments), Fig. 13.2. The segments cannot be fitted together, resulting in interrupted areas. Since a map interrupted in this manner is undesirable, a number of different projections have been devised to meet several diverse map requirements. All map projections have certain limitations, however, since it is impossible to represent a true picture of all areas for any one kind of projection.

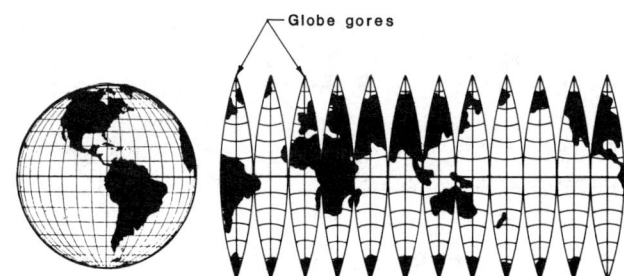

Fig. 13.2 The use of gores to represent the surface of the earth on a flat plane.

13.2 DESIRABLE MAP PROJECTION CHARACTERISTICS

A map, as mentioned earlier, cannot be drawn on a flat surface without some distortion. The amount and character of the distortion varies with the projection system used and the size of the areas covered. On maps of small area the distortion is very small and usually presents no problem. For maps of larger areas, however, the distortion will be very evident. Several properties of concern include shape, area or size, directions, and distances. Since no one projection will have all of these desirable characteristics, the projection system selected will depend on the map requirements and some compromises will be necessary.

Conformity. A map projection is considered *conformal* (orthomorphic) when the scale along meridians and parallels through any point is the same. This requires that the meridians and parallels intersect at right angles since they intersect at right angles on the earth's surface. When all of these conditions are met, the angles between intersecting lines (straight or curved) will be true and the shape of small areas on the map will be the same as the corresponding small areas on the earth. Maps used for navigation, topography, coastal and aeronautical charting are commonly constructed on conformal projections because of the importance of true shape and direction. This type of projection is not too important for world or continental maps due to areas becoming greatly enlarged near the projection margins. The projections most commonly used include Mercator, Transverse Mercator, Lambert's conic, and stereographic.

Equal-area or equivalent. *Equal-area* implies that all areas be represented in their correct relative proportions. A map of the United States and Canada, e.g., should preserve the areas represented in constant ratio to the area on a globe at the same scale. Equal-area projections are used where measurements or comparisons of areas are of prime importance. Small-scale instructional maps depicting geographical areas such as land masses and water commonly employ projections to show equal-area. They are also important in distribution mapping of statistical variables. Several common equal-area projections include Alber's conic, Lambert's equal-area, sinusoidal, Mollweide's, and Eckert's.

True directional. A given direction (azimuth or bearing) from one point to other points on the earth should be correct on the map. Projections that are classified as azimuthal will have this quality. The common azimuthal projections are gnomonic, stereographic, azimuthal equal-area, and azimuthal equidistance.

Equidistance or constant scale. The distance between any two sets of points on the earth should be correctly represented on the map. When representing a curved surface in two directions on a plane surface, a constant scale cannot be maintained. This can be achieved only on selected lines (referred to as standard lines) such as the equator and meridian lines. Equidistance is usually not as important as conformality or equivalence, since it is seldom desirable to measure distances on a map in any one direction. Equidistance maps are used in atlases and other representations of large areas where it is not essential to preserve other properties.

Great circles and rhumb lines. The arcs of great circles should appear as straight lines on a map since they represent the shortest distance between two points. Rhumb lines should also appear as straight lines since they represent lines of constant bearing.

Graticule. The projection system employed should allow the earth's graticule to be easily constructed and of a form to allow geographic coordinates to be found easily and plotted.

It is readily apparent that a map cannot contain all of the desirable characteristics presented. Several of these characteristics are mutually exclusive, e.g., a map cannot be both conformal and equal-area and both great circles and rhumb lines cannot appear as straight lines on the same map. The two different projections illustrated in Fig. 13.3 should be studied carefully to fully understand equal-area and conformality. The attribute of equal-area is obtained by spacing the parallels and meridians so that each square represents the actual area in the same proportional scale, Fig. 13.3 (a). Every square formed by the parallels and meridians has the same area as it has on the earth. Since lines of the graticule do not in-

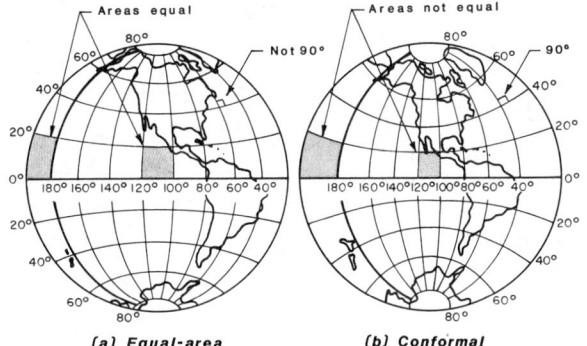

Fig. 13.3 Equal-area and conformality comparison.

tersect at right angles, the angular distortion between these lines increases towards the edge of the map. The map will be conformal when every small area formed by the meridians and parallels has its proper shape, Fig. 13.3 (b). The scale will be the same along both a parallel and meridian, and they will form a right angle.

An area can change shape, in different parts of a map, through stretch and shear and still maintain the same area, Fig. 13.4. Lines may be shorter in one direction and stretch in the other to keep areas equal, Fig. 13.4 (b). An area may shear as shown in Fig. 13.4 (c). If the horizontal lines maintain the same spacing as in Fig. 13.4 (a), the vertical lines must stretch to keep the area equal. In Fig. 13.4 (d) the area is both stretched and sheared.

The graticule on most equal-area projections are curved lines. Since the stretching and shearing will take place along smooth curves, the change in shape will be gradual and not exaggerated as illustrated in Fig. 13.4.

13.3 CLASSIFICATION OF PROJECTIONS

Many systems of map projections have been employed to represent the spherical earth on a flat or plane surface. Each type of projection has characteristics of its own which will have a direct bearing on its use. Individuals involved in the preparation and use of maps should have a rudimentary knowledge of the more commonly used projections. This knowledge is desirable to help make an intelligent choice on the kinds of projection to use and to give an appreciation of the scope of the subject. The chief value in understanding projections is to provide the background for wisely using computer-assisted methods to produce map projections for which the equations can be specified. Software is readily available that will allow automated equipment to draw graticule lines of any scale from any origin or orientation. Before discussing map projections in general non-mathematical terms, it is suggested that Table 13.1 be studied to provide an overview of several commonly used projections.

The spherical earth is not a developable surface and it is not possible to draw straight lines on its surface. The projection systems or transformation of lines from the sphere to a plane may consist of: (1) some form of orthographic or perspective projection from the sphere to a plane or to a developable surface such as a cylinder or cone, Fig. 13.5, (2) the reproduction of the graticule by mathematical computation, or (3) the combination of the two meth-

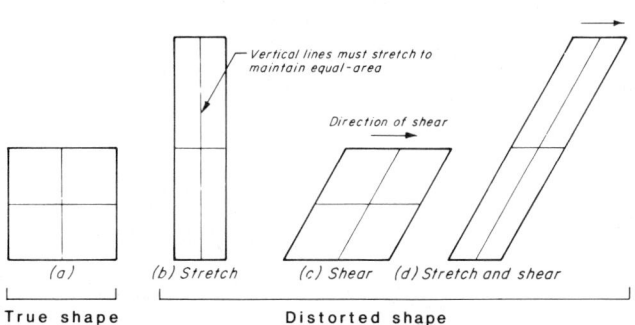

Fig. 13.4 Equal-area diagram.

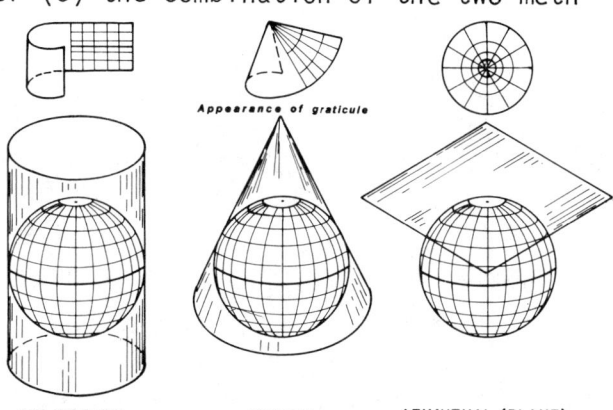

Fig. 13.5 Geometric classification of projections.

Table 13.1 Summary of selected map projections

Characteristics	Mercator	Transverse Mercator	Gnomonic	Lambert conformal	Polar stereographic	Azimuthal equidistant	Azimuthal equal-area
Projection classification	Cylindrical projection (mathematically adjusted) with projectors at center of sphere	Cylindrical projection (mathematically adjusted) with cylinder axis perpendicular to polar axis. Projectors at center of sphere	Tangent plane with projectors at center of sphere	Conical Projection (mathematically adjusted) with projection origin along polar axis	Tangent plane with projectors at opposite pole	Tangent plane (mathematically computed)	Tangent plane (mathematically computed)
Parallels	Parallel straight lines unequally spaced	Elliptical	Curved lines except the equator	Arcs of concentric circles unequally spaced	Concentric circles unequally spaced	Curved lines	Curved lines
Meridians	Parallel straight lines equally spaced	Curved lines	Straight lines	Converging straight lines	Straight lines radiating from the pole	Curved lines	Curved lines
Angle between parallels and meridians	90°	90°	Variable	90°	90°	Variable	
Shape and area	Small areas approximately correct. Not equal area	Small areas approximately correct. Large areas on small scale charts distorted. Not equal-area	Great distortion at edges of projection. Not equal-area	Small areas approximately correct. Large areas on small scale charts distorted. Not equal-area	Small areas approximately correct. Large areas on small scale charts distorted. Not equal-area	Not true shapes. Not equal-area	No true shapes. Equal-area
Conformal	Yes	Yes	No	Yes	Yes	No	No
Distance scale	Good at mid-latitude of course. Scale increases away from equator	True scale along tangent meridian; increases outside tangent meridian	Increases rapidly as distance from point of tangency increases	Correct along standard parallels. Nearly constant on large scale maps	Good	Correct only on lines through center of chart	Variable in order to maintain equal-area
Great circles	Curved lines (except equator and meridians)	Curved lines (except tangent meridian)	Straight lines	Approximated by straight lines	Approximated by straight lines	Straight lines radiating from center	No value
Rhumb lines	Straight lines	Curved lines	Curved lines	Curved lines	Spiral curve	Curved lines	Curved lines
Direction	Good and easily measured	Good between any two points	Azimuthal. Difficult to measure between any two points	Good between any two points	Azimuthal. Good between two points	Azimuthal. Cannot be determined between two points	Very difficult to determine
Geographic coordinates	Excellent. Rectangular grid	Easily plotted or determined	Difficult to plot or determine except polar gnomonic	Easily plotted or determined	Easily plotted or determined	Very difficult to determine	Difficult to plot or determine
Application	Navigation	Grid navigation in polar areas. Military grid (UTM). State plane coordinates for some states	Astronomy. Great circle navigation	Aeronautical charts, meterology, and state plane coordinates for some states	Polar navigation, hemisphere maps, and military grid (UPS)	Aeronautics, earthquake studies, and radio transmission	Hemisphere or continents

ods. The projections most used fall in the last category.

All projections have lines along which distances are shown correctly. These are termed *standard lines* and are usually selected meridians or parallels, Fig. 13.6. The cylinder, cone, and plane may be tangent or secant (intersecting the sphere). It should be noted that a projection does not need to make contact only with the lines illustrated since the sphere can be turned to make contact with lines in other regions.

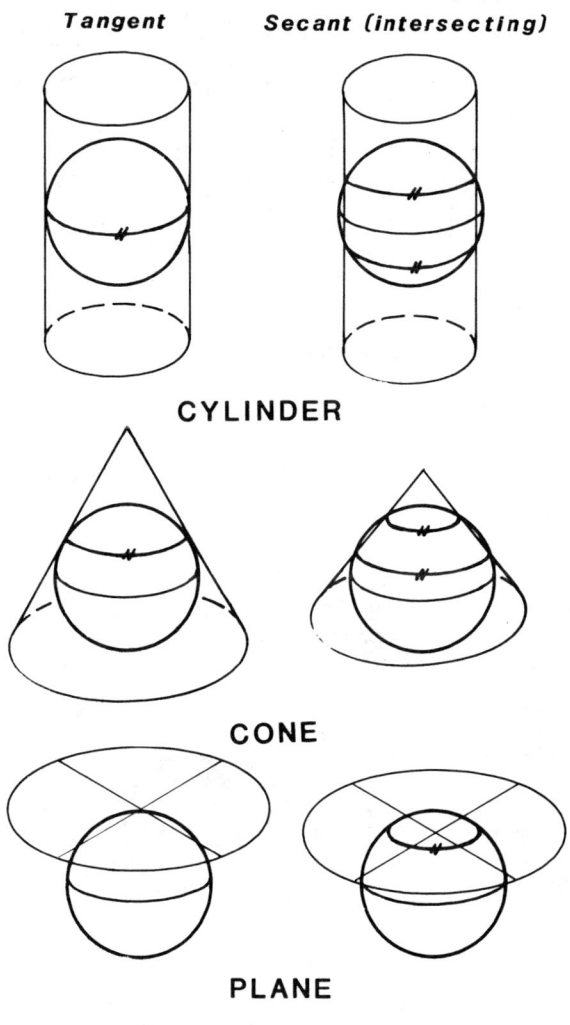

Fig. 13.6 Standard lines.

13.4 CYLINDRICAL PROJECTIONS

The concept of developing an imaginary cylinder, placed tangent or secant to the sphere, in the basis for this group of projections. The lines of tangency can be standard parallels, standard meridians, or any selected great circle. Lines of intersection are standard lines and may not coincide with lines of the graticule. Several properties for these projections are:

(1) Meridians and parallels are straight, parallel, and equal in length.
(2) Meridians are equally spaced.
(3) Meridians and parallels intersect at right angles.

All useful cylindrical projections are mathematically modified to give desired qualities.

Perspective cylindrical projection. In perspective cylindrical projection, the parallels and meridians are projected onto a cylinder tangent to the sphere at the equator, Fig. 13.7. The center of the earth is used as the station point for projecting the parallels onto the cylinder. The poles can not be shown in this projection because the expansion along lines of latitude is infinite. Maps based on cylindrical projections, therefore, are seldom extended beyond 60° of latitude due to the polar distortion that occurs. Since the cylinder is tangent to the equator, true scale can be measured along the equator making the projection best suited to regions elongated in an east-west direction. Although the projection is easy to construct, it has no real value and is seldom used. It is presented here as a basis for understanding other cylindrical projections that follow.

Mercator projection. Mercator projection is a modification of perspective cylindrical projection, Fig. 13.8 (a). The parallels of latitude are expanded using mathematically derived values so that the scale between latitude and longitude is the same. The projection is conformal. Large areas, however, will be distorted by the change in scale from point to point. Table 13.2 indicates the amount of exaggeration for land areas at various latitudes.

Mercator projection is used worldwide for sea navigation and aeronautical charts. On a Mercator map a straight line between any two points is called a *rhumb line* or *loxodrome*. The bearing is constant and crosses all meridians at the same angle. The line is of prime importance to the navigator who can set

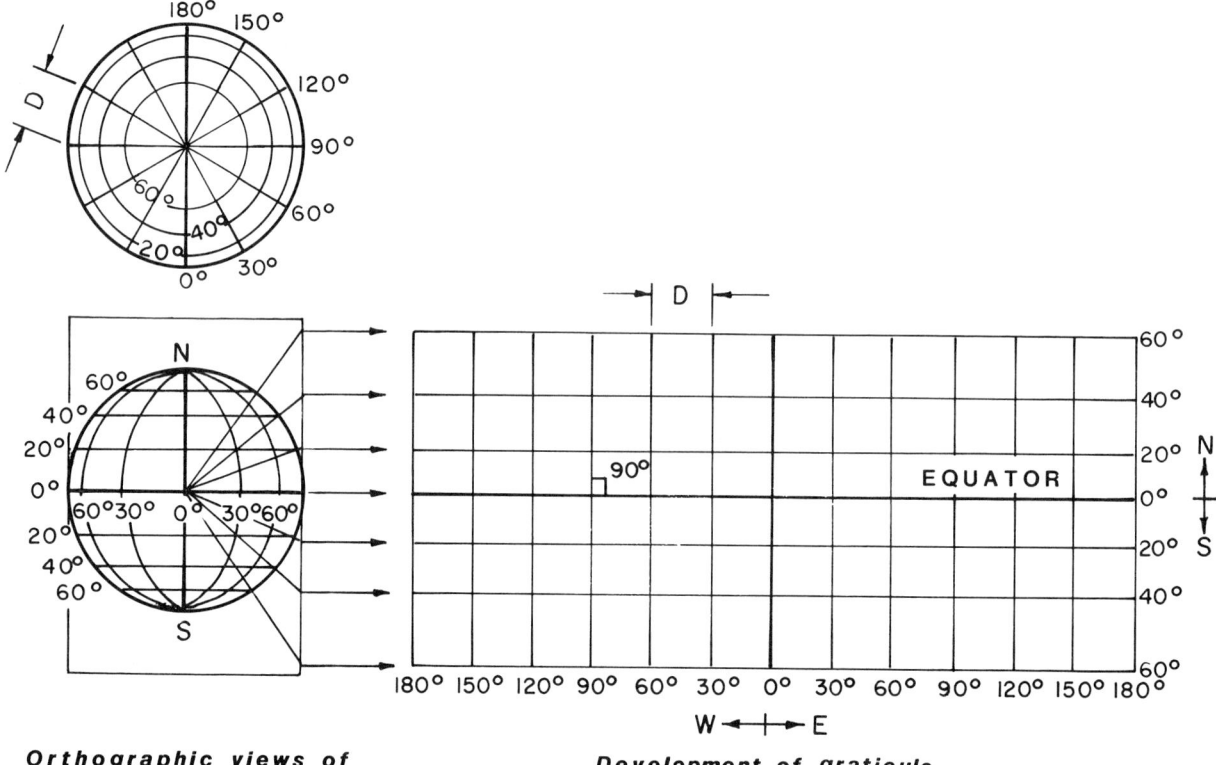

Fig. 13.7 Perspective cylindrical projection.

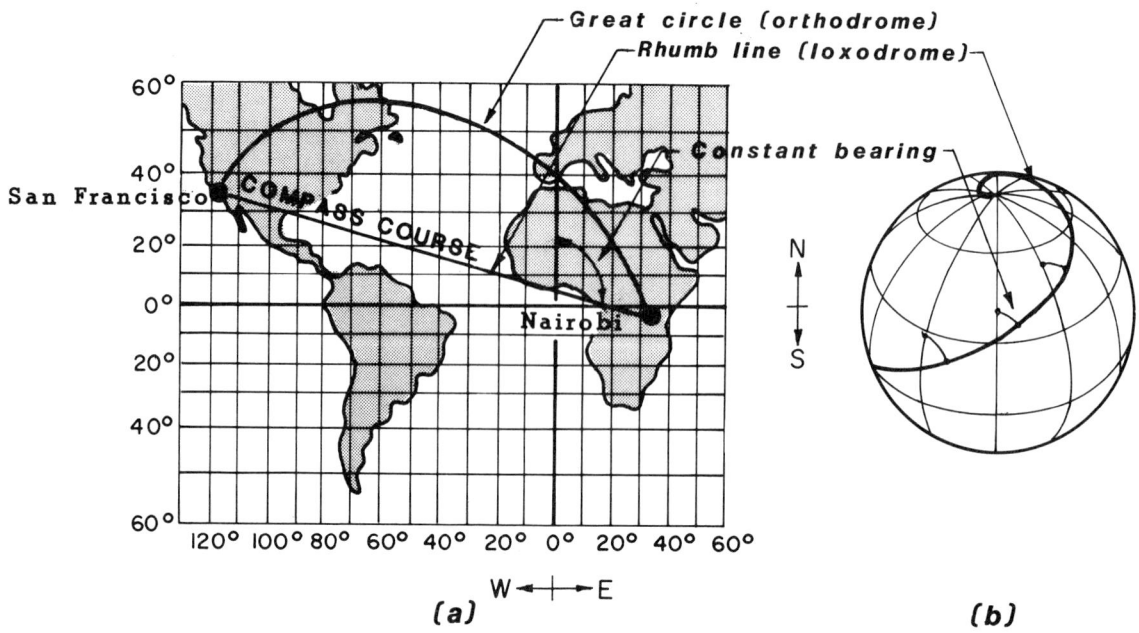

Fig. 13.8 Mercator projection.

Table 13.2 Size of land areas on Mercator maps.

Latitude	Exaggeration of Land Areas
0° (Equator)	None (True scale east-west)
45°	Doubled
60°	Four times too large
75°	Approximately 16 times too large
80°	More than 33 times too large
90° (Poles)	Infinite

the compass direction to follow a straight course, although a curved line is actually followed. The line will appear as a spiral curve on the spherical surface of the earth, Fig. 13.8 (b).

The shortest distance between two points on the earth is a *great circle* or *orthodrome*. Its direction is constantly changing, crossing successive meridians at different angles. A loxodrome will not be the shortest distance between two points on the earth unless the great circle route is along the equator or a meridian. Where short distances occur, there is no practical difference between the rhumb and great circle. In longer distances, however, the great circle may be used to determine the course. All straight lines drawn between any two points on a gnomonic map will show the great circle distance between them. This line can then be plotted on a Mercator map where it becomes a curved line. It can be converted into a series of short rhumb lines to obtain the compass bearings for approximating the great circle course.

Transverse Mercator projection. This is a mathematical variation of the standard Mercator projection and is obtained when the cylinder of projection is tangent to the sphere along one of the meridians passing through the poles. Fig. 13.9 (a) shows a cylinder tangent to the 0°-180° (central meridian) meridians. Except for the central meridian, all meridians are curved lines, Fig. 13.9 (b). All parallels except the equator are ellipses. The projection is conformal and the meridians and parallels are at right angles, although the graticule is not rectangular. True scale exists along the meridian of tangency, making it suitable for mapping areas extended in a north-south direction. It is particularly adaptable to mapping along a selected meridian as variation in scale and azimuth for a small band is negligible. In the United States, the projection is widely used in connection with state-wide plane coordinate systems. It is also extensively used for navigation charts in polar areas and for a vast majority of the world's modern topographical map series. A special application of the Transverse Mercator is the Universal Transverse Mercator (UTM) projection that is used as the basis for the UTM grid (see chapter 15).

Oblique Mercator projection. Oblique Mercator projection differs from the standard Mercator and Transverse Mercator in that the cylinder of projec-

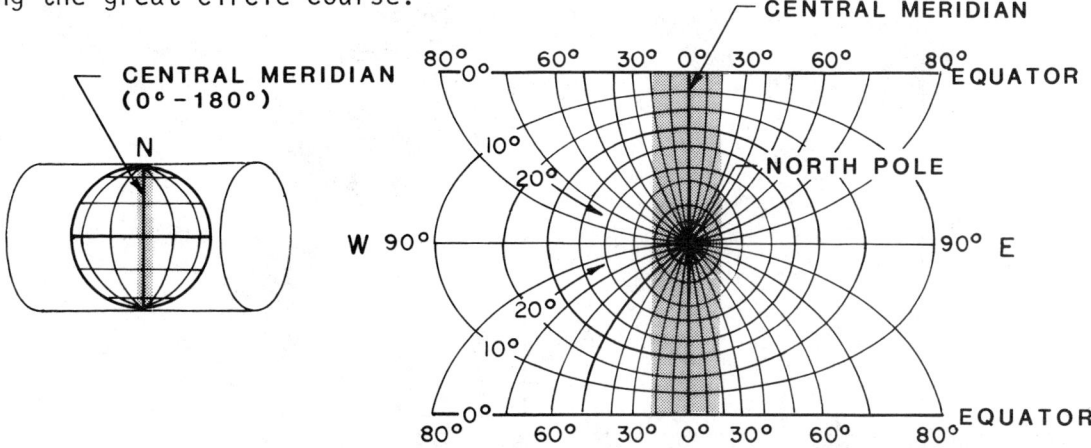

(a) (b) Appearance of graticule in northern hemisphere (southern hemisphere identical)

Fig. 13.9 Transverse Mercator projection.

tion is tangent to the sphere along some great circle other than the equator or a meridian, Fig. 13.10 (a). Some authors do not make this differentiation and refer to any movement of the cylinder as a transverse projection. The projection is derived mathematically and is easy to recognize since the parallels appear as sine curves and the meridians as curved lines, Fig. 13.10 (b). The projection is conformal with true scale occurring along the great circle of tangency. Distortion of shapes and areas increases away from the great circle of tangency. This projection is ideal for mapping a strip of a great circle path.

Miller projection. This projection resembles Mercator projection, but shows less exaggeration of area in the higher latitudes, Fig. 13.11. The projection is neither conformal or equal-area. It is widely used in atlases for climatic maps showing barometric pressures, winds, temperatures, and ocean communications.

Gall's projection. A cylinder is assumed to cut (secant) the sphere at the 45°N and 45°S parallels, Fig. 13.12. The parallels are projected from the antipodal point on the equator. The area between the two 45° parallels (standard parallels) is reasonably good since it exaggerates northern and southern areas less than Mercator projection. It is used by meteorologists in showing distribution of air temperature and pressure. Geographers also find this projection useful for showing the distribution of economic, climatic, and population data. The projection is neither equal-area or conformal and the scale is not constant.

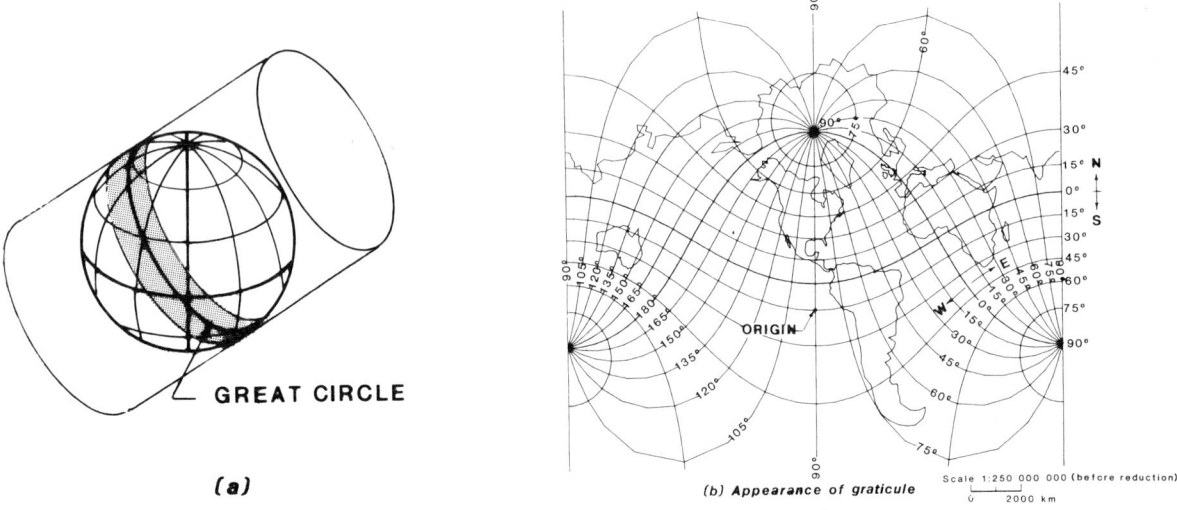

Fig. 13.10 Oblique Mercator projection.

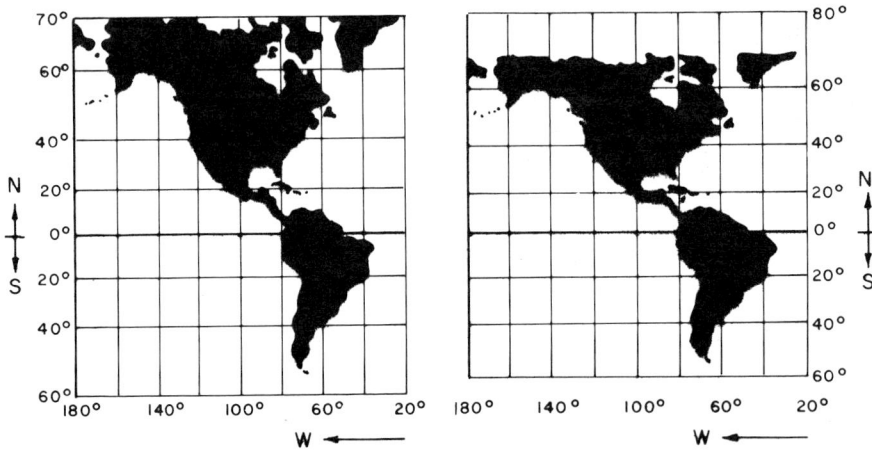

Fig. 13.11 Miller and Mercator projection comparison.

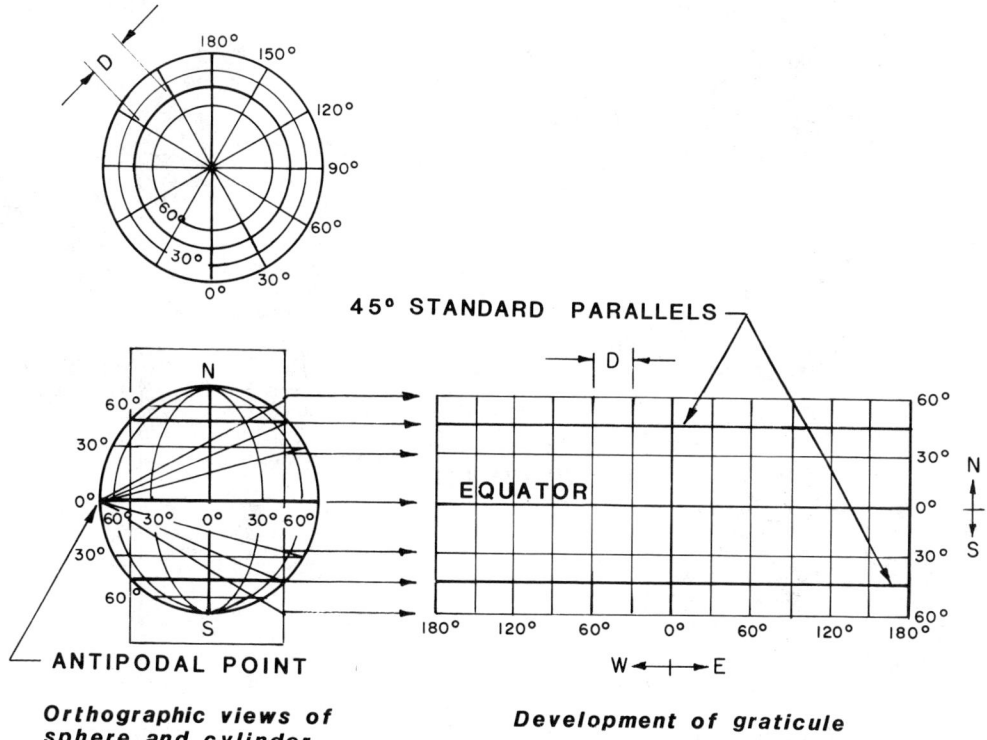

Fig. 13.12 Gall's projection.

13.5 CONICAL PROJECTIONS

Conic projections are derived from a tangent or secant cone that can be developed. The apex of the cone is usually on the earth's polar axis. The sides of the cone can be tangent to the sphere along a selected parallel of latitude or can intersect the earth along two parallels. Parallels of tangency and secancy are called standard parallels. A constant scale is maintained along these lines with distortions increasing away from the standard parallels.

Perspective conic projection. For perspective conic projection a cone tangent to the sphere at the 30° central parallel is used with the apex of the cone falling on the sphere's axis prolonged, Fig. 13.13. This line is called the standard parallel. The center of the sphere is used to project the parallels on the surface of the cone. The meridians appear as straight lines equally spaced apart and coming closer together as they approach the poles. Parallels are concentric arcs whose spacing increases towards the poles. To avoid undue distortion, the area projected upon the cone is limited to that included between the 0° and 60° parallels. True scale exists along the parallel of tangency and increases north and south of the standard parallel. The projection is best suited for mapping areas in the vicinity of the standard parallel. It is used primarily for showing areas in the middle latitudes and is of greater value when modified for several of the other conic projections.

Bonne projection. This is a mathematical modification of the perspective conic projection where one standard parallel and a central meridian are chosen to draw the projection, Fig. 13.14. The cone's slant height, r, is taken as the radius of the standard parallel on the map. Other parallels are drawn as concentric circles that are truly spaced from the standard parallel. True scale can be measured along all the parallels while the central meridian is the only one along which the scale is true. As the distance is increased from the central meridian, the intersection of parallels and meridians depart increasingly from right angles resulting in distortion of shape in areas. This is an equal-area projection and the representation of areas in a narrow zone along the standard parallel and central merid-

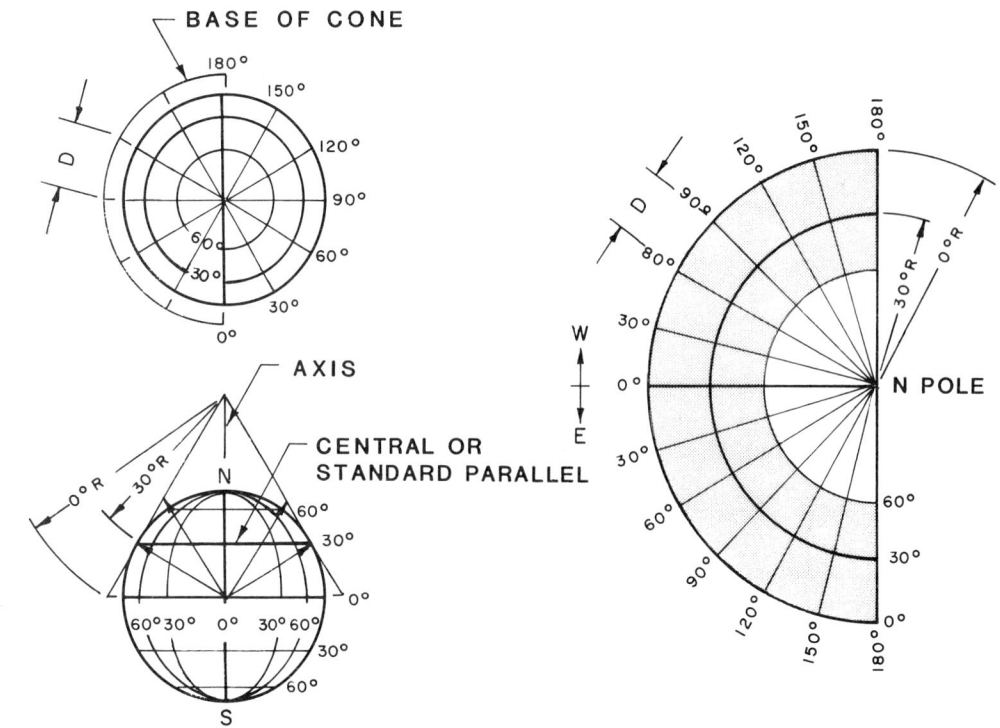

Fig. 13.13 Perspective conic projection with one standard parallel.

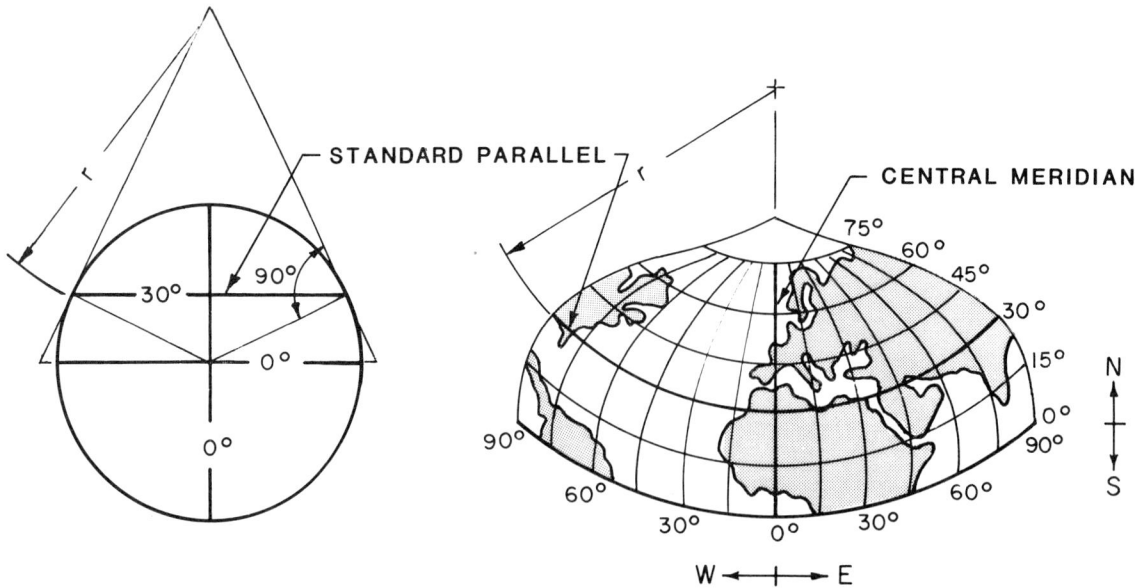

Fig. 13.14 Bonne projection.

ian is reasonably good. Maps of the continents of Europe, Asia, North America, and South America in older atlases frequently used this projectisn since it was well suited for areas in the middle latitudes. This projection, however, is not presently used very extensively for current mapping of large areas.

Lambert conformal conic projection. A tangent cone may be used for this projection, however, a secant cone intersecting the sphere to give two standard parallels is usually the basis for the projection, Fig. 13.15 (a). The graticule for a conic projection with two standard parallels could be developed in a similar manner as illustrated for perspective conic projection with one standard parallel in Fig. 20.13. For Lambert's conformal conic projection, however, the radii of the parallels and their spacing along the meridians are calculated so that the scale errors at any point along parallels and meridians are the same. The graticule will show meridians as straight lines that converge at the poles and parallels as concentric circles that will be nearly equally spaced intersecting the meridians at right angles, Fig. 13.15 (b) and (c). Although the projection is not equal-area, distortions are small. The scale is true along the standard parallels. Between the two standard parallels the scale is smaller than true and outside the standard parallels it is greater than true scale. By carefully selecting the standard parallels, a good representation of limited latitude can be obtained.

The projection is commonly used in military mapping and of areas running predominantly in an east-west direction as in the case of the United States and France. It is widely used for state plane coordinate systems in surveying (see chapter 15) and for aeronautical charts. Great circles will plot as gradual curves departing only slightly from a straight line.

Alber's equal-area conic projection. Very little difference exists between Alber's equal-area and Lambert's conformal conic projection, Fig. 13.16. One or two standard parallels of true scale can be used. The meridians are drawn as in Lambert's but the parallels have their radii calculated so that the areas formed between the parallels is the same as found on the sphere. Equal-areas are thus formed making this projection ideal for mapping mid-latitude countries such as the United States, Canada, and Australia. The USGS uses

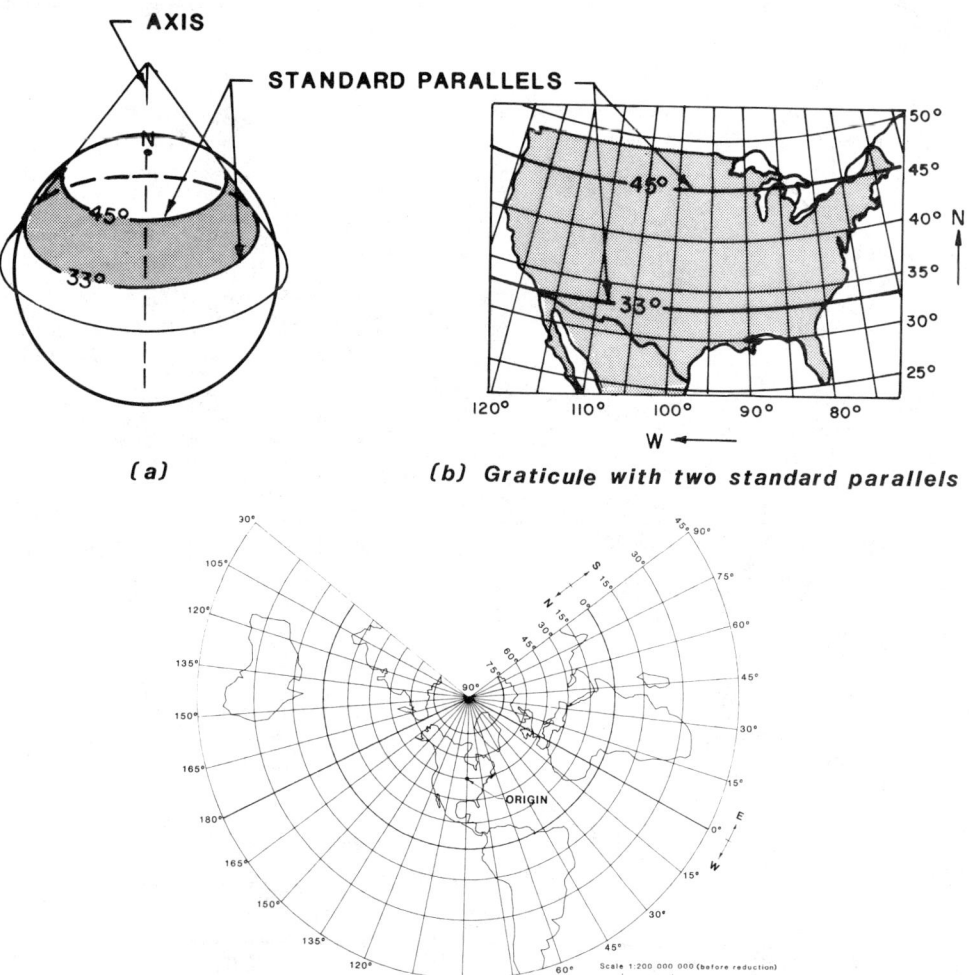

(a)

(b) Graticule with two standard parallels

(c) Graticule with one standard parallel

Fig. 13.15 Lambert conformal conic projection.

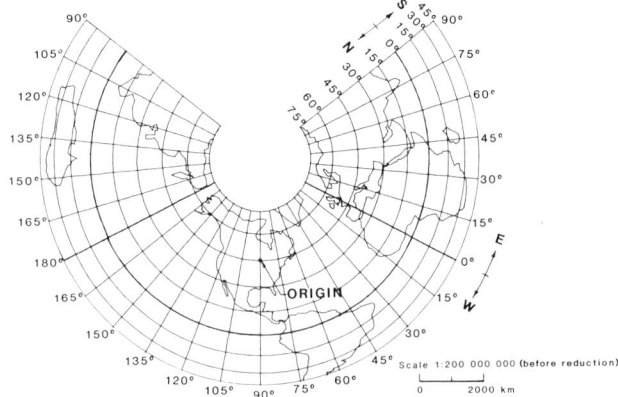

Fig. 13.16 Alber's equal-area conic projection.

this projection with standard parallels at 29.5° and 45.5° for its state based maps at 1:500 000 scale.

Polyconic projection. In polyconic projection a series of cones are used, Fig. 13.17 (a). One tangent cone is drawn for each parallel of latitude with a tangent cylinder drawn at the equator. Each parallel is then developed from center points on an extension of the central meridian. A partial development of each cone 10° on each side of the standard parallel to which the cone is tangent is illustrated in Fig. 13.17 (b). Parallels are non-concentric circles truly divided along the central meridian. The meridians appear as equally spaced lines along each parallel. True scale can be measured along the central meridian and along all standard parallels.

A map drawn on the graticule of Fig. 13.17 (b) would be interrupted and difficult to draw due to the open spaces

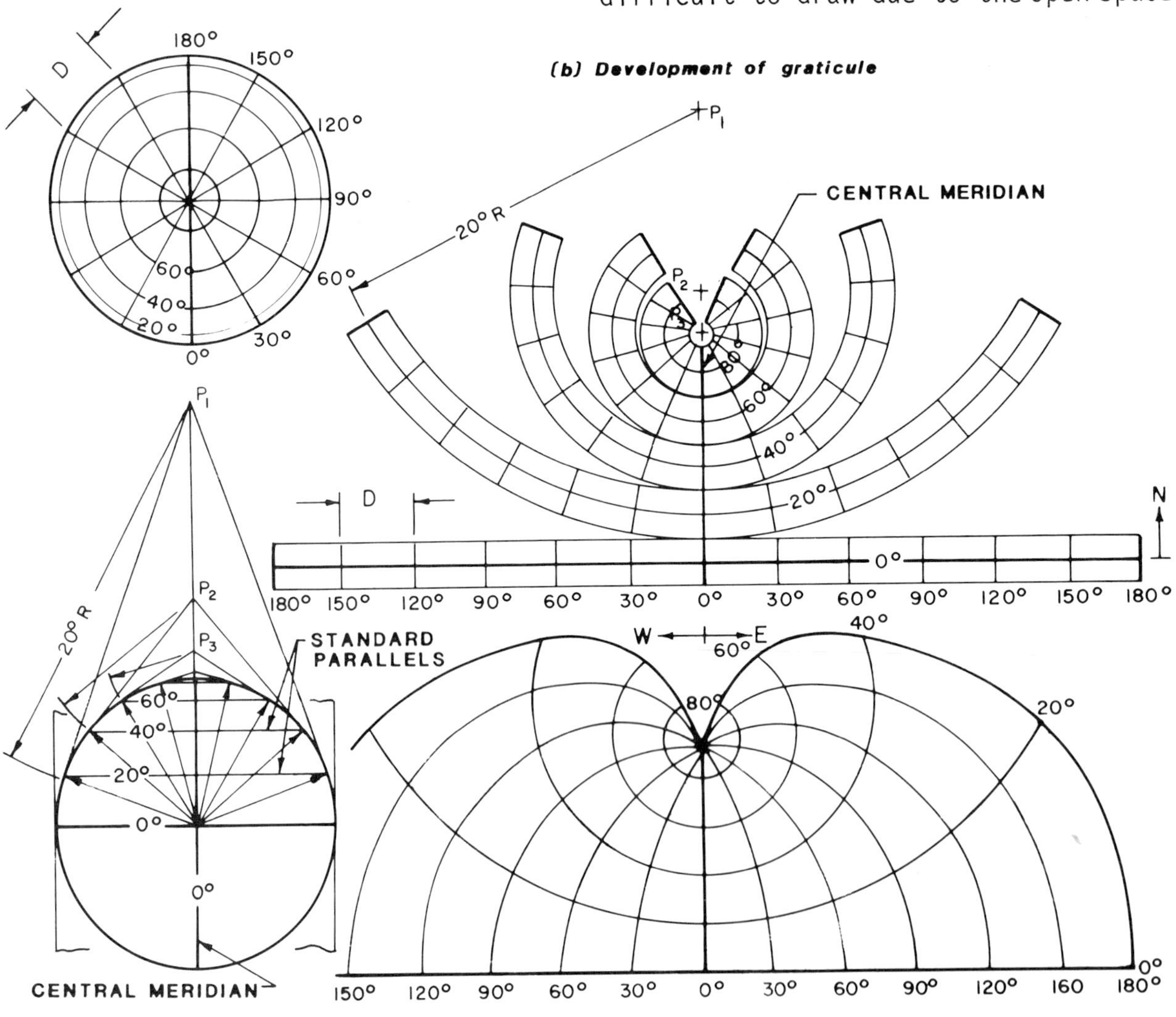

(a) Orthographic views of sphere and cones

(b) Development of graticule

(c) Apperance of graticule with meridians stretched

Fig. 13.17 Polyconic projection.

that exist between the parallels. For a complete map to be drawn, the meridians must be stretched and connected together to form continuous lines. This stretching can be illustrated by drawing the arcs representing the parallels, Fig. 13.17 (c). The points at which the meridians cross the parallels are located by setting off along each parallel the corresponding chord distances between the meridians. The meridians are drawn as curved lines passing through these points. As the distance east or west of the central meridian increases, the scale along the meridians also increases.

The polyconic projection is not conformal or equal-area. Although distortion increases east or west of the central meridian, it is minimal for small areas. Generally the projection is not well adapted to areas of wide longitudinal extent. It is used for maps of areas which are long in a north-south direction and narrow in an east-west direction. The USGS used polyconic projection for many years on the 7.5-minute topographic maps. Since 1953 the Lambert conformal conic and Transverse Mercator projection (state coordinates) have been used for the topographic map series.

13.6 AZIMUTHAL PROJECTIONS

A whole family of projections has been developed based upon a plane surface that is tangent to or intersects the spherical earth at any point. In all cases a part of the sphere's surface is projected upon a plane from some station point which may be different for the various projections in this group. Where it is desirable to determine exact directions from a specific point, azimuthal projections (sometimes referred to as zenithal projections) are used. This class of projections is most usefully applied to the polar regions. All azimuthal projections have the following in common:

1. All great circles passing through the center of projection appear as radiating straight lines of true direction (bearing or azimuth).
2. Points that are equal distance from the center of projection on the sphere are equal distance on the map.
3. The distortion is equal for all places that are equal distance from the center. The center is distortion free.

Orthographic projection. The orthographic projection of a globe onto a flat plane involves the projection of the view using parallel projectors that are at infinity and perpendicular to the plane of projection (picture plane), Fig. 13.18 (a). This projection can show only a full hemisphere. The projection is not equal-area or conformal. There is great distortion in the outer portion of the projection, and the scale changes rapidly along the meridians in the top view and along both the meridians and parallels in the front view. In the equatorial case, the meridians are parts of true ellipses with the parallels appearing as straight horizontal lines spaced more closely near the poles, Fig. 13.18 (b). This graticule form helps to distinguish the projection from other zenithal equatorial type projections that do not have any straight horizontal lines. Since a good three-dimensional pictorial effect can be obtained with orthographic projection, it is used mainly by illustrators to convey a realistic view of the earth. A relatively true picture can be seen of land masses that are located near the center of projection. The projection has also been used for maps of the moon since it shows the moon as seen from an infinite distance.

Gnomonic projection. Gnomonic projection may be considered a perspective projection upon a plane tangent to the sphere at some point. The station point from which the projection lines are drawn is located at the center of the sphere. The projection is limited to less than a hemisphere since the edge of the sphere is an infinite distance from the center of projection. Every great circle is represented by a straight line which is the most important and useful characteristic of this projection.

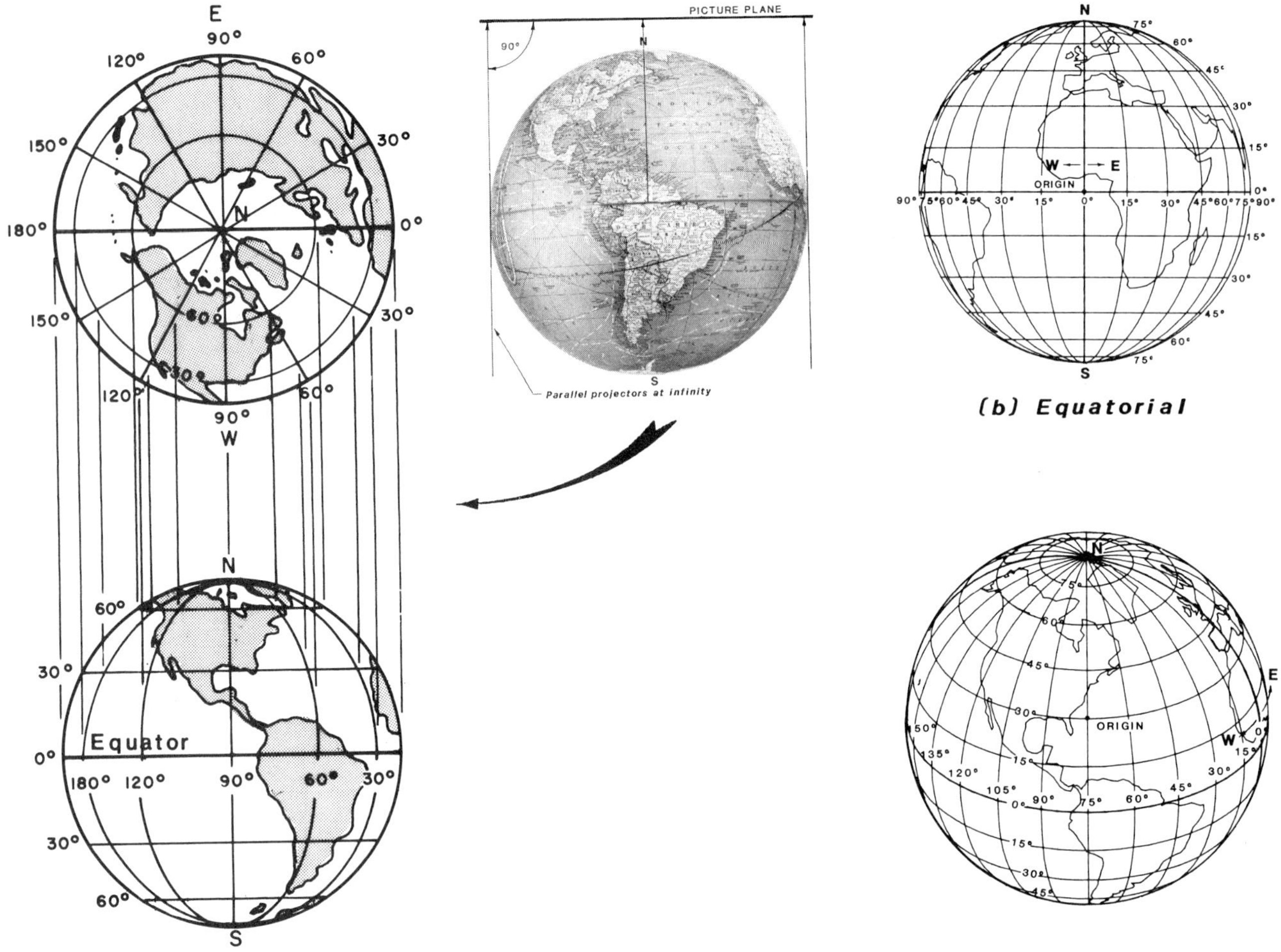

(a) Polar (b) Equatorial (c) Oblique

Fig. 13.18 Orthographic projection of the globe to a plane.

Since the shortest distance between any two points is a great circle, the projection has little value other than its use to plot this course as a straight line for navigation. An increasing scale away from the center of the map results in extreme distortion of areas thus limiting its use for general purpose maps, Fig. 13.19.

In polar gnomonic projection the plane of projection is tangent to one of the poles, Fig. 13.20. Meridians are represented by equally spaced straight lines radiating from the pole and parallels are represented by concentric unequally spaced circles. Distortion increases measurably away from the poles.

In equatorial gnomonic projection the plane of projection is tangent to the equator, Fig. 13.21. The equator and all meridians are represented as parallel straight lines. The parallels are hyperbolic curves concave toward the poles. Spacing of all lines increases

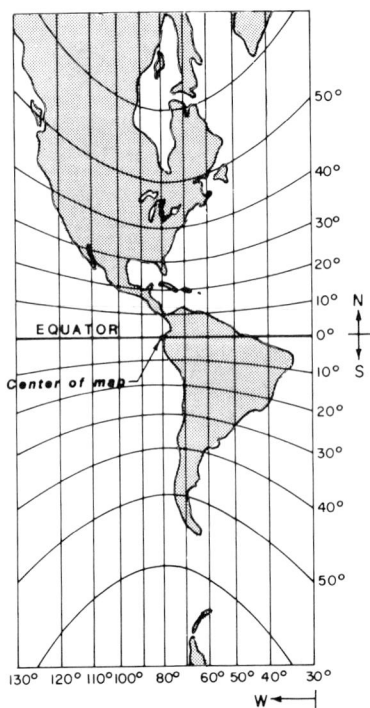

Fig. 13.19 Gnomonic projection distorts areas.

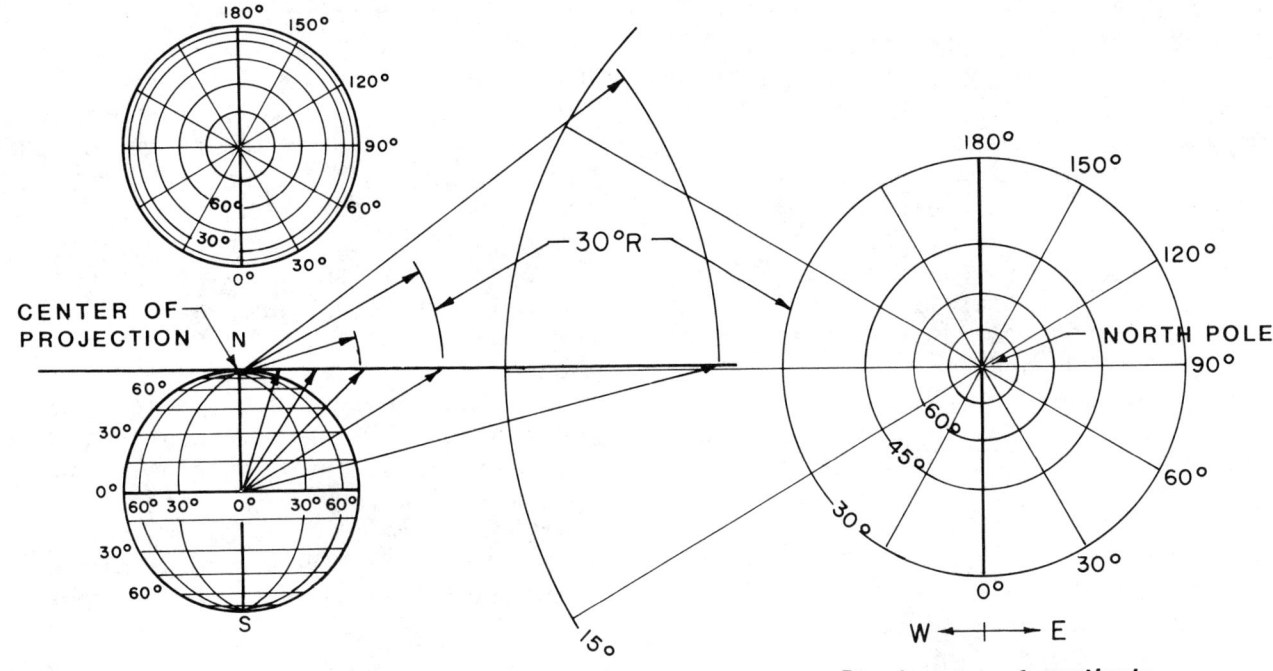

Orthographic views of sphere and plane *Development of graticule*

Fig. 13.20 Polar gnomonic projection.

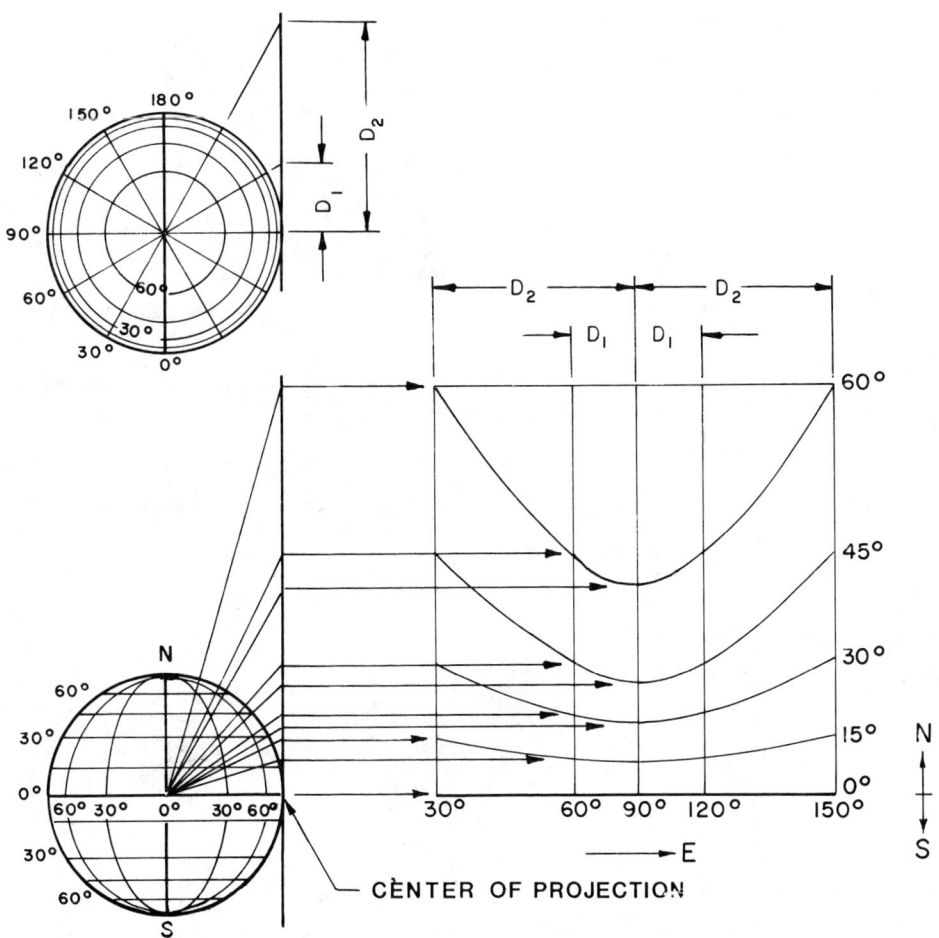

Orthographic views of sphere and plane *Development of graticule*

Fig. 13.21 Equatorial gnomonic projection.

rapidly as the distance is increased from the center of the sphere outwards.

In oblique gnomonic projection the plane on which the projection is drawn may be tangent to the sphere at any point other than the poles and equator, Fig. 13.22 (a). This projection requires a large number of calculations before the graticule can be drawn, Fig. 13.22 (b). Meridians are represented as converging straight lines towards the poles. The parallels are curved lines, concave towards the poles.

Azimuthal equidistant projection. The appearance of the graticule for this projection is determined by mathematical means and depends upon the point selected as the projection center. If the center of projection is one of the poles, polar case, the meridians are represented as straight lines and the parallels as equally spaced concentric circles, Fig. 13.23 (a). All straight lines radiating from the center of the map represent great circles at true azimuth from the center, and distances along these lines are true to scale. Measurements in an east-west direction are not correctly represented and distortion increases in the southern latitudes.

When the center of projection is shifted to the equator, equatorial case, the meridian of the center point and its anti-meridian form a straight line diameter of the circle with the equator, also a straight line diameter perpendicular to it, Fig. 13.23 (b).

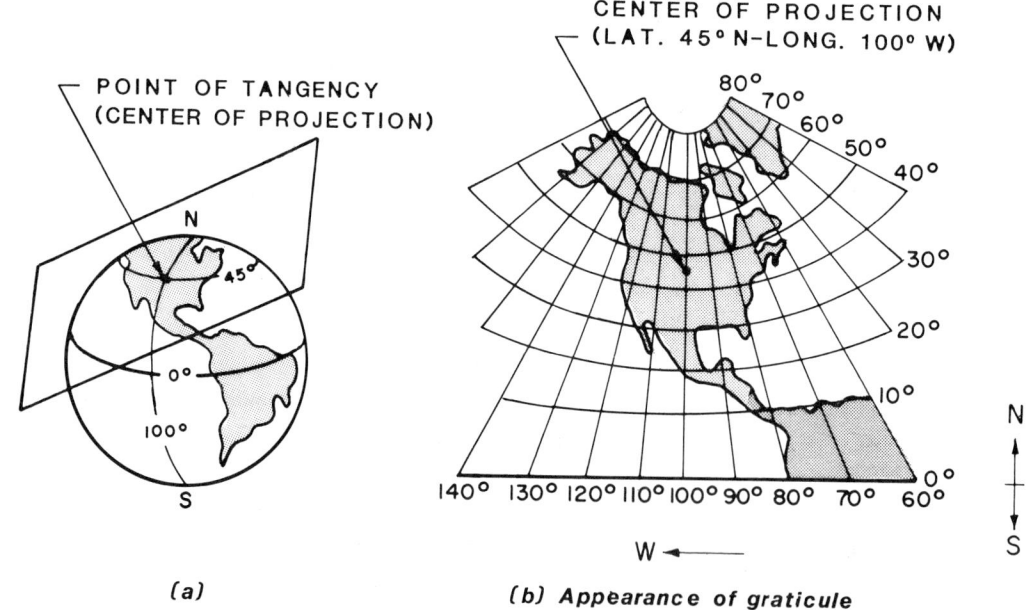

Fig. 13.22 Oblique gnomonic projection.

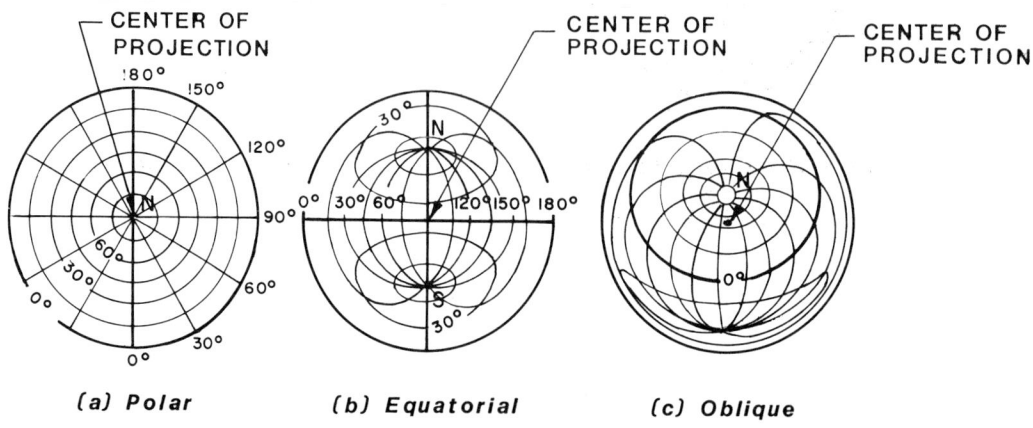

Fig. 13.23 Azimuthal equidistant projection.

The center of projection may be shifted to any point, oblique case, with the graticule appearing as curves, Fig. 13.23 (c). By centering the map projection at any point on the sphere, distortion for a specific region is minimized.

This family of projections is used for maps of the Artic and Antarctic. They are also employed for celestial maps and are particularly useful in communication and aeronautics about any chosen point on which the map is centered.

Azimuthal equal-area projection. Azimuthal equal-area projection is similar in appearance to the azimuthal equidistance projection previously described. Equal-area is obtained by adjusting the meridians and parallels that are projected upon a tangent plane by mathematical methods. Considerable distortion will exist in the shape of countries except near the center of the map. For the polar case, Fig. 13.24 (a), the meridians are represented as straight lines radiating from the center of the map. The parallels are drawn as concentric circles so the area between any two parallels is the same as on the sphere.

The equatorial case, Fig. 13.24 (b), will show the equator and central meridian as straight lines. Meridians and parallels are curved lines, decreasing in spacing from the center outward.

In the oblique case, Fig. 13.24 (c), the north and south poles are enclosed by a complete circle with the remaining parallels appearing as curved lines. The meridians are drawn as radiating arcs from the poles.

This family of projections has found wide use in atlases for the preparation of polar maps and hemisphere maps showing land-water relationships.

Stereographic projection. In Stereographic projection the plane of projection may be placed tangent to the sphere at any point. The station point is taken to be at a point on the surface of the sphere directly opposite the point of tangency. All great circles on the sphere will show as arcs of circles on the map, except those which pass through the center and they are straight lines. The parallels and meridians intersect at right angles and the projection is conformal with small areas formed by the meridians and parallels identical in shape to the corresponding area on the sphere.

In Polar stereographic projection the plane of projection is placed tangent to one of the poles, Fig. 13.25. The meridians appear as radiating straight lines and the parallels as concentric circles spaced further apart from the center outward. The scale increases conformally in all directions and each small section on the map has the true shape of its counterpart on the sphere. The military grid system in the polar regions discussed in chapter 15 utilizes this projection in a system referred to as Universal Polar Stereographic (UPS).

In equatorial stereographic projection, Fig. 13.26, all parallels except the equator are curved with the curvature of each rapidly increasing toward the poles. The spacing of both parallels and meridians increases from the center outward. This projection is often used in atlases to show hemispheres.

Oblique stereographic projection, Fig. 13.27, is easy to recognize since only one meridian occurs as a straight line. All other meridians as well as the parallels are curved lines. As in other oblique projection systems, the

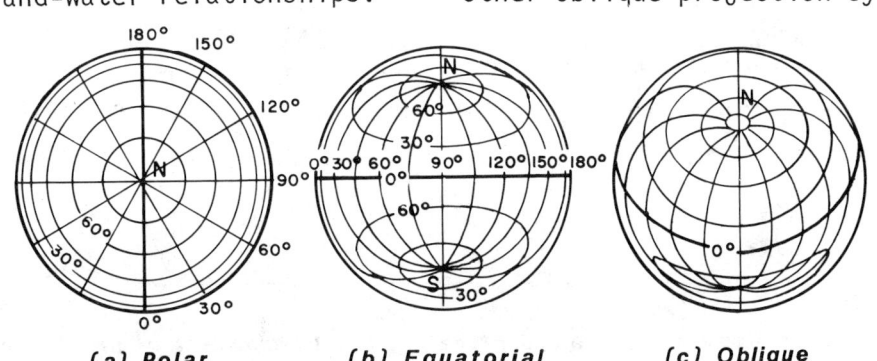

(a) Polar   (b) Equatorial   (c) Oblique

Fig. 13.24 Azimuthal equal-area projection.

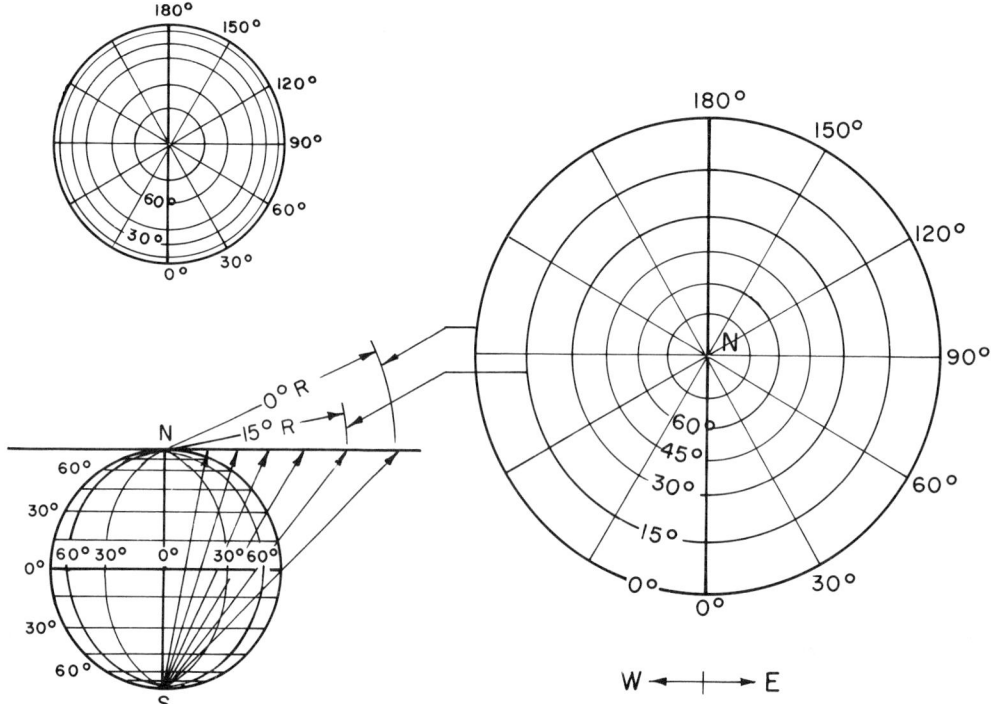

Orthographic views of sphere and plane *Development of graticule*

Fig. 13.25 Polar stereographic projection.

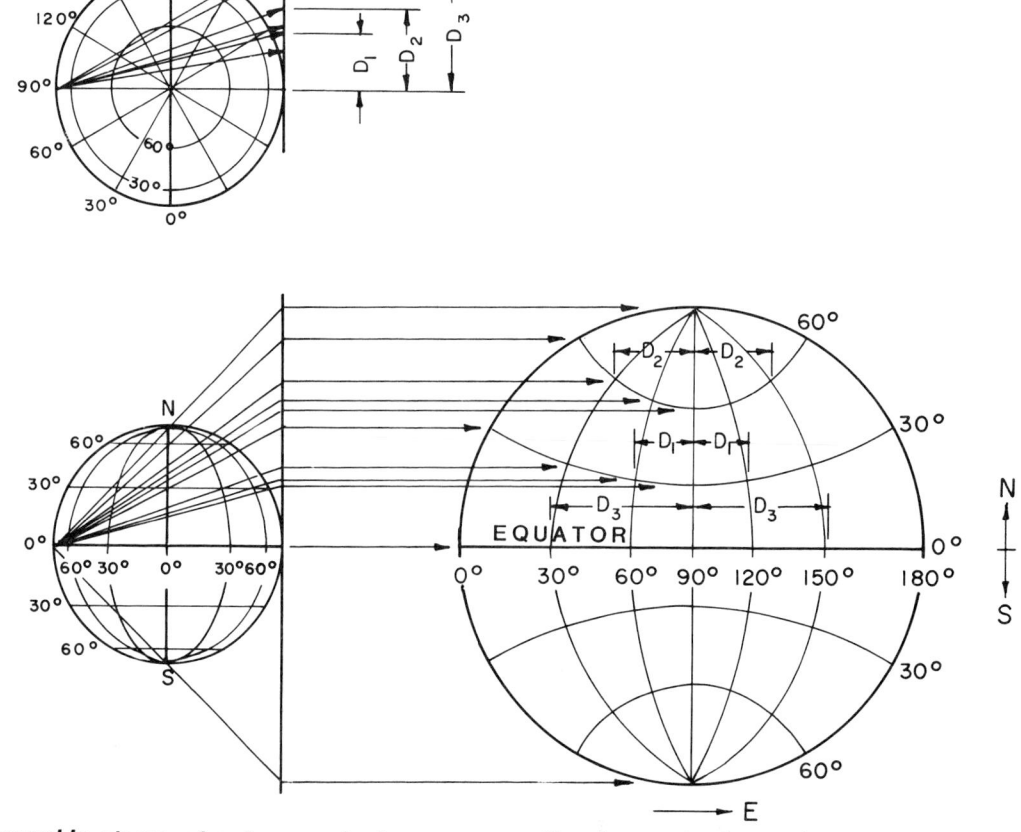

Orthographic views of sphere and plane *Development of graticule*

Fig. 13.26 Equatorial stereographic projection.

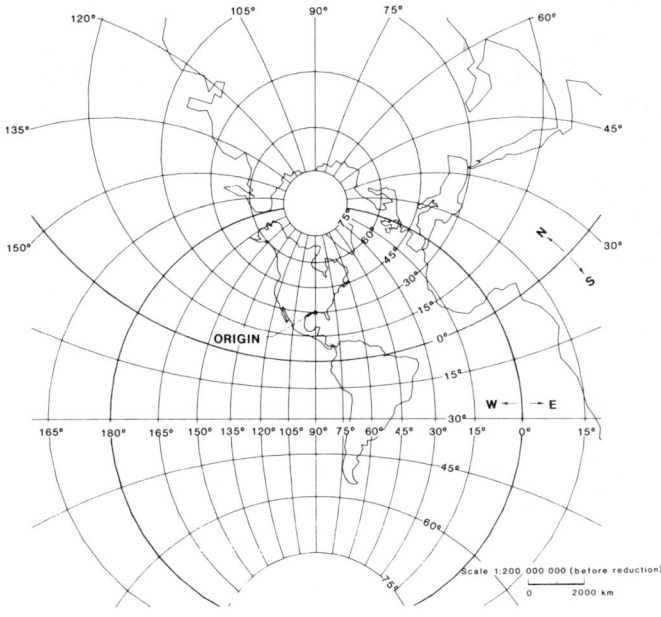

Fig. 13.27 Oblique stereographic projection.

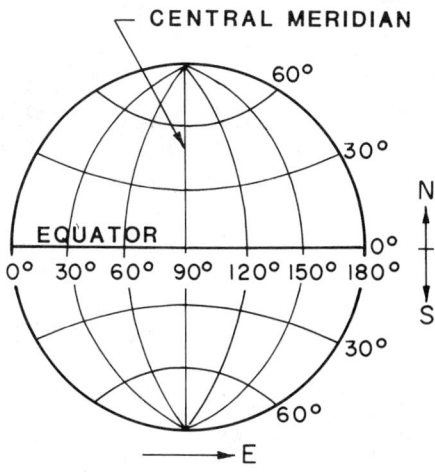

Fig. 13.28 Globular projection.

plane of projection may be centered at any point on the sphere. This class of projection is used extensively for maps of the polar regions and hemisphere maps.

Globular projection. This pictorial projection which is neither conformal or equal-area is easy to understand and construct. Fig. 13.28 illustrates this projection of the eastern hemisphere with the plane of projection taken to be tangent at the equator. The central meridian, equator, and the sphere circumference are divided equally and arcs of circles are drawn through the intersecting points. No scientific properties are attributed to this projection, but it readily affords general information concerning chief land masses since angular distortions and scale changes are not excessive. The projection is popular for hemisphere school maps.

13.7 SPECIAL PROJECTIONS

There are several projections that cannot be readily related to the three general classifications of projections presented. These are purely mathematical and cannot be illustrated graphically.

Sinusoidal projection. This projection (sometimes called Sanson-Flamsteed projection) has straight, horizontal parallels equally spaced at their true distances from the equator, Fig. 13.29. The center meridian is straight, with the others appearing as sine curves. True scale is shown only on the central meridian and on each parallel. Since the remaining meridians are skewed towards the poles, considerable distortion in shape occurs. It is, however, an equal-area projection. To reduce distortion it is common to interrupt the map in the oceans and recenter each continent on its own central meridian, Fig. 13.30.

For maps on the equatorial regions, South America, Africa, Australia, and for smaller countries extended in a north-south direction at high latitudes, this projection is extensively used. Atlases frequently use this projection to portray continental features.

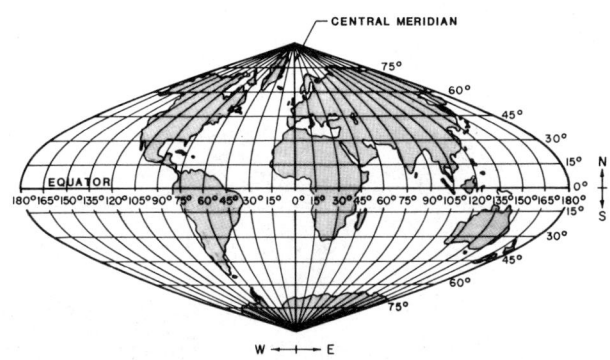

Fig. 13.29 Sinusoidal projection.

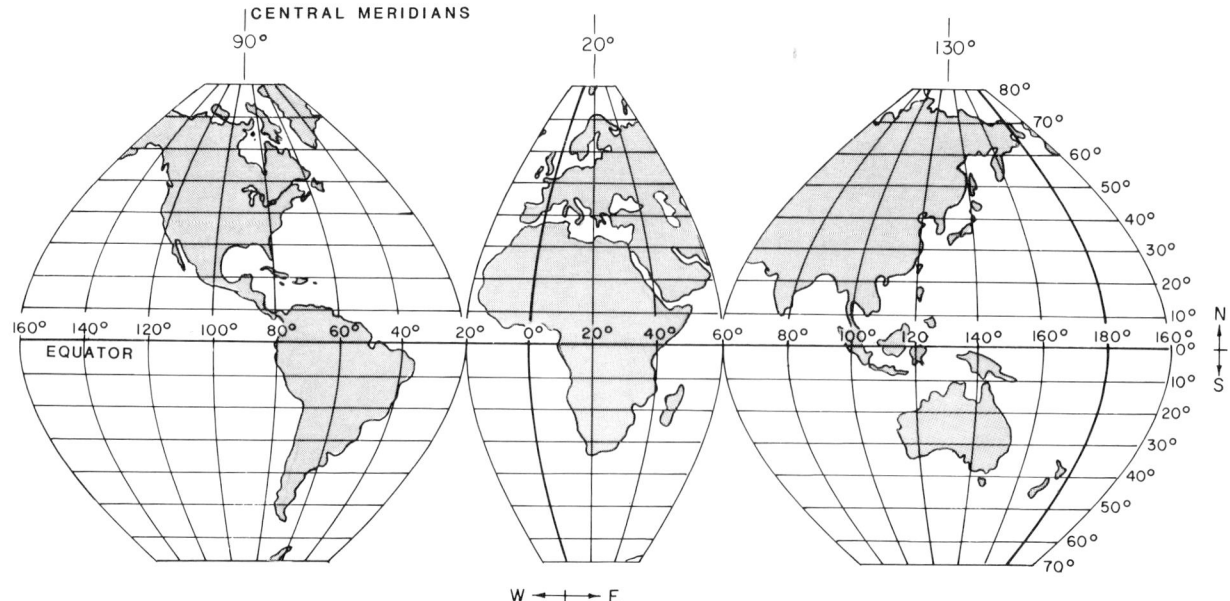

Fig. 13.30 Interrupted sinusoidal projection, polar regions omitted.

Mollweide's homographic projection. One of the well known elliptical projections of the sphere is Mollweide's homographic, Fig. 13.31. The ellipse is drawn to enclose an area equal to that of the sphere at the same scale. The equator is drawn as a straight line and divided equally. A central meridian half the length of the equator is drawn and divided so that the areas of the bands between the parallels are proportional. The parallels are drawn as horizontal straight lines and a central circle of radius, R, is drawn through the poles which is termed the central hemisphere. Each parallel is equally divided and the divisions are connected to form the elliptical meridians.

The projection is equal-area, but there is little uniformity in linear scale except along the major axis (equator) of the ellipse. Distortion in the shape of areas is considerable as the distance increases from the center of the map. The projection may be interrupted and a new central meridian established through each continent so that areas of greater distortion in the outer parts of the projection will be reduced. The projection is used for hemisphere maps in atlases and books on economic geography. It is particularly useful for showing distributions such as population for the whole world.

Hammer-Aitoff projection. This projection is derived by projecting the Lambert equal-area projection into an ellipse, Fig. 13.32. It is similar to Mollweide projection with the exception that the parallels are curved. This reduces angular distortion since the parallels meet the meridians at less oblique angles. The projection is equal-area and provides a realistic map of the

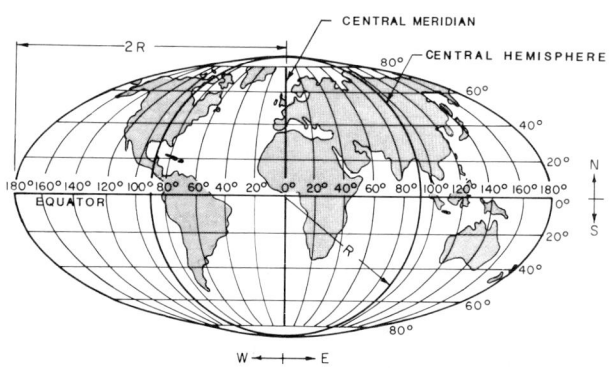

Fig. 13.31 Mollweide's homographic projection.

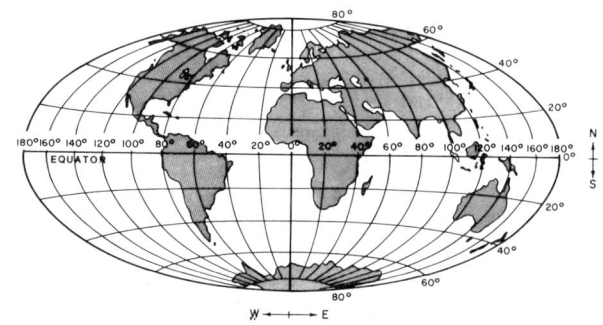

Fig. 13.32 Hammer-Aitoff projection.

world. It is used for showing physical geography as well as physical or cultural distributions similar to Mollweide.

Eckert projection. The Eckert projection makes the poles into lines half the length at the equator, Fig. 13.33. This produces less squeeze than in Mollweide and Aitoff projections, however, stretch in the polar regions is excessive. The parallels are straight lines spaced so as to make the projection equal-area. The meridians will be sinusoidal or elliptical in appearance. As in Mollweide and Aitoff projection, the map may be interrupted and recentered on several central meridians to obtain better land shapes. The projection is commonly used in atlases for showing statistical material.

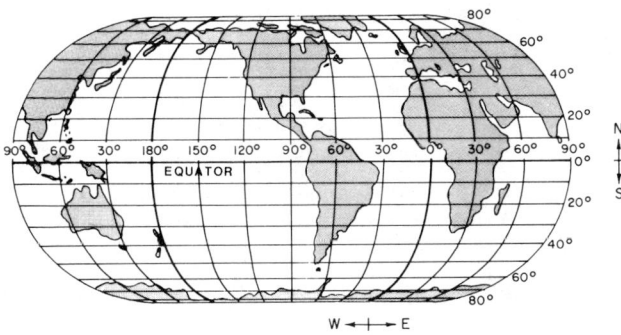

Fig. 13.33 Eckert projection.

Goode's interrupted homolosine projection. This projection is derived by combining several of the better qualities found in sinusoidal and Mollweide projection, Fig. 13.34. For latitudes up to 40° north and south the sinusoidal is used and for areas poleward of these latitudes Mollweide is used. The projection is equal-area and many variations of the projection are possible by using the principle of interruption. The projection may be interrupted anywhere to allow oceans or continents to be shown in the center of projection, resulting in better shape of areas. The parallels are represented by straight lines which enable the map user to study and compare the distributions of climates, population, natural resources, and other world relationships.

The individual parts of objects may be represented and defined clearly by drawing a series of related views. Drawings that are to be used in the production or construction of an object, however, can only be considered half finished if additional information in the form of dimensions, notes, and symbols has been omitted. This information must be thoughtfully placed on the drawing to include all of the descriptive information required to manufacture the object. An understanding of the following fundamentals is essential in determining the best method to use for dimension selection and placement.

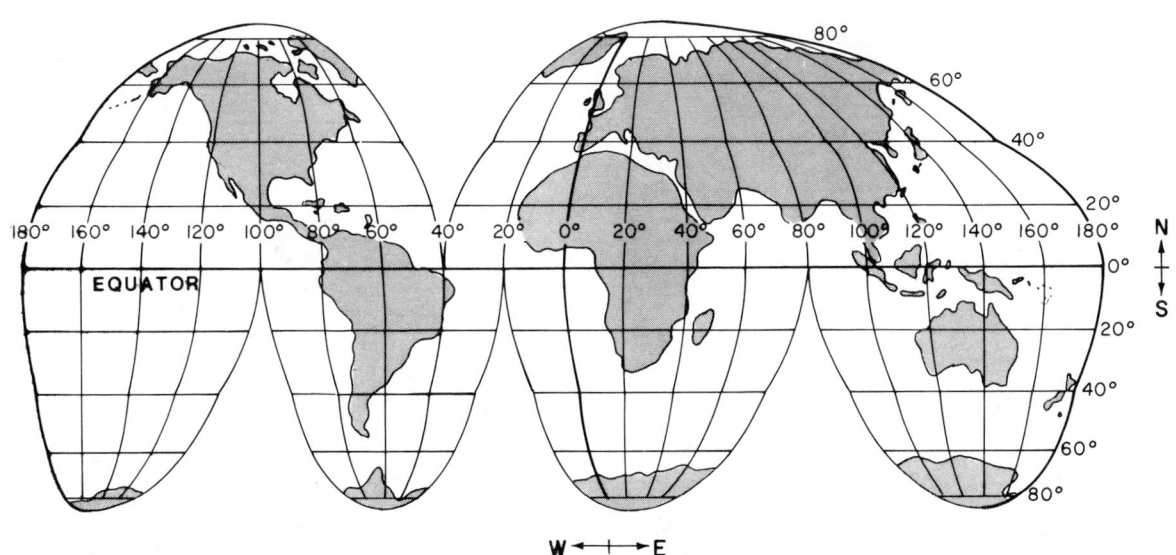

Fig. 13.34 Goode's interrupted homolosine projection.

MAP ORIENTATION

14.1 DIRECTION

A map's value is diminished if it lacks a means of determining direction or of orienting it in respect to some base line. *Bearing* is one method for finding the direction of a line by measuring the horizontal angle it makes with the meridian line running due north and south. An angle 90° or less is given for the bearing, measured either from the north or the south direction. If the bearing of a line, from a given starting point, is said to be N45°E, it means that the line is turned away from the north line 45° toward the east, Fig. 14.1. The bearing of a line is not affected by its slope since all measurements on a map represent only horizontal distances.

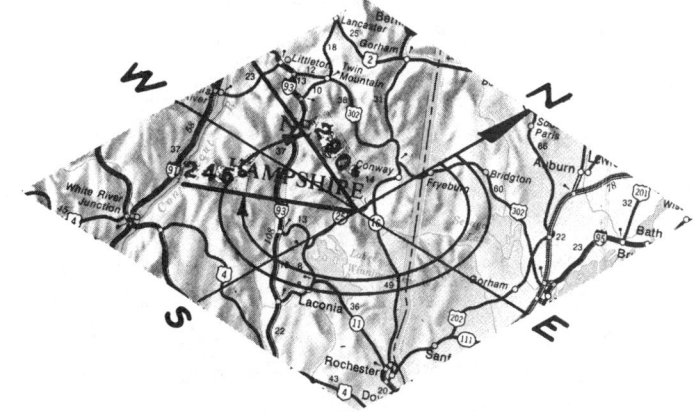

NOTE *
NE bearings are azimuths of 0°-90°
SE   "   90°-180°
SW   "   180°-270°
NW   "   270°-360°

* Bearings and azimuths are identical in NE quadrant

Fig. 14.2 Azimuth.

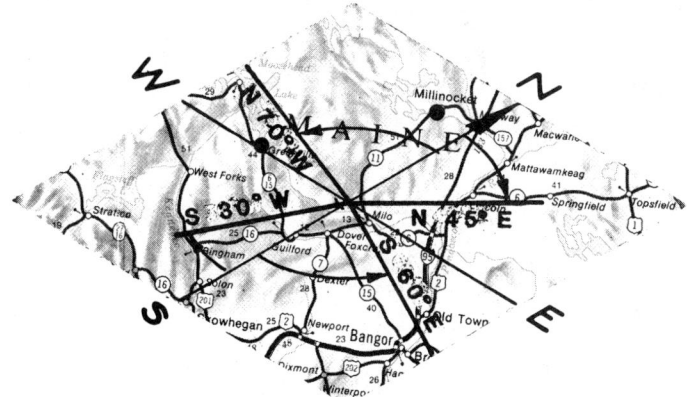

Fig. 14.1 Bearing.

Azimuth is another method of specifying the direction of a line and is defined as the horizontal angle measured 0° to 360° in a clockwise direction from the north-south meridian. Fig. 14.2 illustrates the azimuth of several lines measured from true north. A line actually has two azimuths or bearings since direction may be viewed from either er' The azimuth or bearing looking from observer is a forward azimuth or bea.

ing, while the reverse of this is a back azimuth or bearing. Back bearings retain the same number, but the letters designating the quadrant are opposite. For example the back bearing of a line S30°E is N30°W. Back azimuths are found by adding or subtracting 180° from the forward azimuth, Fig. 14.3.

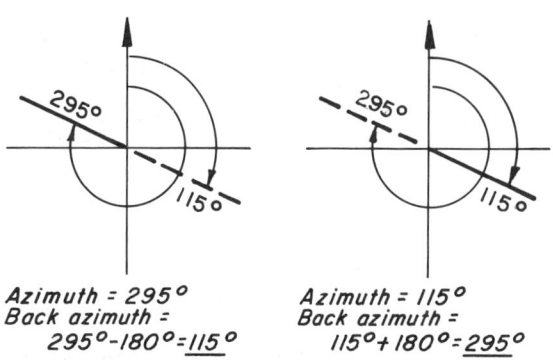

Azimuth = 295°
Back azimuth =
295°-180°=115°

Azimuth = 115°
Back azimuth =
115°+180°=295°

Fig. 14.3 Back azimuth.

The sexagesimal system that is founded on the number 60 (full circle = 360 degrees, 1 degree = 60 minutes, 1 minute = 60 seconds) is used primarily for angular measurements, Fig. 14.4. Other numerical systems in use are based on the radian where 2π radians = 360°, mils where 1 mil = 1/6400 of 360°, and grads where 1 grad = 1/400 of 360°.

for measuring horizontal angles, the use of deflection angles is becoming less frequent. The *interior angle* is the inside angle between two lines in a closed figure. For a closed polygon, if n is the number of sides, the sum of the interior angles will be (n - 2) x 180°. Fig. 14.5 illustrates the relationship of these two angles.

14.2 DIRECTION OF BASE LINES

In determining direction there must be a starting point or zero measurement. Three base lines have been designated for directing measurement. These are true north, magnetic north, and grid north.

True north. The direction of the geographic north pole from any point on the earth's surface is defined as *true north*. On a map, true north is represented by the meridians or lines of longitude. True north is indicated on a map, without longitude lines, by an arrow with a full head or by a star (implying the North Star from which true north determinations are often made), Fig. 14.6. The letters TN for true north are sometimes omitted.

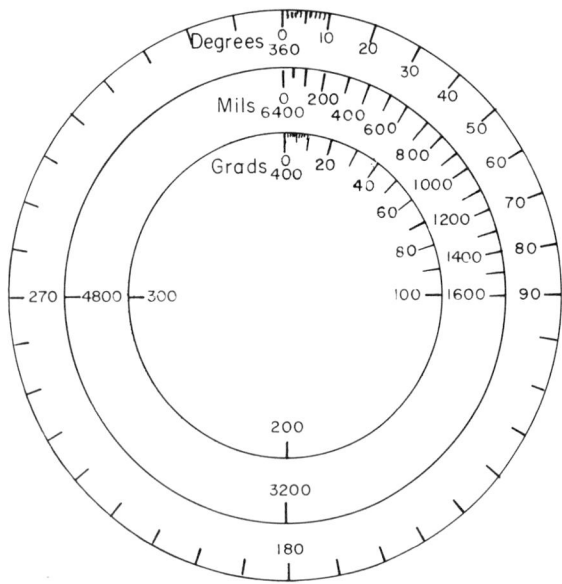

Fig. 14.4 Degree, mil, and grad relationship.

Two angles used in surveying for drawing lines are the deflection angle and the interior angle. The *deflection angle* is the angle between a line and the prolongation of the preceding line. Deflection angles must be designated left or right. With increasing use of the theodolite instead of the transit

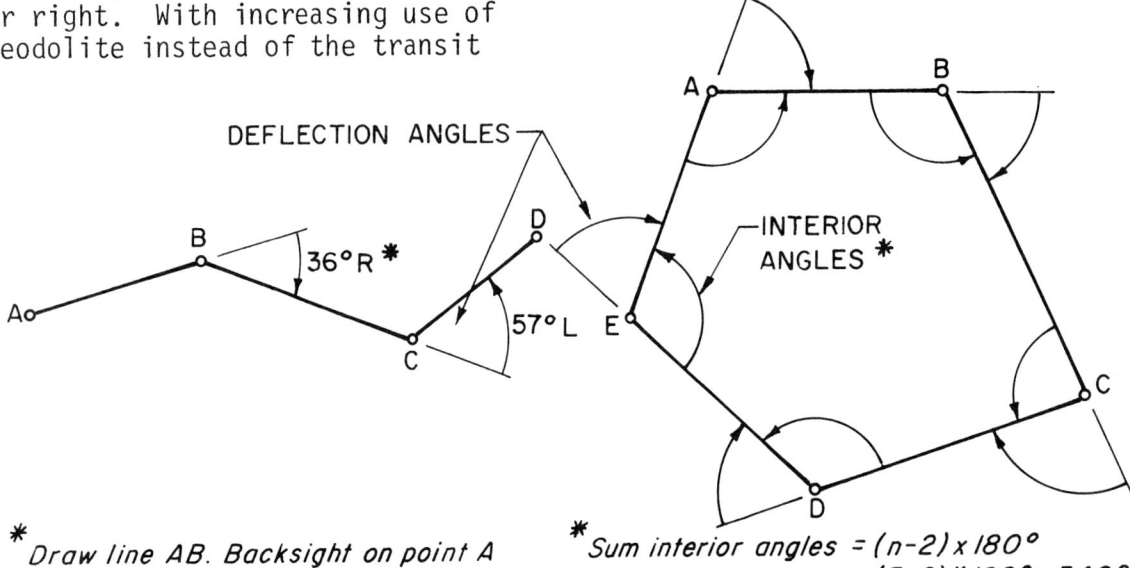

* Draw line AB. Backsight on point A from point B and turn deflection angle

* Sum interior angles = (n-2) x 180°
= (5-2) x 180° = <u>540°</u>

Fig. 14.5 Deflection and interior angles.

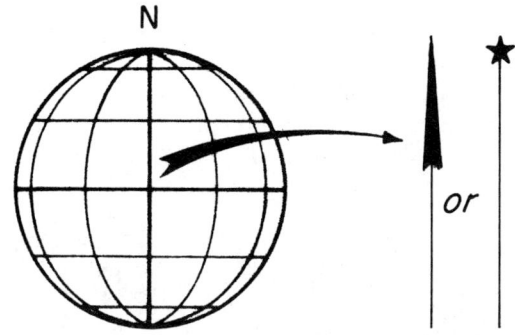

Fig. 14.6 True north.

Magnetic north. The direction toward the *magnetic north* pole is established by the magnetic needle of the compass. It is usually indicated by an arrow with a half head and the letters MN, Fig. 14.7. The year in which magnetic north was established should be given.

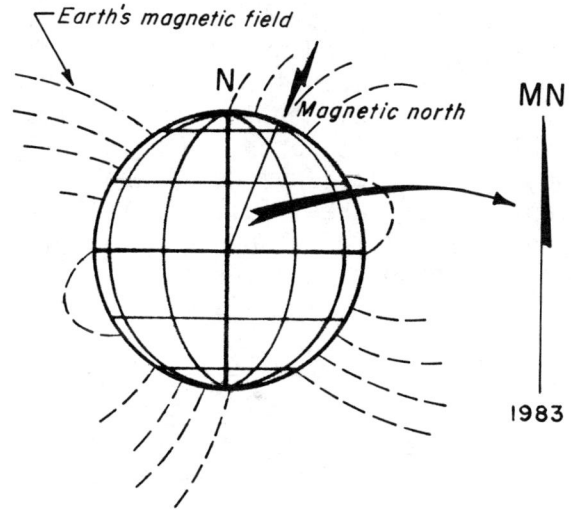

Fig. 14.7 Magnetic north.

Grid north. The meridians of longitude of the earth converge toward the poles and point toward true north. On many maps, however, these lines are constructed parallel to each other. The term *grid north* is used on maps of this kind where direction is represented by north-south grid lines that do not actually point toward true north, Fig. 14.8. It is indicated on a map as a single solid line accompanied by the letters GN. Grid and true north will seldom be the same except where a map is centered on a meridian. For small scale maps there will be little problem. The drawing of large scale maps requires consideration of grid declination since the divergence of true north becomes greater as the distance from the center of the map increases.

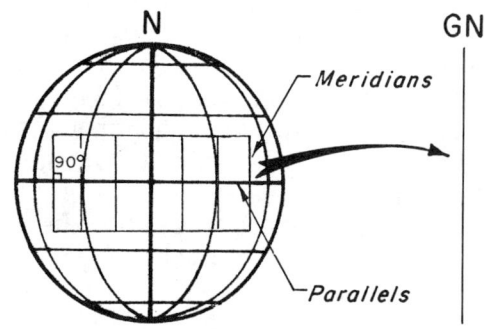

Fig. 14.8 Grid north.

14.3 DECLINATION

Declination shows the relationship between true north, magnetic north, and grid north.

Magnetic declination. Magnetic attractions in the earth's core cause the compass needle to be pulled away from true north. The horizontal angle from true north to magnetic north at any point is known as the *magnetic declination*, Fig. 14.9. The year and magnitude in degrees and direction, east or west, from true north must be given.

Fig. 14.9 Magnetic declination.

The magnetic attraction in one locality is not constant over a period of time but is gradually changing. Isogonic maps, Fig. 14.10, are produced every five years by the National Ocean Survey (NOS) for the use of map makers in determining magnetic variations throughout the United States. Local conditions should be checked since charts of this type indicate the declination but may not provide correct values for all areas. Declination varies from about 25° easterly on the west coast to about 23° westerly in the State of Maine. It is obvious that as the central region of the United States is approached from the coasts, the declination becomes less and less. In the middle of the continent there are a series of points, running in a general north-south direction, at which the needle points due north and the declination is zero. A line drawn through these points is called the *agonic* line. Other lines termed *isogonic lines* have been determined which run generally parallel to the agonic line and on either side of it. These lines have equal declination at all points along the same line.

Since the magnetic declination is always changing, a publication date is usually printed on the isogonic map. Additional lines that indicate the direction and amount of change each year enable the map user to determine the declination from one year to the next. Point A, Fig. 14.10, would have a magnetic declination of 10°E if the publication date were 1980. The declination changes three minutes in a westward direction each year. In 1984 the total change will be 12 minutes westward making the declination 9°48'E.

Grid declination (Grid convergence). The horizontal angle from true north to grid north is termed *grid declination*, Fig. 14.11. The magnitude in degrees and direction, east or west, from true north must be given.

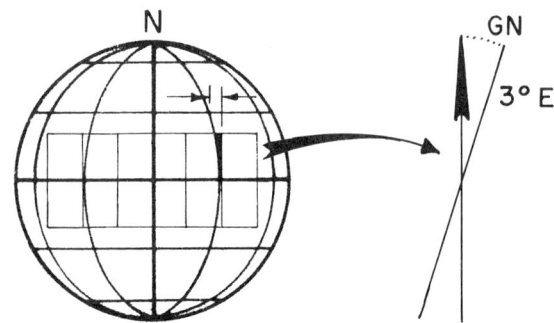

Fig. 14.11 Grid declination.

Declination diagram. A declination diagram representing true north, magnetic north, and grid north is placed on most maps, Fig. 14.12. An arc, drawn as a dotted line, connects true north with the magnetic north and grid north lines. The grid convergence angle is given to the nearest full minute with its equivalent to the nearest full mil. For conversion of degrees to mils, one degree equals 17.777 mils may be used.

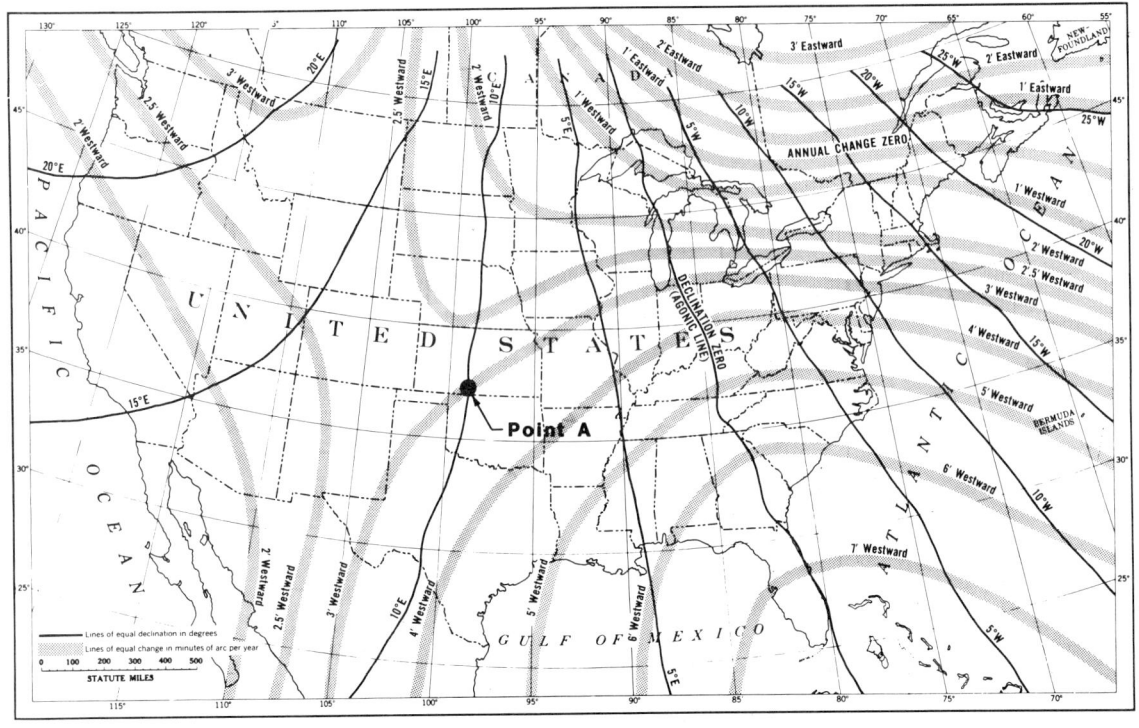

Fig. 14.10 Isogonic map of the U.S. *(Courtesy NOS).*

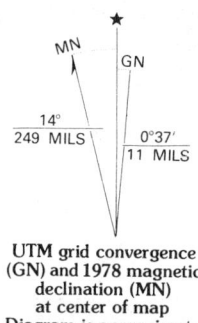

Fig. 14.12 Declination diagram.

The UTM grid convergence (GN) and 1978 magnetic declination (MN) are given at the center of the map. Declination will therefore change from one part of the map to another. The declination diagram may be approximated, although relative proportions are maintained, where the angles between the lines are too small for drawing accurately.

Grid magnetic angle. The horizontal angle from grid north to magnetic north is termed the *grid magnetic angle*, Fig. 14.13. A note is used including the year it was prepared to indicate this value expressed to the nearest one-half degree with its equivalent to the nearest 10 mils. The annual magnetic change reflects the amount by which the grid magnetic angle changes annually due to changes in magnetic north.

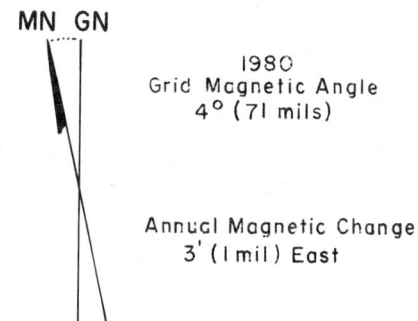

Fig. 14.13 Grid magnetic angle.

14.4 DRAWING DIRECTIONAL POINTS

The choice of whether to use a true or magnetic arrow, or both, is not up to the individual drawing a map. The method in which the control was oriented is required to serve as a guide. If the control was established with a compass, magnetic north or both true and magnetic directional lines giving the angle of declination may be shown. Control done with a transit or theodolite, oriented by celestial observation, will require only the true north line. A few simple rules are suggested for drawing these lines.

(1) In drawing, orient the map so that north is either at the top or at the right-hand side of the sheet.

(2) Draw directional lines, relatively small, in proportion to the size of the map.

(3) Do not draw ornamental arrows and lines or other embellishments. The simpler and the more dignified, the better. A simple arrow, Fig. 14.14 (a), with a clean barb and a short cross-stroke midway across the vertical shaft is adequate. For display maps, plot plans, and architectural drawings the directional arrows may be drawn more creatively as suggested in Fig. 14.14 (b).

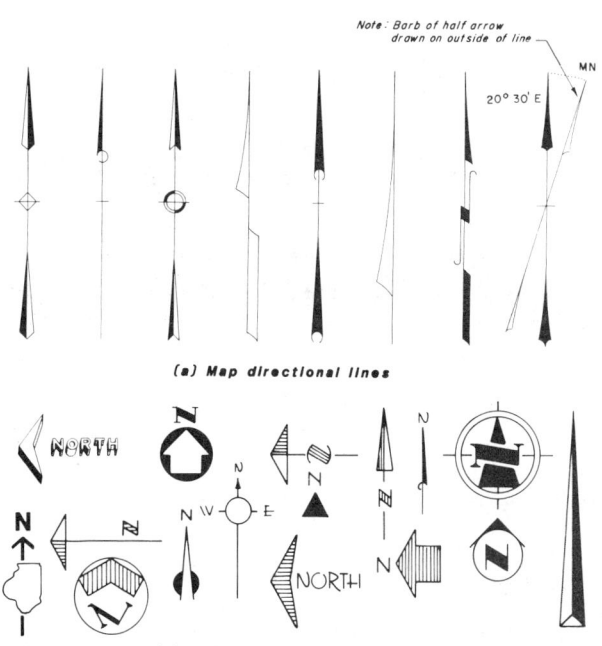

Fig. 14.14 Suggested directional lines and arrows.

14.5 CONVERSIONS

Occasionally it is necessary to convert from one type of direction to another. This is accomplished by adding or subtracting appropriate values. Since confusion may result as to whether to add or subtract values, problems of this type may be simplified if a sketch is used to portray the situation. Problems involving bearing and azimuth in relation to different reference lines are illustrated by the following examples.

Problem 1: Magnetic declination = 18°30'E. What is the magnetic bearing corresponding to a true bearing of S11°E?

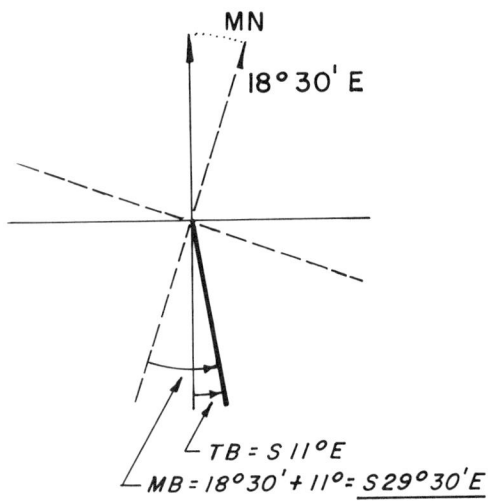

TB = S11°E
MB = 18°30' + 11° = S29°30'E

Problem 2: Magnetic declination = 21°30'E. What is the true bearing corresponding to a magnetic bearing of S8°30'W?

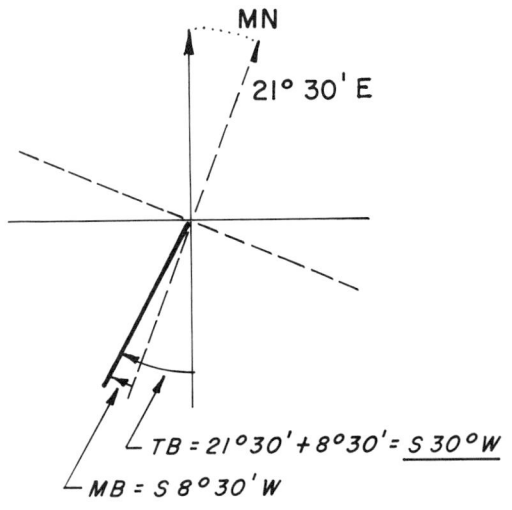

TB = 21°30' + 8°30' = S30°W
MB = S8°30'W

Problem 3: What is the true bearing corresponding to an azimuth of 260°?

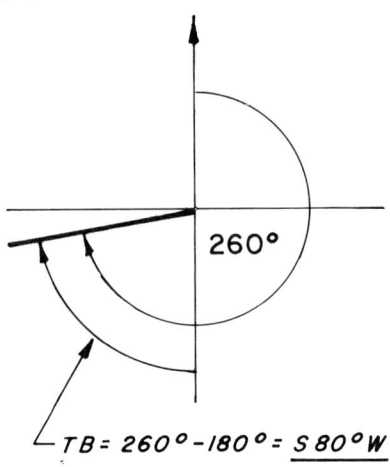

TB = 260° - 180° = S80°W

Problem 4: Grid declination = 6°W. What is the grid azimuth corresponding to a true azimuth of 72°?

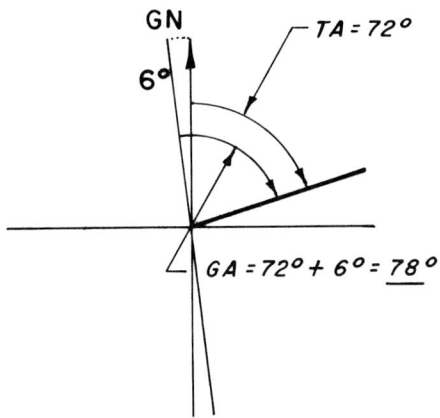

GA = 72° + 6° = 78°

Problem 5: Grid magnetic angle = 8°E. What is the magnetic azimuth corresponding to a grid azimuth of 250°?

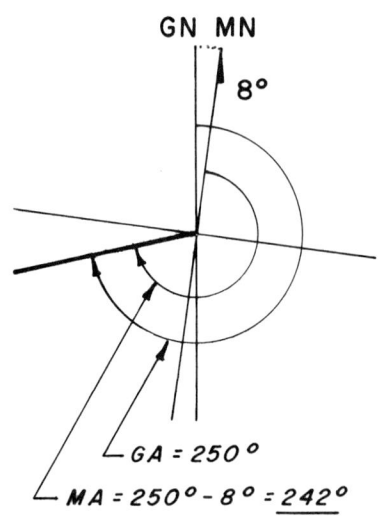

GA = 250°
MA = 250° - 8° = 242°

Figure 14.15 illustrates the calculation of bearings from interior angles. The solution of the problem is simplified by again using a sketch. A line of known bearing such as line AB having a bearing of N15°E is drawn and then extended to the opposite quadrant as denoted by line AB', Fig. 14.15 (b). The bearing of line BC is found by laying off the interior angle of 112° from AB'. The back bearing AB' was drawn since the interior angle of 112° must be laid off counter-clockwise to obtain its true direction. Other angles may easily be determined from the sketch for calculating the bearing of line BC of N83°E. The bearing of line CD is determined in the same way as shown in Fig. 14.15 (c) by prolonging the line of known bearing (BC, N83°E) into the opposite quadrant, laying off the interior angle at station C, and calculating the bearing of line CD.

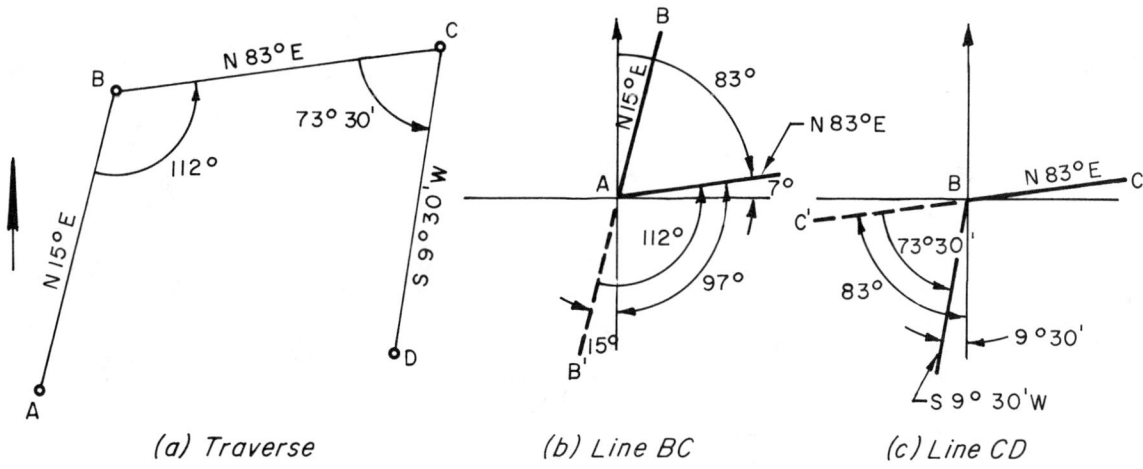

(a) Traverse   (b) Line BC   (c) Line CD

Fig. 14.15 Calculation of bearings from interior angles.

LAND LOCATION AND DESCRIPTION

15.1 INTRODUCTION

The location and description of points, lines, and areas on maps representing the earth's surface must be determined accurately. Many different ways have been developed for their systematic location and designation. Each method serves a different purpose which is explained in subsequent sections of this chapter.

15.2 THE EARTH

The study of *geodesy* concerns itself with the determination of the exact size and shape of the earth and the precise location of points on the earth's surface. The topographic surface of the earth is made up of relief features on which actual measurements are made. Since mathematical computations are difficult to make on an irregular surface, the shape or "figure of the earth" may be considered to be a sphere, Fig. 15.1 (a). This shape represents an easy surface to work with in making approximate computations involving spherical trigonometry for astronomical and navigation problems.

The earth is not spherical but is slightly flattened (oblate spheroid at the poles and bulged around the middle). This is attributed to the centrifugal force of the earth's rotation which deforms it into a form in equilibrium with respect to the forces of gravity and rotation. In geodesy the earth's shape is more precisely represented mathematically by selecting an ellipsoid to define the figure of the earth, Fig. 15.1 (b). An ellipsoid is a surface of revolution that is generated by rotating an ellipse about its minor axis. A number of ellipsoids have been computed and adopted worldwide for precise work in surveying, mapping, and science, Table 15.1. These have been used in different areas and the ellipsoid employed as a reference surface is one selected as a mathematical convenience to represent the figure of the earth. The size of an ellipsoid is usually designated by the semi-major axis or equatorial radius, designated by the letter "a" in Fig. 15.2. The semi-minor axis or one-half the length of the polar axis (minor axis) is designated by the letter "b". In expressing the flattening of the earth ellipsoid, designated by the letter "f", the expression $F=(a-b)/a$ is used. The earth's flattening is about 1/300. This size difference between the ellipsoid and a sphere to represent the earth is actually very small.

The earth's figure for measuring and describing its form is referred to still another surface termed a *geoid* (meaning "earth shaped"), Fig. 15.3. This surface form may be considered the sea-level surface of the oceans extended in an imaginary way under the land to form a continuous figure. Under the continents the sea-level surface or geoid lies somewhat higher than the ellipsoid due to less gravitational force. Under

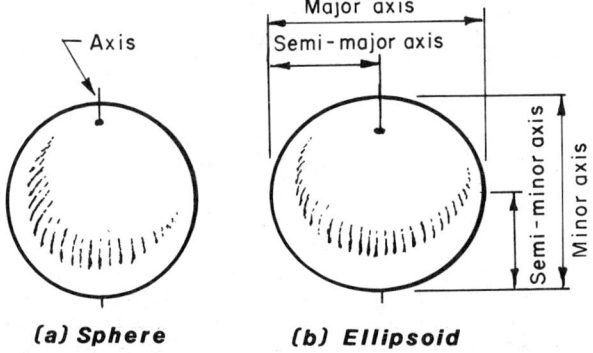

Fig. 15.1 The figure of the earth.

Table 15.1 Reference ellipsoids.

Name	Semi-major axis (Equatorial radius)*	Flattening	Where Used
Geodetic reference system (1980)**	6 378 137	Derived	World-wide science computation
Geodetic reference system (1967)**	6 378 160	Derived	World-wide science computation
Krassowsky (1940)	6 378 245	1/298	Russian datum
International, Hayford (1924)	6 378 388	1/297	European datum
Helmert (1907)	6 378 200	1/298	Egypt
Clarke (1880)	6 378 249	1/293	New French, Africa
Clarke (1866)	6 378 206	1/295	North American datum
Bessel (1841)	6 377 397	1/299	Tokyo datum
Everest (1830)	6 377 276	1/300	Indian datums

* metres
** Computed from earth-satellite data.

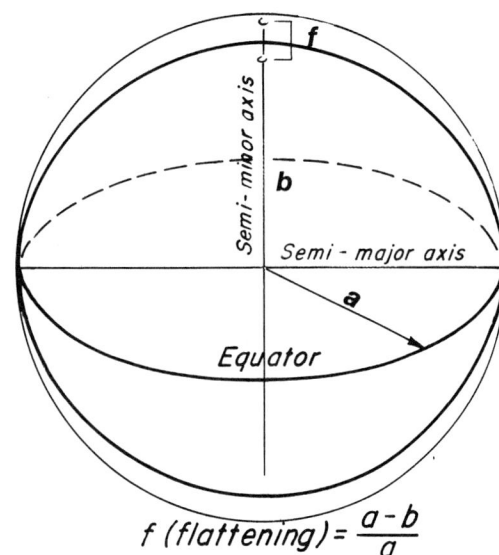

Fig. 15.2 The earth's flattening.

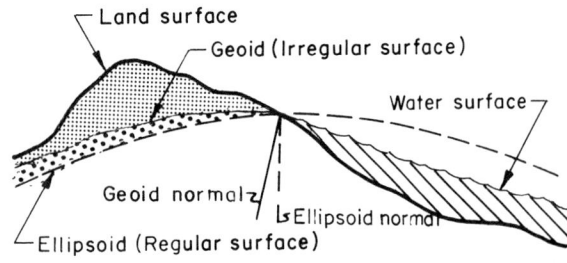

Fig. 15.3 The geoid.

earth's crust. The geoid is consequently considered to be the actual shape of the earth's surface at which gravity potential is the same everywhere and where the surface is normal or perpendicular to the direction of gravity or plumb line. This undulating surface will not be as uneven as the earth's actual surface. Uneven but less pronounced areas will still occur since the forces acting on the earth's mass produce an irregular surface when compared to that of an ellipsoid which is a regular surface.

The earth's statistics. The earth is constantly in motion traveling through space at a speed of approximatley 107 000 kilometres (66,700 miles) per hour and at the same time spinning upon its axis counter-clockwise (west to east) once every 23 hours, 56 minutes, and 4.09 seconds, Fig. 15.4.

the oceans where less dense water occurs, the force of gravity at sea-level is greater, depressing the surface of the geoid beneath that of the ellipsoid. The forces acting on the oceans would include the actual attraction of the earth's mass, centrifugal force due to the earth's rotation, and other attractions due to density differences in the

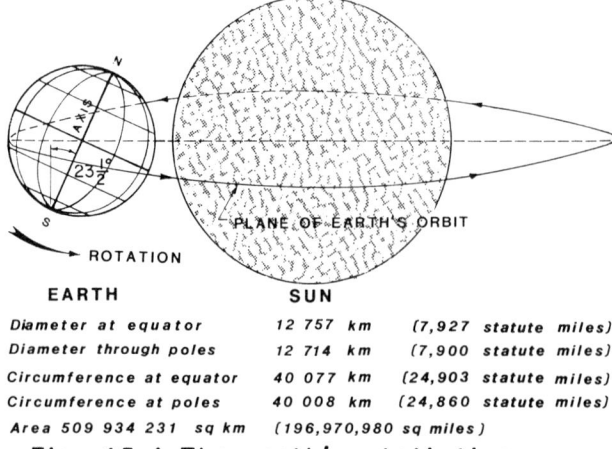

EARTH		SUN
Diameter at equator	12 757 km	(7,927 statute miles)
Diameter through poles	12 714 km	(7,900 statute miles)
Circumference at equator	40 077 km	(24,903 statute miles)
Circumference at poles	40 008 km	(24,860 statute miles)
Area 509 934 231 sq km	(196,970,980 sq miles)	

Fig. 15.4 The earth's statistics.

The earth's axis is an imaginary line passing through the center of the poles. It is tipped or inclined approximately 23½° from the perpendicular and this slant is termed the angle of inclination. The angle remains constant as the earth rotates upon its axis in a fixed elliptical path around the sun once a year. Day-night relationships and seasons of the year are due to the earth's rotation and revolution around the sun.

The sun's declination or latitude for each day of the year influences the length of the day. On many globes an indicator referred to as the *Analemma*, Fig. 15.5, shows where the sun's rays are vertical on any day of the year. The Analemma, in the form of a figure 8, is divided into months and days and extends between the points 23½° north, *Tropic of Cancer*, and 23½° south, *Tropic of Capricorn*. The sun's vertical rays do not fall outside the region between these lines known as the tropics. By examining the Analemma, the days length and seasons are easily understood. *Equinox* meaning "equal nights," occurs on March 21 (vernal equinox) and September 23 (autumnal equinox). On these two dates the days and nights are of equal length. The word *solstice* means "sun stand still." June 21 (summer solstice) and December 22 (winter solstice) are the dates the sun's vertical rays have reached their northern and southern limits. In the northern hemisphere spring occurs from March 21 to June 21, summer from June 21 to September 23, autumn from September 23 to December 22, and winter from December 22 to March 21. The seasons are just the opposite in the southern hemisphere.

The earth's coordinates. A systematic method for locating points on the earth's surface based upon two sets of lines requires the use of geographical coordinates. One set of lines termed *meridians* are formed by passing a series of imaginary planes through the earth's poles, Fig. 15.6. If an imaginary plane is passed through the center of the earth halfway between the poles and perpendicular to the axis, the circle it forms on the surface is the equator. When the plane is passed through the earth, parallel to the equator without passing through the center, small circles called *parallels* are formed. Any circles formed by planes that pass through the center of the earth but are not perpendicular to the axis are called *great circles*. All meridians may be considered great circles since they converge at the poles. Only the parallel representing the equator is a great circle, with all others small circles. The great circle is of prime importance in navigational work since it indicates the path to be followed to reach a destination in the shortest distance and time. The distance between two points on the earth, Fig. 15.7, is measured along the arc formed by the great circle passing through them. This great circle on the earth's surface is termed an *orthodrome* by navigators.

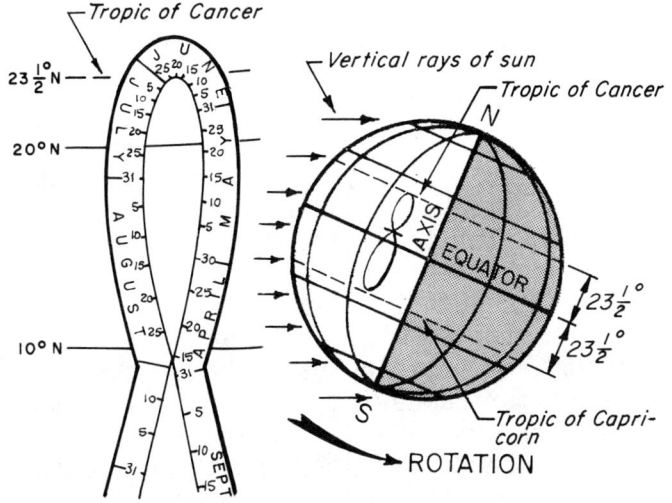

Fig. 15.5 The Analemma.

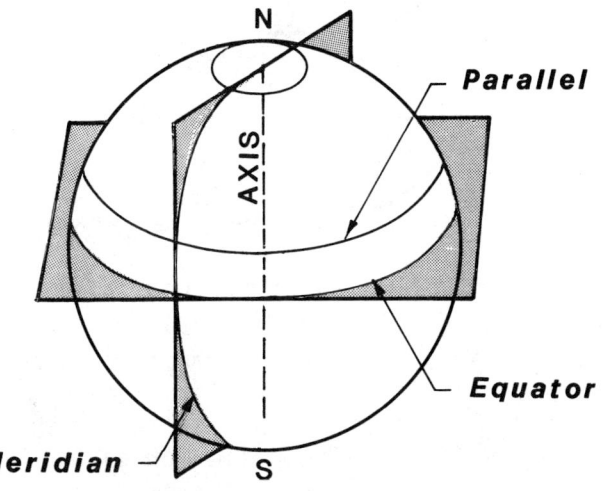

Fig. 15.6 Parallels and meridians.

Fig. 15.7 Great circle route between New York and Moscow.

The earth's meridians and parallels form a useful network of geographical or spherical coordinates, Fig. 15.8. These may be used to locate any point on the earth's surface. There are 360° around the earth and one numbered meridian for each degree. Starting with 0°, they are numbered to 180° eastward and to 180° westward. The 0° and 180° meridians together form a great circle through the poles which divides the earth into two halves. One half is called the Eastern Hemisphere and the other the Western Hemisphere.

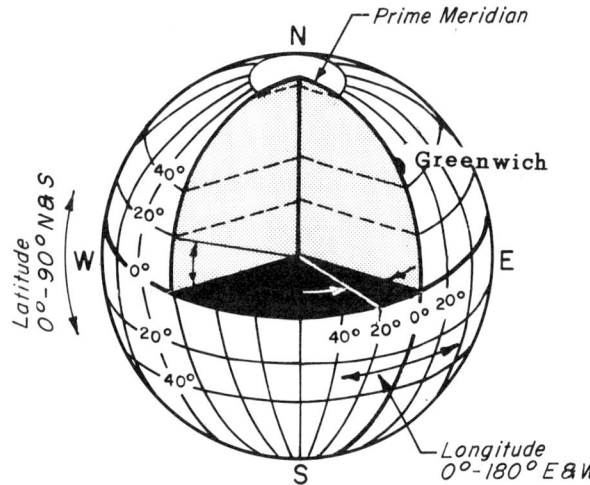

Fig. 15.8 The earth's coordinates.

On most charts the meridian designated as 0° is the one which passes through the Royal Observatory near Greenwich, England. This is termed the *prime meridian*. The prime meridian, however, may be located anywhere and the maps of some foreign countries use a different prime meridian. The 180° meridian, opposite the prime meridian, is the international date line or the Sunday-Monday line. The date line passes near the center of the Pacific Ocean, avoiding inhabited lands where it might cause confusion. The date line deviates from the 180° meridian sufficiently to pass through the Bering Strait, west of the Aleutian Islands, and east of certain islands near New Zealand.

Figure 15.9 illustrates that longitude and time are closely related. In traveling from west to east, standard time advances one hour for each 15° or 24 hours (one day) in passing completely around the earth. An east to west traveler would therefore find it necessary to set a watch back one hour for each 15° of longitude. New York City near the 75th meridian is five hours earlier than the time in Greenwich.

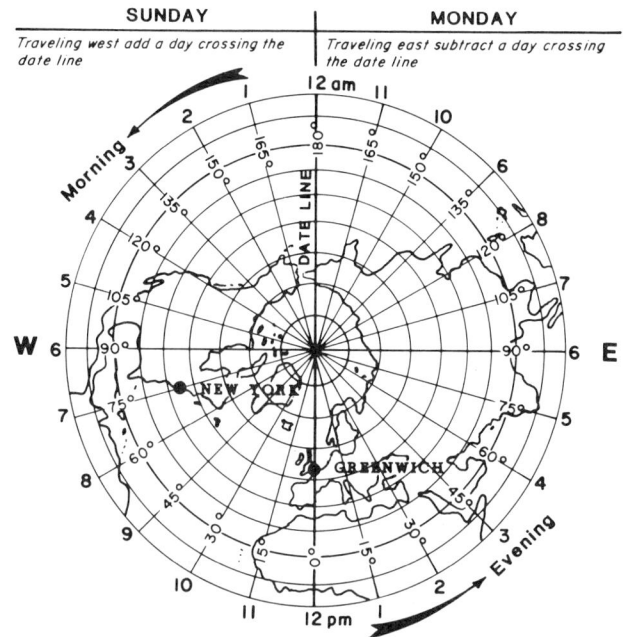

Fig. 15.9 Longitude and time relationship.

The equator divides the earth into two halves, the Northern Hemisphere and the Southern Hemisphere. Starting at the equator there are parallels for each degree both north and south. The parallels are actually a series of circles which grow smaller as they approach the north and south poles. The starting point for numbering the parallels is the equator, which is the 0° parallel. They are numbered 0° to 90° north and south of the equator and every parallel runs true east and west.

The meridians and parallels are used to express distances in degrees, minutes, and seconds. In expressing this distance, the terms *latitude* (the arc distance measured in degrees from the equator) and *longitude* (the arc distance measured in degrees from the prime meridian) are used instead of parallel and meridian. The small circles are called circles of latitude, or parallels of latitude and the great circles, meridians of longitude. Expressed in another way, latitude is used for north-south measurements and longitude for east-west measurements. In locating a point, latitude is given first. For example Orono, Maine is latitude 44°53'N - longitude 68°41'W, Fig. 15.10.

Fig. 15.10 Latitude and longitude.

The ground distance covered by one degree of longitude is 111.321 kilometres (69.172 statute miles) at the equator but decreases as the poles are approached until it becomes zero. The approximate length of one degree of longitude at different latitudes may be found by using the formula illustrated in Fig. 15.11.

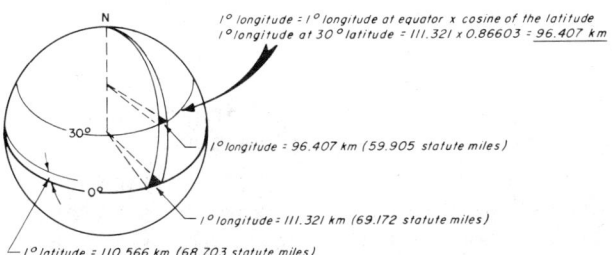

Fig. 15.11 Computation of longitude.

The degrees of latitude are approximately the same length on the earth. One degree of latitude at the equator varies from 110.566 kilometres (68.703 statute miles) to 111.699 kilometres (69.407 statute miles) at the poles.

15.3 SPHERICAL TRIANGLES

Since the earth is approximately a sphere, the relationship between the sides and angles of a spherical triangle must be understood when distances and directions on the earth are in question. The solution of problems involving the spherical triangle are common in areas such as navigation, geodesy, and astronomy.

A spherical triangle is the figure on the surface of a sphere bounded by the arcs of three great circles, Fig. 15.12 (a). There are six parts to the spherical triangle ABC, three sides and three angles. Each side is an arc of a great circle and is defined by the subtended angle originating at the center of the sphere. Side AB is defined by angle AOB, side AC is defined by angle AOC, and side BC is defined by angle BOC. The angles are formed by the intersection of the arcs of the great circles and are defined by the dihedral angles between intersecting planes. Angle A is the dihedral angle between plane AOB and AOC, angle B is the dihedral angle between planes BOA and BOC, and angle C is the dihedral angle between planes COA and COB.

Navigation problem. Problems involving great circle distances may be solved by the use of spherical triangles. The shortest distance along the earth's surface between two points (point A, lat-

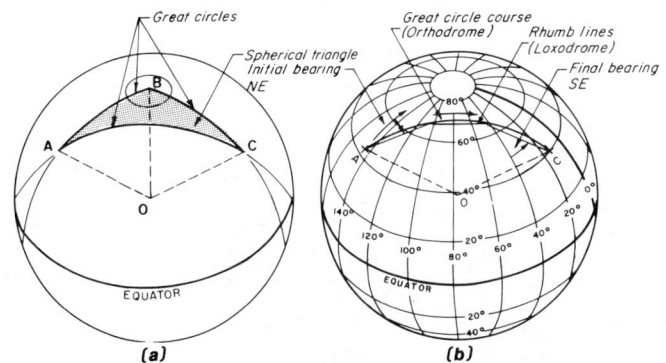

Fig. 15.12 Shortest distance between two points using spherical triangles.

itude 45°N - longitude 145°W and point C, latitude 40°N - longitude 10°W) is illustrated in Fig. 15.12 (b). To determine the shortest distance AC, it is necessary to find the true size of plane angle AOC. The initial bearing and azimuth of AC may be determined by finding the dihedral angle between planes AOB and AOC, and the final bearing and azimuths of AC may be determined by finding the dihedral angle between planes COA and COB, Fig. 15.12 (a).

The graphic solution of the problem, Fig. 15.13, is as follows: To locate points ABC on the surface of the earth, the front and top views of the earth are drawn with point B taken to be at the north pole. Point C_F is found by laying off the angle of latitude 40°N in the front view. The meridian of longitude 10°W is the outside circular arc. This arc will appear as a frontal line in the top view for locating point C_T. Point A_{F_r} is found by laying off the angle of latitude 45°N in the front view. Since this point falls on the outside meridian, it again must be frontal in the top view to locate point A_{T_r}. The actual location of point A_T on the spherical earth is found by revolving point A_{T_r} about the axis $O_T B_T$ in the top view until it intersects the 145°W meridian. Point A_{F_r} in the front view must move in a direction perpendicular to the axis $O_F B_F$. By projecting point A_T in the top view, point A_F in the front view may be precisely located.

The true size of a plane is found by finding an edge view in which any line in that plane will appear as a point and drawing another view at right angles to the edge view to obtain the true size. For plane angle AOC, line $O_F C_F$ is true length in the front view since the line is frontal or parallel to the image plane. Auxiliary view 1 is drawn using a line of sight parallel to the true length line $O_F C_F$. The plane AOC appears as an edge in this view since line $O_1 C_1$ in the plane appears as a point. Auxiliary view 2 is drawn looking perpendicular to the edge view of plane AOC to obtain the true size. The arc $A_2 C_2$ represents the shortest great circle distance and its length is found by measuring the angle 87° at point O_2 and multiplying by the appropriate factor where:

1 nautical mile = 1.516 statute miles = 1.855 35 km = 1' arc

Arc distance AC = 87° × (1.855 35 × 60) = 9684.9 km

In traveling from one point to another along a great circle course, the bearing changes constantly. Since navigating would be difficult and impractical, the shortest distance course is traveled along a series of short lines termed *rhumb lines* or loxodromes, Fig. 15.12 (b), which approximate the great circle and cross every meridian at the same angle. The initial bearing of AC is determined by finding the dihedral angle formed by planes AOB and AOC and measuring the plane angle cut out of the given planes by a plane which is perpendicular to their line of intersection. The auxiliary view 3 is drawn showing the line of intersection $A_3 O_3$, which is true length in view 2, as a point. Since the line of intersection is on planes AOB and AOC, both planes appear as edges to allow the true size of the

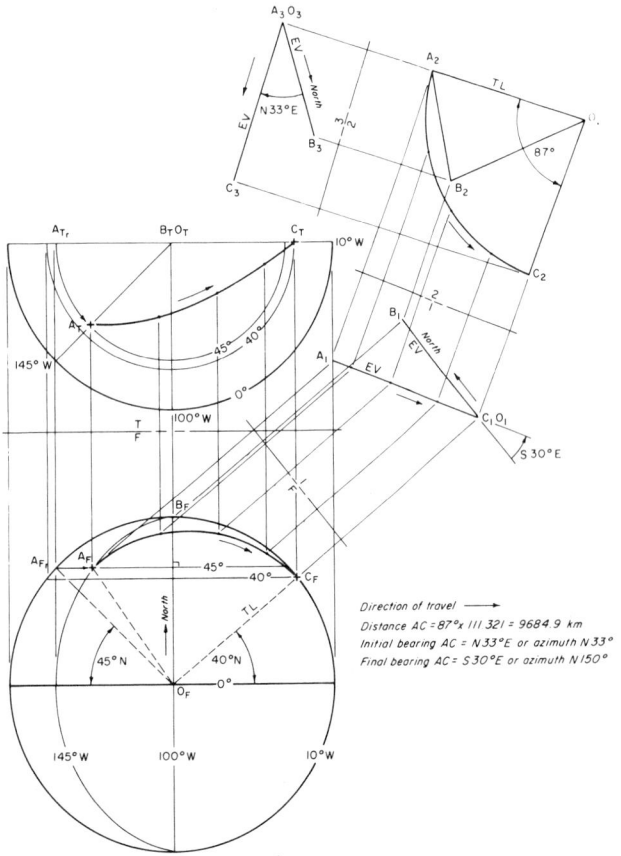

Fig. 15.13 Solution of navigation problem.

dihedral angle of N33°E (azimuth N°33) to be measured. Arrows placed along the lines will help to visualize the direction of travel. View 3 must be drawn looking in the direction of the line of intersection A_2O_2. If the view were drawn looking along O_2A_2, a reverse or mirror image would result. Views must be taken looking at the earth's surface from the outside instead of from within the earth. Correct numerical answers will be obtained either way but the correct quadrant for the bearing cannot be obtained by viewing O_2A_2 from within the earth.

The final bearing of AC is determined by finding the dihedral angle formed by planes COA and COB as was done for the initial bearing. Since the line of intersection C_FO_F is true length in the front view, view 1 will show this line as a point and both planes as an edge. The true size of the dihedral angle of S30°E (azimuth N°150) is measured in this view.

The distance between any two points on the sphere can be calculated by the formula:

cos D = (sin a x sin b) + (cos a x cos b x cos P)

where: D = arc distance between points A and B
        a = latitude of point A
        b = latitude of point B
        P = degrees of longitude between points A and B

The product of the sines will be negative if points A and B are on opposite sides of the equator. The product of the cosines will be negative if P is greater than 90°.

<u>Spherical triangle problem given three sides.</u> The problem in Fig. 15.14 illustrates the solution on finding the angles A, B, and C of a spherical triangle given the sides AB = 65°, BC = 40°, and AC = 50°. The top and front views of the spherical triangle ABC are found first. Plane BOC of the spherical pyramid if placed in a frontal position will appear as an edge in the top view. The side BC of 40° is laid off from the true length line O_FB_F in the front view. Point C_F will fall on the outside circular arc and point C_T is located by pro-

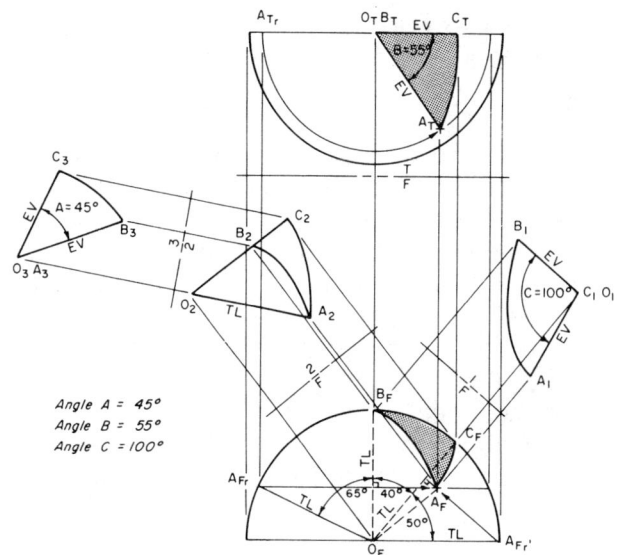

Fig. 15.14 Spherical triangle problem given three sides.

jection. The given angle of side AB = 65° is laid off from the true length axis O_FB_F to locate point A_{F_r}. The top view of point A_{T_r} must be in a frontal plane since this indicates its revolved position. Using the true length line O_FC_F as the axis, angle AC = 50° is laid off. Point A_F is found by counter-revolving A_{F_r} about the true length axes O_FB_F and O_FC_F. Both A_{F_r} and $A_{F_{r'}}$ must move in a direction perpendicular to the axes. By projecting point A_F in the front view, point A_T in the top view may be precisely located.

The dihedral angle at B is measured in the top view since planes BOA and BOC appear as an edge. The dihedral angle at C is determined by finding an edge view of planes COB and COA. Since the line of intersection O_FC_F between the two planes is true length in the front view, auxiliary view 1 drawn looking in the direction of O_FC_F will show both planes as an edge to allow the dihedral angle at C to be measured. To find angle A, an edge view is required of planes AOB and AOC. The auxiliary view 2 is drawn using a line of sight perpendicular to the line of intersection O_FA_F to obtain its true length. View 3 is drawn looking in the direction of the line of intersection O_2A_2 to obtain an edge view of both planes for the dihedral angle at A.

In all views the elliptical great circle arcs have been approximated to aid visualization of the problem.

Spherical triangle problem given two sides and included angle. The problem in Fig. 15.15 illustrates the solution of finding the angles A and C and the side AC of a spherical triangle given side AB = 43°, side BC = 45°, and the included angle at B = 60°. The top and front views of the spherical triangle ABC are found first. Point C is located by laying off the given angle for side BC of 45° in the front view. Plane BOC is placed in a frontal position to locate points C_F and C_T. The given angle of side AB = 43° is laid off from the true length axis $O_F B_F$ to locate point A_{F_r} and A_{T_r} in its frontal position. In the top view the line of intersection $O_T B_T$, between planes AOB and COB, appears as a point. The included angle for point B = 60° is laid off in this view. Point A_{T_r} is counter-revolved until it intersects the edge view of plane AOB to locate point A_T. In the front view point A_{F_r} must move in a direction perpendicular to the true length axis $O_F B_F$. The point where this line intersects a vertical projection line from point A_T locates point A_F.

The dihedral angle at point A = 67° and point C = 66° is found by drawing auxiliary views following the principles explained in the previous problem. Since plane AOC is frontal, point A_F is revolved about the true length axis $O_F C_F$ to locate the true length line $O_F A_{F_r'}$. The side AC = 41° may be measured in this view.

15.4 METES AND BOUNDS

In the United States, early original land grants in the colonies were of irregular shape. It was common practice to describe property boundaries based on surrounding natural and artificial features. This type of survey was known as Rural Land Description. As the country developed and the value of land increased, it became necessary to determine the boundaries of land by measurements. Surveys are made using the transit or theodolite and distance measuring equipment to obtain the bearings and distances of land having no regular pattern. A tract of land meas-

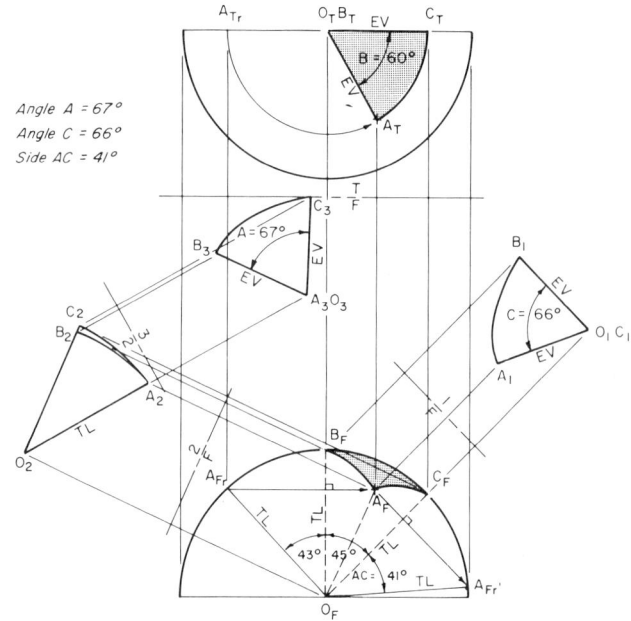

Fig. 15.15 Spherical triangle problem given two sides and included angle.

ured in this way is described by *metes and bounds*. The corners of the property are marked or described with the calculated area of the tract usually given along with all known easements and encroachments. Additional information required to be shown consists of existing buildings, fences, lakes, and streams. Using the metes and bounds description of a land parcel makes it possible to relocate the boundaries provided one original corner can be identified and the true direction of one of the boundary lines can be determined. The following is a metes and bounds description of a tract of land, with Fig. 15.16 representing the same area in map form. Either the bearings or distances may be lettered outside the traverse as long as they are consistent.

A certain lot or parcel of land situate, lying and being in the Town of Orono, County of Penobscot, State of Maine, bounded and described as follows:

Commencing at an iron pin lying on the southwesterly right of way line of State Route 2, leading from Bangor to Orono, said iron pin is located approximately one thousand three hundred forty-four feet (1344') northwesterly from the intersection of the centerline of Gilbert Road, so called, and the southwesterly sideline of State Route 2, as measured along the said southwesterly sideline of Route 2, said iron pin also marks the northerly corner of the herein described parcel of land; thence running S36°-20'E by and along the southwesterly sideline of said State Route 2 a distance of two hundred fifty-one feet (251'±) more or less to an iron pin; thence running S53°-20'W by and along the land of Ron E. Smith a distance of three hundred feet (300'±) more or less to an iron pin; thence running N35°-45'W by and along said Ron E. Smith's land a distance of two hundred twenty-eight feet (228'±) more or less to an iron pin; thence running N48°-58'E by and along said Ron E. Smith's land a distance of two hundred ninety-eight feet (298'±) more or less to the iron pin at the point of beginning. All courses are referenced to magnetic north in the year 1973. The above described parcel of land contains approximately one and six tenths (1.6) acres.

Meaning and intending to describe a portion of the premises as deeded to Harold R. and Donald H. Moneypenney by Ellis W. Simpson by his deed dated September 13, 1970, recorded in Vol. 2018, page 32 of the Penobscot Registry of Deeds.

15.5 UNITED STATES PUBLIC LAND SURVEY

The public land areas acquired by the United States require a system to subdivide the land into smaller parcels of useful shape that can be described and marked on the ground. The start of a program of land disposal, ownership recording, and actual on-the-ground boundary marking occurred on May 20, 1785, when the Continental Congress called for the rectangular system of cadastral survey. *Cadastral survey* is defined as the art of creating, recreating, marking, and defining of boundaries of tracts of land. The Bureau of Land Management is responsible for the township and range system of public lands in the United States. Fig. 15.17 indicates the areas surveyed under this system. The laws regulating the subdivision of public lands and the surveying methods employed are described more fully in the "Manual of Instructions

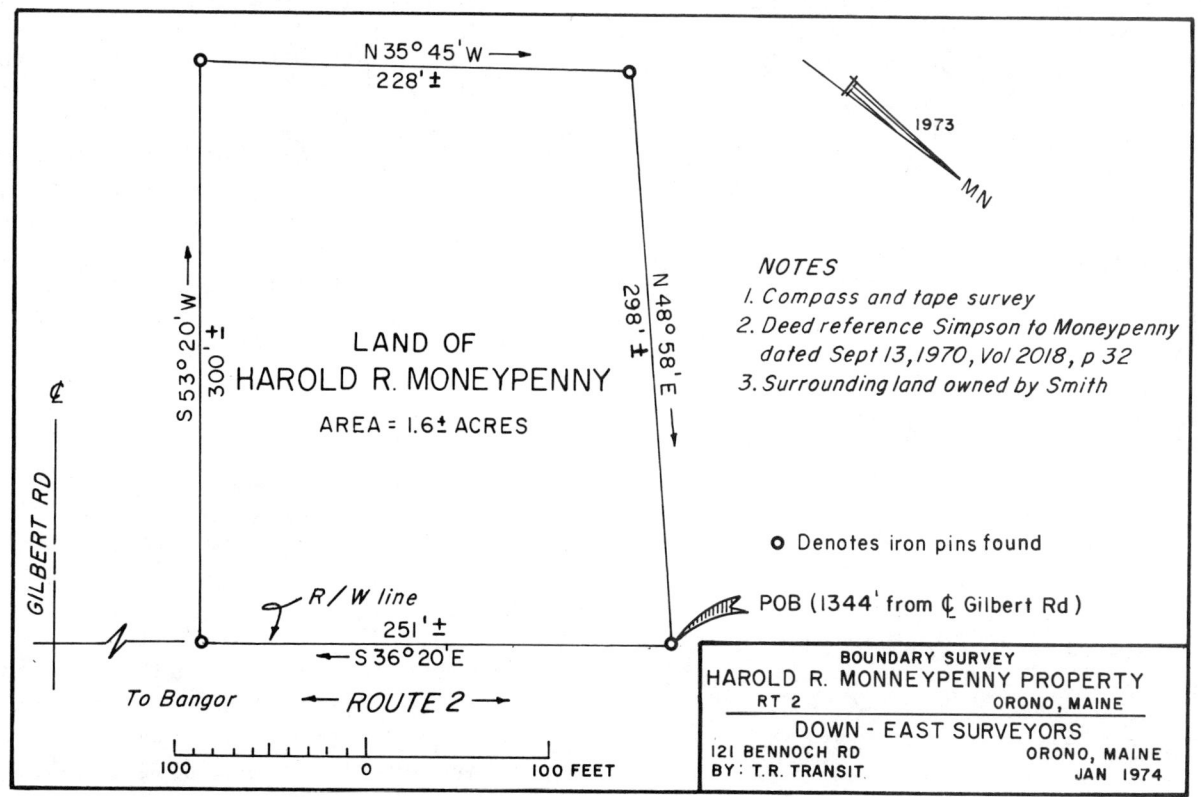

Fig 15.16 Map of metes and bounds land description.

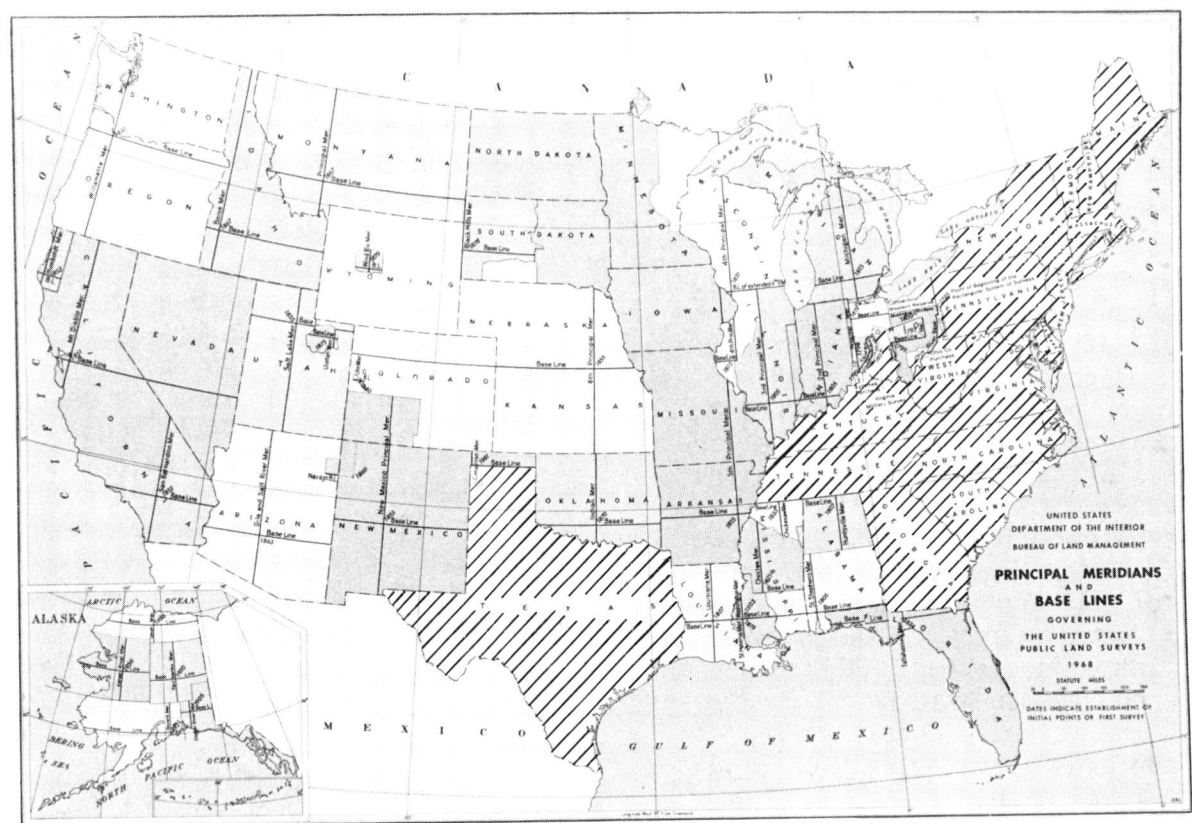

Fig. 15.17 Principal meridians and base lines governing the United States Public Land Surveys. Areas excluded are cross-hatched *(Courtesy Bureau of Land Management)*.

for the Survey of the Public Lands of the United States, 1973," published by the Bureau of Land Management, U.S. Department of the Interior.

The Public Land Surveys are based on separate surveys in relation to principal meridians and base lines that have been established throughout the United States for use as a reference system, Fig. 15.18. Each principal meridian is given a name to which all subdivisions are referred. The points where the principal meridians intersect the base lines are termed *initial points* for beginning a survey to subdivide the land. At intervals of 24 miles north and south of the base line, additional lines referred to as *standard parallels* are drawn and numbered, Fig. 15.18 (a). From the principal meridian, lines termed *guide meridians* are drawn and numbered at intervals of 24 miles in an east and west direction. The standard parallels and guide meridians divide the land into quadrangles of approximately 24 miles square. Each square will measure slightly less than 24 miles at the top since the meridians converge. To adjust for this, correction lines are run after every fourth township or every 24 miles. The base of the township above the correction line is allowed to extend over the boundary of the township below. This offset allows for the convergence of the north-south lines and keeps the acreage of the townships approximately the same.

The areas formed by the standard parallels and guide meridians are furthur divided into 16 parts that are called *townships*, Fig. 15.18 (a) and (b). The townships measure six miles on each side except at the top, again due to meridian convergence. Vertical lines are called *range lines* and the horizontal lines are called *township lines*. A row of townships extending north and south is called a *range*, and a row extending east and west is called a *tier*. The township rows are numbered according to their direction from the base line as T1N, T2N, etc., and T1S, T2S, etc.. Range columns are numbered according to their direction from the

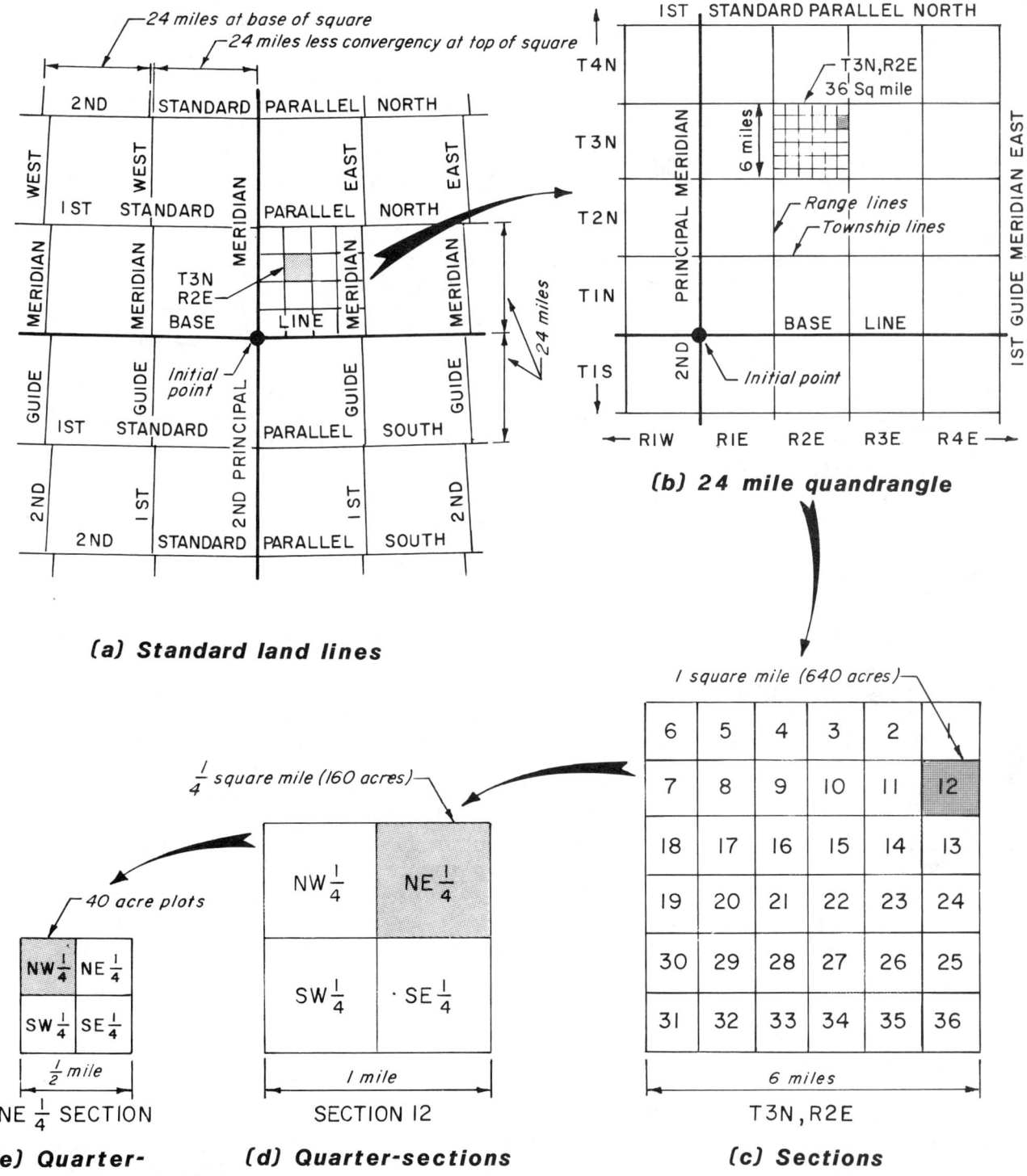

Fig. 15.18 Public land divisions.

principal meridian as R1E, R2E, etc., and R1W, R2W, etc.. Each township is designated by the number of its tier and range, followed by its principal meridian to identify the administrative or subdivision location in the system, e.g., T2N, R2E of the Second Principal Meridian.

Each township consists of 36 sections approximately one-mile square (640 acres) that are numbered as shown in Fig. 15.18 (c). The sections are further subdivided into quarter sections of 160 acres which become the basic unit under the Homestead Act of 1862, Fig. 15.18 (d). Additional subdividing into quarter-

quarter sections or 40 acre plots is possible, Fig. 15.18 (e). The quarter and quarter-quarter sections are designated by their compass positions.

The description of any specific land area must begin with the smallest unit. An example of this, for the shaded area, Fig. 15.18, would be NW¼ of NE¼ of Section 12 of T3N, R2E, of the Second Principal Meridian. This same description could be expressed as NW 40, NE¼, Section 12, T3N, R2E, Second Principal Meridian.

Smaller USPLS land divisions, in addition to the major divisions presented, are illustrated in Fig. 15.19. The description of any of the fractional units again must be given starting with the smallest unit.

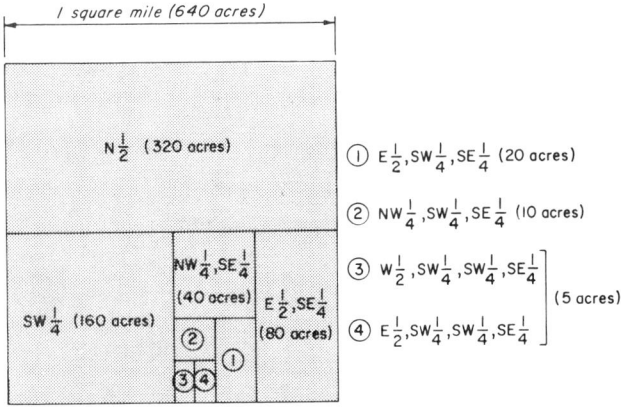

Fig. 15.19 Fractional public land divisions for a section.

15.6 GRIDS

The use of grids on a map to locate any point is widespread. A *grid* is a numbered network of two series of regularly spaced straight parallel lines intersecting at right angles. This system of rectilinear lines is easier to use than spherical coordinates where parallels and meridians frequently do not appear as straight lines. For small scale maps, points are usually located by their spherical or geographic position while grids of some type are frequently employed on maps of large scale. A map may have more than one grid system with the grid lines drawn as fine solid lines across the surface of the map or indicated by grid ticks along the edge or "neatline" of the map. Each line is given a number to allow points to be identified. The values may be expressed in units such as yards, feet, metres, or kilometres, depending on the type of grid and the map scale. The number is usually printed outside the border of the map directly opposite the grid line it indicates. The vertical grid lines are usually numbered on the bottom border and the horizontal grid lines on the right border. Some maps may have them printed on all four sides. The grid reference numbers increase from left to right and from bottom to top, since a rule when using grid references is to "read right up."

15.7 RECTANGULAR COORDINATES

The grid consists of a network of rectilinear numbered lines forming areas of identical dimensions. The position of any area or point on the grid may be determined by using the coordinate lines. One of these distances, known as the X coordinate, gives the position in an east-west direction. The other known as the Y coordinate gives the position in a north-south direction. In reading coordinates the X value is given first and is termed an *easting*; the Y value is termed a *northing*. Grid references to locate points may be expressed in an even number of digits (e.g., 104107), without any punctuation separating the values. These values could also be written X 104, Y 107. In Fig. 15.20 the shaded area is designated 104107,

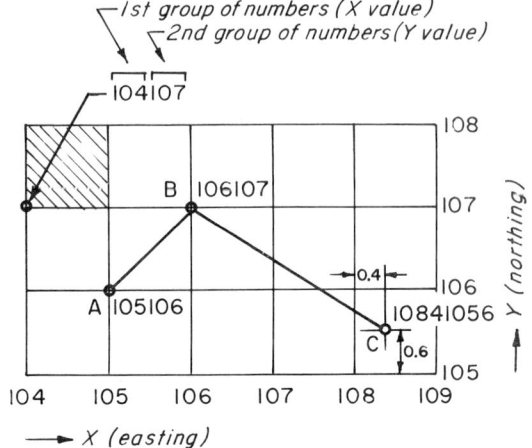

Fig. 15.20 Rectangular coordinates.

point A is located at 105106, and point B is located at 106107. The grid numbers given are those representing the SW corner of the square in which the area or point lies. A point falling within a square may be located with greater precision by estimating the distance in tenths from the SW corner in which the point lies. Point C is 0.4 of the way from 108 to 109 and 0.6 of the way from 105 to 106. Its coordinate position would be stated 10841056.

Coordinate plotting. In plane surveying, points may be located by plotting their Y and X coordinate values or latitudes and departures or northings and eastings from the origin, Fig. 15.21 (a). For latitude, the values will be positive in a north direction and negative in a south direction. Departures are positive in an east direction and negative in a west direction. It is obvious from this statement that for a line having a bearing of N60°W, latitude would be positive in the north direction and departure negative in the west direction,

Fig. 15.21 (b). The measurements for laying off the distances for latitude and departure may be found by using the following formulas:

Latitude: Length of Line X Cosine of the Bearing

Departure: Length of Line X Sine of the Bearing

The line AB in Fig. 15.21 (c) has a bearing of N30°W and is 100.00 m (328.08') in length. Using the equations above, the latitude is 86.603 m (284.13') and the departure is -50.000 m (-164.04'). Since the bearing is NW, the latitude is positive and the departure is negative. To locate point C for line BC having a bearing of S75°W and length of 120.00 m (393.70'), the latitude and departure are calculated and the values laid off in relation to point B. Successive lines may be calculated and drawn in a similar manner. The position of point C, however, must be given in respect to the origin (0,0). These values may be found by summing the latitudes and departures of lines AB and BC, Fig. 15.21 (c).

In the surveying profession, it is common practice to use the terms northing (Y) and easting (X) when working with coordinate values. The northing value is usually given first for plotting.

A closed traverse will usually not close due to unavoidable small random errors in making linear and angular measurements. The distance between where the traverse starts and ends is known as the *error of closure* (misclosure), Fig. 15.22, and is used to determine the accuracy of a survey. The error of closure is calculated as the square root of the algebraic sum of the latitudes squared plus the sum of the departures squared. For any closed traverse, these values should be equal to zero.

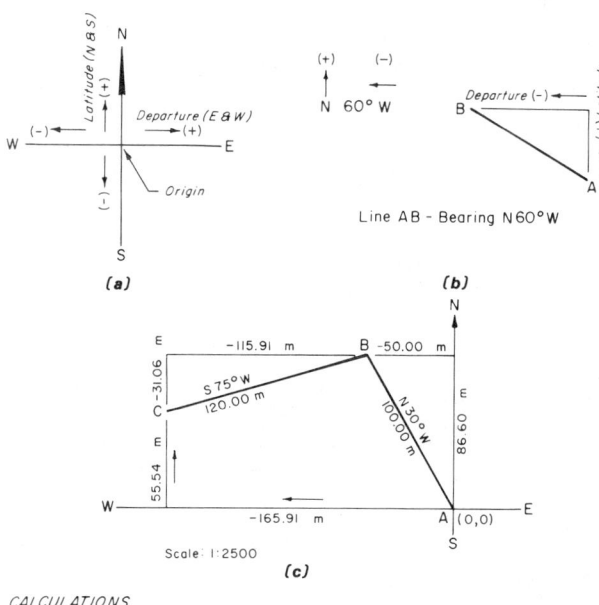

CALCULATIONS
Line AB, N30°W, length 100.00 m   Line BC, S75°W, length 120.00 m
  Latitude = length of line x cosine of the bearing
  Lat AB = 100.00 x .86603 = 86.60 m
  Lat BC = 120.00 x .25882 = -31.06 m

  Departure = length of line x sine of the bearing
  Dep AB = 100.00 x .50000 = -50.00 m
  Dep BC = 120.00 x .96593 = -115.91 m

Latitude and departure for point C
  Latitude = Lat line AB + Lat line BC
           = (86.60) + (-31.06) = 55.54 m
  Departure = Dep line AB + Dep line BC
            = (-50.00) + (-115.91) = -165.91 m

Fig. 15.21 Plotting latitudes and departures.

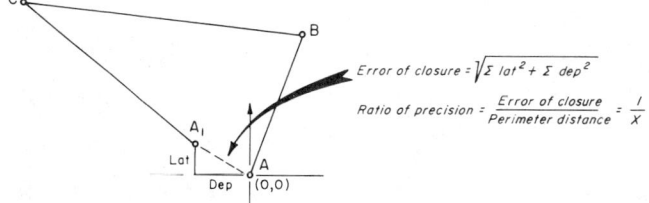

Fig. 15.22 Error of closure.

The *ratio of precision* or error is the reciprocal of the error of closure divided by the total perimeter distance around the traverse. As a guide a ratio of precision greater than 1/10,000 indicates a more precise survey, while ratios less than 1/2,000 indicate a less precise survey (1 in 10,000 implies a greater precision than 1 in 2,000).

For a traverse to close, the bearings and/or the lengths of lines must be adjusted. The Bowditch rule (the compass method) is widely used for making adjustments and is based on the assumption that errors are attributed proportionately to bearings and lengths. The error of closure and the ratio of precision should be calculated before a traverse is adjusted.

The traverse problem in Fig. 15.23 and the steps outlined below explain how adjustments are made. For additional methods in making adjustments, a standard surveying text should be used. To avoid negative values the origin could be assigned an arbitrary value, such as 1000,1000. The origin of 0,0 would then be moved 1000 metres west and south of its present position.

1. Check to see if all interior angles are equal to (N-2) 180°, where "N" equals the number of sides in the traverse. If the sum of the measured angles do not total zero, the angles must be adjusted to the proper geometric total.

   (N-2) 180° = (5-2) 180° = 540°

2. Calculate bearings from interior angles as explained in chapter 14, Fig. 14.15.

3. Tabulate data as illustrated by listing the points, length of lines, and bearings.

4. Calculate latitudes and departures using the trigonometric formulas previously discussed, see table 15.2.

5. Sum the columns algebraically for latitude and departure. The algebraic difference represents the error of closure for the latitude direction and departure direction of the traverse.

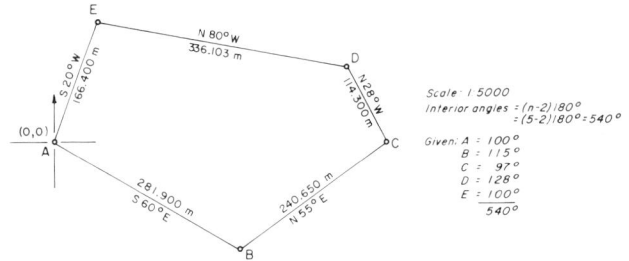

Point	Length	Bearing	LAT	DEP	ADJUSTED LAT	ADJUSTED DEP	COORDINATES Y (Northing)	COORDINATES X (Easting)
A							0.000	0.000
	*281.900	*S60°E	-140.950	244.134	-140.951	244.210		
B							-140.951	244.210
	*240.650	N55°E	138.032	197.128	138.031	197.193		
C							-2.920	441.403
	*114.300	N28°W	100.921	-53.660	100.921	-53.629		
D							98.001	387.774
	*336.103	N80°W	58.364	-330.998	58.363	-330.907		
E							156.364	56.867
	*166.400	S20°W	-156.364	-56.912	-156.364	-56.867		
A								
Σ	1139.353		0.003	-0.308	0.000	0.000	0.000	0.000

* Given values

$$\text{Closure} = \sqrt{\Sigma \text{ lat}^2 + \Sigma \text{ dep}^2} = \sqrt{(0.003)^2 + (-0.308)^2} = 0.308$$

$$\text{Ratio of precision} = \frac{\text{Closure}}{\text{Perimeter}} = \frac{1}{X} : \frac{0.308}{1139.353} = \frac{1}{X} : 3699 \text{ or } \frac{1}{3700}$$

Fig. 15.23 Balancing latitudes and departures for a closed traverse by the Bowditch rule.

6. Calculate the error of closure.

   $$\text{Closure} = \sqrt{\Sigma \text{ lat}^2 + \Sigma \text{ dep}^2} =$$
   $$\sqrt{(0.003^2) + (-0.308^2)} = 0.308$$

7. Calculate the ratio of precision.

   $$\frac{\text{Closure}}{\text{Perimeter distance}} = \frac{1}{X}$$

   $$\frac{0.308}{1139.453} = \frac{1}{X} = 3699 \text{ or } 1/3699$$

   (Usually rounded off to the nearest 10 or 100 value - 1/3700)

8. Calculate Bowditch rule corrections for latitude and departure using the following formula: (see table 15.2)

   $$\frac{-\text{Length of line}}{\text{Perimeter distance}} \times \text{Closure in lat or dep}$$

9. For this example the adjusted latitude and departure columns are found by adding or subtracting the Bowditch rule corrections. Corrections are added for S latitudes and E departures since their total value is smaller. Corrections are subtracted for N latitudes and W departures since their total value is larger.

10. The coordinates for Y (northing) and X (easting) are determined by algebraically summing the balanced departure column. The sum of both columns should equal zero.

Table 15.2 Calculations for Fig. 15.23.

Latitudes and Departures		
Line	Latitude	Departure
AB	281.900 x .50000 = 140.950 m	281.900 x .86603 = 244.134 m
BC	240.650 x .57358 = 138.032 m	240.650 x .81915 = 197.128 m
CD	114.300 x .88295 = 100.921 m	114.300 x .46947 = 53.660 m
DE	336.103 x .17365 = 58.364 m	336.103 x .98481 = 330.998 m
EA	166.400 x .93969 = 156.364 m	166.400 x .34202 = 56.912 m

Bowditch Rule Corrections		
Line	Latitude	Departure
AB	(-281.900 ÷ 1139.353) 0.003 = -0.0007	(-281.900 ÷ 1139.353) -0.308 = 0.0762
BC	(-240.650 ÷ 1139.353) 0.003 = -0.0006	(-240.650 ÷ 1139.353) -0.308 = 0.0650
CD	(-114.300 ÷ 1139.353) 0.003 = -0.0003	(-114.300 ÷ 1139.353) -0.308 = 0.0309
DE	(-336.103 ÷ 1139.353) 0.003 = -0.0009	(-336.103 ÷ 1139.353) -0.308 = 0.0909
EA	(-166.400 ÷ 1139.353) 0.003 = -0.0004	(-166.400 ÷ 1139.353) -0.308 = 0.0450

15.8 STATE PLANE COORDINATE SYSTEM

The National Geodetic Survey (formally the U.S. Coast and Geodetic Survey) has been engaged for over one hundred years in determining precise geographic positions of thousands of monumented points throughout the United States. In 1933, a system of state plane coordinates for use as a common datum of reference was established for specifying positions. This grid system is used by the states to draw accurate continuous maps that cover large areas, without the direct use of geographical coordinates. The system is widely used by land surveyors, engineers, regional planners, and in photogrammetric mapping. Advantages in using state plane coordinates are that measurements may be made in decimal units of feet uniformly in both the north-south and east-west direction and the dependence upon the magnetic needle may be forgotten.

In using state plane coordinates it will be necessary to refer to publications related to the projection system used, location of zones, assigned values, and tables for the conversion of geographic positions to plane coordinates for each state. The two basic projections employed in the state plane coordinate system are the transverse Mercator projection and the Lambert conformal conic projection. In states larger in an east-west direction, the Lambert conformal conic projection is used, Fig. 15.24 (a). The transverse Mercator projection is used for states larger from north to south, Fig. 15.24 (b). The state grids are developed mathematically from these projections, whereby the latitude and longitude of a position on the earth is transformed into plane rectangular coordinates for locating the same position on a flat plane. Each state usually has two or more zones. The zones follow the county boundaries so that each county lies wholly within one zone. By dividing the states into zones and limiting the width to approximately 158 miles, the grid distortions are held to a level of ± 1 in 10,000 feet or less. Each zone has a central meridian passing

(a) Lambert conformal conic projection used for states with larger dimension running east and west.

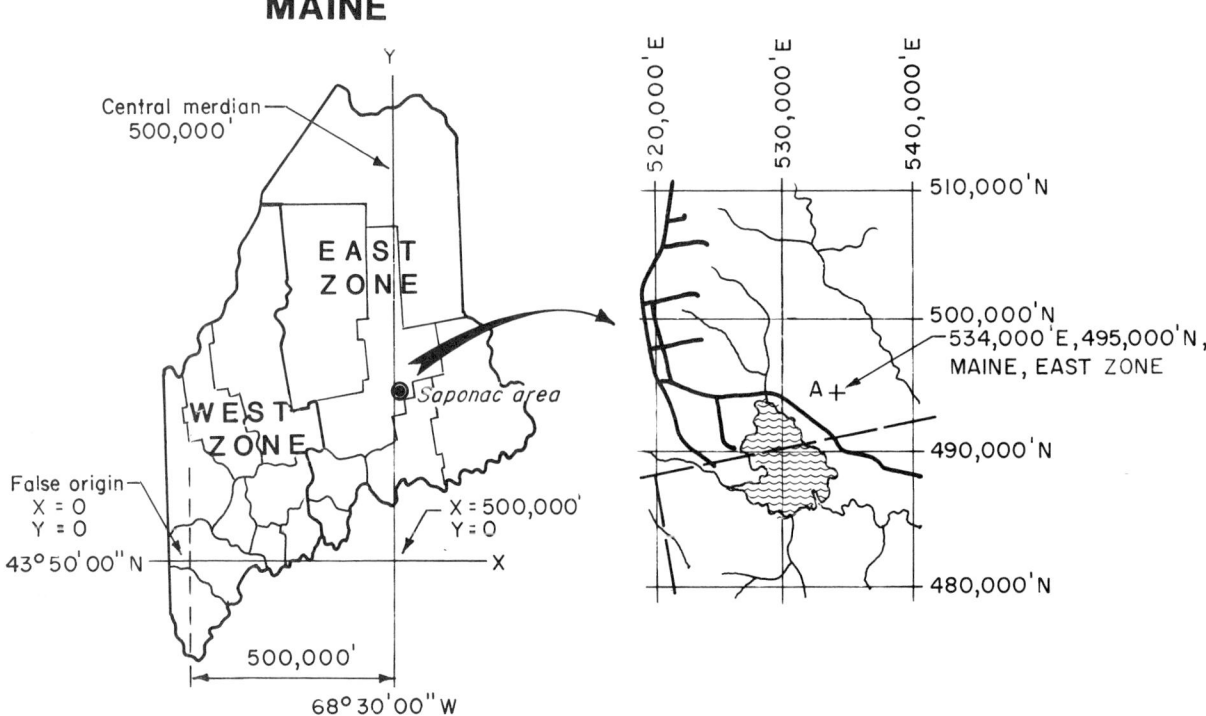

(b) Transverse Mercator projection used for states with larger dimension running north and south.

(c) Grid lines and designation for Saponac area, Penobscot County, Maine.

Fig. 15.24 State plane coordinate system.

approximately through the center of the zone. The central meridian for the Maine west zone is at 70°10'00"W longitude and 68°30'00"W longitude for the east zone. A value of 500,000 is assigned the central meridian in both zones. In the east zone, the point where the 43°50'00"N parallel of latitude intersects the central meridian has a value of X = 500,000 and Y = 0. A false origin is established by moving this point 500,000 feet west of the central meridian. The value of the false origin is now X = 0 and Y = 0. This origin is used to lay off a grid (e.g., 10,000 feet) for the Maine east zone. A portion of this grid for the Saponac area of Penobscot County is illustrated in Fig. 15.24 (c). The coordinates to locate any point are first read to the east (X) of the false origin and then to the north (Y) of the false origin followed by the state and the zone in which a point is located. The location of point A would be given as 534,000 feet E, 495,000 feet N, Maine, east zone.

Published maps frequently show more than one grid system. The State grids on most maps of the Geological Survey, e.g., are indicated by ticks along the edges or "neatlines" and may be drawn on the map by connecting corresponding ticks on opposite edges.

15.9 UNIVERSAL TRANSVERSE MERCATOR GRID

The Universal Transverse Mercator (UTM) grid is used extensively in mapping, surveying, and by the military throughout the world to locate points on the surface of the earth using plane coordinates.

<u>Grid zones</u>. The world is divided into zones each covering 6° of longitude and 8° of latitude extending from 80°S to 72°N, Fig. 15.25. The most northerly zone covers 12° of latitude from 72°N to 84°N. The longitude rows are numbered 1 through 60 starting at the 180°W meridian and increasing eastward around the earth. There are 20 columns of latitude lettered from C through X with I and O omitted. Each of the grid zones may be designated by reading the grid zone column followed by the row

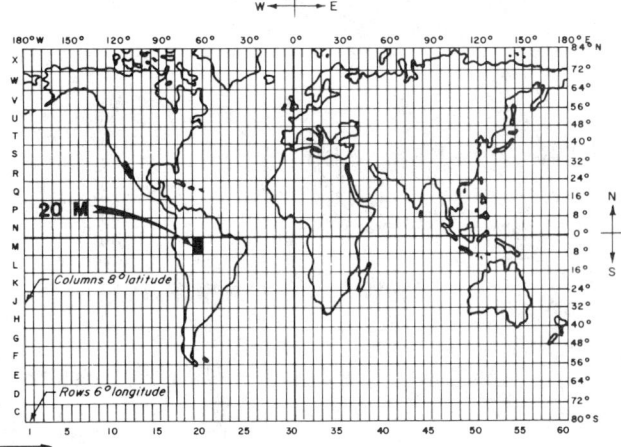

Fig. 15.25 UTM grid zones for the world.

letter. The grid zone of the shaded area located in South America, Fig. 15.25, is 20M.

Each zone has a central meridian passing through its center, Fig. 15.26 (a). The intersection of the central meridian and the equator is used as the origin or starting point for designating longitude. To avoid negative values, an arbitrary value of 500 000 metres is assigned to the central meridian. The values increase from west to east and are known as false eastings (X coordinate). The latitude of origin in all zones is the equator. In the northern hemisphere, the equator is given a value of 0 metres, and in the southern hemisphere, the equator is given a value of 10 000 000 metres. The values, known as false northings (Y coordinate), increase towards the north pole and decrease towards the south pole.

A 100 000 metre rectangular grid is superimposed on the UTM grid zones illustrated in Fig. 15.26 (b). All grid lines have been labeled with their false easting or northing values as previously explained. The coordinates to locate a point are designated giving the false easting, the false northing, the zone number, and the zone half (north or south). The location of point A in the 100 000 metre square would be given as 670 000 m E, 9 750 000 m N, 20M, S.

A large number of published maps indicate full UTM grids or just the grid ticks along the edge or "neatline." The grid interval varies and is usually

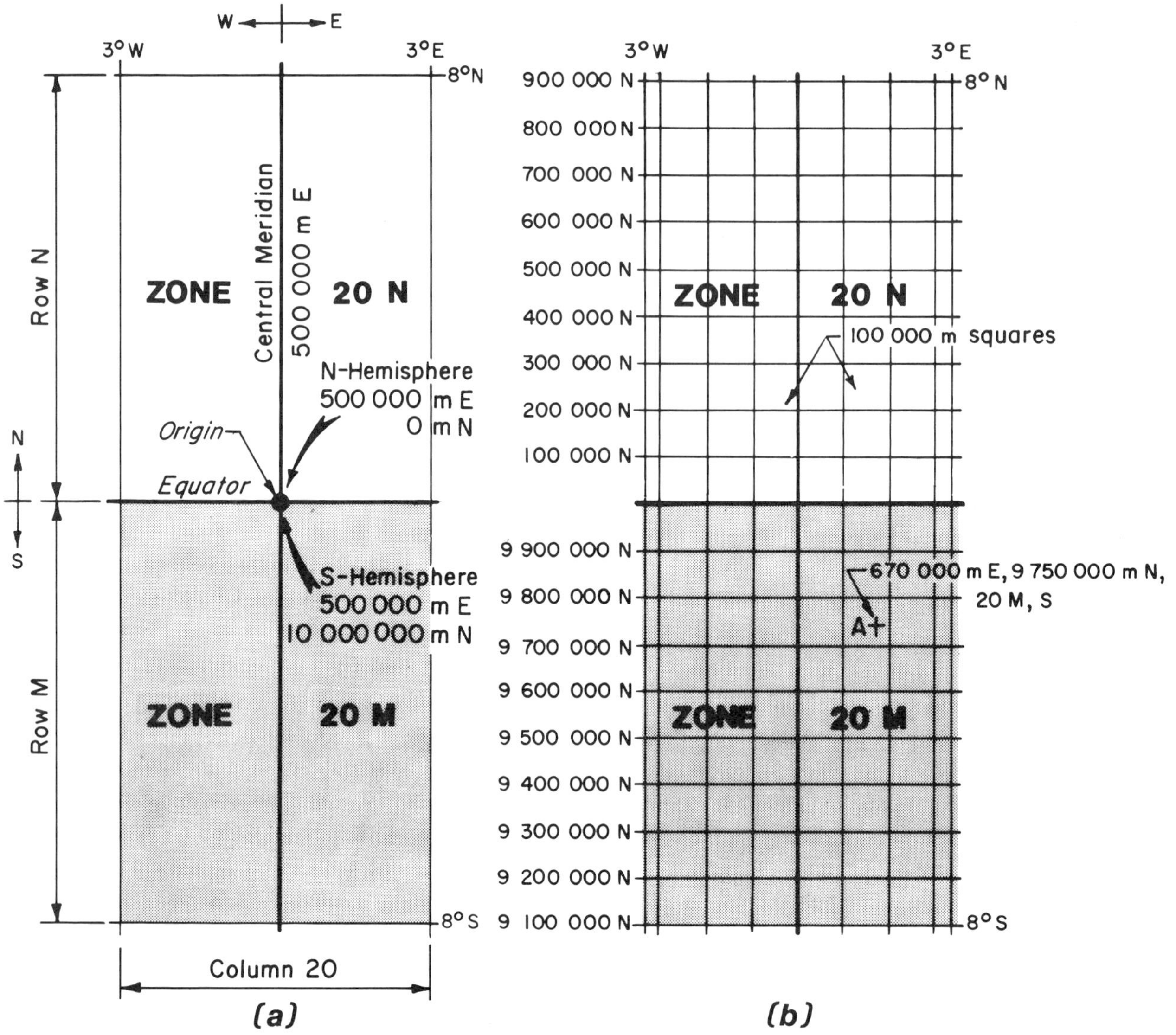

Fig. 15.26 Origin and false coordinates for UTM grid zone.

1000 metres for large scale maps, 10 000 metres for medium scale maps, and 100 000 metres for small scale maps. In some instances the full values of the grid lines may not be labelled. It is customary to print one or two of the digits in larger type than the others and to omit the zeros except for the values labelling the first grid line in each direction. A map with a 1000 metre grid would be labelled with the last three digits omitted, Fig. 15.27 (a). On a map with a 10 000 metre grid the last four digits are omitted, Fig. 15.27 (b).

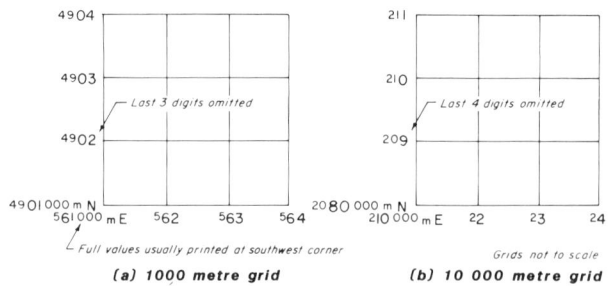

Fig. 15.27 UTM principal digits.

15.10 MILITARY GRID REFERENCE SYSTEM

The military grid reference system is the method for referencing points worldwide on maps with either Universal Transverse Mercator or Universal Polar Stereographic grids. It uses grid zones and 100 000 metre squares to divide the earth's surface for identifying points.

The 100 000 metre squares. The UTM grid zones are subdivided into 100 000 metre squares, Fig. 15.28. The intersection of the 180° meridian and the equator is the origin for subdividing and lettering the 100 000 metre squares. Each 100 000 metre column, including partial columns along grid junctions, is lettered A to Z with I and O being omitted, and repeated every 18°. The 100 000 metre rows are lettered A to V from south to north and repeated every 2 000 000 metres.

In each 100 000 metre square, two letters of designation are printed following the basic rule of read right up. The row letter depends on the zone in which the 100 000 metre square is located. In odd numbered grid zones, 1N, 3N, etc., the letter is that of the 100 000 metre row letters, A through V, beginning at the equator. The even numbered grid zones, 2N, 4N, etc., are lettered using the 100 000 metre row letters beginning 500 000 metres south of the equator (row F). This staggering lengthens the distance between 100 000 metre squares having the same two identification letters.

Each of the 100 000 metre squares may be designated by reading the column letter followed by its row letter. The 100 000 metre square that is shaded is XG. It's full grid reference is designated 3QXG. The southwest corner of the square is 600 000 m E, 2 600 000 N. This corner is assigned coordinate values of zero, however, for locating a point that falls within a 10 000 metre square. The conventional method to list these values is to give the easting and northing, both of which must have the same number of digits, and to omit the zeros. All punctuation and spaces between values should be omitted. The location of several points, falling in the 10 000 m square 3QXG, with complete notation, is illustrated in table 15.3.

15.11 UNIVERSAL POLAR STEREOGRAPHIC GRID

The Universal Polar Stereographic (UPS) grid is used, in conjunction with the UTM grid, for the north and south polar regions. Both polar areas are divided into two zones each by the 0° and 180° meridian, Fig. 15.29. The region covered extends from 84° north latitude to the north pole and 80° south latitude to the south pole. Each of the two zones, for both polar areas, are designated by using the letters Y (west), and Z (east) in the northern hemisphere and A (west), B (east) for the southern hemisphere. The origin for the north and south pole is assigned an arbitrary false easting and false northing of 2 000 000 metres.

A 100 000 metre rectangular grid is superimposed on the UPS south polar region, Fig. 15.30. Columns are lettered, west to east, J through Z in the west zone and A through R in the east zone. The letters I and O are omitted along with the letters D, E, M, N, V,

Table 15.3 Military grid reference designations.

Coordinates	Complete Notation	Reference to Nearest
10 000 m E, 90 000 m N	3QXG19	10 000 m
15 000 m E, 84 000 m N	3QXG1584	1 000 m
14 800 m E, 64 400 m N	3QXG148644	100 m
12 280 m E, 78 120 m N	3QXG12287812	10 m
18 121 m E, 81 724 m N	3QXG1812181724	1 m

* Zone designation   ** 100 000 m square   *** Coordinates

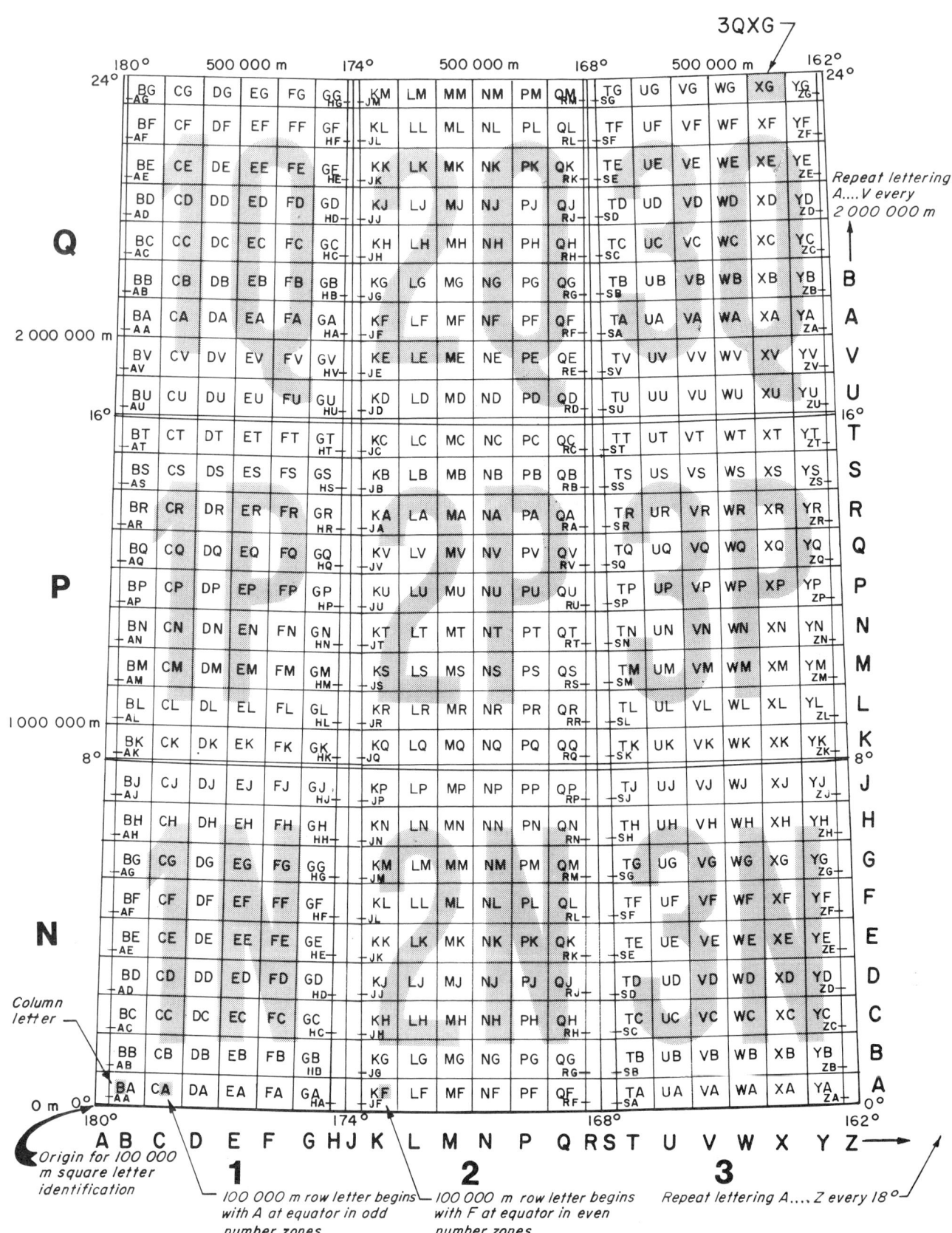

Fig. 15.28 UTM 100 000 metre square identification.

269

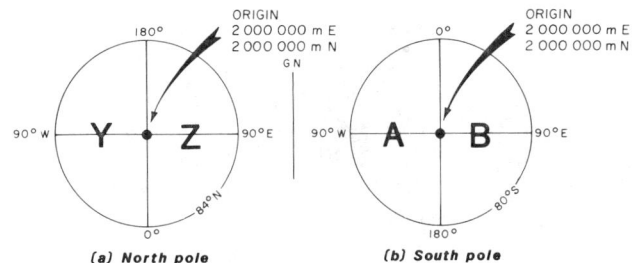

Fig. 15.29 UPS grid zone designations.

and W to avoid confusion with 100 000 metre squares in adjoining UTM zones. The 100 000 metre rows are lettered A to Z with I and O being omitted, in the direction of grid north. Each of the 100 000 metre squares is identified by giving the column letter followed by its row letter.

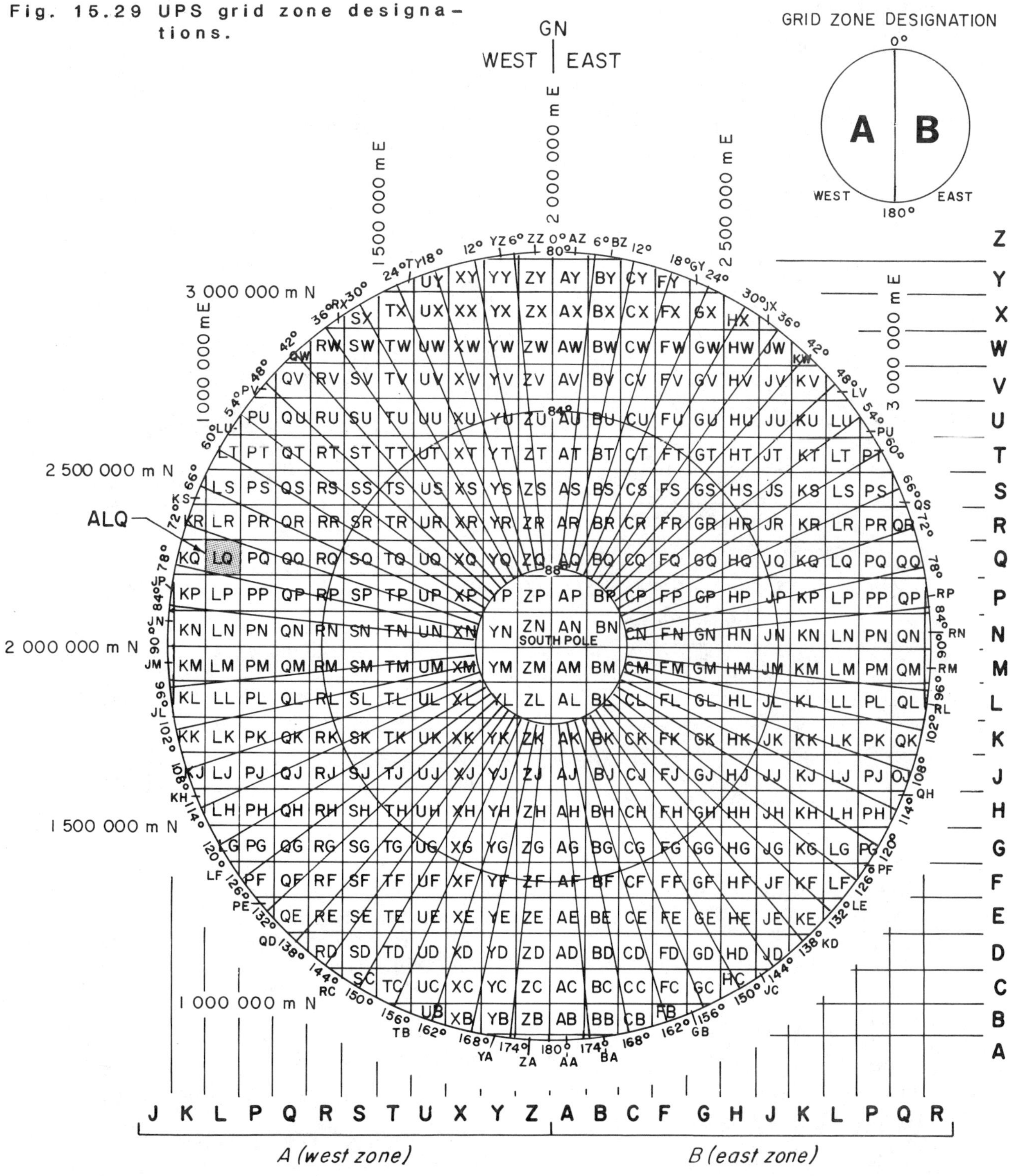

Fig. 15.30 UPS 100 000 metre square identification for the south pole.

The referencing or location of a point is similar to that used in the UTM grid system. The 100 000 metre squares are designated by giving the zone designation followed by its column and row letter. The 100 000 metre square that is shaded is ALQ. A more precise location of a point falling within a 100 000 metre square would consist of three letters followed by the necessary digits, e.g., ALQ6449. The identification of 100 000 metre squares and points is similar for the north polar region, Fig. 15.31.

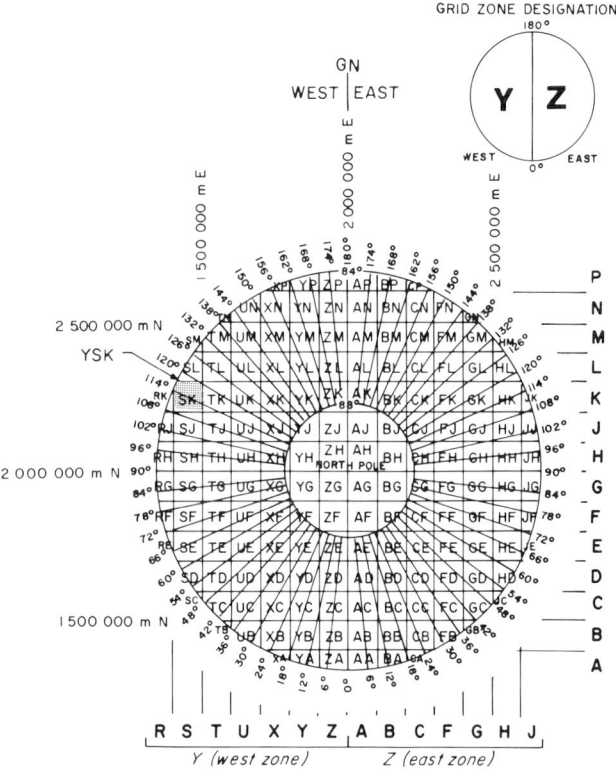

Fig. 15.31 UPS 100 000 metre square identification for the north pole.

15.12 WORLD GEOGRAPHIC REFERENCE SYSTEM

The World Geographic Reference System (GEOREF) is used primarily by the United States Air Force (USAF) to designate the location of a point anywhere in the world. It is based on latitude and longitude divisions that appear normally on all charts and small scale maps. GEOREF has worldwide application and may be applied to any chart or map regardless of the projection system used.

The surface of the earth is divided into 15° quadrangles, Fig. 15.32. There are 24 columns of longitude of 15 degrees each that are lettered A through Z, with I and O omitted, in a west to east direction from the 180 degree meridian. There are 12 rows of latitude of 15 degrees each lettered A through M, omitting I, from the south to the north pole. The surface of the earth is divided into 288 basic 15 degree quadrangles, each of which is identified by two letters. The column is given first, followed by the row. In Fig. 15.32, a portion of the eastern United States is in the 15 degree quadrangle designated HJ.

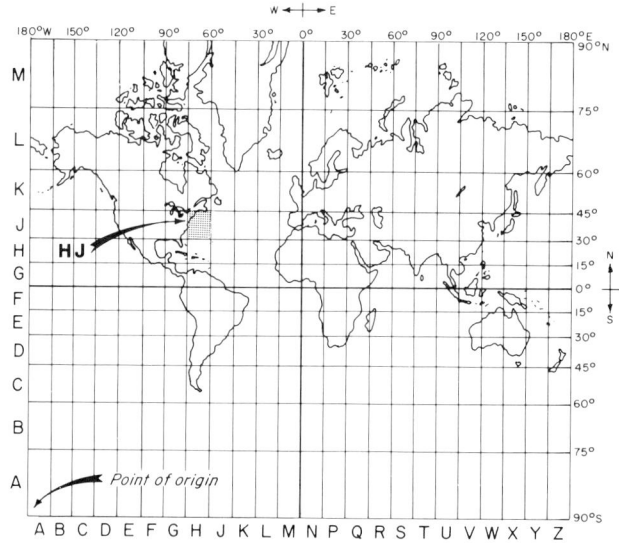

Fig. 15.32 GEOREF 15° quadrangles.

The 15 degree quadrangles are further divided into 15 one degree quadrangles, Fig. 15.33. These are lettered A through Q, omitting I and O, from west to east for the columns and from south to north for the rows. A one degree quadrangle anywhere on the earth's surface may now be identified by two additional letters. Atlantic City is located in the one degree quadrangle AK and the 15 degree quadrangle HJ. Its GEOREF reference is designated HJAK. The designator for the 15 degree quadrangle HJ is lettered in the southwest corner.

Each of the one degree quadrangles is divided into 60 one minute columns and rows, Fig. 15.34. The columns are

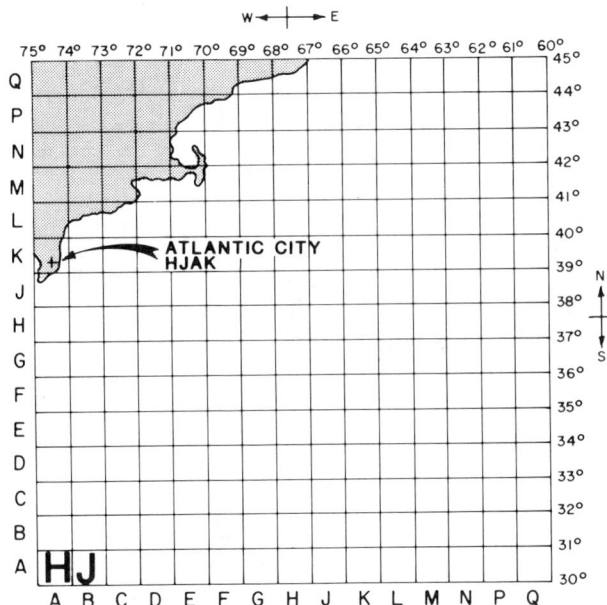

Fig. 15.33 GEOREF one degree quadrangles.

numbered from west to east and the rows from south to north. This reference system will identify the position of a point to an accuracy of approximately one minute, 2 kilometres (1.2 miles) by using four letters and four numbers. The GEOREF location of Atlantic City to one minute is HJAK3422. The designator for the one degree quadrangle AK is lettered in the southwest corner. On occasion it may be necessary to define the location of a point with greater accuracy than one minute. Each one minute quadrangle may be divided into 60 seconds of longitude and 60 seconds of latitude to allow additional coordinate values to be expressed.

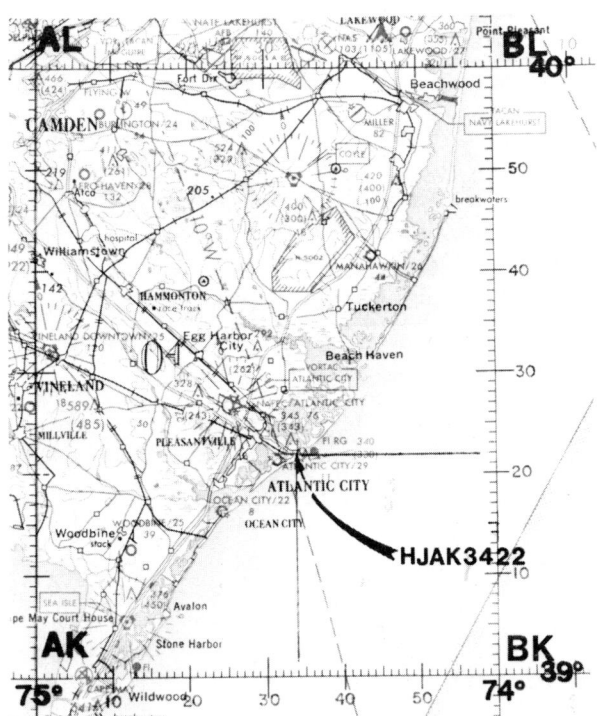

AREA AND VOLUME

16.1 INTRODUCTION

Land areas and earthwork volumes are quantities that frequently must be calculated. The determination of area may be thought of as a two dimensional analysis. Volume is three dimensional and is readily obtained by adding the third dimension.

In the metric system units to be used for area are square metre (m^2), hectare (ha), and square kilometre (km^2). Most land areas are shown in hectares, except those that are a small part of a hectare (small land surveys, city lots, plots, plans, and deeds) which should be shown in square metres. In general, areas greater than 1000 hectares (geopolitical divisions, states, countries, and continents) are expressed in square kilometres.

The cubic metre (m^3) is the fundamental unit for expressing volume. This unit is used for volumes involving engineering surveys, construction excavation, and cut-and-fill.

Area and volume may be calculated by a variety of methods depending on the area itself and the data available. Most surveying texts discuss these methods thoroughly and the reader should refer to other resource materials for additional information or procedures not described in the following sections.

16.2 DETERMINATION OF AREA

The calculation of area involving several simple geometric figures is illustrated in Fig. 16.1. Many of the methods for finding the area of a tract of land are based on the principle of dividing the land into some geometric shape.

Method of squares. The area of a figure or tract of land may be estimated by drawing or superimposing the figure on graph paper, Fig. 16.2. The total

SQUARE
Area = S^2

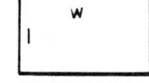
RECTANGLE
Area = $l \times w$

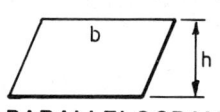
PARALLELOGRAM
Area = $b \times h$

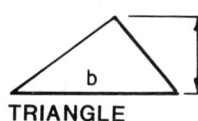
TRIANGLE
Area = $\frac{1}{2} b \times h$

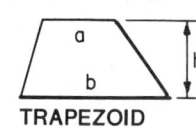
TRAPEZOID
Area = $h \frac{(a+b)}{2}$

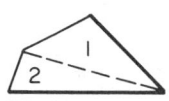
TRAPEZIUM
Divide figure into two triangles and find area using equation for triangle

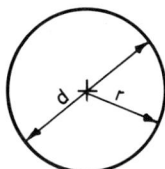
CIRCLE
Area = $\frac{\pi d^2}{4} = \pi r^2$
Circumference = $2\pi r = \pi d$

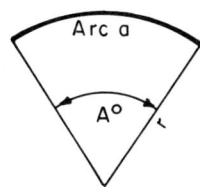
CIRCULAR SECTOR
Area = $\frac{1}{2}$ length of arc $a \times r$
    = area of circle $\times \frac{A°}{360}$
    = $0.0087 \, r^2 A°$
Arc $a = \frac{\pi r A°}{180°} = 0.0175 \, r A°$

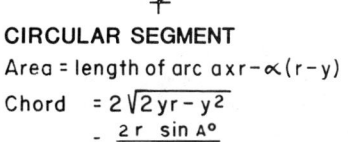
CIRCULAR SEGMENT
Area = length of arc $a \times r - \alpha(r-y)$
Chord = $2\sqrt{2yr - y^2}$
     = $\frac{2r \sin A°}{2}$

REGULAR POLYGON
Area = $\frac{n r_i}{2}$ (n = number of sides)
Any side = $S = 2\sqrt{r_o^2 - r_i^2}$
$R = \frac{S}{2 \sin \emptyset}$ $R = \frac{S}{2 \tan \emptyset}$

Fig. 16.1 Computation of area.

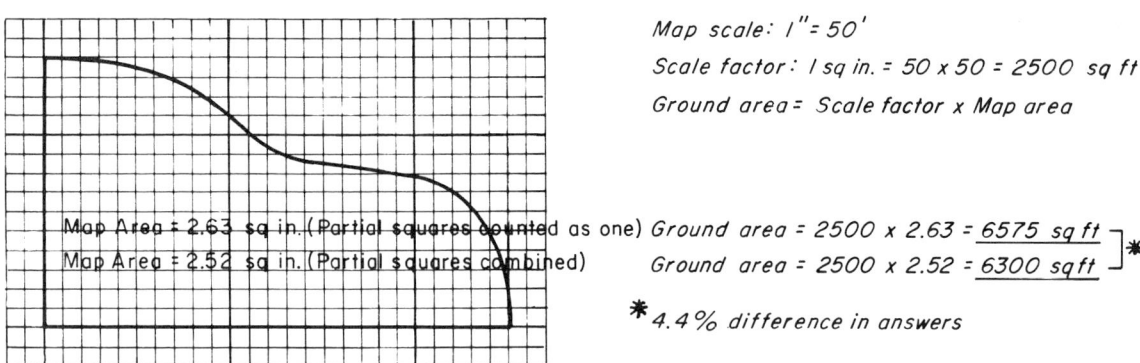

Fig. 16.2 Area by method of squares.

area is found by counting both the large and small squares enclosed by the figure. To convert the area of the map into the area on the ground, a scale factor must be applied. Multiplying the area of the enclosed figure by the scale factor gives the total area. Table 1 gives conversions for several units of area measurements.

The degree of accuracy depends on counting the partial squares as one or combining the partial squares. A more accurate answer is obtained by combining the smaller squares when counting to give full squares. In counting the partial squares, those that are overestimated or underestimated will tend to cancel out.

Area by dot grid. A dot grid pre-prepared on a transparent overlay may be used instead of squares to calculate area with reasonable speed and accuracy, Fig. 16.3. The dot grid, composed of alternate black and colored dots,

Table 16.1 Conversions for several units of area measurements.

Square feet	Square chains	Acres	Square miles	Square metres	Hectares	Square kilometres
4,356	1	0.1	0.000156	404.687	0.040 469	0.000 405
43,560	10	1	0.0015625	4 046.87	0.404 687	0.004 047
27,878,400	6,400	640	1	2 589 998	258.999 8	2.589 998
107,638.7	24.7104	2.47104	0.003861	10 000	1	0.01
10,763,867	2,471.04	247.104	0.386101	1 000 000	100	1

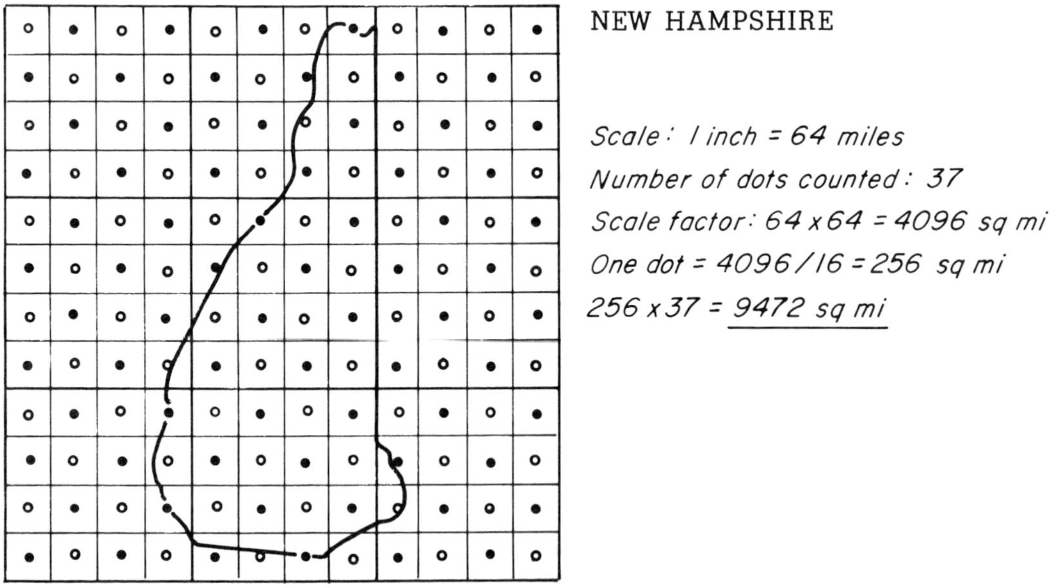

NEW HAMPSHIRE

Scale: 1 inch = 64 miles
Number of dots counted: 37
Scale factor: 64 x 64 = 4096 sq mi
One dot = 4096/16 = 256 sq mi
256 x 37 = 9472 sq mi

Fig. 16.3 Area by dot grid.

(shown as open circles) may be made on a sheet of graph or tracing paper. Dot grids of different densities are also commercially available. The dots are equally spaced and the accuracy of the calculation depends on the number of dots. The dot grid is laid over the area to be measured and the dots counted. A low or high dot count may be avoided by counting all dots falling within the boundary with dots touching the boundary counted only for the black or colored dots. In counting the dots, it is faster if a line grid encloses blocks of dots of 10, 16, 20, etc.. The number of dots is multiplied by the area factor each dot represents to obtain the total area.

Strip method. An area rectangular in shape is easily determined as illustrated in Fig. 16.4 (a). For an area enclosed by an irregular boundary, Fig. 16.4 (b), the unknown area may be divided into a series of narrow strips a unit distance apart. Since the degree of accuracy depends on the distance between the lines of the strips, a convenient small unit should be used. Each strip should be converted into rectangles by drawing lines at right angles to the vertical lines forming the strips. These lines are drawn as "give and take lines" in forming rectangles so as to exclude as much area as they include. The strips may be numbered for identification purposes and their lengths determined. The accuracy of the final results will depend on how well the length of each strip is measured. The lengths of all the strips are added and multiplied by the width of strip using the formula:

$$A = (L_1 + L_2 + L_3 \ldots) \times B$$
(B = width of strip)

The total area will yield square units and a scale factor will have to be applied to yield square miles or any other units that are desirable.

A variation of the method described above that is often used in practice and referred to as slipping areas may be illustrated by referring to Fig. 16.4 (b). A strip of paper may be used to add the lengths of the strips. Starting at zero the average length of strip number one is laid off followed by strip number two, etc. The area is determined by multiplying the total length of all the strips by the width of strip.

Triangle method. An area as illustrated in Fig. 16.5 is divided into triangles occupying as much of the figure as possible. The area is calculated using one of two formulas:

1. $A = \dfrac{\text{Base} \times \text{Perpendicular Height}}{2}$

2. $A = \sqrt{S(S-a)(S-b)(S-c)}$

where a, b and c are the three sides of the triangle and $S = \dfrac{a+b+c}{2}$.

The area of the remaining portions may be found by a procedure termed the Mean Ordinate Rule. Equally spaced perpendicular lines are drawn from the bounding lines of the various triangles to the margin of the area. The closer the offsets, the more accurate the results. The area of the irregular portion is found by the formula:

$$A = \dfrac{L(O + O_1 + O_2 + \ldots)}{n}$$

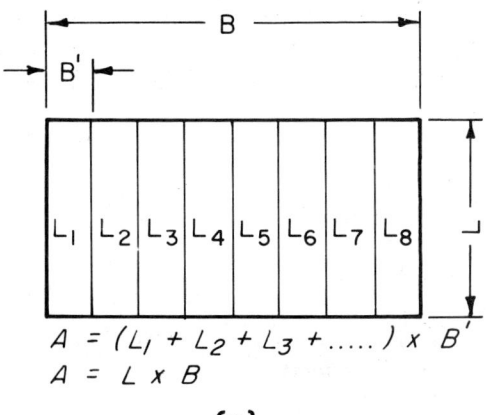

(a)

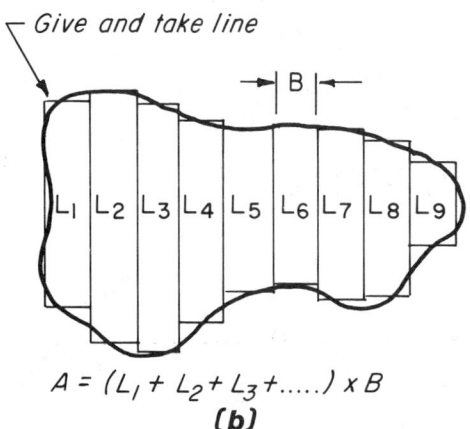

(b)

Fig. 16.4 Area by strip method.

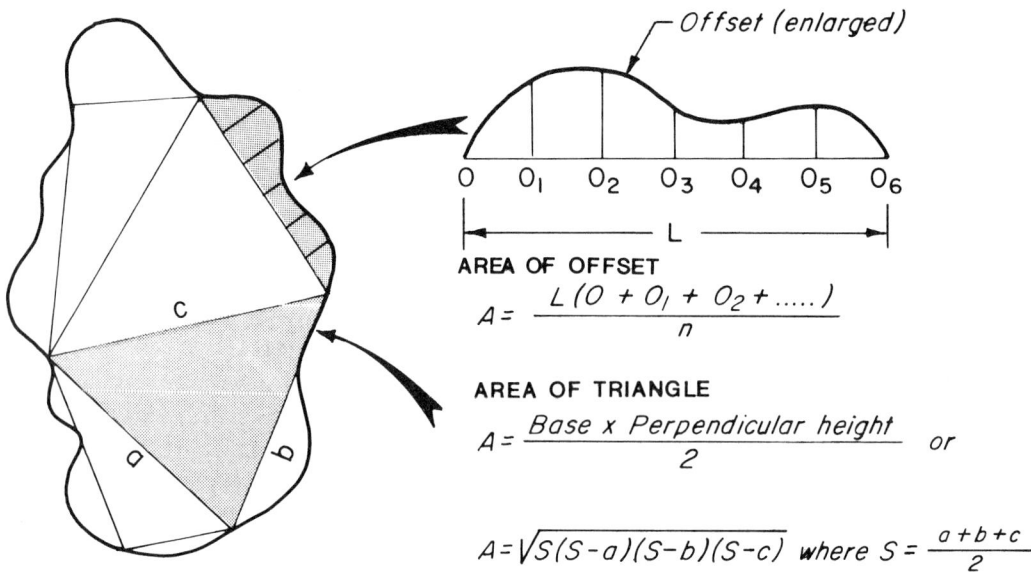

Fig. 16.5 Area by triangle method.

where L is the length of the line, 0, $O_1, O_2 \ldots O_6$ are the lengths of each offset, and n the number of offsets.

Trapezoidal rule. Often an irregular boundary as illustrated in Fig. 16.6 delineates a tract of land and the area must be determined within this boundary and a traverse line. The area lying between the boundary and traverse line is divided into equal strips of width d and successive points on the curve are connected by straight lines. The trapezoidal rule then consists in approximating the total area by a summation of the narrow trapezoidal areas. To improve the accuracy of the approximations, additional increments may be taken. The equation may be expressed conveniently in the form of the following rule:

Add the average of the end offsets to the sum of the intermediate offsets. The product of the quantity thus determined and the common interval between offsets is the required area.

For example, if rectangular offsets are taken at intervals of 10 metres, and the values of the offsets in metres are $h_1 = 4.1$, $h_2 = 4.4$, $h_3 = 3.7$, $h_4 = 4.9$, and $h_5 = 5.7$, the total area by the trapezoidal rule would be:

$$A = 10 \left(\frac{4.1 + 5.7}{2} + 4.4 + 3.7 + 4.9\right) = $$

179 square metres

Simpson's rule. The trapezoidal rule approximates the area under a curve by summing the areas of uniform width trapezoids formed by connecting successive points on the curve by straight lines. A more accurate approximation of the area under a curve is obtained by Simpson's rule since it consists of connecting successive groups of three points on the curve by second degree parabolas and summing the area under the parabolas to obtain the approximate area. The area under the curve in Fig. 16.7 may be found by dividing the area into an even number of strips of uniform width d. The total area is the

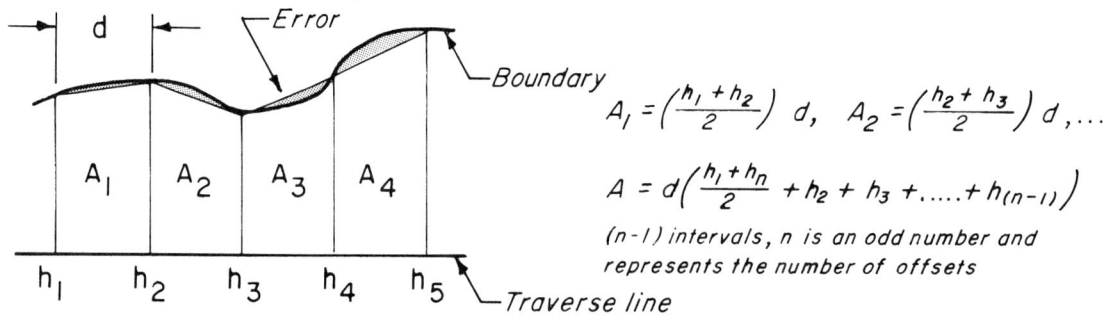

Fig. 16.6 Area by trapezoidal rule.

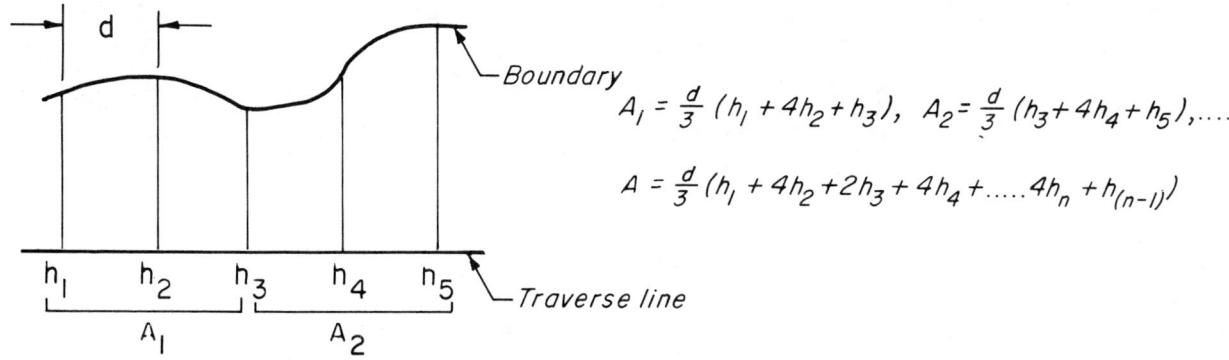

$$A_1 = \frac{d}{3}(h_1 + 4h_2 + h_3), \quad A_2 = \frac{d}{3}(h_3 + 4h_4 + h_5),\ldots$$

$$A = \frac{d}{3}(h_1 + 4h_2 + 2h_3 + 4h_4 + \ldots 4h_n + h_{(n-1)})$$

Fig. 16.7 Area by Simpson's rule.

sum of all the parabolic areas in the region between the two limits as expressed in the given equation. This is called Simpson's One-third Rule and may be expressed conveniently by the following rule, which is applicable for an odd number of offsets with a uniform interval between offsets.

Find the sum of the end offsets, plus twice the sum of the off intermediate offsets, plus four times the sum of the even intermediate offsets. Multiply the quantity thus determined by one-third of the common interval between offsets, and the result is the required area.

The area by Simpson's One-third Rule for the example used under the discussion trapezoidal rule is:

$$A = \frac{10}{3}[4.1 + 5.7 + 2.(3.7) + 4(4.4 + 4.9)] = \underline{180 \text{ square metres}}$$

Area by planimeter. The polar planimeter, Fig. 16.8, is an instrument designed to make area measurements of either straight or irregular boundaries. To measure an area the tracer point is manually run around the periphery of the figure in a clockwise direction. The dial is graduated into 10 parts and indicates the number of revolutions of the measuring wheel. The measuring wheel is numbered one to ten and each number represents 1/10 of a revolution of the measuring wheel. Each of the 10 parts of the measuring wheel is further divided into 10 parts representing 1/100 of a revolution of the measuring wheel. A vernier scale is also divided into 10 parts and represents 1/10 of each measuring wheel graduation (1/100 revolutions) or 1/1000 of a revolution

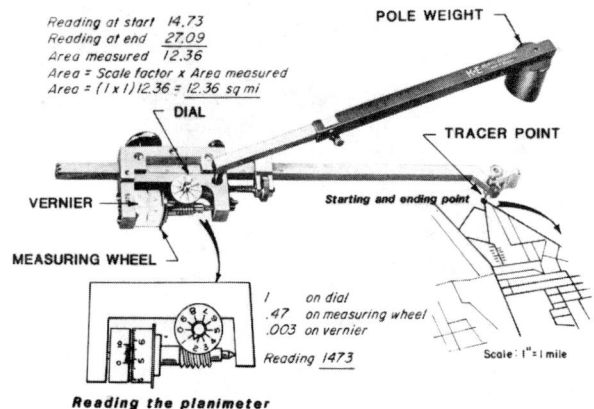

Fig. 16.8 Area measurement with the polar planimeter.

of the measuring wheel. Any planimeter may be read to four places very quickly and without error by understanding this principle.

The area of a figure is determined by establishing a starting point and taking an initial reading. Many planimeters are available to give dual readings in English (square inches) or metric units (square centimetres). When the boundary of the figure has been traced carefully in a clockwise direction and the tracer point returned to the starting point, a final reading is made. The difference in the two readings is used to indicate the area. Some planimeters have a zero setting device that eliminates the need to record an initial reading. It is best to repeat the operation at least once to serve as a check. The average of two or more tracings may be made to increase the accuracy of the results. Areas usually may be measured within a range of from .05% to 1% of the correct area. For areas too large to circumscribe in one operation, divide the area into measureable segments, measuring each separately and adding the results.

For convenience, most planimeters include factors to aid the user in converting planimeter readings to commonly used area units.

A digital readout planimeter, Fig. 16.9, is advantageous to use where a large number of areas are to be measured. Valuable time is saved with readouts made in an instant on a readout dial for any scale ratio or measuring system. The correct scale area is automatically computed while the figure is being measured. A switch allows the

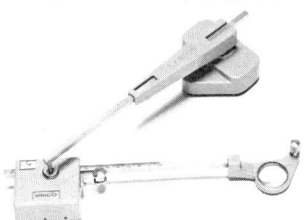

Fig. 16.9 Digital readout computing planimeter *(Courtesy The Ben Meadows Co)*.

planimeter to be set at zero and consecutive areas may be added or subtracted from each other with an accumulative switch.

The following examples will serve to illustrate the calculation of area in English and metric units using the expression:

Area = Scale Factor x Area Measured

Example using English units

Map scale: 1" = 100'

Planimeter reading: Final 83.45

Initial 71.09

Area measured = 12.36 in.^{2}

Scale factor: 1 in^2 = 100 x 100 = 10,000 ft^2

Area = 10,000 x 12.36 = 123,600 ft^2 / 43,560 = 2.837 acres

Example using metric units

(a) Map scale: 1:2500 m

Planimeter reading: Initial 0

Final 213.29 cm^2

Area measured = 213.29 cm^2 = 0.021 329 m^2

Scale factor: 1 m^2 = 2500 x 2500 = 6 250 000 m^2

Area = 0.021 329 x 6 250 000 = 133 310 m^2

(b) Map scale: 1:1000 mm

Planimeter reading: 2.75 cm^2

Area (cm^2) = ?

Change map scale from 1 mm: 1000 mm to 0.1 cm: 100 cm

Scale factor: $\frac{1}{10}$ x $\frac{1}{10}$ = 100 x 100

$\frac{1}{100}$ = 10 000

1 cm^2 = 1 000 000 cm^2

Area = 1 000 000 x 2.75 = 2 750 000 cm^2

Area by coordinates. The computation of area from known coordinates may be used for a closed polygon traverse, Fig. 16.10 (a). This is accomplished basically by finding the areas of trapezoids, e.g. DEed, CDdc, etc., formed by projecting lines from the traverse corners upon one of a pair of coordinate axes, usually the y-axis. To avoid negative values, the origin for the axes could be moved so that all traverse corners lie in the northeast or first quadrant, Fig. 16.10 (b).

The area of the tract is computed by summing algebraically the areas of the trapezoids falling both within and outside the traverse. The equation for this calculation may be written:

$$\text{Area} = \frac{-\left[Y_A(X_B - X_E) + Y_B(X_C - X_A) + Y_C(X_B - X_D) + Y_D(X_C - X_E) + Y_E(X_D - X_A)\right]}{2}$$

A general rule used to express the calculation of area by coordinates is:

To determine the area of a tract of land when the coordinates of its corners are known, multiply the parallel distance, or ordinate, of each corner by the difference between the meridian distances, or abscissas, of the following and the preceding corners, always algebraically subtracting the following from the preceding. The area of the tract is one half of the algebraic sum of the resulting products.

In Fig. 16.10 (a) the resulting area of the traverse is negative since the calculations were made proceeding in a counter-clockwise direction. The reverse of this would occur by calculating the area proceeding in a clockwise direction. A result identical except for sign would also be obtained by always subtracting the preceding from the following. In any case, the sign of the area is not significant.

The equation used in Fig. 16.10 may be reduced to an easier form by listing the X and Y coordinates of each point in succession in two columns, Fig. 16.11. The coordinates of the starting point, A, are repeated at the end. Products are determined as indicated by the arrows, with the solid arrows considered plus and the dashed arrows minus. The algebraic signs of coordinates must be watched carefully. A positive product, dashed arrows, having a negative coordinate will be minus while a negative product, solid arrows, having a negative coordinate will be positive. The total area is obtained by dividing the algebraic summation of all products by two.

$$\text{Area} = \frac{-\left[Y_A(X_B - X_E) + Y_B(X_C - X_A) + Y_C(X_B - X_D) + Y_D(X_C - X_E) + Y_E(X_D - X_A)\right]}{2}$$

$$= \frac{-\left[0(-187.343) + (-140.951[-441.403]) + (-2.920[-143.564]) + 98.001(384.536) + 156.364(387.774)\right]}{2}$$

$$= \frac{-187.343 - 62\,216.194 - 419.207 - 37\,684.912 - 60\,633.894}{2}$$

$$= \frac{-161\,141.550}{2} = -80\,570.775 \text{ sq m} \quad \text{or} \quad -80\,570.775 / 10\,000 = -8.1 \text{ ha}$$

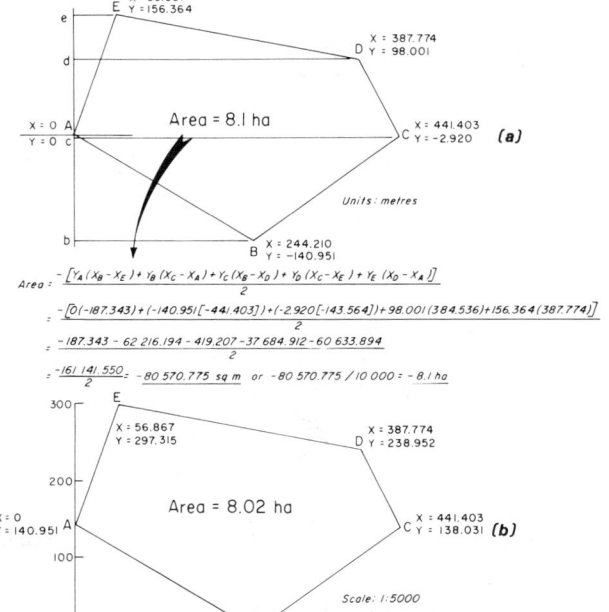

Fig. 16.10 Area by coordinates.

Point	X	Y	(+) ⟶	(-) - - -→
A	0	0		
B	244.210	-140.951	0	0
C	441.403	-2.920	-713.093	62 216.194
D	387.774	98.001	43 257.935	1 132.300
E	56.867	156.364	60 633.894	5 573.023
A	0	0	0	0
			Σ 103 178.736	57 775.471

160 954.207

Area = 160 954.207 / 2 = 80 477.104 sq m or 80 477.104 / 10 000 = 8.05 ha

Fig. 16.11 Computation of area by coordinates.

Area by double meridian distance (DMD) method. The area computation of a closed figure by the double meridian distance method is convenient when latitudes and departures of boundary lines are known. A reference meridian, Fig. 16.12 (a), is assumed to pass through the most westerly traverse station to avoid negative values. The meridian distance of a straight traverse line is the perpendicular distance from the mid-point of a line to the reference meridian. The double meridian distance of a straight traverse line is the sum

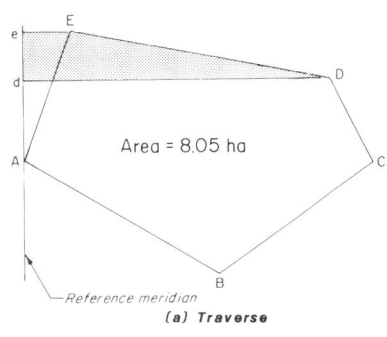

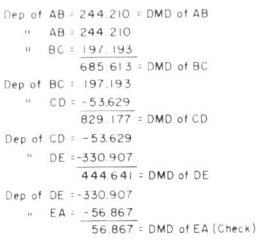

(b) Computation of DMD's   (c) Tabulation of DMD's

Fig. 16.12 Area by DMD method.

of the meridian distance of the two ends, e.g., the double meridian distance of line DE is Ee + Dd. For the trapezoid, DEed, the double area is the product of the double meridian distance of the course and the latitude of the course:

Double area = DMD x latitude

The algebraic summation of all double areas gives twice the area of the tract inside the traverse. The signs of the products of DMD's and latitudes must be taken into account. If the reference meridian is passed through the most westerly corner, all DMD's are positive. The products of DMD's and north latitudes are plus and the products of DMD's and south latitudes minus.

Several convenient rules of determining DMD's are:

1. The DMD of the first traverse course (where the reference meridian passes through the beginning station) is equal to the departure of that course.

2. The DMD of any other traverse course is equal to the DMD of the preceding course, plus the departure of the preceding course, plus the departure of the course itself.

3. The DMD of the last traverse course is equal to its departure, but of opposite signs.

The computation of area using DMD's is shown in Fig. 16.12. All latitudes and departures of the traverse lines have been determined and the survey balanced by the Bowditch rule. Computations and procedures for these steps are described in chapter 15, Fig. 15.24. The DMD's are calculated using the preceding rules, Fig. 16.12 (b). The DMD of the last line, EA, in the traverse is equal to its departure, but of opposite sign, to serve as a check that the computations are correct.

A convenient method for tabulating the DMD's is illustrated in Fig. 16.12 (c). The double positive and negative areas are computed by multiplying each DMD by the corresponding adjusted latitude. These values are summed algebraically. The result is divided by two to obtain the area 80 477.103 square metres, and by 10 000 to obtain the number of hectares, 8.05.

Area of cross sections. Many of the methods for determining land area are appropriate for use on cross sections involving cut or fill earthwork calculations. The three-level section, Fig. 16.13, is used where the ground conditions are not too rough. Three points, one point on each side of the centerline, are required to determine the ground surface. The area is found using the equation:

$$A = \frac{c(d_l + d_r)}{2} + \frac{w(h_l + h_r)}{4}$$

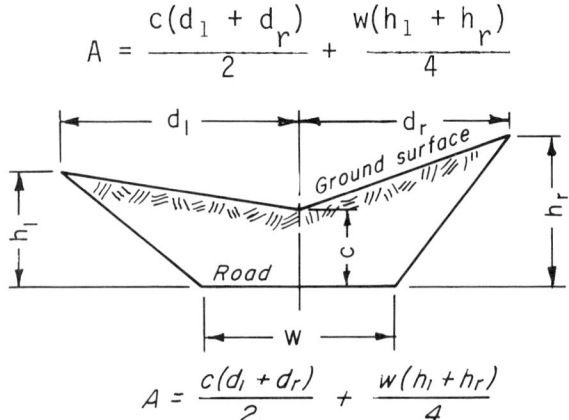

Fig. 16.13 Area of three-level section.

16.3 DETERMINATION OF VOLUME

The amount of earthwork in the form of cut and fill is often required and must be calculated in construction work involving roads, railroads, dam sites, the grading of areas, etc. It is sometimes necessary to be able to find the volume involving different shaped solids, Fig. 16.14. Computation methods for obtaining the volume of irregular shaped figures are discussed in subsequent sections.

Volume by profiles and cross sections. A profile may be used to obtain sufficient data for estimating earthwork quantities involving roadways, Fig. 16.15. The profile is taken along the proposed road centerline and a "give and take line" representing the road gradeline is drawn bisecting the high and low terrain so the amount of cut will approximately be equal to the amount of fill. The total area of cut and fill is determined separately along the road, using methods previously described. The total volume of earthwork may be estimated by the formula:

V = A x w (w = width profile represents)

Ideal conditions exist when the cut and fill balance or are equal, requiring no material to be hauled away or onto the site.

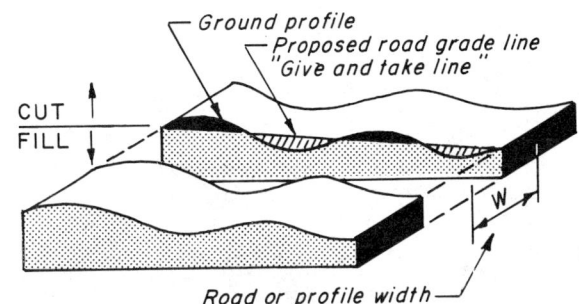

Fig. 16.15 Volume of profile by "give and take line".

A more accurate determination of earthwork quantities may be obtained by the use of cross sections. A series of equally spaced cross sections are drawn, Fig. 16.16 (a). The closer together the cross sections are taken, the more accurate the final results. An assumption is made that the slope from one cross section to the next is uniform. This is normally not true, however the results are reasonably accurate. The volume of each trapezoid is found by finding the average of the two parallel sides and multiplying by the distance between them. A total volume

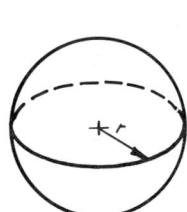

SPHERE

Vol = $\frac{4\pi r^3}{3}$ = 0.5236 d^3

Surface = $4\pi r^2$ = πd^2

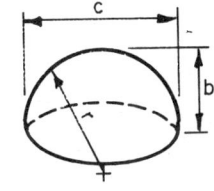

SEGMENT OF SPHERE

Vol = $\frac{l\pi b^2 (3r-b)}{3}$

Surface = $2\pi r b$
(Excluding circular base)

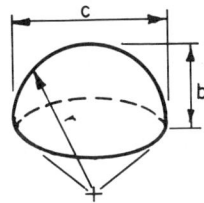

SECTOR OF SPHERE

Vol = $\frac{2\pi r^2 b}{3}$

Surface = $\frac{\pi r (4b+c)}{2}$

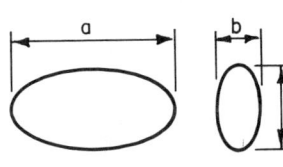

ELLIPSOID

Vol = $\frac{\pi a b c}{6}$

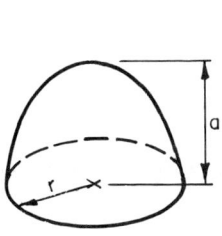

PARABOLOID

Vol = Area of circular x $\frac{1}{2}$ Altitude

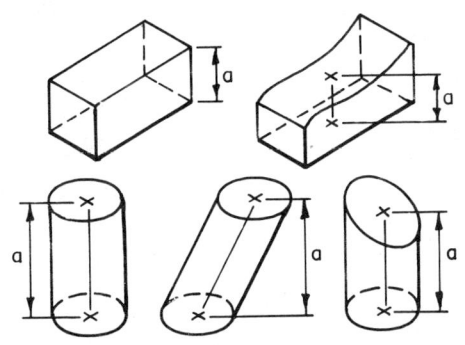

PRISMS AND CYLINDERS

Vol = Area of base x Altitude

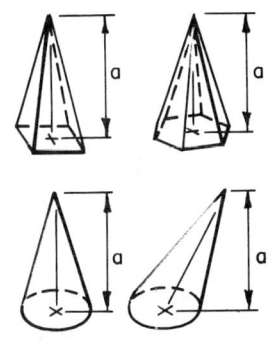

PYRAMIDS AND CONES

Vol = Area of base x $\frac{1}{3}$ Altitude

Fig. 16.14 Computation of volume.

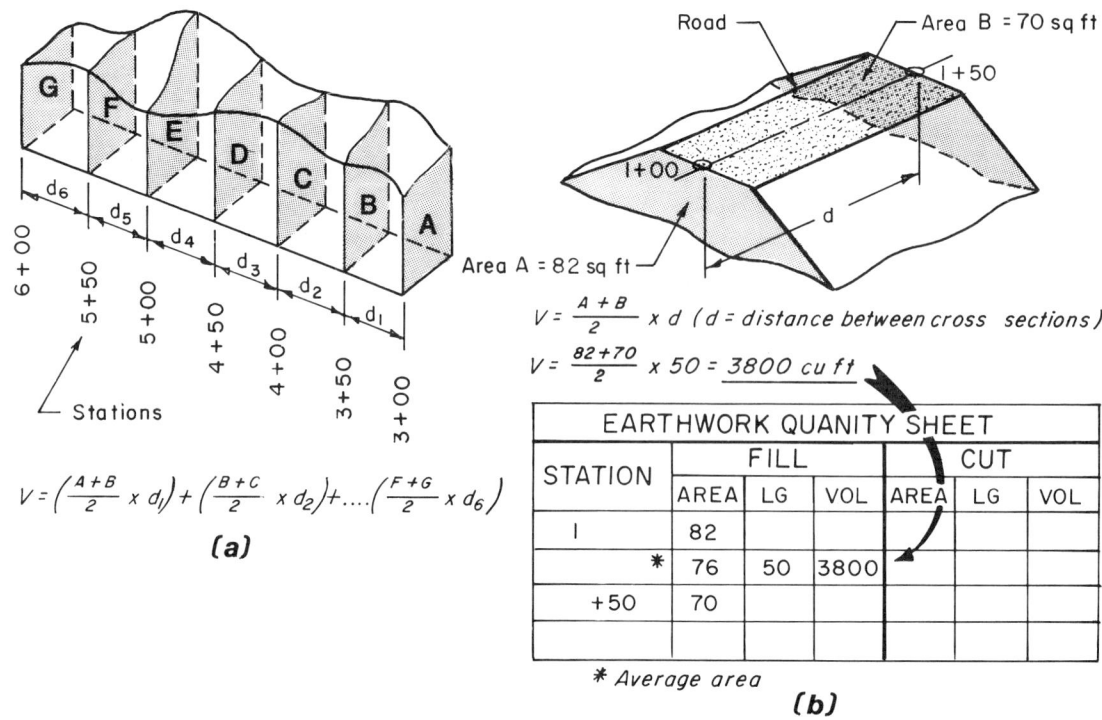

$$V = \left(\frac{A+B}{2} \times d_1\right) + \left(\frac{B+C}{2} \times d_2\right) + \ldots \left(\frac{F+G}{2} \times d_6\right)$$

(a)

Fig. 16.16 Volume of cross sections.

that is accurate for estimating and other purposes is found by the following summation.

$$\text{Volume} = \left(\frac{A+B}{2} \times d_1\right) + \left(\frac{B+C}{2} \times d_2\right) + \ldots \left(\frac{F+G}{2} \times d_6\right)$$

The estimation of quantities of earthwork involving cut and fill may be conveniently tabulated using an earthwork quantity sheet, Fig. 16.16 (b). The cut and fill areas are easily obtained by drawing a series of cross sections for filling in and computing the volume.

<u>Volume by prisms</u>. The volume of earthwork for an area to be excavated or filled may be calculated by dividing the figure into vertical truncated prisms, Fig. 16.17. The elevation of the four corners of each prism is determined and the volume found by multiplying width by depth by average height. For prism A the elevation of each corner at a, b, c, and d is determined. If the area in square feet is indicated by A and h_1, h_2, h_3, and h_4 are the corner heights, the volume in cubic yards may be determined by the formula:

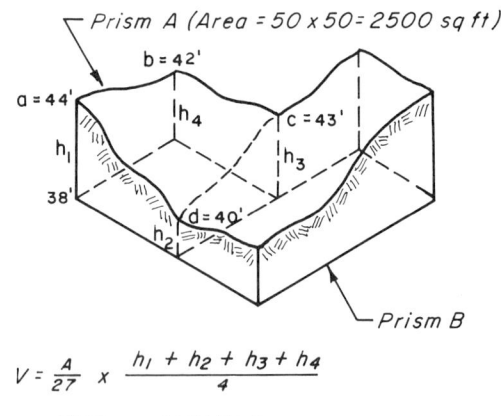

Fig. 16.17 Volume by use of vertical truncated prisms.

$$V = \frac{A}{27} \times \frac{h_1 + h_2 + h_3 + h_4}{4}$$

The total volume is found by computing the volume for the remaining prism B in a similar manner.

The volume of an irregular figure may be found by constructing a section midway between the bases and calculating the volume using the prismatoid equation:

V = (Area of base A + Area of base B + 4) (Area of midsection C) × $\frac{1}{6}$ Perpendicular distance between bases (D)

The formula is quite accurate for any solid with two parallel bases connected by a surface of straight line elements, Fig. 16.18 (a), or smooth simple curves, Fig. 16.18 (b).

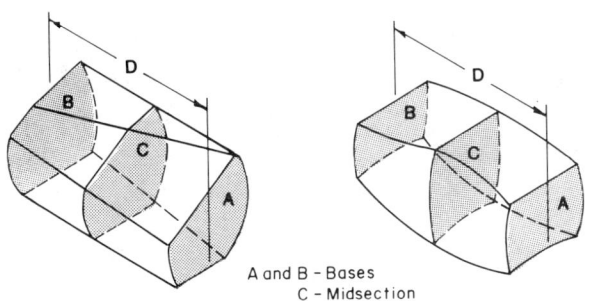

A and B – Bases
C – Midsection

V = (Area of base A + Area of base B + 4)(Area of midsection C) × $\frac{1}{6}$ Perpendicular distance between bases (D)

(a)   (b)

Fig. 16.18 Volume by prismatoid formula.

<u>Volume by contours.</u> Volumes may be obtained from contours by finding the area enclosed by each contour. The area of the first and last contours plus two times the area of the remaining contours is multiplied by one-half the contour interval to find the volume.

$$V = A_1 + 2(A_2 + A_3 + A_4) + A_5 \times \tfrac{1}{2} CI$$

The contours in Fig. 16.19 represent a loose pile of four-foot wood. Since one cord of loosely piled four-foot wood will equal approximately 170 cubic feet, the calculated volume is divided by this value to give 46,470 cords of wood.

16.4 PROGRAMMABLE CALCULATORS AND DIGITAL COMPUTERS

The computation of area and volume may be expedited by using programmable calculators or digital computers. Many kinds of hardware devices are manufactured to help save time in making calculations and to reduce the chance of error, Fig. 16.20. Where lengthy or repetitive computations are required, the use of this type of equipment is recommended. The user, however, must be able to understand the different methods presented for the computation of area and volume.

The cost of programmable calculators has been reduced substantially. Both desk top and pocket models are available with keyboard programming and card reading programs. The cards may be punched and read into the calculator or libraries of card programs for different types of problems may be purchased.

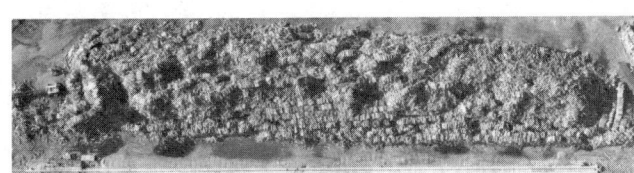

Pulp piles may be contoured from aerial photographs to find volume.

$$V \text{ (cords*)} = \frac{[A_1 + 2(A_2 + A_3 + A_4) + A_5] \times \tfrac{1}{2} CI}{170 \text{ ft}^3}$$

$$= \frac{[203,125 + 2(143,750 + 91,875 + 48,750) + 18,125] \times 10}{170}$$

$$= 46,470 \text{ cords}$$

*1 cord = approximately 170 ft³ loosely piled 4' wood

Contour	Area (sq ft)
0	203,125
20	143,750
40	91,875
60	48,750
80	18,125

Fig. 16.19 Volume by contours.

Electronic self-contained mini-
computers have become common place at
an affordable price. Most models may
be easily expanded by adding peripheral
equipment as the need increases. A program or software package is required to
translate the language of the program
into the language of the computer.
There are many kinds of ready-to-run
programs available from equipment manufacturers. Some exposure to programming methods and procedures is necessary, however, before any program will
be meaningful.

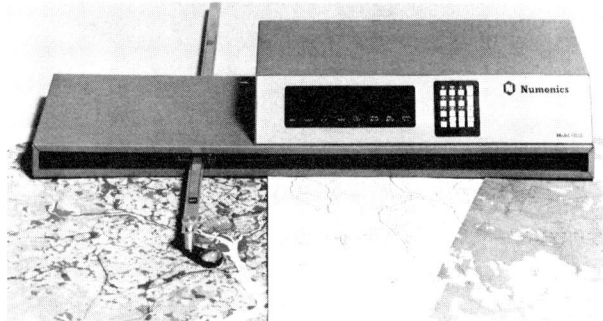

Fig. 16.20 Area-length calculator
(Courtesy Numonics Corp).

REPRESENTATION OF RELIEF

17.1 INTRODUCTION

It is necessary to have an understanding of the various shapes and forms of the earth's surface to assist individuals in solving the many problems of land use. Relief information is needed on maps to furnish coordinated data for computing problems involving the terrain and to present a 3D graphic picture of the ground surface. This type of data is used in the planning of large projects such as buildings and highways in addition to a myriad of problems in engineering, geology, soils, and forestry.

The earth's physical features have been shaped into many forms by the action of forces such as erosion, water, freezing, thawing, and wind. The cartographer commonly uses the term *topography* when discussing land shapes and forms making up the configuration of the earth's surface. This surface is not flat but an endless series of ups and downs and the term *relief* is used when referring to variations in land heights. The term *elevation* is used to refer to vertical distances measured above or below a datum plane. The shapes of features such as depressions, ridges, hills, and valleys are given the general name of *land forms*.

17.2 RELIEF METHODS

A number of different methods have been developed and used for portraying relief. Each of these methods has its own advantages and limitations. Several ways of showing relief to add to the maps third dimension are relief models, hachures, shading, layer tints, and contours.

17.3 RELIEF MODELS

Terrain models are excellent to show relief for individuals that find it difficult to visualize a contour map or lack the training to interpret them properly. Essentially, *relief models* are small three-dimensional maps showing the configuration of the earth's surface as it appears on the ground. Early relief models were made of plaster, sponge rubber, or layers of cardboard. More recently, lightweight plastic relief models have been developed. The map is printed on a plastic sheet, Fig. 17.1, and then deformed by heat and vacuum over a plastic mold. These can be manufactured economically in great quantities. The vertical scale for most models must be exaggerated to accentuate the relief with the extent of exaggeration stated in the marginal information.

Plastic relief models are finding increasing use in fields like geological structures, landscape architecture, military operations, city and regional planning, etc. The user, in many instances, finds it easier to access the vertical elevation data as compared with looking at contours on the usual topographic map. Many models will show contours as well as other man-made features These are excellent for studying different land forms to help improve an individual's ability to visualize features shown by contours on a two-dimensional surface.

17.4 HACHURES

In 1799 a symbol termed a *hachure*, Fig. 17.2 (a), was introduced and widely used to portray slopes. This was one of the first attempts to represent relief

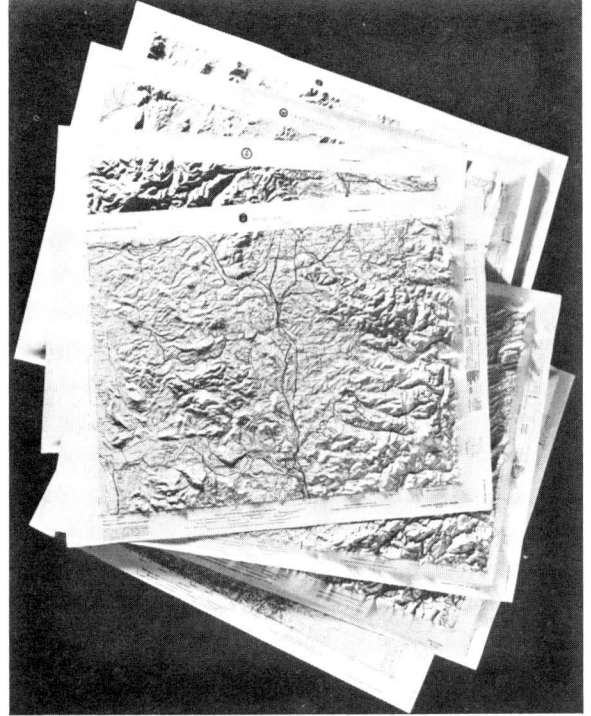

Fig. 17.1 Plastic raised relief topographic maps *(Courtesy Hubbard Scientific Co)*.

in plan. Hachures are drawn as short straight lines of varying widths always running downhill in the direction of greatest slope. The steeper the slope, the shorter the hachures and the closer their spacing. The thickness of the hachures is varied in proportion to the angle of slope with the steeper slopes drawn with thicker lines. A very skilled drafter is required to obtain a good visual impression of ridges, peaks, and drainage systems, Fig. 17.2 (b). The use of hachures will not show exact relief and their use is limited in general to showing relative relief and terrain shapes. For this reason, hachures are no longer used to represent relief. Their use today is as a supplemental device for showing small features such as pits, crags, rock ledges, and levees that cannot be shown adequately or feasibly by contour lines alone.

17.5 SHADING

The pictorial effect of a map may be emphasized by relief shading techniques, Fig. 17.3. The appearance of solid three-dimensional topography, similar to that of an aerial photograph, is simulated by the degree of light and shadows on the terrain. On many maps the area to be shaded is illustrated obliquely from the northwest. The north

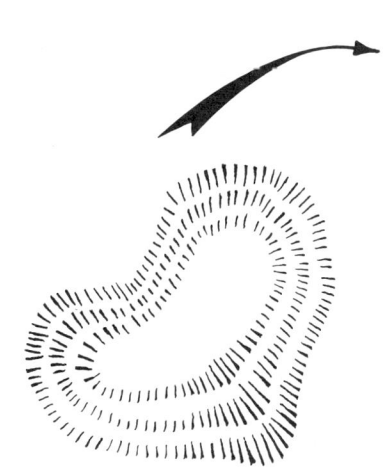

(a)

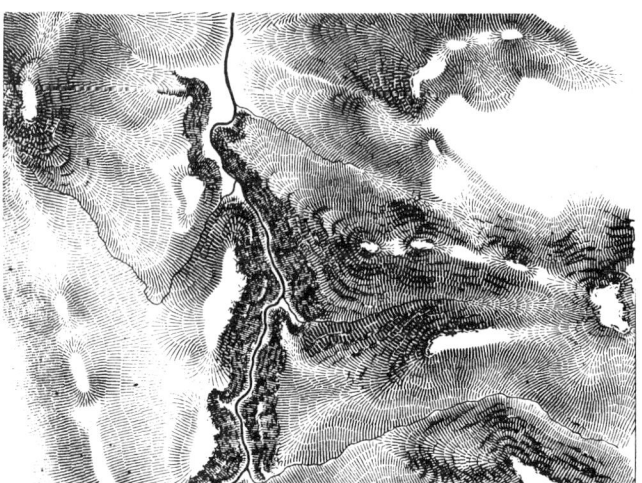

Portion of a map showing hachures. Exact source unknown, probably Dufour. The Dufour map of Switzerland was started in 1833 and completed in 1866.

(b)

Fig. 17.2 Hachures.

Fig. 17.3 Relief shading (Portion of Maine, USGS).

and west slopes would be in light and the south and east facing slopes in shadow. A three-dimensional effect is achieved through the artistic use of various color tonal values. The degree of slope is indicated by the density of shading. It is sometimes necessary to shift the light source direction to show slopes that run parallel to the light source.

Relief shading is best suited to maps of mountainous areas since distinguishing one tone from another is infeasible in areas of low relief. While this method can show relative relief clearly, it does not show exact elevations. Used in conjunction with contours, relief shading helps the map user to interpolate terrain patterns.

17.6 LAYER OR ALTITUDE TINTING

Various colors or tones of the same hue can be used to show different zones of elevation. The map is divided into bands of elevation, e.g., 0-50, 50-100, 100-150 metres, etc., and a color is assigned to each band. Usually the lower elevation bands are a lighter color and the higher ones darker with the bands in between becoming progressively darker as the elevation is increased. The sequence of colors from the lowest to the highest elevation on many maps is green, yellow, brown, and red. A key to the colors is always printed in the margin.

Layer tinting is used primarily for presenting land surface information on wall maps, atlases, and other physical maps of small scale. It should be emphasized that layer tinting portrays little about the land surface since each color band must cover a wide vertical elevation range. Each zone shows a sharp change from one elevation level to the other while abrupt differences rarely exist. This problem may be reduced by vignetting, the use of colors that gradually merge one tint into the next. More success may be realized when layer tints are used in conjunction with contour maps to add to the third dimension.

17.7 CONTOURS

Throughout the history of mapmaking, the depiction of relief above and below the water of the earth has been a major problem for cartographers. Several methods of expressing relief have been discussed and work is still going on to improve existing techniques. The principal method of showing relief is by the use of *contours* or *isarithms*. Contouring was first introduced by the Dutch surveyor Cruquis in 1729 and is considered the most effective method for showing topography. Through the use of contours, elevation information as well as a three-dimensional terrain picture is readily obtained. This one advantage over other relief methods makes its use invaluable for many engineering and related projects where more precise information is required concerning ground relief.

A *contour*, Fig. 17.4, is an imaginary level line of constant elevation on the surface of the ground that connects all points of equal elevation. It may be thought of as a line formed by the intersection of the shore line of a body of water with the ground surface. A *contour line* is the actual line drawn on the map to represent the corresponding contour on the ground. No distinction will be made in the use of these two terms throughout the remainder of the text.

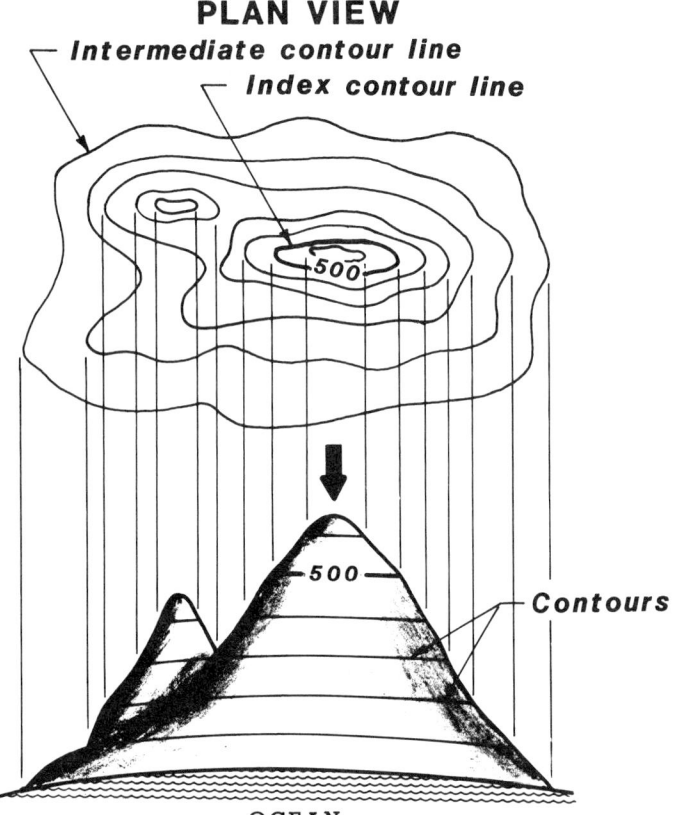

Fig. 17.4 Contour terminology.

ject line weight and the intermediate lines center line weight. The intermediate contour lines are usually not labelled except where the terrain is relatively flat and their elevations are not readily obvious.

In places where the intermediate contours are so closely spaced as to be coalescing, or nearly so, it is standard practice for readability to drop the lines for short distances, Fig. 17.5. This treatment is called *feathering*.

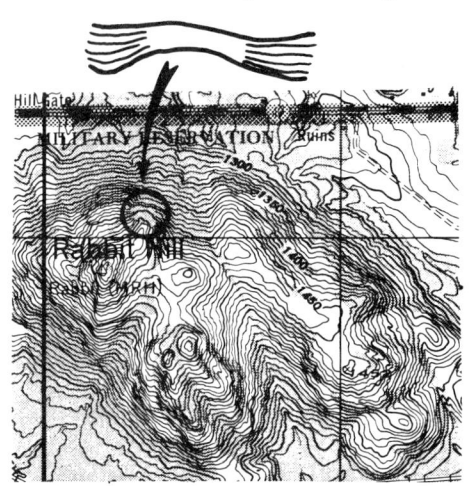

Fig. 17.5 Feathering contour lines (Portion of Fort Sill, Oklahoma area, Defense Mapping Agency).

There are several types of contours that are used to express the ground relief and they may be classified into the following types:

Index contours. Usually every fifth contour line, depending on the contour interval, is an *index contour*, Fig. 17.4. For example, if the contour interval is 10 metres, every multiple of 50 metres will be an index line. These lines are never omitted and should be drawn thicker in relation to the other contour lines. The line is broken and the elevation lettered following procedures described in chapter 6. Contour numbers basically should be lettered and spread on every index line in such a way that will enable the value of other lines or points to be easily read or interpreted. The contour number may have to be repeated several times for long contour lines that meander over the map.

Intermediate contours. The thinner lines drawn between adjacent index contour lines are termed *intermediate contour lines*, Fig. 17.4. It is recommended that the index lines be drawn ob-

Supplementary contours. On flat areas where contour lines are spaced too far apart, *supplementary contour lines*, Fig. 17.6, drawn as dashed lines are often used to help depict the relief. These normally are drawn where relatively flat areas appear on the same map alongside rugged areas. A more complete picture of the terrain is presented by drawing supplemental contour lines usually at one-half the contour interval of the map. The conformation of the flatter ground as well as the rise and fall of the more rugged areas can be more easily assessed by using this technique.

Approximate contours. In some instances it may not be feasible to obtain accurate contour lines. The area may be inaccessible on the ground or it may be difficult to interpret contours from aerial photographs where heavy shadows occur. *Approximate contour lines*, drawn as dashed lines, are used in these instances to give a reasonable idea of relative elevations.

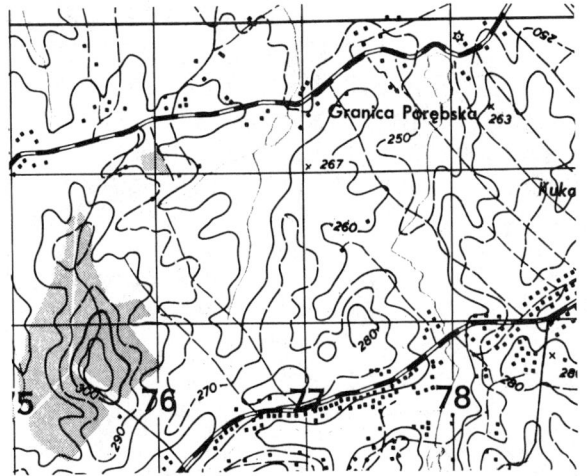

Fig. 17.6 Supplemental contour lines (Portion of Biala-Bielsko, Poland area, Defense Mapping Agency).

Form lines. Information is sometimes insufficient, sketchy, or not available to draw contour lines. For representing the general shapes of terrain features without reference to a datum plane, *form lines* are used to show the approximate representation of relief. The lines resemble contour lines but do not always connect points of equal elevation and should not be labelled.

17.8 CONTOUR INTERVAL

The vertical distance between two adjacent contour lines is termed the *contour interval*, Fig. 17.7. Neighboring contour lines may be close together or far apart on the map to show changes in slope and relief variations. The contour interval, obviously, will affect the degree of detail for showing this information. A small contour interval will allow slight changes in slope and small features like depressions and hills to be drawn. A large contour interval will frequently mask out irregularities of this kind. Regardless of the slope or feature represented, the vertical distance from one contour line to the next represents a change of height of the ground equal only to the contour interval.

The contour interval must always be consistent throughout any one map but may be varied between map sheets to best portray the terrain. In planning and determining the contour interval to be used, the appropriate interval should be based on such factors as map scale, characteristics of the terrain, density of ground cover, relative costs, map use requirements, available information, and

SUGGESTED CONTOUR INTERVALS

Scale	metres	feet
1: 2 000	1	5
1: 5 000	2	10
1: 10 000	5 or 10	25
1: 25 000	10	25
1: 50 000	20	50
1: 100 000	25	100
1: 250 000	50	200

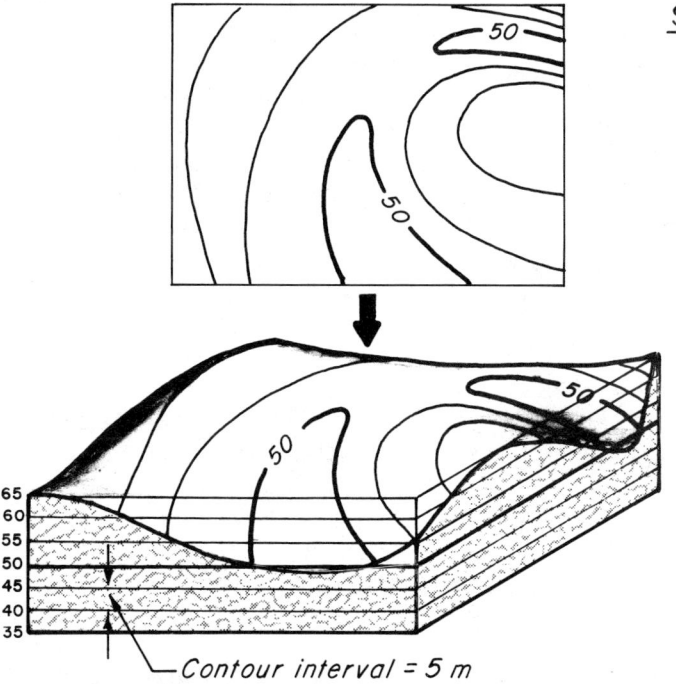

Fig. 17.7 Contour interval.

difficulty in obtaining this information. The map scale and slope of the terrain affect the contour interval selected since the minimum practical separation between contours to avoid coalescence is about 1.3 mm (0.05"). Where the ground surface cover is extremely heavy, it may be difficult to see the ground for drawing contour lines derived by photographic methods. In these instances the contour lines are commonly drawn as dashed instead of solid lines. Many map users are interested in the general slope and form of the terrain and do not require maps having a small contour interval. In general, the contour interval is a matter of economy with the smallest interval that can be afforded desired. The cost of the map will increase rapidly as the contour interval is reduced. This cost increases roughly in proportion to the square of reduction, e.g., reducing the contour interval from four to two increases the map cost almost four times.

The contour interval is almost invariably fixed at multiples of 2, 5, 10, 20, 50, and 100. The metre will become the standard unit of vertical measurement for all maps. New maps such as the topographic series maps produced by the USGS are now all metric. Many existing maps, however, are contoured in feet and this unit will continue to be used for a number of years. The contour interval used for general types of terrain may be summarized as follows: (a) flat to gently rolling - 1 to 5 metres (2 to 20 feet), (b) hilly - 5 to 20 metres (5 to 50 feet), (c) mountainous - 25 metres (100 feet).

17.9 GEODETIC CONTROL DATA

The National Geodetic Survey is an agency of the National Ocean Survey within the Department of Commerce, National Oceanic and Atmospheric Administration. It maintains the nation's geodetic control system that allows accurate modern maps to be produced. In the United States a network of precisely known geodetic control points have been established that locate the horizontal position (latitude and longitude) and elevation for each of these points. The geodetic networks consist of approximately 250,000 horizontal and 500,000 vertical control stations. The control points are indicated by bronze markers or monuments fixed in the ground. Other Federal organizations, such as the U.S. Geological Survey, the Bureau of Reclamation, the Tennessee Valley Authority, the U.S. Army Corps of Engineers, and many state agencies, also place markers in the course of their work. These markers are incorporated into the national system. In carrying out mapping and surveying activities, individuals rely on NOS for geodetic control and upon the USGS for base maps.

The location of geodetic control points can be found by referring to geodetic control diagrams, Fig. 17.8. These diagrams are produced at a scale of 1:250,000 that cover 1° of latitude and 2° of longitude. Vertical control lists, Fig. 17.9 (a), contain the description, location, and elevation of bench marks. Horizontal control lists, Fig. 17.9 (b) contain the description, geodetic and/or geographic position of transit traverse, triangulation, and electronic traverse stations. Both types of lists are assembled in 15-minute quadrangle units. The National Geodetic Survey vertical and horizontal control data are published separately in 30-minute quadrangles.

Horizontal control. The national network of horizontal control is known as the North American Datum of 1927. This network of control has deteriorated over the years because of ground movement and the destruction of monuments. The National Geodetic Survey is redefining and readjusting this network to produce the North American Datum of 1983. The latitude and longitude of almost all points will be changed slightly to reflect the correct position in relation to the entire world.

Vertical control. A map showing elevations must have a reference or *datum* surface from which heights are indicated and determined. In the United States elevations on most maps are usually measured with respect to mean sea level, Fig. 17.10, known as the National Geodetic Vertical Datum (NGVD) of 1929. This datum is based on mean sea

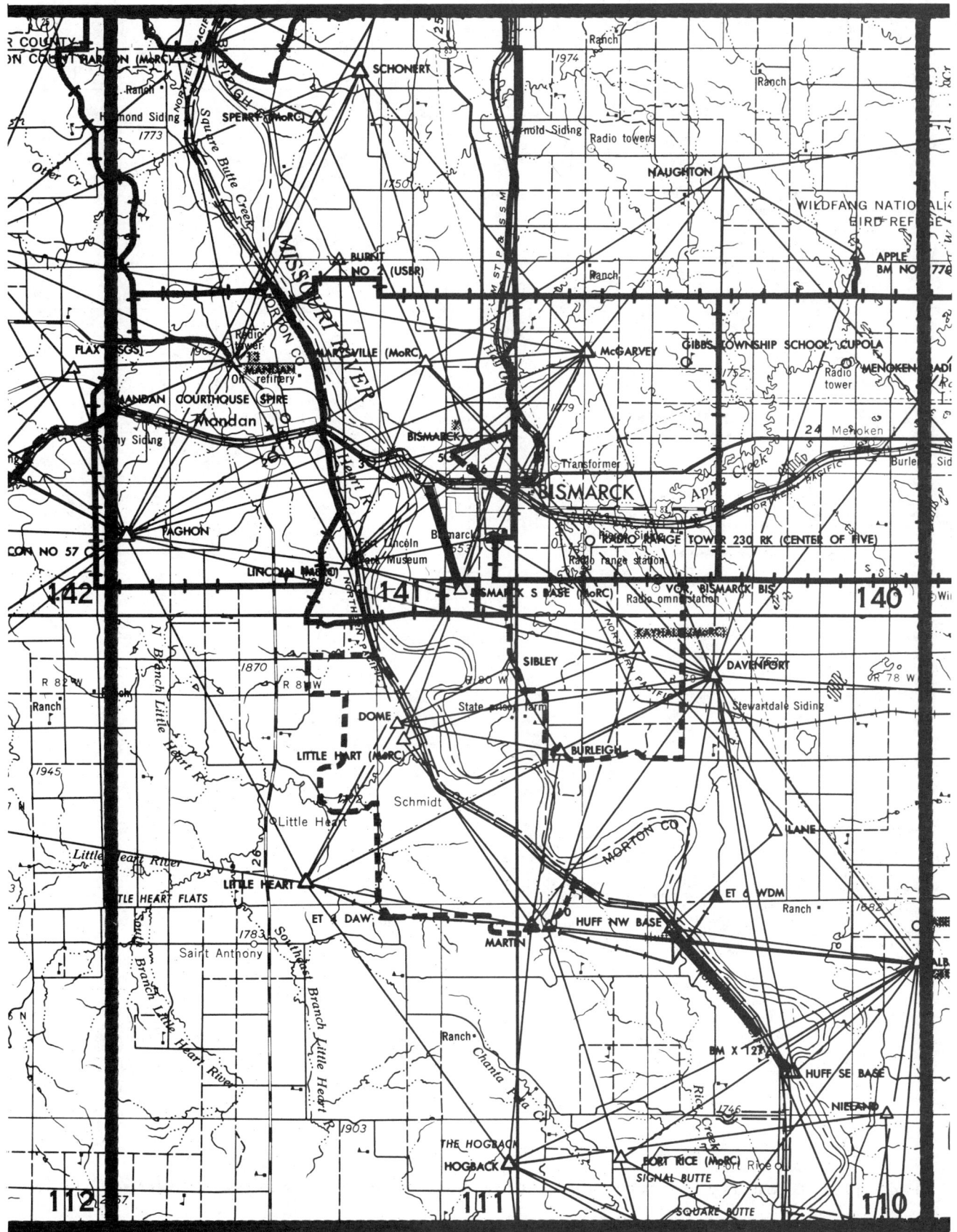

Fig. 17.8 Geodetic control diagram for a portion of Bismarck, North Dakota (*Courtesy USGS*).

MARCH 1969
U.S. DEPARTMENT OF COMMERCE
ENVIRONMENTAL SCIENCE SERVICES ADMINISTRATION
COAST AND GEODETIC SURVEY

VERTICAL CONTROL DATA
by the
Coast and Geodetic Survey
SEA-LEVEL DATUM OF 1929

QUAD 311081 PAGE NO. 1
NEW MEXICO
LATITUDE   31°30' TO   32°00'
LONGITUDE 108°00' TO 108°30'
DIAGRAM DOUGLAS NH 12-3

LINE 101
(Second-order)

The field work (L-4605) was done in January 1935 by a party supervised by G.R. Fish.

These elevations are based on a supplementary adjustment of 1958.

BENCH MARK	ADJUSTED (Meters)	ELEVATION (Feet)
E 132	1376.017	4514.482
4553 (USGS)	1387.611	4552.520
RV 191 (SP)	1395.371	4577.980
F 132	1402.163	4600.263
G 132	1430.664	4693.770

LINE 102
(Second-order)

The field work (L-4385) was done in December 1934 by a party supervised by G.R. Fish. The line was releveled (L-8545) in March 1939 by a party supervised by R.A. Earle.

These elevations are derived from a combination of both levelings, and are based on a supplementary adjustment of 1941.

BENCH MARK	ADJUSTED (Meters)	ELEVATION (Feet)
Y 130	1373.620	4506.618
4514 (USGS)	1375.770	4513.672
Z 130	1371.876	4500.896

LINE 101

DEPARTMENT OF COMMERCE
U.S. Coast and Geodetic Survey
Form 685 A

RECOVERY NOTE, BENCH MARK

Designation E 132   State New Mexico   County Hidalgo
Nearest town Hachita   County Grant   Jan. 1953
Distance and direction from nearest town 12.7 miles west
Character of mark A bench mark disk   Stamping E 132 1934
Established by U. S. Coast & Geodetic Survey
Present condition Good
Detailed report 12.7 miles west along the Southern Pacific Railway from the station at Hachita, 0.1 mile west of the east end of a side track, 97 feet north of the north rail of the main track, 3.0 feet south of a fence, 3.0 feet east of a witness post, about 4 feet higher than the track, and set in the top of a concrete post projecting 1.0 foot above the ground.

Note: This bench mark is 1.8 miles east of bench mark "4432 (U.S.G.S.)".

(a)

JULY 1966
U.S. DEPARTMENT OF COMMERCE
ENVIRONMENTAL SCIENCE SERVICES ADMINISTRATION
COAST AND GEODETIC SURVEY
REVISED JAN 1972

HORIZONTAL CONTROL DATA
by the
Coast and Geodetic Survey
NORTH AMERICAN 1927 DATUM

QUAD 311081 STATION 1008
N MEX-MEX
LATITUDE   31°30' TO   32°00'
LONGITUDE 108°00' TO 108°30'
DIAGRAM NH 12-3 DOUGLAS

Near (Luna County, J. S. Hill, 1910; 1936).—The station is about 5 miles S. 33° W from the town of Victorio, about one mile west-northwest of the old International Mine and tank, and on a high, rocky peak which is the higher and more eastern one of two peaks in the vicinity. The station is marked with a standard disk station mark set in rock according to note 2. The reference mark, a cross cut in the rock, is 4.11 meters (13.5 feet) from the station in azimuth 116°30'. In 1936 the station and reference marks were recovered and a new reference mark, a standard disk set in rock according to note 12a, was established on a ridge about one foot higher than the general level and 10.265 meters (33.68 feet) from the station in azimuth 288°13'. An air beacon northwest of Hermanas, which can be used as an azimuth mark, is about 10 miles (estimated) from the station in azimuth 249°08'41". Elevation: 1,649.4 meters (5,411 feet).

ADJUSTED HORIZONTAL CONTROL DATA

NAME OF STATION: NEAR
STATE: NEW MEXICO   YEAR: 1909   THIRD —ORDER
LOCALITY: SOUTHWESTERN NEW MEXICO TEXAS CALIFORNIA ARC
SOURCE: G-14699   FIELD SKETCH: NEW MEXICO 4

GEODETIC LATITUDE: 31 47 23.44738
GEODETIC LONGITUDE: 108 11 02.61675
ELEVATION: 1649.4 METERS / 5411 FEET

STATE COORDINATES (Feet)

STATE & ZONE	CODE	X	Y	θ or Δ α ANGLE
N. MEX. W.	3003	391,028.17	287,461.89	- 0 11 05

*PLANE AZIMUTH HAS BEEN COMPUTED BY THE θ or Δ α FORMULA NEGLECTING THE SECOND TERM.

TO STATION OR OBJECT (From south)	GEODETIC AZIMUTH (From south)	PLANE AZIMUTH*	CODE
US AND MEXICO BNDRY MON NO 39	16 08 51.7	16 19 57	3003
HERMANAS	266 10 33.7	266 21 39	3003

(b)

Fig. 17.9 Geodetic control lists for a portion of New Mexico *(Courtesy USGS)*.

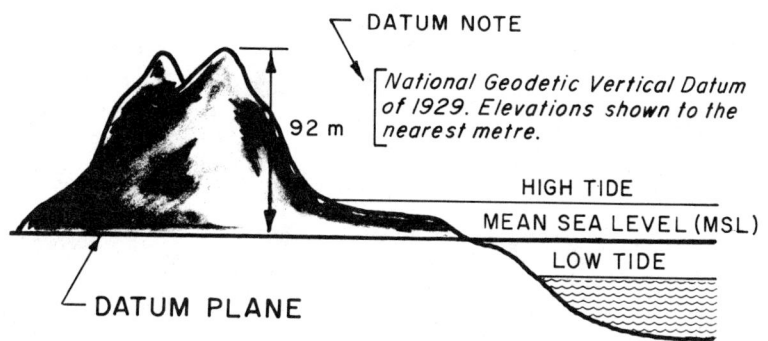

Fig. 17.10 Datum plane-mean sea level.

level readings taken at 26 tide stations in the United States and Canada. A datum note is placed in the map margin to establish the elevation datum and the elevation units.

An assumed reference or datum elevation, other than mean sea level (MSL), may be used for mapping small areas where no known elevation with respect to sea level exists. A convenient point such as the top of a hill or a road intersection may be used for the datum. The datum selected should be one easily found or determined at some future time. Any value may be assigned to the datum. A large value is usually used to avoid negative elevations for all nearby points.

In studying the ground relief on a map, certain points are marked and their values given to indicate the height of the ground at that exact spot. A *bench mark*, Fig. 17.11, is a definite permanent point of known elevation and is commonly identified by standard inscribed brass or copper markers set in concrete or stone. Bench marks have been established throughout the country and they can be found in such varied places as the tops of remote mountains and the sidewalks of major cities. Only the bench marks identified by standard markers are indicated on published maps by a cross symbol (x) together with the letters "BM" and value of elevation (BM x 1340). The elevation value may or may not appear on the markers and should be checked for the correct value with the agency setting the marker. The elevations are checked periodically since they may be changed by natural occurring events such as frost action.

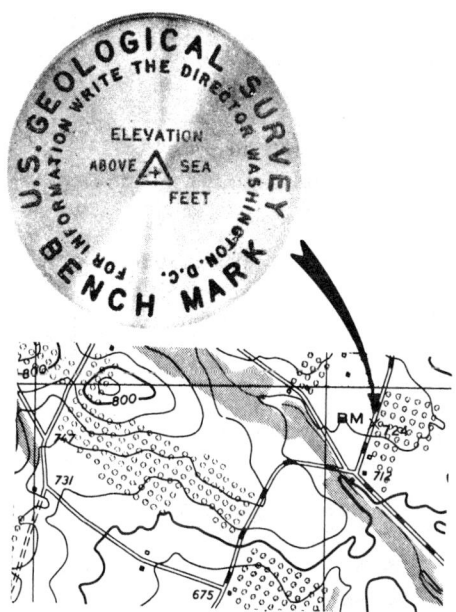

Fig. 17.11 Bench marks.

Spot elevations are shown on maps to help the user in visualizing the relief by supplementing the information provided by the contour lines, Fig. 17.12 (a). Since many places on a map may not show relief exactly, a spot elevation can be useful in showing the height of a crossroad, hill, saddle, etc. Maps show the elevation symbolized by a small cross (x) and value of elevation (x 2681). The elevations are usually determined by field surveys or photogrammetric methods and are not normally marked on the ground.

Spot elevations may be estimated with reasonable accuracy on even surfaces with fairly close contouring, assuming the slope is uniform between adjoining contours. Where contours are widely spaced, close estimation will

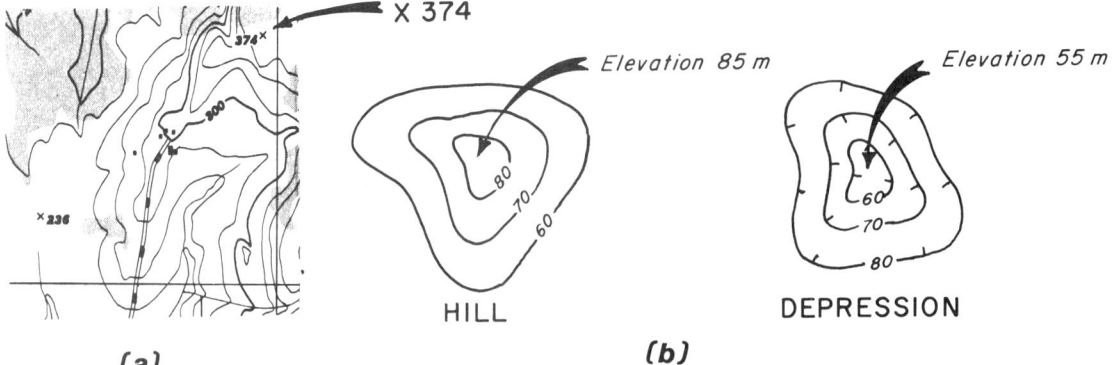

Fig. 17.12 Spot elevations.

probably not be possible. For the elevation of hills and depressions that are unmarked, one-half the contour interval should be added to the highest contour around the hill and subtracted from the lowest contour around the depression, Fig. 17.12 (b).

17.11 CONTOUR CHARACTERISTICS

Drawing accurate contour lines that faithfully represent the terrain is viewed as both an art and science. It is a science in the sense that the lines must be positioned within certain accuracy limitations and an art because within this limitation the cartographer must draw the lines to represent the typical characteristic of the ground surface. A map is said to have good topographic expression when the contours portray the terrain reliably. This is achieved by shaping and spacing contour lines in relationship to each other so that topographic features are represented with exactness and can be interpreted with ease. The following points should be understood and followed closely in drawing contour lines so as not to give a misleading picture of the landscape.

1. All contour lines are horizontal and every point along the line has the same elevation.

2. Contour lines are drawn perpendicular to the slope or dip of the land.

3. Contour lines are closer together on steep slopes and more widely spaced on gentle slopes, Fig. 17.13. The shape of the contour line will show slope irregularities and land forms like valleys, ridges, and hills.

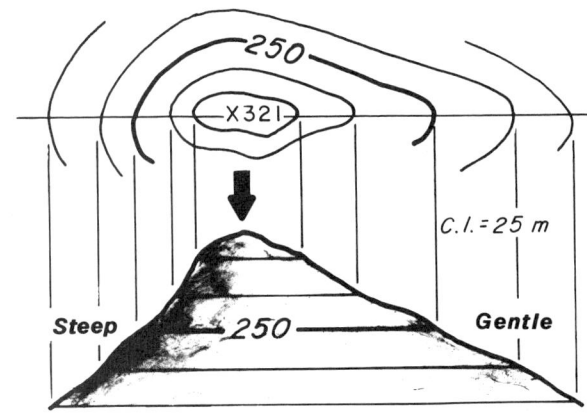

Fig. 17.13 Steep and gentle slopes.

4. The regular spacing of contour lines indicates a uniform slope while irregular spacing denotes an irregular slope, Fig. 17.14. On uniform slopes where the contours are closely spaced, it is common practice to plot and draw

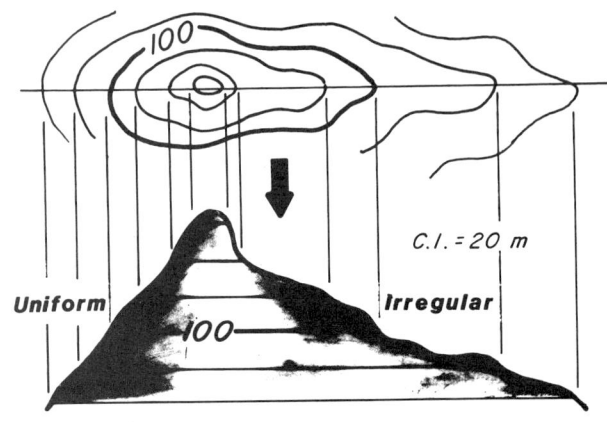

Fig. 17.14 Uniform and irregular slopes.

only the index contour lines and to interpolate the intermediate contour lines. The even spacing of these lines may be attained as accurately and more quickly by eye as by instrument. Where slopes are not uniform, all contour lines must be plotted and drawn independently.

5. The contour lines for a convex slope are closely spaced near the bottom, gradually widening toward the top. On a concave slope this arrangement is reversed, Fig. 17.15.

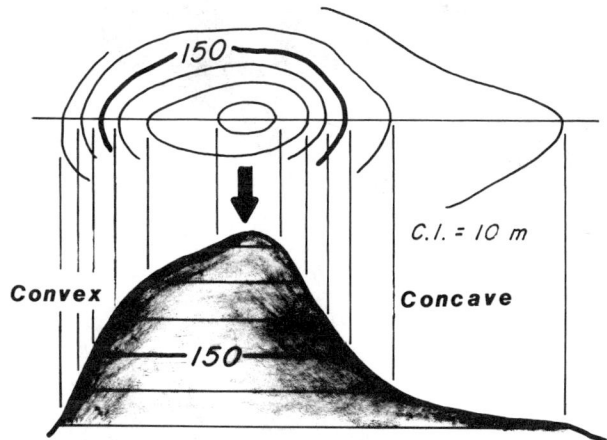

Fig. 17.15 Convex and concave slopes.

6. Contour lines do not split or branch, Fig. 17.16 (a) and seldom merge except in instances where vertical or overhanging cliffs are shown. The contour lines are stacked one on top of the other, and do not actually touch although they will appear this way on the map.

A vertical cliff is shown by drawing a *carrying contour* which is formed by bringing together into a single line all the contour lines contained within the cliff, Fig. 17.16 (b). If any of the contour lines representing the cliff are index lines, the carrying contour should be drawn as an index contour. Carrying contours are also used to show other vertical or near-vertical topographic features such as cuts and fills.

An overhang is shown by an upper contour line crossing a lower one in the form of a protrusion, Fig. 17.16 (c). Normally this contour is drawn as a dashed line.

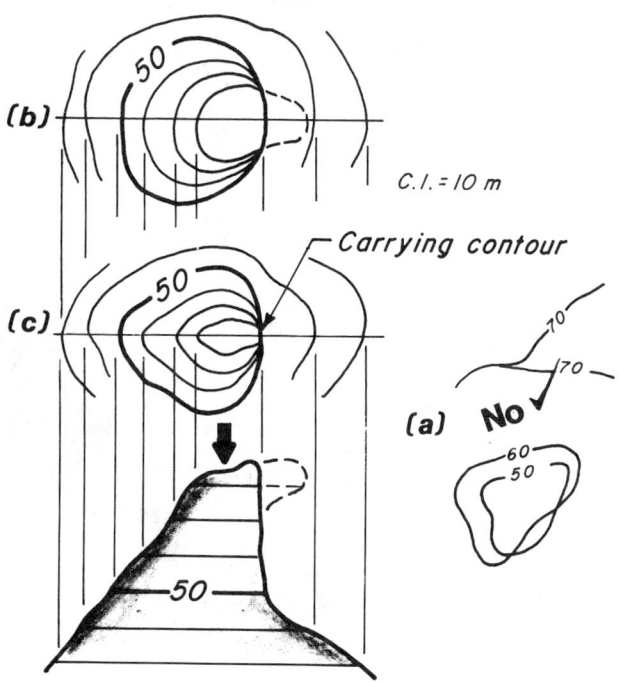

Fig. 17.16 Vertical and overhanging cliffs.

7. A valley is indicated when a recession of the contours from the low land pressing back into a hill or the higher land occurs. A ridge is represented when the reverse of this occurs. The contour lines for valley and ridge lines are generally "U" shaped, Fig. 17.17. The contour lines should point uphill for a valley and downhill for a ridge. The parts of the contour lines representing the valley that recurve are commonly called *reentrants*. They are very common on topographic maps and slight changes in how they are drawn will greatly affect the appearance of a map.

8. Contour lines are drawn crossing streams and rivers forming an inverted "V" or "U" reentrant that curve toward the higher elevation, Fig. 17.18. The reentrant part of the contour lines should not be too pointed or blunt when crossing single line streams and the stream course should pass through the center of the reentrant. For a double line stream symbol, the

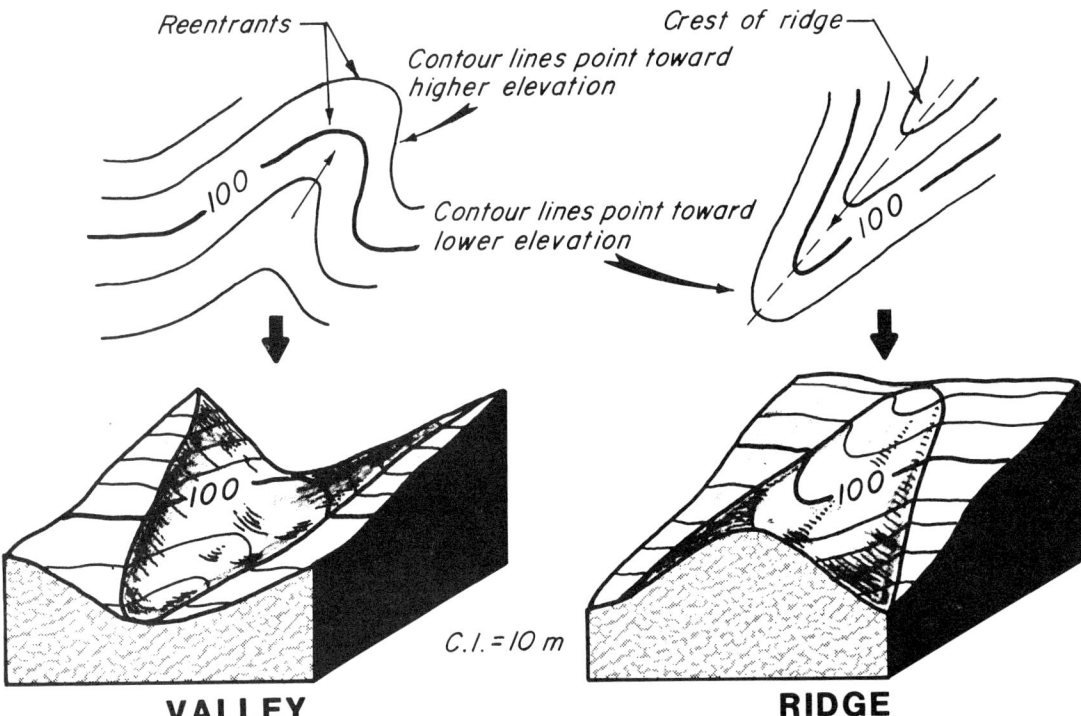

Fig. 17.17 Valley and ridge lines.

contour lines cross the water approximately at right angles to the two sides. The contour lines shaping this land form may be considered analogous to a valley since streams and rivers mark the most depressed line in a valley.

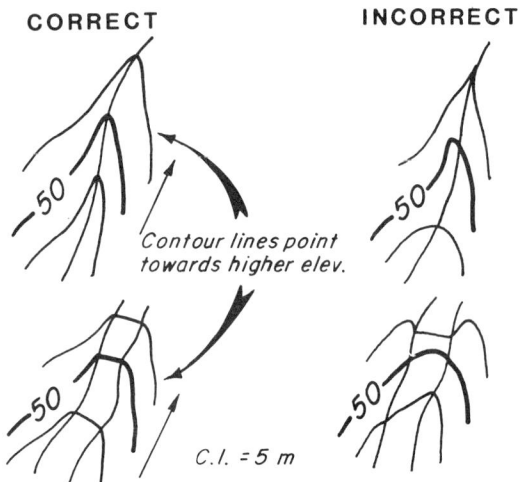

Fig. 17.18 Contour lines crossing streams and rivers.

9. Contour lines are drawn across railroads and other transportation routes as straight lines at right angles to the alignment, Fig. 17.19.

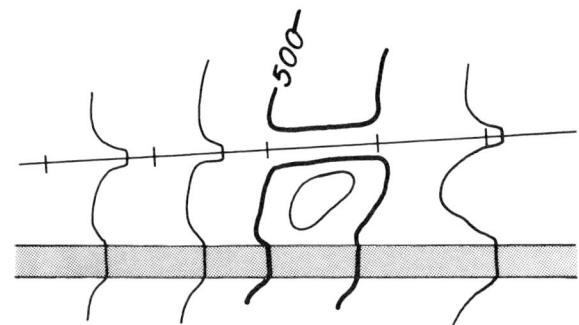

Fig. 17.19 Contour lines crossing transportation arteries.

10. A contour line always closes on itself, either on or off the limits of the map. If a contour line closes on itself within the map boundary, it indicates a hill unless there are short tick marks (hachures) drawn pointing downhill perpendicular to the direction of the contour for representing a depression, Fig. 17.12. Depressions are areas lower in elevation than all the surrounding terrain. Several contour lines may be needed to depict a very deep depression, one contour line inside another.

11. Cuts and fills are normally plane surfaces appearing adjacent to roads, railroads, and other man-made features. The contour lines usually appear as straight equally spaced lines parallel to each other, Fig. 17.20. A cut passes through hills and ridges while a fill will pass over small streams, gullies, or depressions. Fills are distinguished from cuts by hachures drawn on the downhill side of the contours. Where the face of the slope is not too steep, all the contour lines are shown; but for a steep-faced or near vertical slope the interrupted contours are connected by a carrying contour.

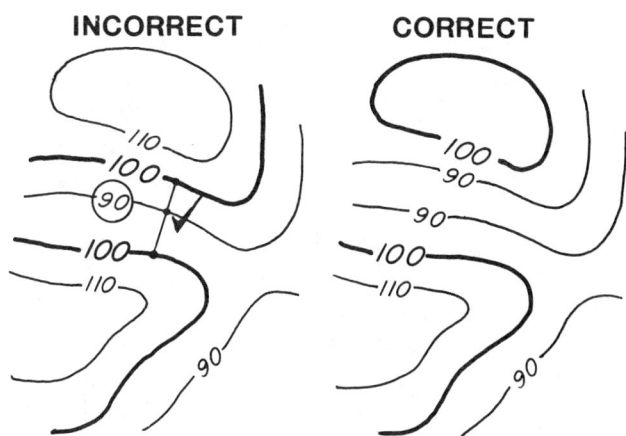

Fig. 17.21 Elimination of single contour lines between those of higher or lower elevation.

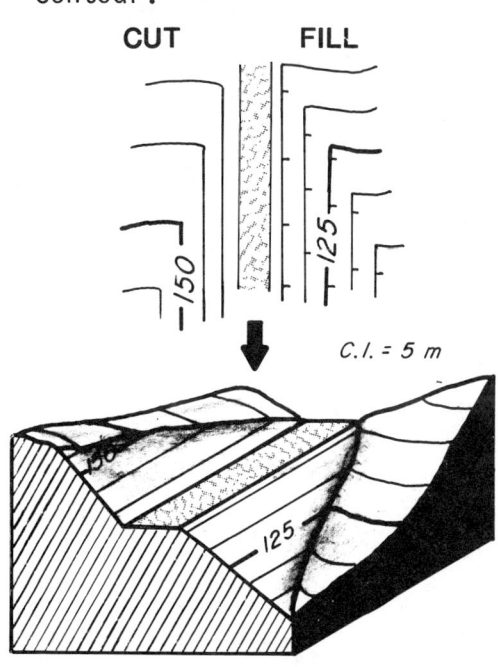

Fig. 17.20 Cuts and fills.

12. A single contour line cannot lie between two contour lines of higher or lower elevation, Fig. 17.21. In land formations a level knife-edge condition would probably never be encountered in natural formations. This is a common mistake found on contour maps and elevations of all contour lines should be checked for this mistake. The map will probably require that several adjacent contour lines having the same elevation be drawn to eliminate the error.

13. Contour lines should be drawn as smooth curves and tend to parallel streams and each other. The contour lines should also more or less parallel each other throughout any particular geologic formation. The principal function of contour lines is to represent the kinds of land patterns involved rather than show every small undulation in the grounds surface. As the map scale becomes smaller or the contour interval becomes larger, it will be necessary to generalize the contours.

14. A contour line is never skipped and may only be succeeded by the next contour immediately above or below. Given a contour interval of 10 metres, the 50 metre contour line must lie next to the 40 or 60 metre contour line.

17.11 LAND FORMS

For those involved in the preparation and use of topographic maps, an understanding of the characteristics of a few common land forms, *geomorphology*, is necessary. The cartographer must exercise great care to make sure the area mapped is portrayed accurately. The pattern of contour lines will vary considerably and lines must be drawn carefully for land forms to be recognized by persons using the map. Since a relief map is a composite representation of many different land forms, it may be difficult to recognize features based on a single contour characteristic. The recognition of topography will depend to a considerable extent on the interpretation of all the related features such as drainage lines, valleys, and ridges. A study of Fig. 17.22 will show how various features are depicted on a topographic map.

A thorough understanding of orthographic projection discussed in chapter 8 is essential to understand and visualize the different land forms represented on contour maps. Several geometric figures with contours drawn on the faces of the object are shown in Fig. 17.23. The orthographic top or plan view with contour lines added to the view is sufficient to describe the object fully. The front view would be required to describe the object completely if the contour lines were omitted.

Since all measurements made on a map are *horizontal*, it is necessary to determine the horizontal distance (HD) between adjacent contour lines on each surface using the equation given in Fig. 17.23.

After the direction of the contour lines has been determined, all lines will be parallel and equally spaced on each plane surface. The contour lines will be parallel to any horizontal line on the object since all contours represent level lines on the surface of the ground.

The representation of some common land forms is illustrated in Fig. 17.24. These land forms should be studied carefully to enable various relief features to be quickly recognized. In addition, contour lines on maps should be compared with the corresponding area on the ground to see how different shapes are represented.

A contour map showing several distinctive land forms is shown in Fig. 17.25. The drawing should be examined and all questions answered as indicated.

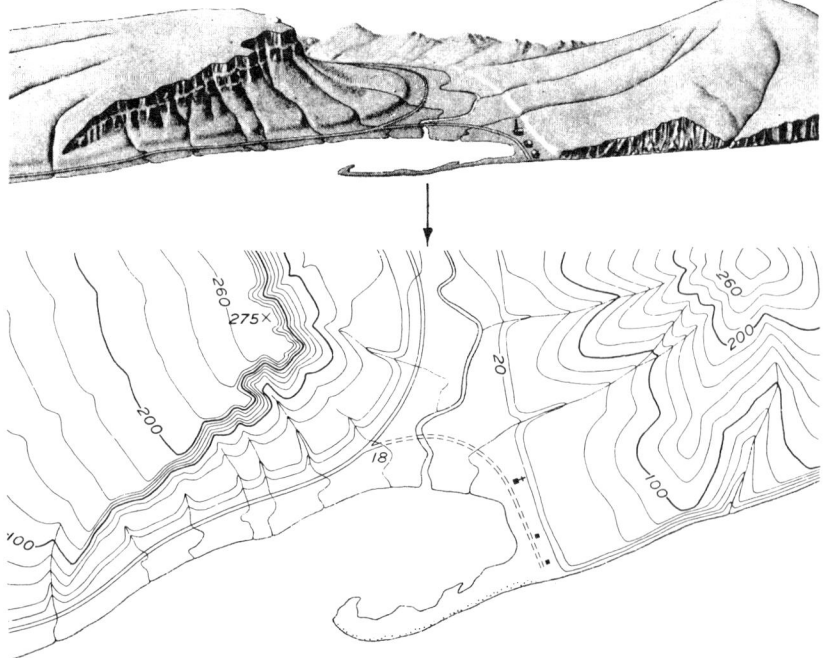

THE USE OF SYMBOLS IN MAPPING

These illustrations show how various features are depicted on a topographic map. The upper illustration is a perspective view of a river valley and the adjoining hills. The river flows into a bay which is partly enclosed by a hooked sandbar. On either side of the valley are terraces through which streams have cut gullies. The hill on the right has a smoothly eroded form and gradual slopes, whereas the one on the left rises abruptly in a sharp precipice from which it slopes gently, and forms an inclined tableland traversed by a few shallow gullies. A road provides access to a church and two houses situated across the river from a highway which follows the seacoast and curves up the river valley.

The lower illustration shows the same features represented by symbols on a topographic map. The contour interval (the vertical distance between adjacent contours) is 20 feet.

Fig. 17.22 Symbols and their uses in portraying various map features *(Courtesy USGS)*.

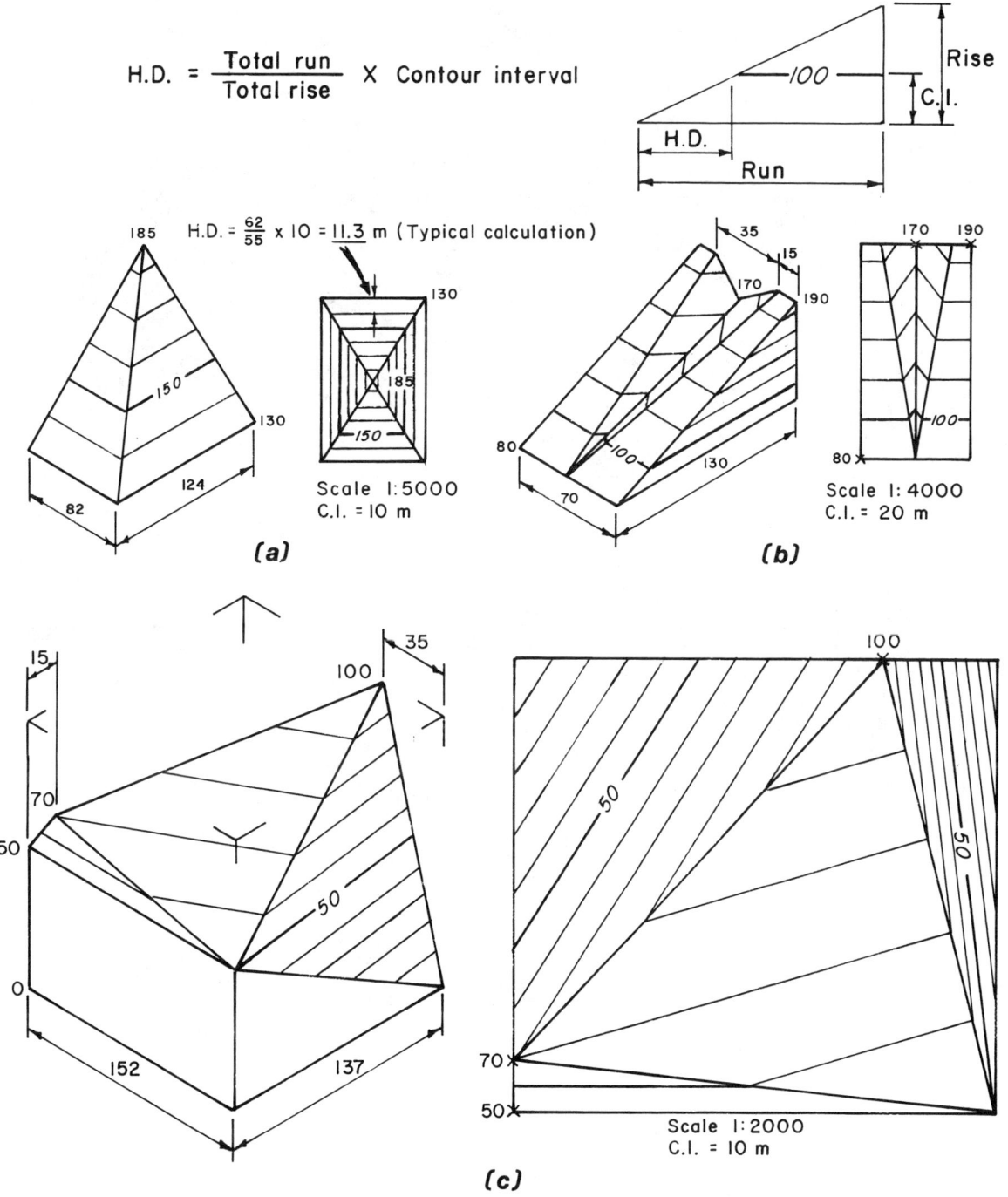

Fig. 17.23 Geometric figures described by contour lines.

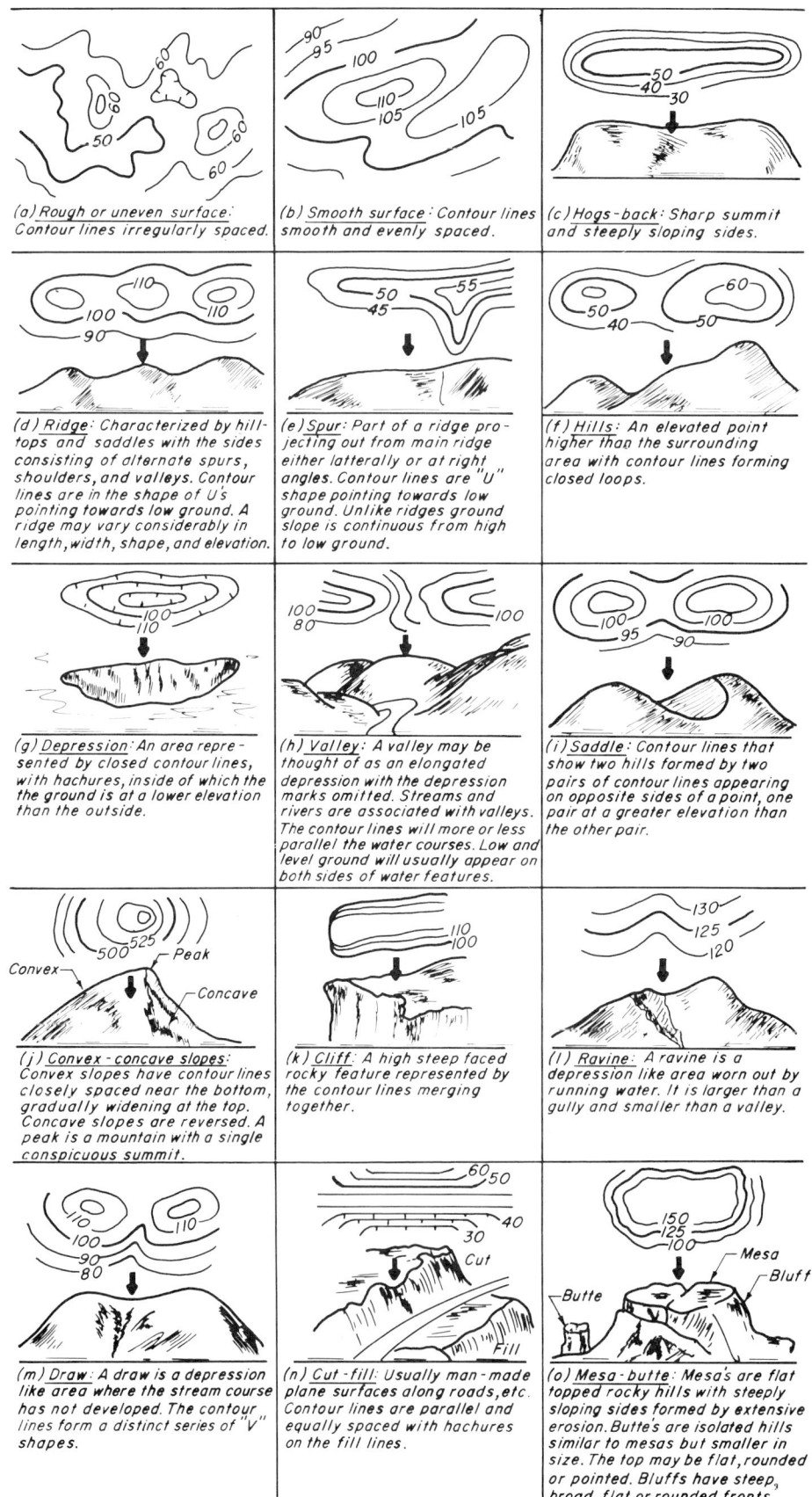

Fig. 17.24 Typical land forms.

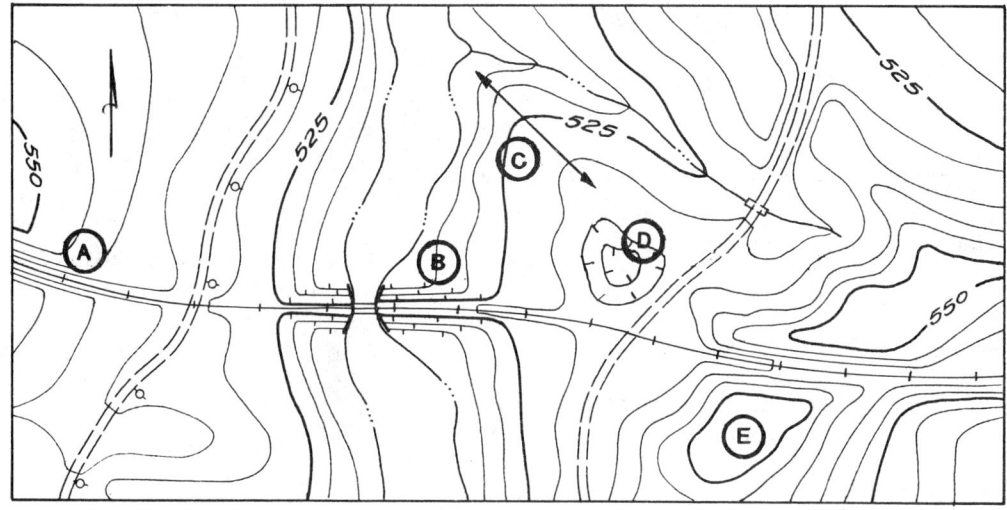

IDENTIFY THE FOLLOWING LANDFORMS — ANSWERS

A ______ Cut
B ______ Fill
C ______ Ridge
D ______ Depression
E ______ Hill

Lowest point on map = ______ 512.5
Highest point on map = ______ 552.5
The stream flows in a generally ____ direction. NW
The area having the most gentle slope is in the ____ part of the map. NW
Contour interval = ______ 5

Fig. 17.25 Contour lines expressing relief.

17.12 METHODS OF OBTAINING TOPOGRAPHY

The data required for the representation of topography is usually obtained from stadia surveying, plane table surveying, and photogrammetric methods.

Stadia surveying. In stadia surveying, distances, directions, and elevations for many points are obtained using the transit or theodolite. A typical set of field book stadia notes appears in Fig. 17.26.

One method commonly used to obtain topography for small gentle sloping areas is the use of coordinate squares

STADIA SURVEY - BOARDMAN

⊼ @ A B.S. Sta C Plates 0°00'

C = 0 K = 100 hi = 5.16' Elev A = 115.30'

Sta	Stadia Interval	Horiz Angle	Vertical Angle	Horiz Dist (Ft)	Elev (Ft)
1	0.81'	132°30'	00°-10'	81	115.54
2	0.45'	231°-03'	00°-15'	45	115.50
3	0.87'	233°02'	00°-12'	87	115.60
4	1.19'	241°25'	00°-04'	119	115.44
5	1.50'	255°09'	00°-01'	150	115.34
6	0.73'	256°-06'	00°-06'	73	115.43

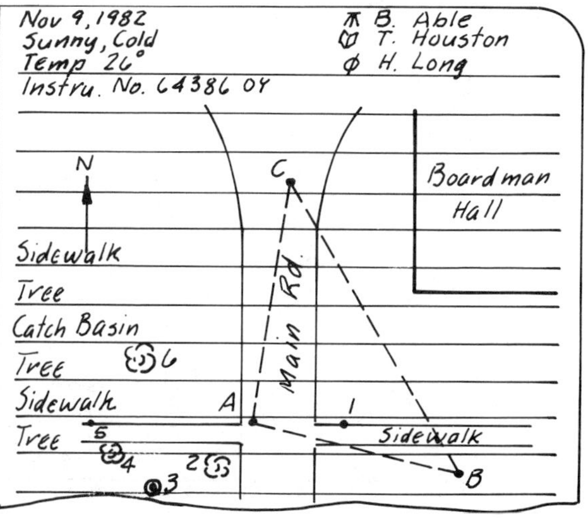

Fig. 17.26 Typical fieldbook stadia notes.

(grid method). An area may be staked into squares, Fig. 17.27, using sides equal to 2, 5, 15, or 30 metres (10, 20, 50, or 100 feet). The size of the squares depends on such factors as terrain involved and the desired accuracy required. The corner elevations and elevations at other critical points where slopes suddenly change, e.g., valley and ridge lines, are obtained by survey methods.

The contour lines may be located and drawn for the sides of each grid by interpolating between known points of elevation using the following methods.

1. Estimation -- On small scale maps where the ground form is not too irregular, contour lines may be spaced between known points of elevation by estimation. This method for finding points to sketch the contour lines is fast and can produce reasonable results if care is exercised.

2. Proportion -- Where high accuracy is desired, the spacing of the contour lines is calculated. This method is well suited for large scale maps. Distances between points of known elevation are measured and the location of contour points located by proportion, Fig. 17.27.

3. Contour finder -- Graphical devices are frequently used and recommended for finding points to draw contour lines between plotted elevations, assuming a uniform slope. These devices allow many interpolations to be made accurately and quickly. The device in Fig. 17.28 (a) consists of a series of equally spaced parallel lines that may be drawn on transparent media using any specified scale. The elevations of the lines are labelled according to the contour interval of the map.

A converging line device of any size may also be prepared and graduated similar to the example shown in Fig. 17.28 (b) for interpolating contour lines between plotted elevations. Any number of converging and horizontal lines having equal spacing may be drawn.

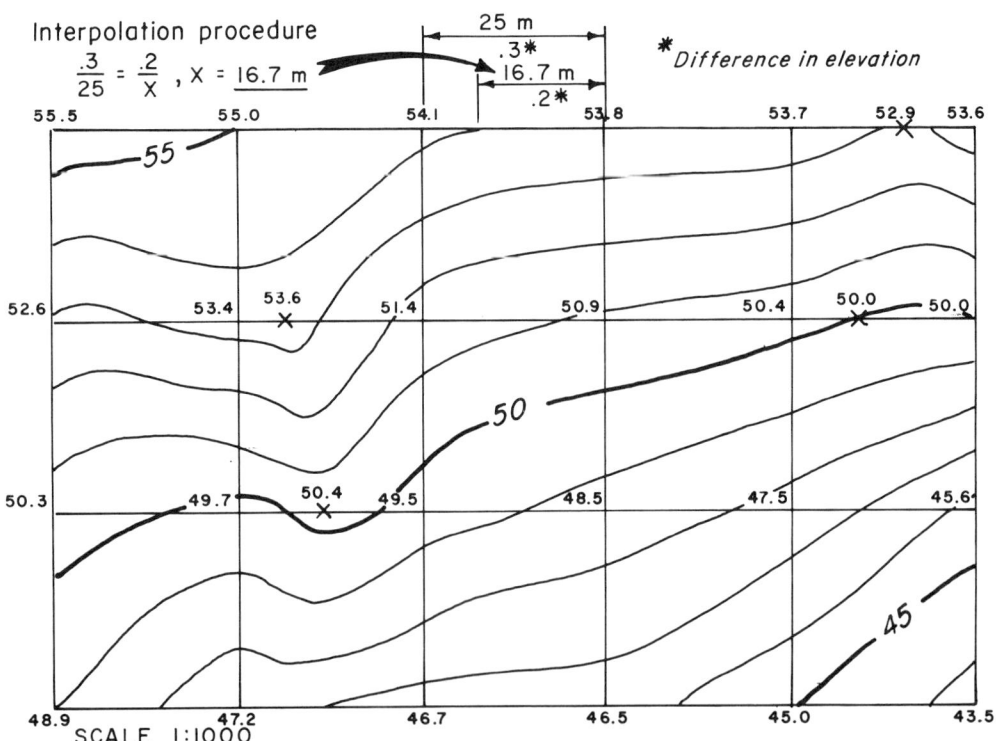

Fig. 17.27 Coordinate squares for drawing contour lines.

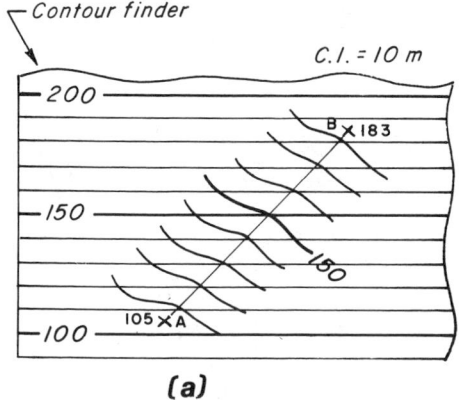

PROCEDURE

1. Place the contour finder over the map until the elevations of points A and B coincide with the same elevations indicated along the parallel lines of the finder.
2. The points where the parallel lines of the finder intersect line AB determine where the contour lines are to be drawn.

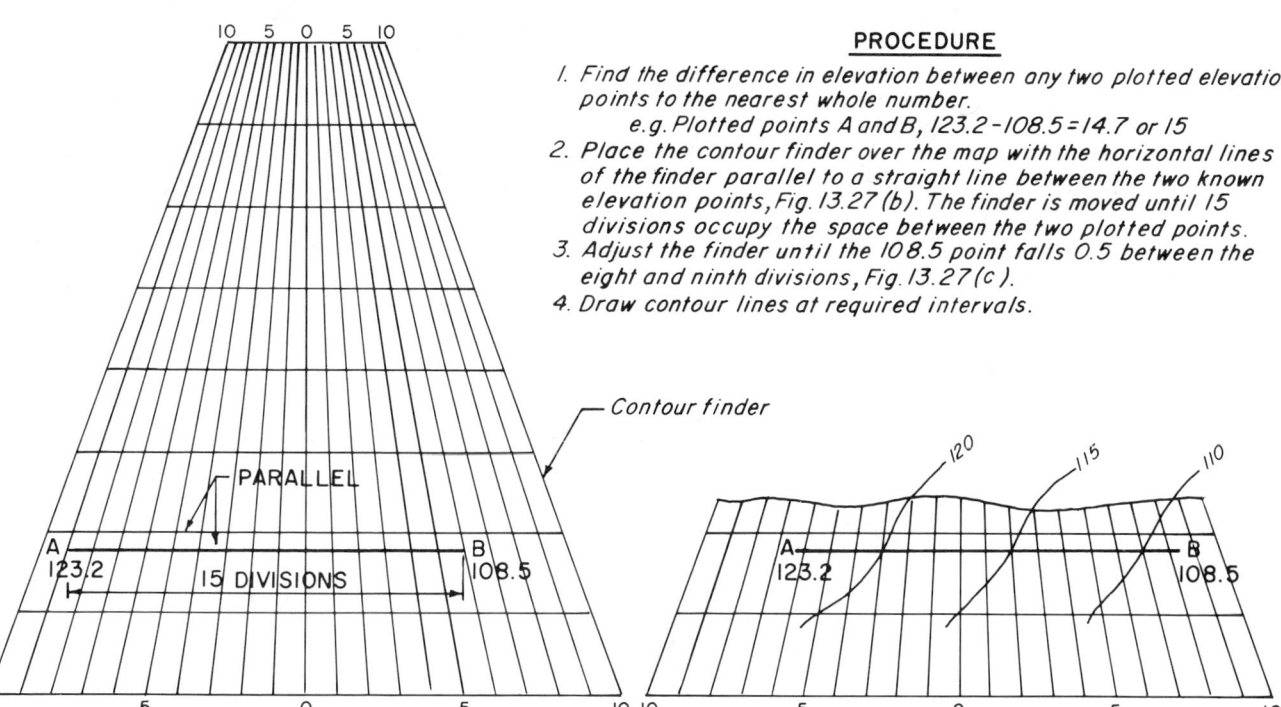

PROCEDURE

1. Find the difference in elevation between any two plotted elevation points to the nearest whole number.
   e.g. Plotted points A and B, 123.2 - 108.5 = 14.7 or 15
2. Place the contour finder over the map with the horizontal lines of the finder parallel to a straight line between the two known elevation points, Fig. 13.27 (b). The finder is moved until 15 divisions occupy the space between the two plotted points.
3. Adjust the finder until the 108.5 point falls 0.5 between the eight and ninth divisions, Fig. 13.27 (c).
4. Draw contour lines at required intervals.

Fig. 17.28 Contour finder.

A very common method for selecting points and interpolating to find contour lines for known elevations is the logical or controlling point method. In this procedure the contours are drawn from field notes showing the drainage pattern and certain critical elevations. The elevation points are the tops of hills, the bottoms of valleys depressions, and other points where important changes in slope occur. Contours may be interpolated accurately by spacing them proportionately between the given elevations. The procedure in interpolating for contours at 10 metre intervals where the drainage system and spot elevations have previously been determined is illustrated in Fig. 17.29. A systematic approach should be followed and this may be divided into five steps:

1. Determine the elevation of all stream junctions, Fig. 17.29 (b). Streams are important on any topographic map since they help to emphasize the low ground. The elevation of the stream junction at A is determined by noting the difference in elevation between the known elevations above and below point A and finding the proportional distance for point A between the two points of known elevation. Since the total difference in elevation is 15 metres between the known elevation of 25

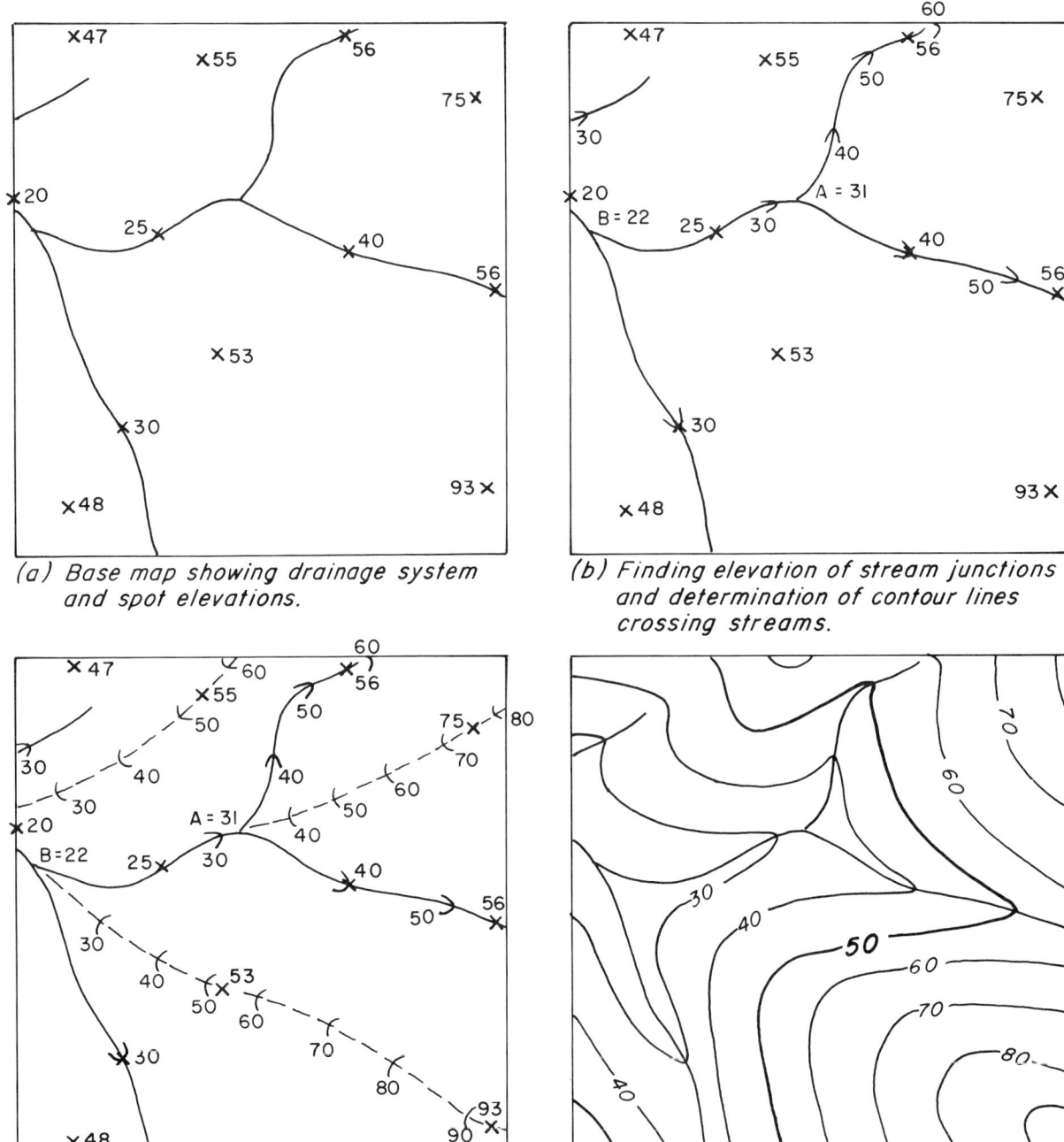

Fig. 17.29 Contour line interpolation.

and 40, the elevation at point A would be 31 metres. It is approximately 2/5 the distance or 6 metres higher than the known elevation of 25 metres. The remaining stream junctions are found in a similar way.

2. <u>Locate the points where contour lines cross the streams</u>, Fig. 17.29 (b). Since the slope of a stream may be considered uniform between any two points of known elevation along the stream, contour lines crossing the stream at 10 metre intervals may be determined. The procedure is the same as outlined in step one. The characteristic V-shaped marks with the points upstream are used to indicate the locations where the lines cross.

3. <u>Sketch in the lines to represent the crest of the higher land,</u> Fig. 17.29 (c). Higher ground always exists between streams and terminates in hills and ridges. With this in mind, lines (shown as dashed lines) may be drawn through the spot elevations which fall between streams and down into the stream junctions. They should be curved where necessary and kept roughly midway between the streams. Since spot elevations are given where a change in slope occurs, the crest of higher ground represents one place where slope changes. Minor lines may also be drawn with the slope to aid in the drawing of the contour lines but should not cross each other or a stream.

4. <u>Locate the points where contour lines cross the crest of the higher ground,</u> Fig. 17.29 (c). These points are determined in the same way as the locations where contour lines cross the streams. The contour lines are in the shape of a U with the curve of the U pointing downhill to indicate where the lines cross the ridges or higher ground.

5. <u>Draw the contour lines connecting points of equal elevation,</u> Fig. 17.29 (d). The contour lines should be sketched in lightly first as smooth curves which follow the shape of the drainage system. The map should be gone over one or two more times smoothing out and forming the contour lines to faithfully express the terrain. This is perhaps the most difficult thing for an individual to accomplish. In forming the contour lines remember the map provides no direct information of the land between contours. Height may grade evenly from one contour line to the next or there may be a series of irregularities. If these are rough, the land surface is probably uneven, etc. The contour irregularities are smoothed out to some extent as the scale of the map is reduced.

Plane table surveying. A plane table and alidade, Fig. 17.30, may be used to draw a map in the field. The plane table is mounted on a tripod that can be set up and centered over a station point using a plumb bob. The alidade consists of a telescope mounted on a brass ruler that is aligned with the line of sight of the telescope. A small compass for measuring direction, a bubble tube for leveling the table, and a vertical arc for measuring vertical angles to determine elevation differences complete the equipment. Good quality drawing paper should be attached to the drawing surface. The plane table is set up with the position of a point on the paper corresponding to the same point on the ground. The drawing may be oriented since the table may be rotated. Any number of desired points may be sighted on for determining the distance and elevation difference by stadia. All information is plotted on the paper in the field making it possible to correlate the drawing with the actual features. For large scale maps of small areas and the filling in of details between survey control points, plane table surveys are convenient.

Photogrammetry. The drawing of topography covering large areas is usually accomplished using this method. The subject of aerial photography is discussed in chapter 18.

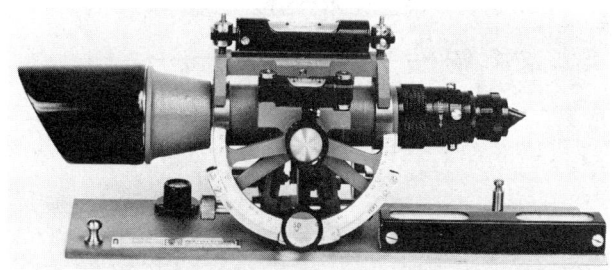
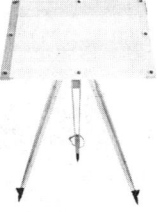

Fig. 17.30 Plane table and alidade (Courtesy Warren-Knight Instrument Co).

17.13 SLOPE

Topographic maps provide all kinds of information on slopes and it is necessary to be familiar with several methods for its expression. Slope is always measured relative to the horizontal and depends on the relationship of vertical distance to horizontal distance between any two points. The most important fact to remember is that all measurements, regardless of direction, made on a map are horizontal. Distances measured on the ground and the map will not be the same when slopes are encountered. In Fig. 17.31 the horizontal map distance A_TB_T and the corresponding ground distance A_FB_F are not the same. Since one must walk on the surface of the ground, distances measured on a map may be misleading unless slopes are considered.

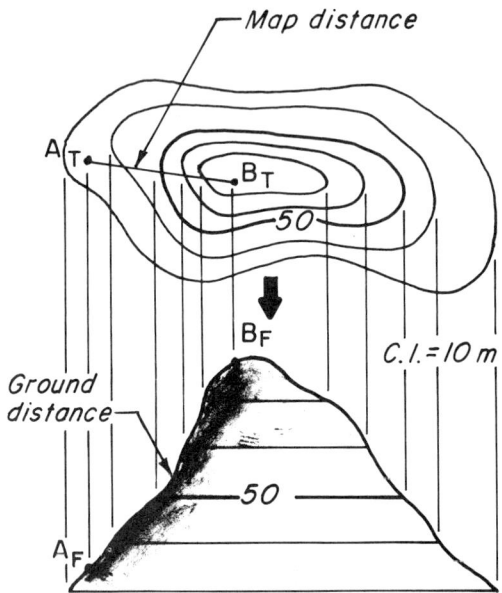

Fig. 17.31 Map-ground distance relationship.

Methods for expressing slope. In discussing the following methods for expressing slope, the same drawing is used throughout for comparison of the slope values.

1. Percent (Grade). The grade between points A and B is -40%. A plus or minus sign must be given to indicate if the slope is rising or falling. All level lines will have a slope of zero (0) percent and a 45 degree line with the horizontal will have a 100 percent slope. A slope in excess of 100 percent is possible since lines may be drawn greater than 45 degrees. Percent grade is used for plot grading and roadways.

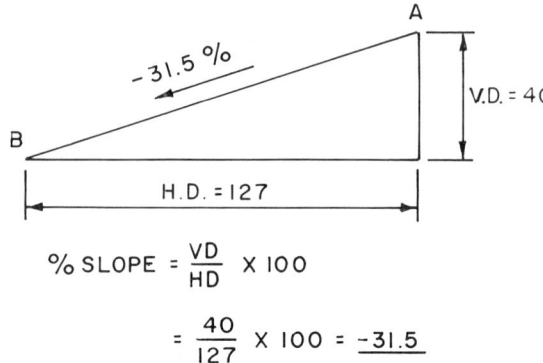

$$\% \text{ SLOPE} = \frac{VD}{HD} \times 100$$

$$= \frac{40}{127} \times 100 = \underline{-31.5}$$

2. Degrees. Slope may be expressed as the angle in degrees that a line makes with the horizontal. In mathematics the tangent of an angle is expressed as:

$$\tan A = \frac{\text{opposite leg}}{\text{adjacent leg}} \text{ or } \frac{VD}{HD}$$

Where precise answers in degrees are required to express slope, the following equation should be used:

$$\text{arc tan } \frac{VD}{HD}$$

A reasonable answer may be obtained for slopes under 20 degrees by using the constant value 57.3 as illustrated. This value is based on the radian where:

(1) 2π radians = 360° or
    π radians = 180°

(2) 1 radian = $\left(\frac{180°}{\pi}\right)$ = 57.2958°
    = 57°17'45"

(3) 1° = $\frac{\pi}{180}$ radian = 0.017453 radian

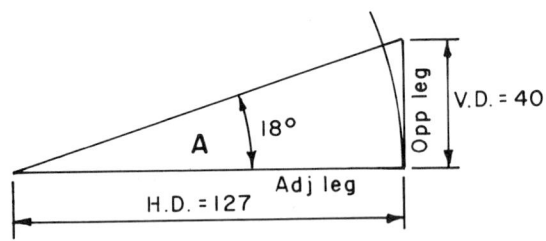

SLOPE IN DEGREES
For precise determination
arc tan $\frac{VD}{HD} = \frac{40}{127} = 17.4824 = \underline{17°28'57''}$

For approximation (slopes under 20°)
$\frac{VD}{HD} \times 57.3 = \frac{40}{127} \times 57.3 = \underline{18°}$

3. Gradient. Slope for gradient is expressed as a simple fraction or a ratio showing the relative value of the vertical and horizontal distances that are always in the same unit of measurement. The vertical distance is always given as one (1) unit.

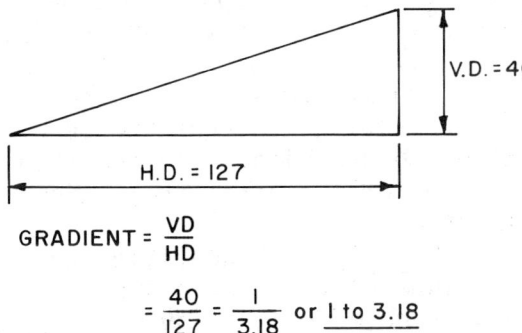

GRADIENT = $\frac{VD}{HD}$

= $\frac{40}{127} = \frac{1}{3.18}$ or $\underline{1 \text{ to } 3.18}$

4. Mils. A mil is a unit of measurement that equals 1/6400th of a circle. A degree equals 17.8 mils. The word mil, meaning one-thousandth, as a unit came about as the result of using the angle that subtends an arc equal to 1/1000 of the radius. It is the angle subtended at the center of a circle having a radius of 1000 units by an arc length of one (1) unit on the circle.

The military commonly uses the mil for expressing slope for computing field artillery firing data, sketching, and reconnaissance mapping.

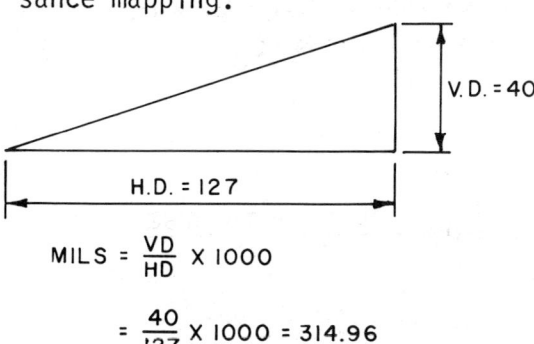

MILS = $\frac{VD}{HD} \times 1000$

= $\frac{40}{127} \times 1000 = \underline{314.96}$

5. Grads. A grad is a unit of angular measurement based on the centesimal system. One (1) grad equals 1/400th of a circle. The slope is found similar to that of slope in degrees with the value 63.7 used instead of 57.3. Where other map measurements are in metric units, the grad is frequently used. Many foreign maps use this unit.

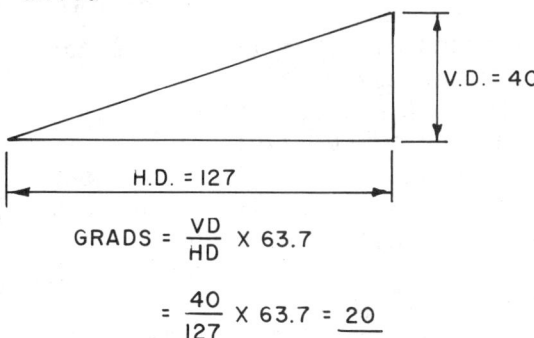

GRADS = $\frac{VD}{HD} \times 63.7$

= $\frac{40}{127} \times 63.7 = \underline{20}$

6. Slope ratio. Slope ratio is expressed as rise/run where the rise is always one (1) unit. The slope ratio value is normally written as 1½ to 1 where the first number in the ratio is the run (horizontal distance) and the second number the rise (vertical distance). In engineering work this method of slope expression is used when working with cuts and fills.

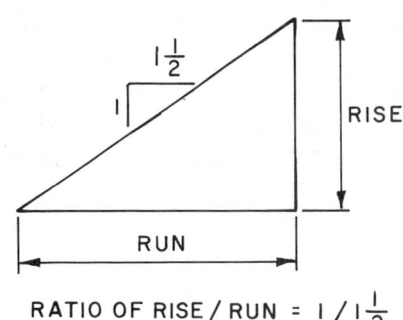

RATIO OF RISE / RUN = $1 / 1\frac{1}{2}$

RISE IS ALWAYS "1" UNIT

NORMALLY WRITTEN $1\frac{1}{2}$ to 1

The slope between any two points on a topographic map may easily be determined as illustrated in Fig. 17.32. The bottom of the slope at point A has an elevation of 79 metres and the top of the slope at point B has an elevation of 82.5 metres. Subtracting the difference between the two values gives a vertical

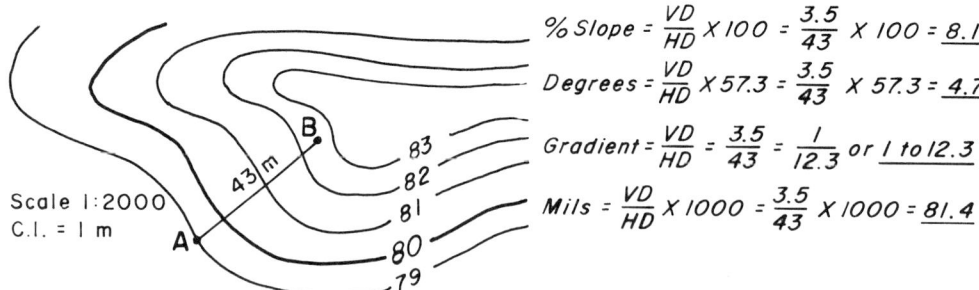

Fig. 17.32 Calculation of slope from a map.

distance of 3.5 metres. The horizontal distance scales 43 metres. Using these values, the slope between points A and B may be calculated. This represents the average slope since the spacing of the contour lines is not uniform. There will normally be some variance in the ground either greater or less than the average slope.

In some instances the horizontal distance is required when the slope (ground) distance and the vertical angle between the horizontal ground is known. The vertical distance (rise) of the slope and the slope (ground) distance may be known in other situations. The methods for making this calculation in these two cases are shown in Fig. 17.33.

In land use work slope maps may be used to represent different degrees of slope, Fig. 17.34. The range of slope percentage is shown by various shades of gray or color. Maps of this kind may be drawn by finding the distance between contour lines on a topographic map, or they may be generated by computer programs that raster scan the contour lines.

17.14 CONTOUR MAP APPLICATIONS

In areas such as geology, mining, and engineering, topographic maps are commonly used to solve problems involved with preliminary work estimates and construction projects of all kinds. Many of the problems encountered lend themselves to graphical solution techniques based on orthographic projection principles and concepts.

Profiles. Visualizing the vertical slope of the ground surface is best understood after a study of slope profiles. The *profile* is a vertical section taken along a given line showing the elevation of the ground. The line may be either straight or curved, but the length of the profile is the same as the true length of the line. A curve on a graph is similar to a profile and one or more curves may be drawn to depict a specific slope or several slopes. Profiles are used for the center line of roads and railroads, for establishing grade lines, and for determining visibility.

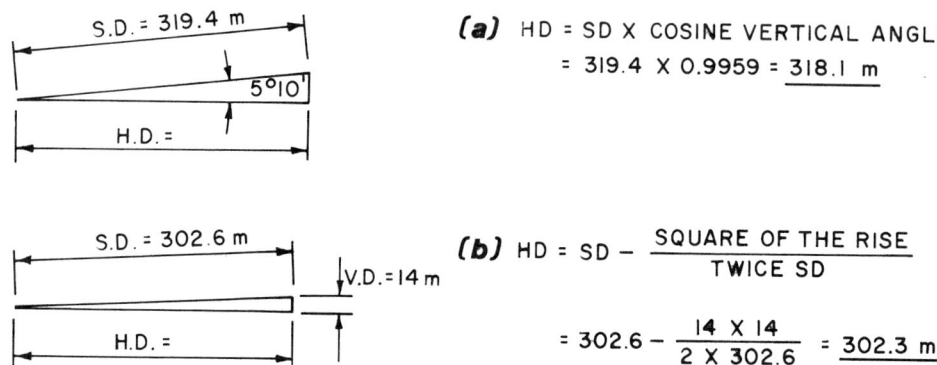

Fig. 17.33 Calculation of horizontal distance.

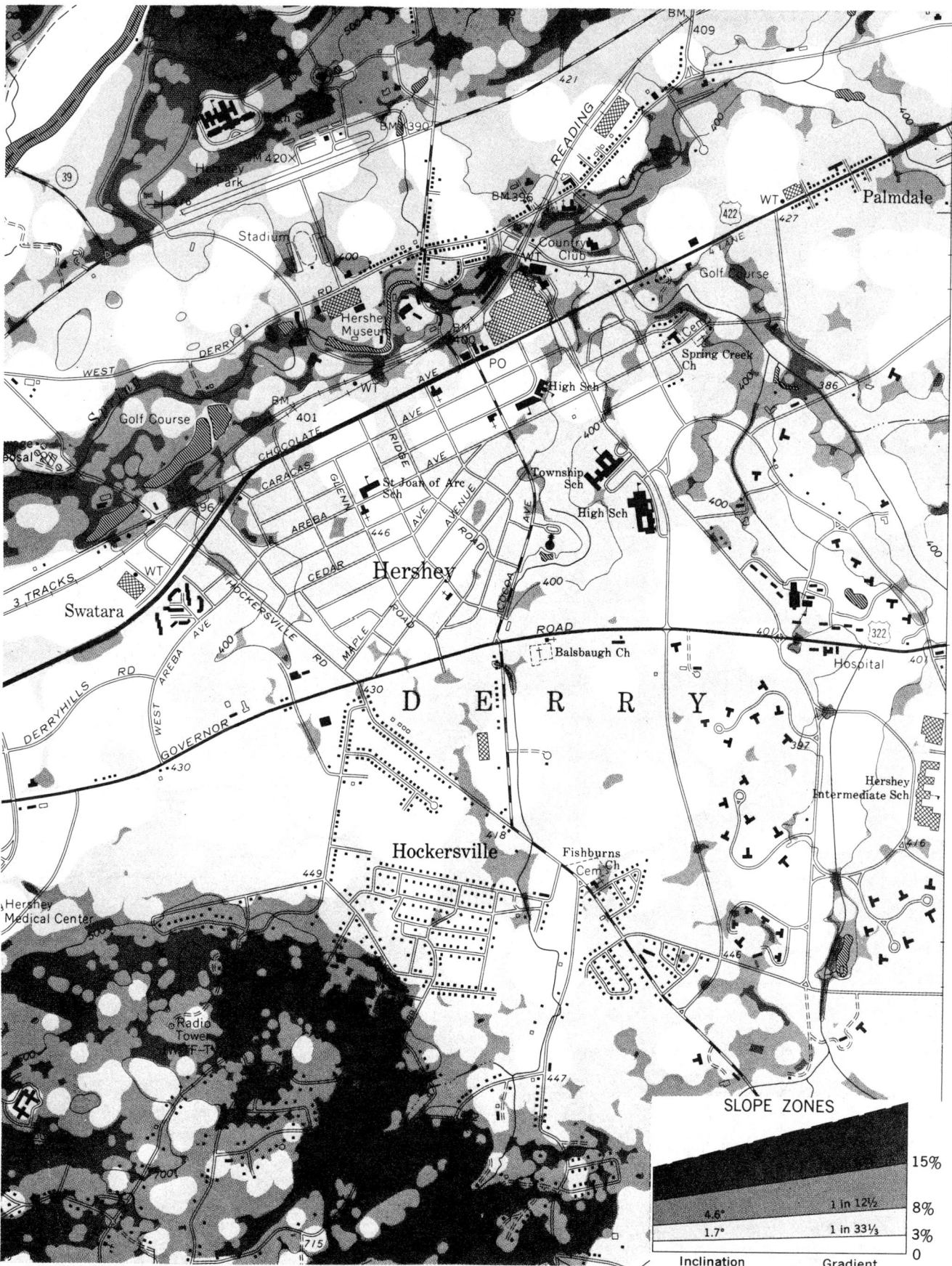

Fig. 17.34 Slope map of Hersey, PA., at a scale of 1:24,000 *(Courtesy USGS)*.

A profile is drawn by plotting horizontal distances as abscissas (X axis) and elevations as ordinates (Y axis), Fig. 17.35. The vertical axis in the profile view is marked at frequent intervals with each horizontal line normally representing one contour interval. To plot the profile between points X and Y, the highest (1680) and lowest (1580) elevation of the contour lines that cross or touch line XY in the plan view are noted to determine the scale values to assign to the vertical axis. The vertical scale should be extended slightly beyond the highest and lowest values to allow the tops of hills and the bottoms of valleys to be rounded with smooth curves. Elevations are plotted by dropping perpendiculars from each point where a contour line crosses or touches line XY in the plan view to the horizontal lines of corresponding elevation in the profile view. The elevation value for plotting hilltops and other critical points is determined by interpolating between contours. Where it is not possible to project horizontal distances as illustrated, the dividers may be used to transfer these distances. The plotted points are connected with a smooth freehand curve or as a straight line between points for engineering work to represent the profile of the land.

The vertical scale is generally exaggerated to better portray the ground surface since valleys, peaks, or other features are more clearly shown and may be compared. The amount of vertical exaggeration should be stated below the

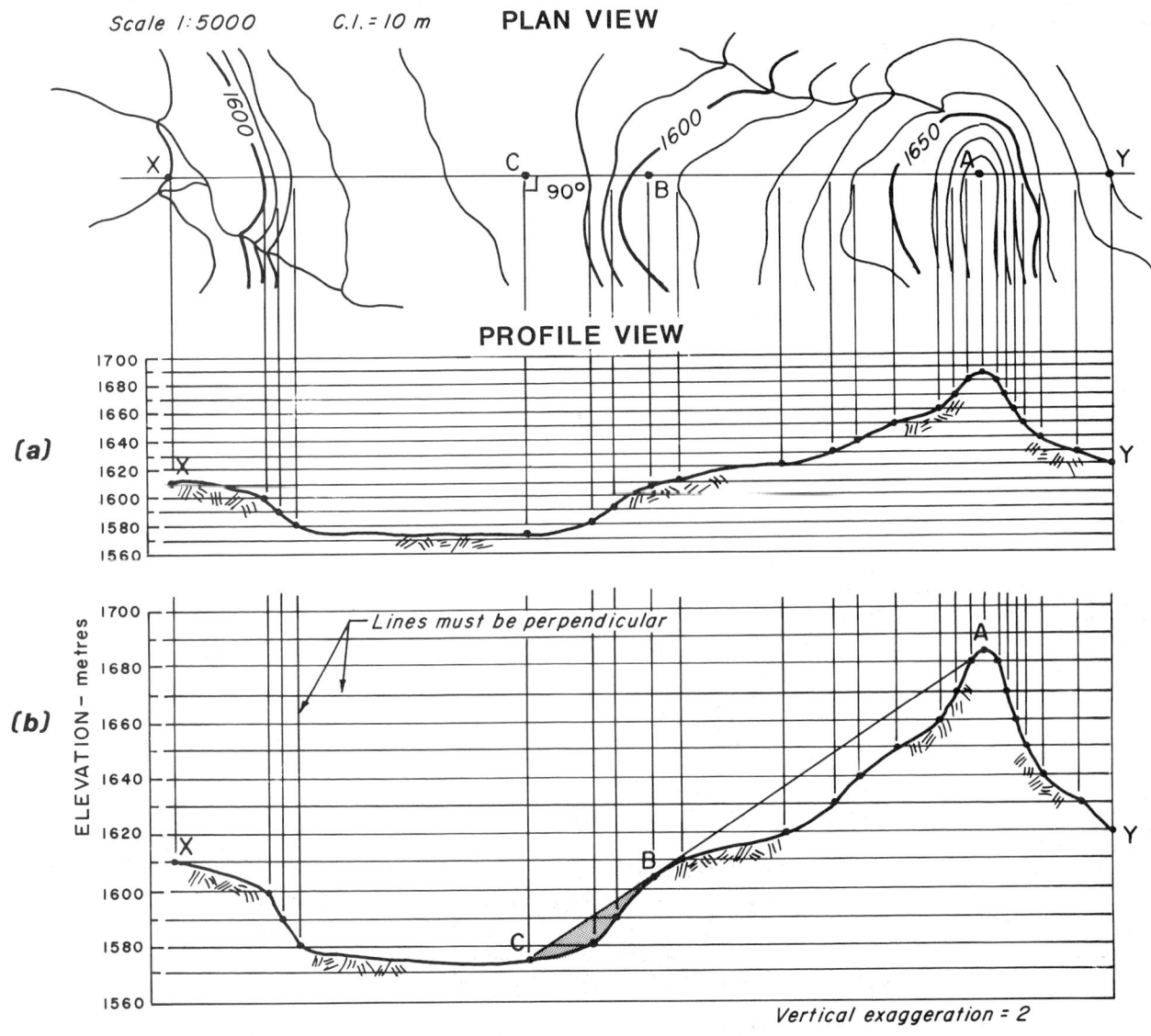

Fig. 17.35 Profile plotting.

profile. For example, given a horizontal scale of 1:5000 and a vertical scale of 1:2500 the vertical exaggeration is as follows:

Vertical exaggeration = $\frac{\text{Vertical Scale}}{\text{Horizontal Scale}}$ = $\frac{1}{2500} \times \frac{5000}{1} = 2$

The vertical and horizontal scales are shown the same for the profile drawn in Fig. 17.35 (a). In Fig. 17.35 (b) the vertical scale has been exaggerated two times for comparison.

In visibility studies, profiles may be used to determine points that are visible from a given point. An area or object hidden from sight is said to be sight defiladed. The area or object that prevents a sighted point to be seen is called the mask. Standing at point A on the ground, Fig. 17.35 (b), the mask at point B forms the defiladed area between points B and C.

Profiles taken along fixed routes usually indicate distances by stationing, Fig. 17.36. The stations are usually taken at intervals of 50 to 100 feet. The full stations are those having even multiples of 50 to 100 feet such as 0+00, 1+00, and 2+00. Any intermediate point between two stations, e.g., 1+28.57, 2+32.48, is referred to as a plus station.

Commercial paper is usually used for plotting profiles. The paper is ruled with vertical lines at intervals of ¼ or ½ inch and with horizontal lines at intervals of 1/10 or 1/20 inch.

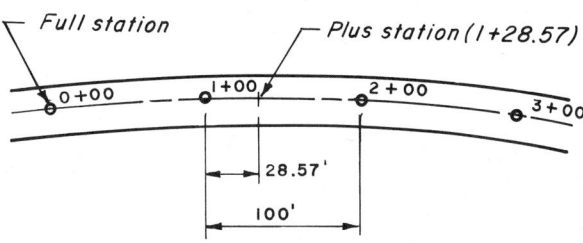

Fig. 17.36 Stationing.

Cross sections. A *cross-section* is used primarily to obtain the areas of cut and fill for computing earthwork quantities. Any number of cross-sections may be drawn by establishing a perpendicular to the centerline, e.g., at each station of a road and drawing a vertical section. The section should be viewed in a direction from which the stationing was laid out (17, 18, etc.) in order to transfer measurements left or right as required. The manner of drawing a cross-section is similar to that for a profile. In Fig. 17.37 the map scale is 1"=200' and the profile is drawn having a horizontal and vertical scale of 1"=50'. Since this represents a scale exaggeration of four, all horizontal measurements where a contour line crosses the dashed line on the map must be scaled or quadrupled to transfer measurements.

Cross-sections for earthwork are commonly drawn to scale on regular cross-section paper which is ruled usually with 10 divisions to the inch in both directions.

Mining and geology terms. Problems related to mining and geology involve a number of technical terms that are illustrated in Fig. 17.38 (a) and defined as follows:

1. *Strike*. The bearing of a level line in a plane.

2. *Dip*. The angle measured downward from a horizontal plane and at right angles to the strike.

3. *Stratum*. A sheetlike mass of sedimentary rock or earth of one kind usually in layers between beds of other kinds.

4. *Vein or lode*. Any mass of ore, coal, etc., more or less sharply separated from the inclosing rock.

5. *Fault*. A fracture in the earth's crust with displacement of one side of the fracture with respect to the other and in a direction parallel to the fracture. The surface along which the dislocated masses have moved is called, when not notably curved, a fault plane.

6. *Outcrop*. The part of a stratum or vein exposed to the surface of the ground.

7. *Bedding plane*. The bounding surface of a stratum or vein.

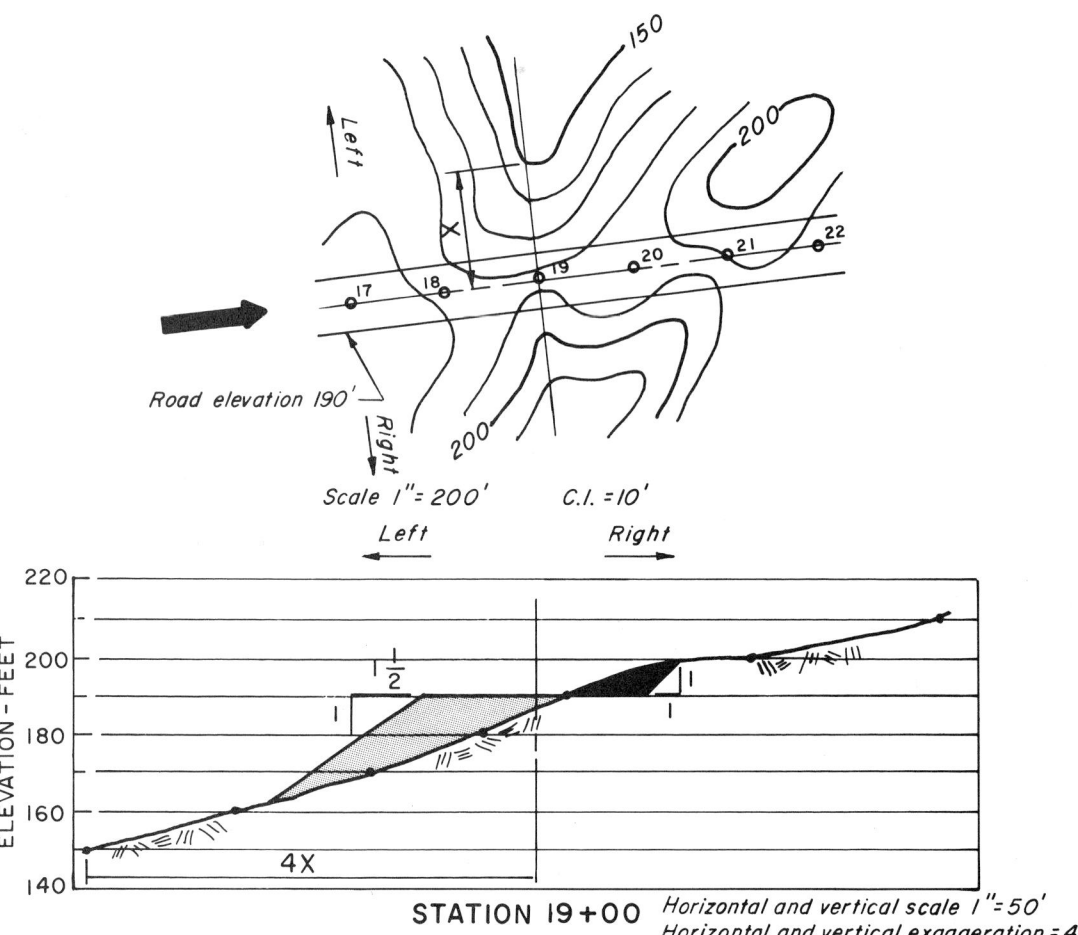

Fig. 17.37 Cross-section plotting.

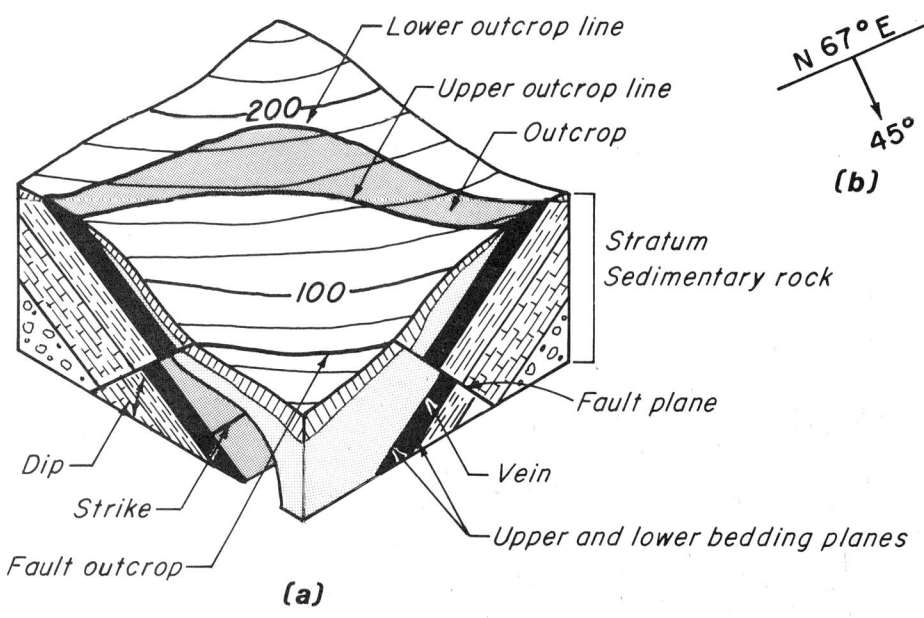

Fig. 17.38 Mining and geology terminology.

The symbol to indicate strike and dip on a map is shown in Fig. 17.38(b). In the field an outcrop of a vein is located and a clinometer, operating on the principle of a spirit level and a compass, is used to record the exact location of the strike and dip on a map.

Strike, dip, thickness, and outcrop from contour maps. A bedding plane may be defined by any three points using the outcrop as one point with other points found by drilling. Point A is on the outcrop of the upper bedding plane of a vein of ore shown in Fig. 17.39. Vertical test holes bored at B and C reach the vein at elevations of 310 and 243 metres respectively. The hole at B is extended until it reaches the lower bedding plane at 282 metres.

The front elevation view of plane ABC is established from the given elevations. A level line $A_F E_F$ is drawn on the vein in the front elevation and projected to the top view at $A_T E_T$. The strike of this true length line is N54°E. The elevation view 3 is drawn by obtaining a point view, $A_3 E_3$, of the strike line. Points $B_3 C_3$ on the upper bedding plane are located in this new view to show the vein as an edge and its dip at 25°. The dip is shown on the map by drawing an arrow perpendicular to the strike line pointing in a downward direction with the value of the dip lettered on it. In view 3, point F_3 on the lower surface is located and the lower bedding plane drawn parallel to the upper one. The distance 8.3 metres be-

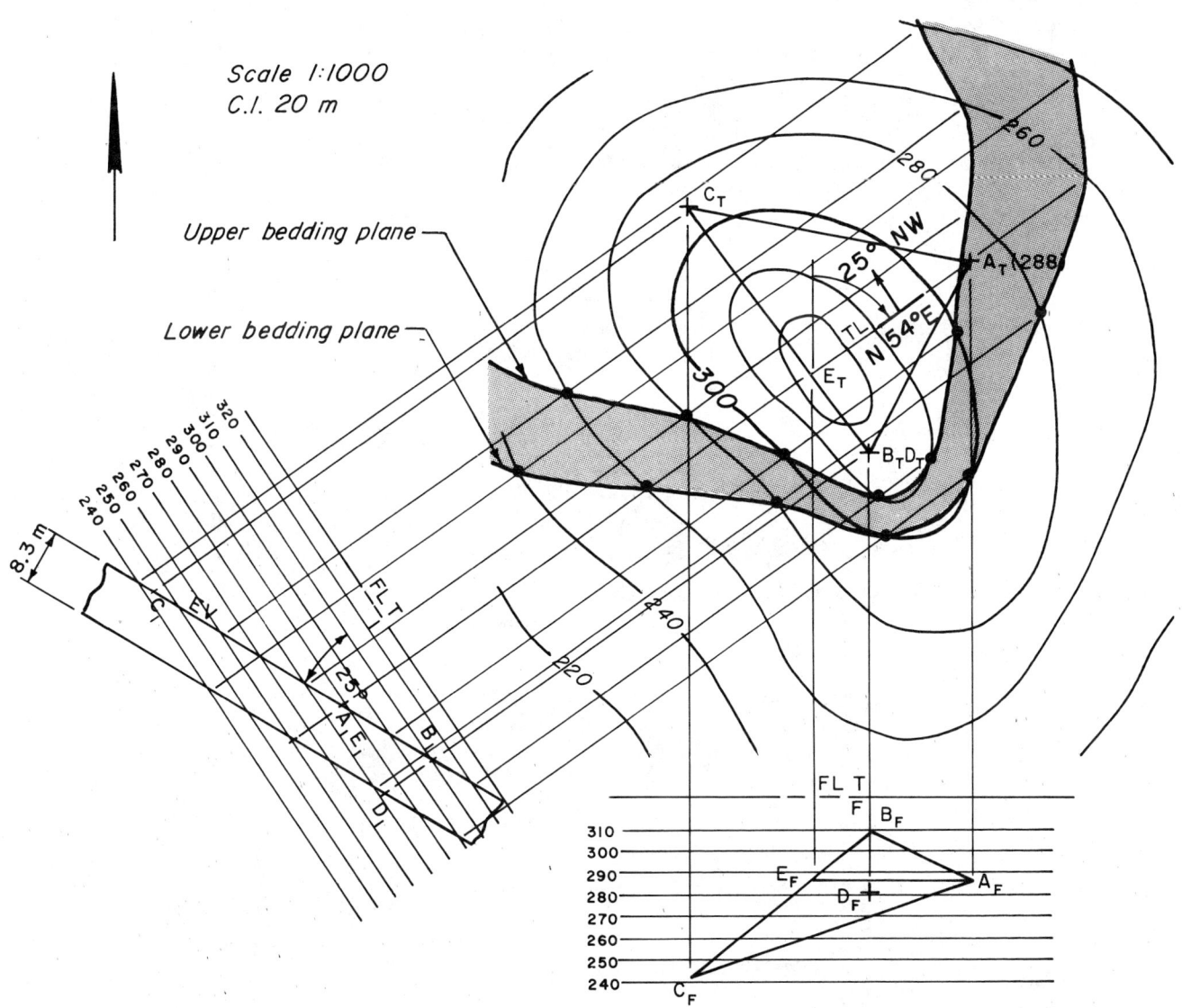

Fig. 17.39 Strike, dip, thickness, and outcrop from a contour map.

tween these two planes is the thickness of the vein.

The line of outcrop is the line of intersection of the bedding plane of the irregular ground surface. Each point on this line may be found by intersecting a level line on the vein with a contour line on the ground at the same elevation. Several points must be found to determine the irregular line of outcrop for both upper and lower bedding planes.

Cut and fill boundaries from contour maps. The boundaries of cut and fill are required in highway construction and other operations involving earthwork. To find the outline of cut and fill, the proposed contour lines are superimposed on the existing contour lines. The points where the two contour systems intersect are connected to delineate the extent of cut and fill.

The proposed contour lines for the level building site, Fig. 17.40, are drawn as dashed lines parallel to the sides of the site. Where the 135 metre contour line crosses the level site, no cut or fill is required. This contour line also marks the cut from the fill area. The spacing of the proposed contour lines depends on the slope. Since the slope for the cut area is 1 to 1, the spacing of the lines will be the same as the contour interval (5 metres) of the map. The slope for the fill is 1½ to 1 and the proposed contours are laid off 7.5 metres apart. Where the proposed contour lines cross the existing contour lines of equal elevation, boundary points for the cut and fill are established. The two contour lines may intersect at several points. All points should be connected with a smooth freehand curve.

A proposed level 15 metre wide road at an elevation of 160 metres is shown in Fig. 17.41. The boundary for the cut and fill lines is drawn similar to the previous example. Every point where the 160 metre contour line crosses both sides of the road is marked. Areas between these points will require a cut where the topography is higher than the road and a fill where the land is lower than the road. The

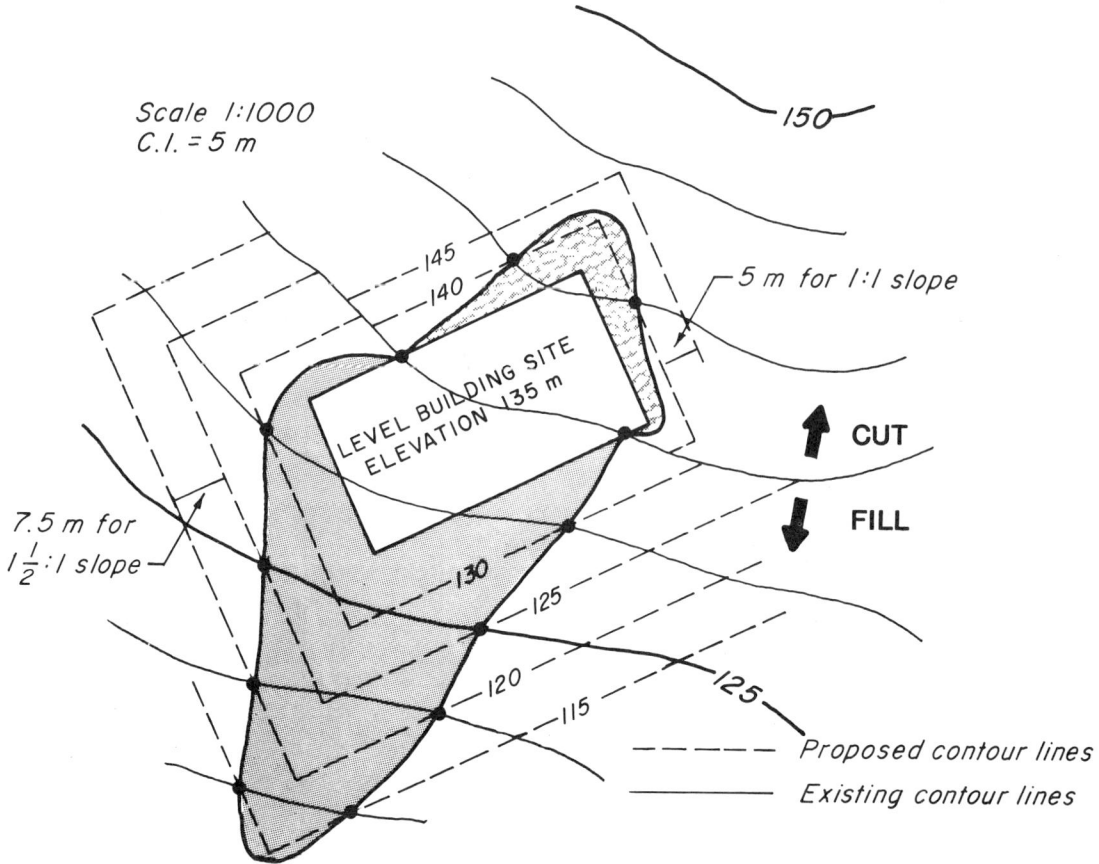

Fig. 17.40 Cut and fill boundary for level building site.

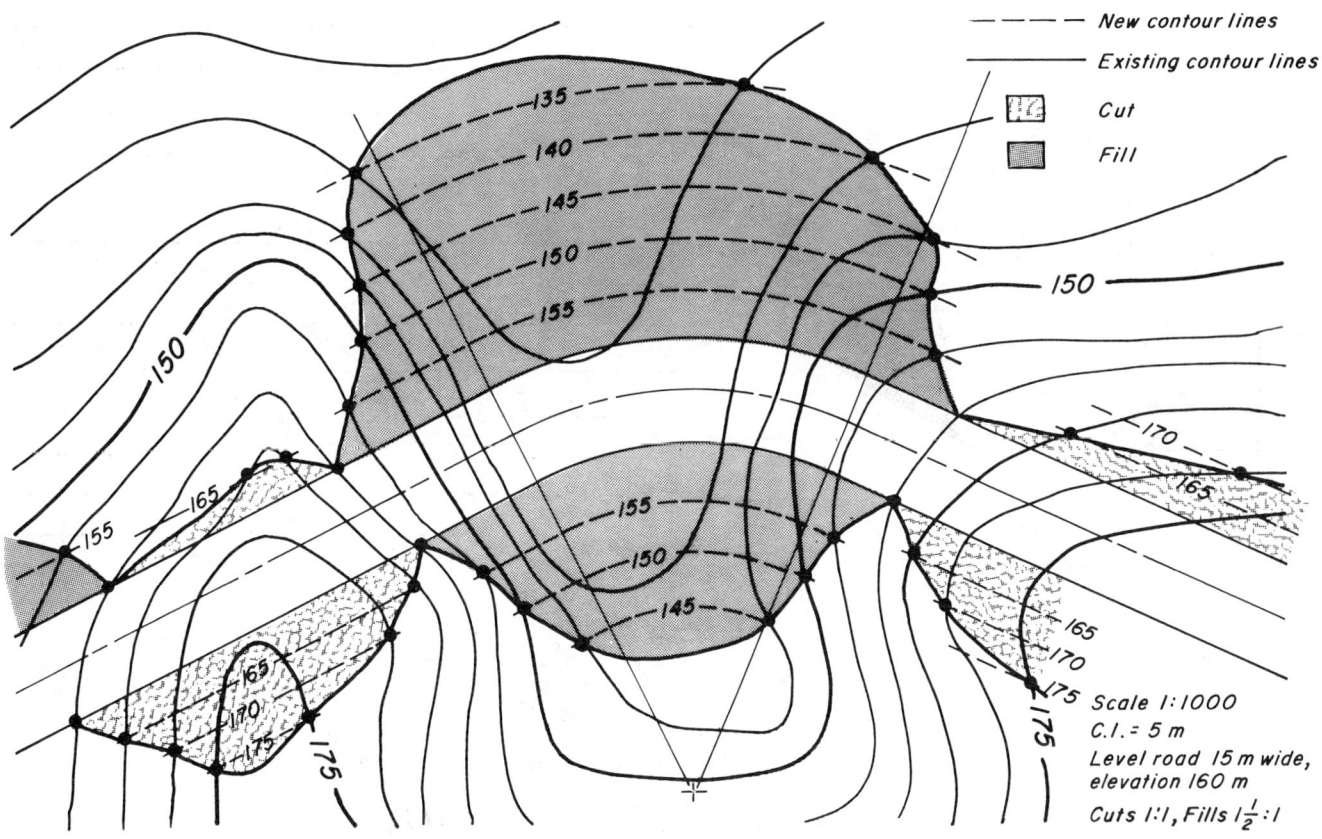

Fig. 17.41 Cut and fill boundaries along a level road.

contour lines for the cut and fill areas are drawn, using the specified slope, at the same elevations as the map contours. Since the roadbed is level, the contour lines will be parallel to the edge of the road. The sides of the cut and fill areas are plane surfaces and all contour lines will be parallel and equally spaced. The boundary lines for cut and fill are found by drawing a line through the points where the new contour and existing contour lines having the same elevation intersect.

A roadbed built on a grade will not have contour lines parallel to the edge of a road. For a road built in the uphill direction, the contour lines for a cut converge toward the edge of the road and for a fill the contour lines diverge from the edge of the road, Fig. 17.42 (a).

The spacing and direction of the contour lines for the uniform sloping road illustrated in Fig. 17.42 (b) may be found by drawing a semi-circle with a radius one-half the distance from the bottom to the top of the slope. Since the side slopes are 2 to 1, the spacing of the contour lines will be 8 metres.

A chord, XY, 80 metres in length is drawn with increments of 8 metres laid off along the line. This length is used since five contour lines will be required from the bottom to the top of the sloping road. The direction of the contour lines is found by drawing parallel lines through the points on line XY. Contour lines may be established along the level road by drawing a series of lines parallel to the edge of the road 8 metres apart to reflect a 2 to 1 slope. The line of demarcation between the plane surfaces along the level and sloping road is found by drawing a line where contour lines having the same elevation intersect.

The boundaries of cut (slope 1 to 1) and fill (slope 1½ to 1) along a road of 14 percent grade may be found as shown on the map in Fig. 17.43. All contour lines will cross the road at intervals of 35.7 metres. At point A on the edge of the road a series of concentric circles are drawn and labelled at 5 metre intervals of elevation. Since fill will be required at this point, the radii will increase in increments of 7.5 metres to represent a slope of 1½ to 1. The fill

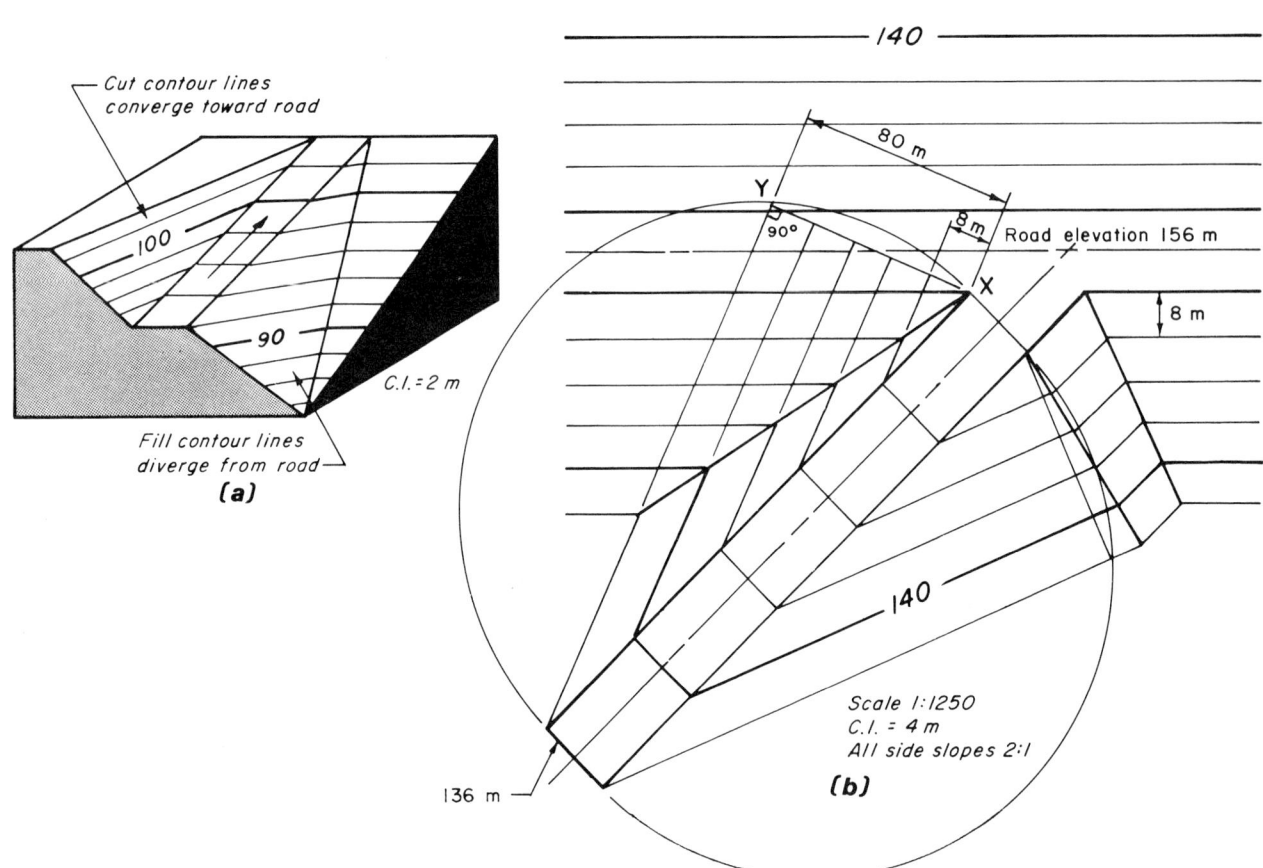

Fig. 17.42 Fill along a grade road.

$$HD = \frac{CI \times 100}{SLOPE} = \frac{5 \times 100}{14} = \underline{35.7} \text{ m}$$

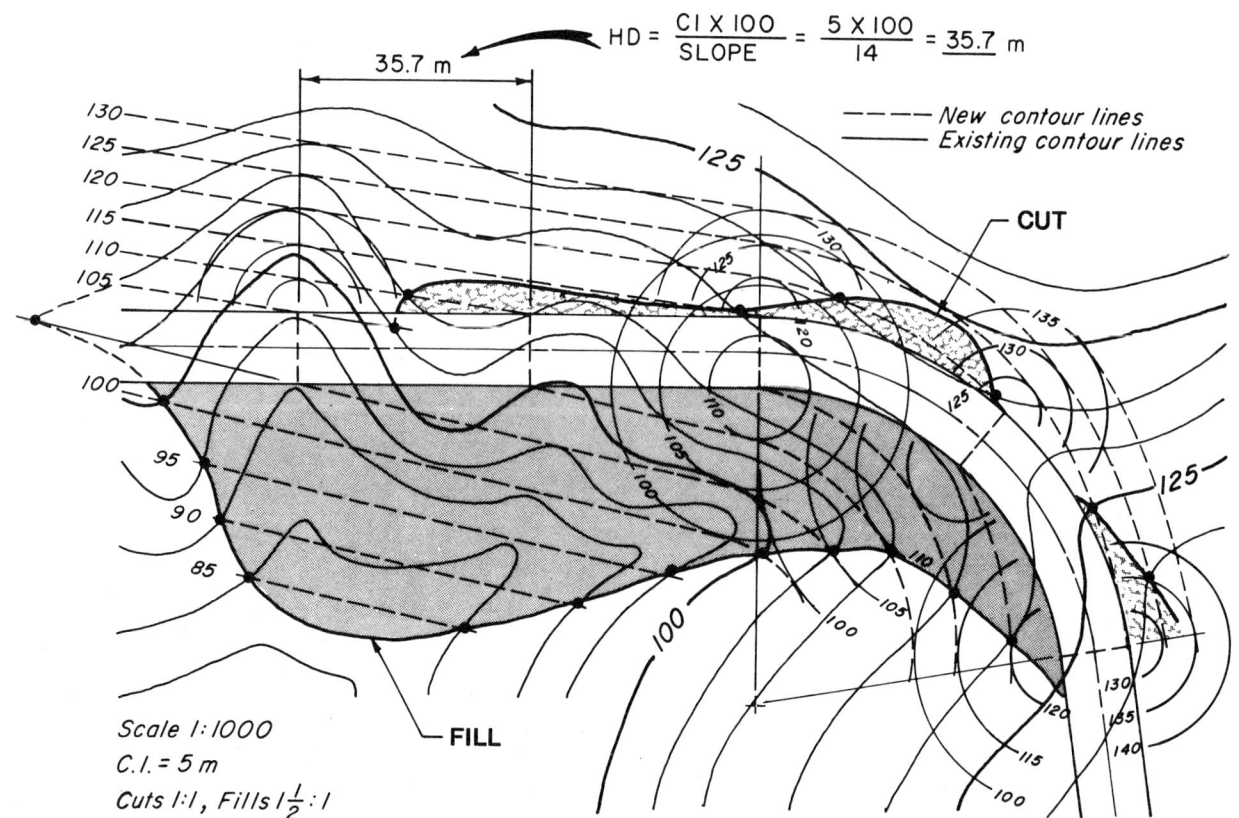

Fig. 17.43 Cut and fill along a grade road on a contour map.

contour lines are drawn from the points where the contour lines cross the road tangent to the concentric circles having the same elevation. These contour lines will be parallel and equally spaced. Additional parallel contour lines may be drawn where required.

The procedure described is repeated at various points where contour lines cross the road to establish the remaining cut and fill contour lines. The concentric circles for cuts will have radii increasing in increments of 5 metres to represent a slope of 1 to 1. It is not essential to draw the complete circles. Along the curved edge of the road it is necessary to draw the concentric circles at regular intervals as at the 120 and 125 metre contour lines. The contour lines for the curved sections of the road are drawn as parallel spiral curves tangent to the successively larger circles.

After all contour lines have been found, the boundary for cut and fill is established by connecting the points where the existing contour lines intersect a fill or cut contour line having the same elevation.

REMOTE SENSING AND AERIAL PHOTOGRAMMETRY

18.1 INTRODUCTION

The first maps made from photographs, taken from high buildings, were exhibited in Paris, France, in 1867. One of the first patents involving aerial photographs was issued to C. B. Adams in 1893. This patent used the radial line principle to locate a point on a map using aerial photographs taken from a balloon. The earliest record of aerial photographs actually taken and used for map-making was reported in a paper presented at the International Society of Photogrammetry in Vienna, Austria, in 1913. It was a mosaic of the city of Bengazi, Libya, at a scale of approximately one inch to 333.3 feet (RF 1:4000).

In 1922, the U.S. Geological Survey issued its first quadrangle sheet made from aerial photographs, and since 1939 all of these sheets have been made by the use of aerial photography. Every large government or private map agency uses air photos almost exclusively in the preparation of maps. Air photos are more widely used for this single activity than for any other purpose. Since it is reasonably quick, maps of large areas may be made to pre-determined standards of accuracy more economically than by ground techniques, and information for inaccessible areas is easy to obtain.

Since the early 1960's, frontiers in aerial photography have been pushed back with the development of mapping via space photography and other remote sensing imagery.

Remote sensing. As defined by the American Society of Photogrammetry, *remote sensing* is the measurement or acquisition of information of some property of an object or phenomenon by a recording device that is not in physical or intimate contact with the object or phenomenon under study. In a broader definition, remote sensing is the use of indirect methods to gather information which is difficult or impossible, either physically or economically, to gather directly. While the aerial camera and conventional photographs are the most commonly used remote sensing methods, other systems relying on different portions of the electromagnetic spectrum are used for gathering information to which cameras are not sensitive.

18.2 ELECTROMAGNETIC SPECTRUM

All objects with a temperature above 0° Kelvin (-273° Celsius or -459.4° Farenheit) emit electromagnetic energy. This energy is in the form of waves of varying frequency in what is called the *electromagnetic spectrum*, Fig. 18.1 (a).

When energy emitted from the sun strikes the earth, it is absorbed or reflected. The energy absorbed is converted to heat or other energy forms by photo-chemical reactions. Energy which is reflected by any object, if it is in the visible part of the spectrum, gives that object its color, as received by the human eye. This "visible" part of the spectrum is in the 0.4 to 0.7 (blue to red) micrometre wavelengths. Films for cameras were developed to be sensitive to these same wavelengths.

Other systems have been developed by technology to receive energy from various parts of the spectrum and convert it to forms usable by man. These other parts of the spectrum include X-rays, ultra-violet, visible light, infrared (both reflected or near and thermal), radar (microwave), and radio. Those portions of the electromagnetic spectrum used in remote sensing of the earth's resources fall in the regions of visible light, infrared (both reflected and thermal), and radar or microwave bands, Fig. 18.1 (b).

18.3 SENSORS USED FOR REMOTE SENSING

Remote sensing systems, or instrumentation, utilizing various bands of electromagnetic energy fall into two categories: (1) passive systems, including cameras and infrared sensors which receive reflected or emitted energy from objects, and (2) active systems, which send out a signal and receive the reflected echo as in a radar system.

Specific systems, or types of instrumentation, developed to utilize visible and non-visible portions of the spectrum in electromagnetic remote sensing of earth resources are: (1) cameras and film (conventional photography), (2) multi-spectral cameras (photography in narrow wavelength bands), (3) line-scanners (thermal and multi-spectral), (4) side-looking airborne radar (SLAR), and (5) satellite scanners or imaging systems, Fig. 18.2.

Cameras. The cameras most commonly used in aerial photography for mapping and photo-interpretation of resources are single-lens precision cameras that operate in the visible, infrared, or ultraviolet portions of the electromagnetic spectrum. Conventional aerial cameras in general have four basic components: (1) film magazine, (2) drive mechanism, (3) cone assembly and (4) lens, Fig. 18.3 (a). The cameras are designed to expose a large number of photographs in rapid succession. The length of time between individual exposures is normally governed by flight altitude, aircraft speed, camera focal length, and film format. Most aerial photography for mapping and the inventory of natural resources use focal lengths of 6 inch (152 mm), 8¼ inch (210 mm), or 12 inch (305 mm). The focal length (f), Fig. 18.3 (b), of a camera is the distance along the optical axis from the rear nodal point of the lens to the focal plane (film).

(a) The electromagnetic spectrum

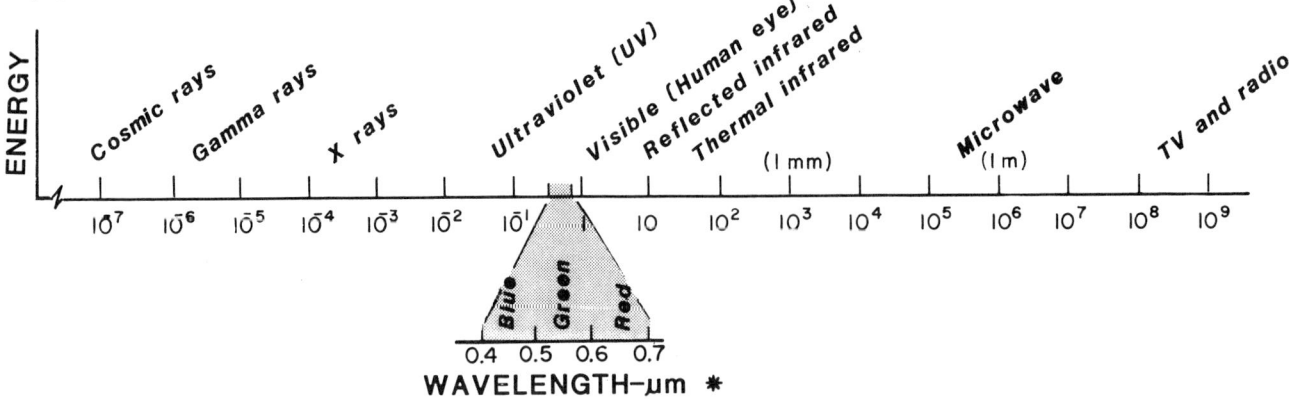

(b) Bands of spectrum used for remote sensing systems

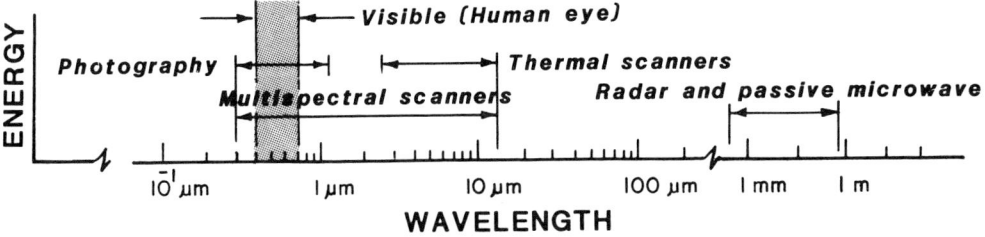

*µm - micrometre (one millionth of a metre)

Fig. 18.1 The electromagnetic spectrum and remote sensing bands.

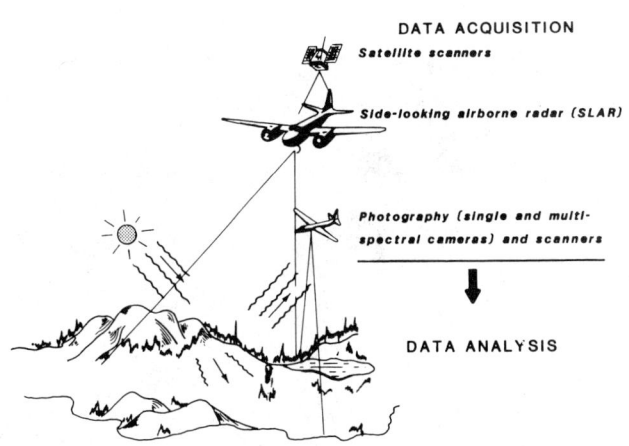

Fig. 18.2 The remote sensing process.

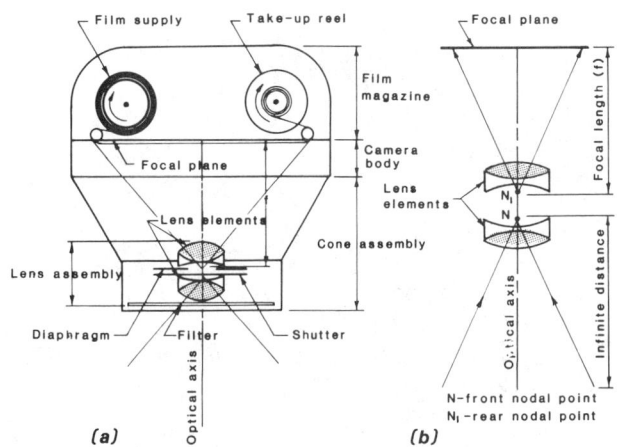

Fig. 18.3 Principal components of a single-lens frame mapping camera.

Black and white films. A variety of film types can be employed, with the commonly used black and white panchromatic film being the most economical, Fig. 18.4. This film is sensitive to all wavelengths of the visible portion of the electromagnetic spectrum. Since it is necessary to block out certain light rays of the visible spectrum, light filters of different kinds are always used with aerial cameras to filter out the scattered blue light or haze of the atmosphere and/or unwanted colors which will degrade the images recorded on the film. The resulting images are shown in various shades of gray. High resolution black and white aerial photographs are used for topographic, engineering, and land-use mapping.

In addition to panchromatic black and white film, infrared black and white film is used for the sensing of images in the infrared portion of the electromagnetic spectrum, Fig. 18.5. The image is frequently much clearer than the same image produced on panchromatic film. A clearer picture results from infrared radiation reflected from an object being scattered less by dust and haze particles than visible wavelengths.

resulting in lakes and rivers appearing dark with wet soils also appearing darker than dry soils. Another example is that deciduous vegetation will appear lighter than coniferous vegetation since it is a better reflector of infrared light.

Color films. The use of color films increases the value of photography for many interpretation uses, as in identifying features such as soils, vegetation, and water conditions. Two types of color films are widely used: (1) conventional or true color film, which gives color prints and/or transparencies with "normal colors" which appear natural to the human eye, and (2) color infrared or false-color film, which records green, red, and near infrared wavelengths. True color films are especially valuable in analyzing terrain geology, distribution and condition of soil types, and hydrologic features. Photographs taken at high altitudes, however, may lack color contrast and appear in a monotone tint because of atmospheric effects.

Color infrared films are used in the management of renewable resources such as agriculture, range, and timber crops. Healthy green vegetation will appear in various shades of red and pinks while areas that are less healthy will usually appear in progressively darker colors with man-made surfaces often in a bluish-gray. When used with the correct filters and developed properly, both types of color films produce fine-grained pictures of high resolu-

Fig. 18.4 Black and white panchromatic aerial photograph of Langenburg, West Germany (*Courtesy Carl Zeiss, Inc*).

Fig 18.5 Black and white infrared aerial photograph of Langenburg, West Germany *(Courtesy Carl Zeiss Inc)*.

tion suitable for magnification by several diameters without degradation of the images.

Multi-spectral camera system. This is an array of smaller frame cameras (usually four or six) with various film-filter combinations in each camera, Fig. 18.6. All cameras are sychronized to record imagery of the same scene in a narrow band of visible color or infrared part of the electromagnetic spectrum, Fig. 18.7. The same type of films and filters are used as in the larger frame single lens cameras. The chief value of using this type of photography is that each camera produces an image of the same area with a different tonal value. A comparison of these different images will provide more information regarding certain objects and conditions than would be possible from a single photograph.

Line scanners. Scanning devices collect and record data by detecting released energy. In line scanning the collecting device, a rotating or oscillating mirror, repeatedly sweeps over an area along a series of scan lines that is perpendicular to the flight line. Only a small portion of the surface is viewed at a time. The electromagnetic energy is received and focused on special detectors which react by instantaneously changing electrical resistance, resulting in a stronger or weaker signal being recorded on magnetic tape for later processing.

In a multi-spectral scanner system, Fig. 18.8, that utilizes the visible and infrared light portions of the electromagnetic spectrum, the energy is split by prisms and/or grating into several narrow bands, each focused on a different detector and recorded on tape.

Fig. 18.6 Multi-spectral camera system *(Courtesy Itek Optical Systems)*.

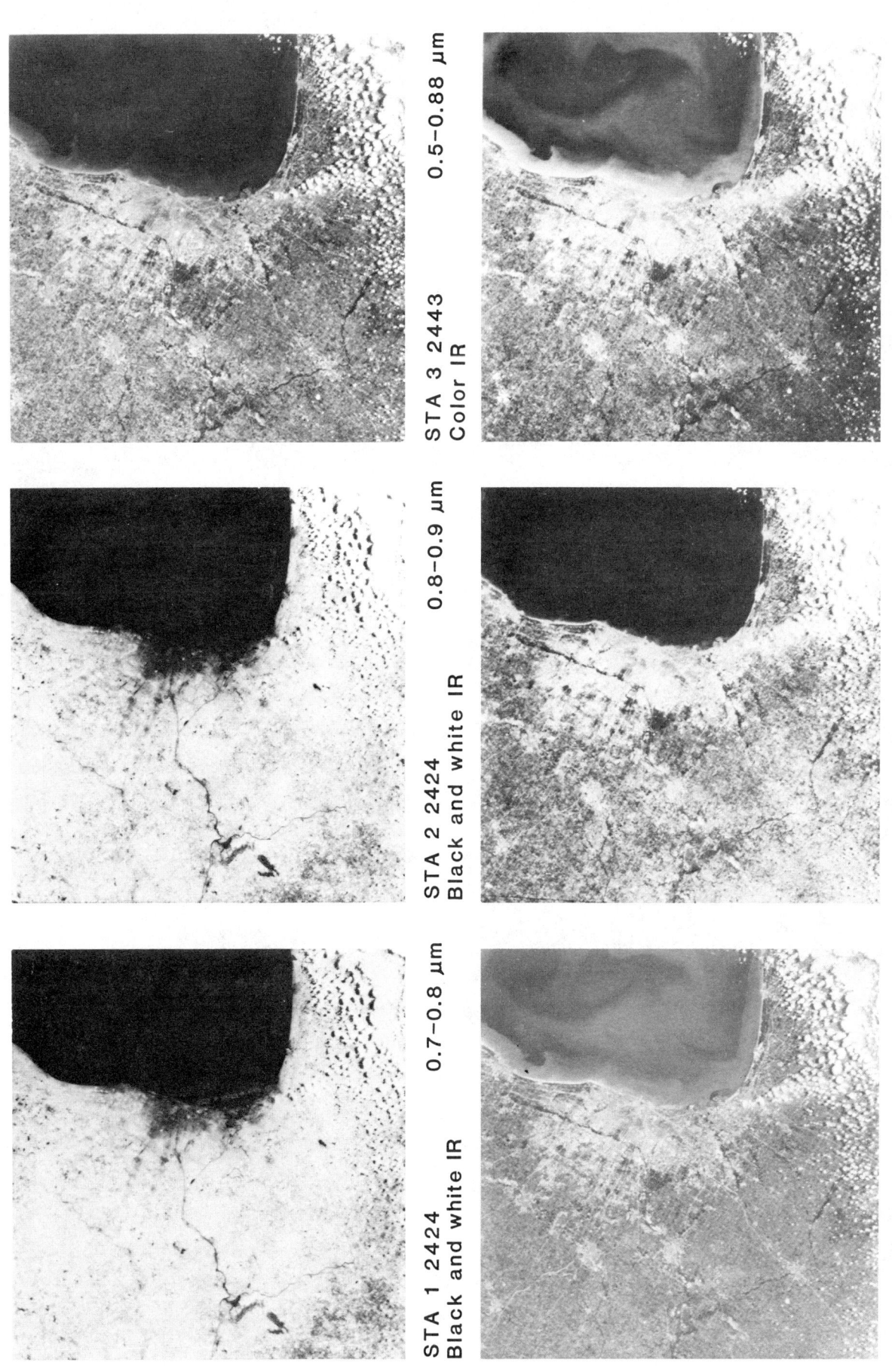

STA 1 2424   0.7–0.8 μm
Black and white IR

STA 2 2424   0.8–0.9 μm
Black and white IR

STA 3 2443   0.5–0.88 μm
Color IR

STA 4 SO-356   0.4–0.7 μm
True color

STA 5 SO-022   0.6–0.7 μm
Red filter   Black and white

STA 6 SO-022   0.5–0.6 μm
Green filter   Black and white

Fig. 18.7 Multi-spectral skylab photography of Chicago area and Lake Michigan. Film types and spectral regions are noted on photos *(Courtesy Itek Optical Systems).*

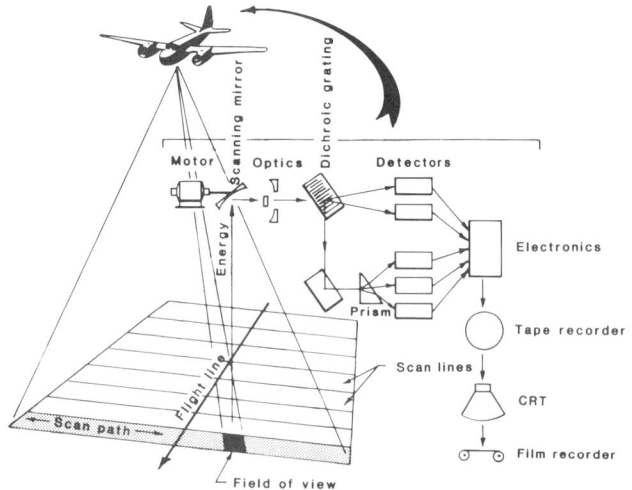

Fig. 18.8 Multi-spectral scanning system.

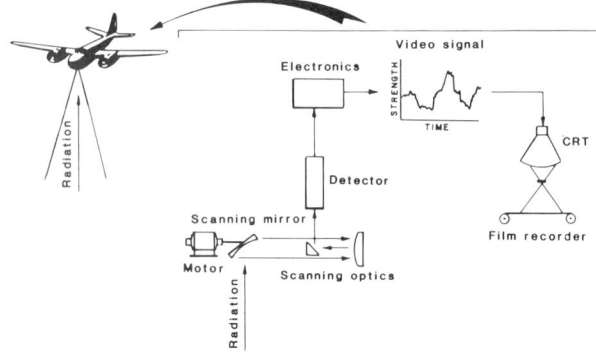

Fig. 18.9 Thermal scanning system.

tector, with its varying electrical resistance, records the energy on tape, as in the multi-spectral scanner, or it may be recorded directly on film passing by a slit in a CRT display, Fig. 18.9.

The images from each detector can then be displayed on a CRT (cathode ray tube terminal) display and photographed or film copy can be made directly.

In a thermal line scanner system, the energy is from the far infrared part of the electromagnetic spectrum (known as heat or thermal infrared). The de-

The quality of scanner images will vary depending on the system utilized. Although the image is similar to a photograph, it will usually be poor in detail resolution and have a fuzzy appearance, Fig. 18.10. The main advantages of using scanners are (1) no film is needed

Fig. 18.10 Thermal scanning image for a portion of the University of Maine at Orono campus. Building and steam line heat loss indicated by white areas (*Courtesy James W. Sewall Co*).

as in a camera and information beyond the visible part of the electromagnetic spectrum can be recorded and (2) recording of the information on tape allows data to be converted to digital form and fed into a computer for storage, analysis, and printout. The measuring of surface temperatures and the mapping of their distributions are valuable in resource and environment studies. Infrared scanners are finding an increased use in specific areas such as surface temperature measurements to indicate volcanic activity, disease in plants, burning in abandoned coal mines, heat loss in buildings, and water currents. It is also being used to reveal surface temperature distributions in water for discovering sources of water and the discharge of pollutants.

Side-looking airborne radar (SLAR). The side-looking airborne radar system is the only active system used in earth resources remote sensing. This system utilizes pulsed radar energy signals emitted from a transmitter, using a directional antenna and a receiver to record echo returns along scan lines on one side of the aircraft, Fig. 18.11. The imagery recorded is a combination of time and strength of the echo returns. The return signals may be recorded on tape for later processing or displayed instantaneously and recorded on film. The main advantage of SLAR is that it can use frequencies that penetrate clouds, smog, and precipitation. As an active system it can be used at night as well as in the daytime. It is usually used at high altitudes to give coverage of large areas at reasonably low cost.

The imagery, Fig. 18.12, produced by SLAR systems is used for broadscale subjective interpretation in evaluating many natural resources, such as geological formations and mineral evaluation, vegetational studies, water resources, and agricultural crop evaluations.

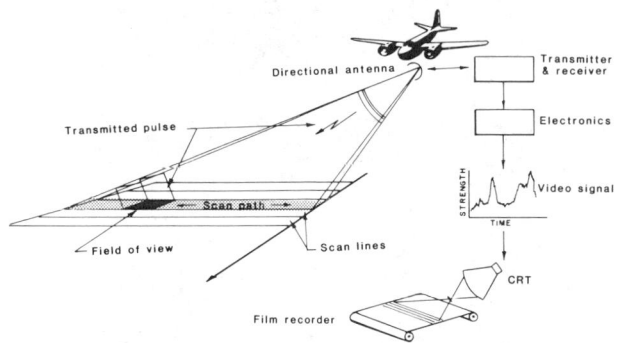

Fig. 18.11 Side-looking airborne radar (SLAR) system.

Satellite imaging systems. The first satellite imaging system, on an operational basis, was the TIROS-1, launched in 1960. This was a weather satellite, sending back images of cloud formations and occasional land-water features. Several Skylab missions, both unmanned and manned, took place in the early 1970's. The spacecraft traveled in orbits 430 km (270 miles) above the Earth acquiring photography, imagery, and selected data from latitudes 50°N to 50°S.

The Earth Resources Technology Satellite (ERTS), launched in 1972, was the first spacecraft, however, designed

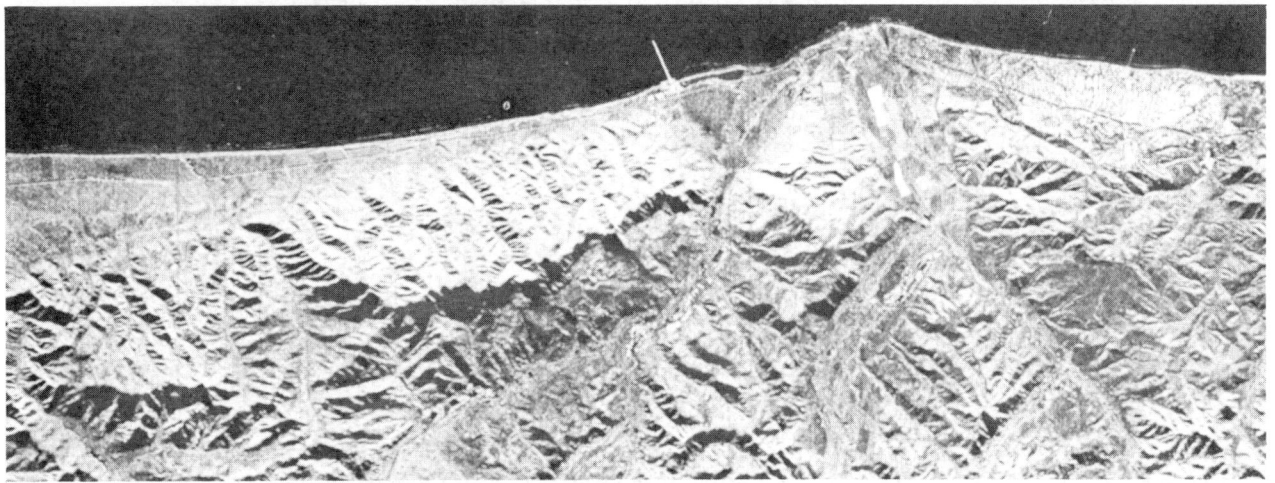

Fig. 18.12 Side-looking airborne radar (SLAR) imagery of coastal California (Courtesy Westinghouse).

to provide systematic coverage of the Earth's surface, except for the poles. With the launching of two additional satellites in 1975 and 1978, the mission was renamed Landsat, Fig. 18.13 (a).

Landsat satellites orbit the earth (north to south) at an altitude of 920 km (570 miles) every 103 minutes or 14 times each day, Fig. 18.13 (b). Pictures are created continuously during the daytime

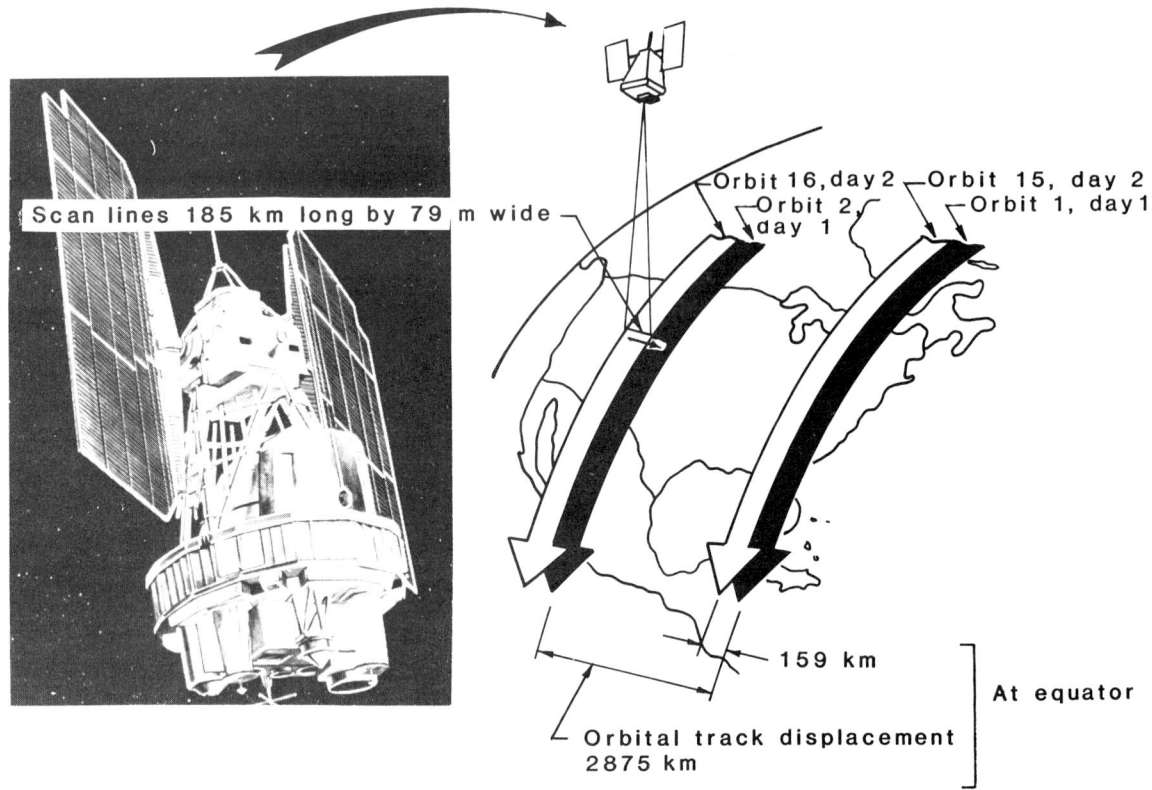

(a) Landsat (Courtesy NASA)   *(b) MSS ground coverage pattern*

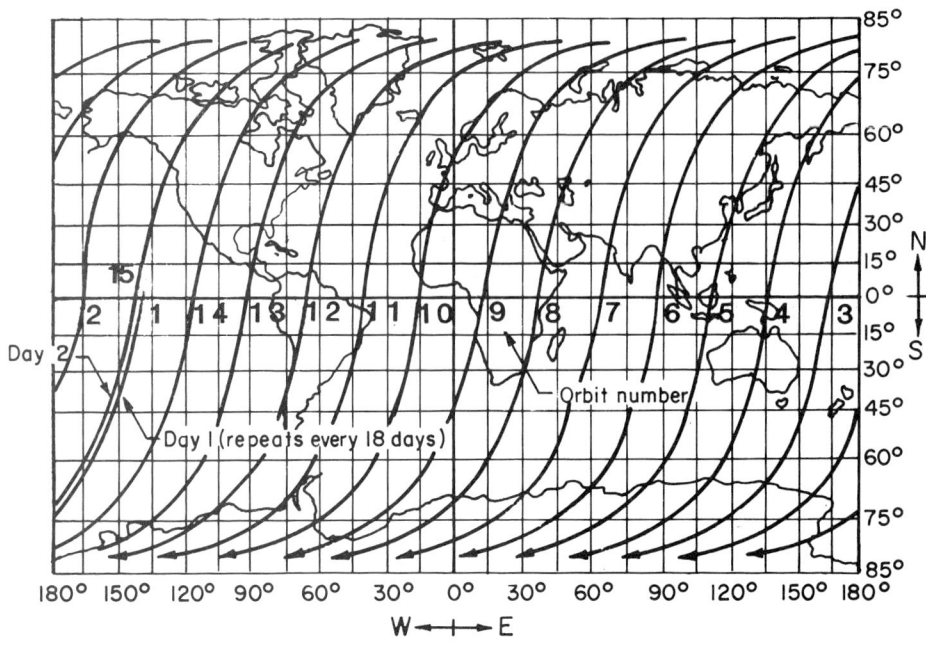

(c) Daily daylight orbit pattern

Fig. 18.13 Landsat.

of the entire globe, except areas north and south of 82° with the same areas receiving repetitive coverage every 18 days, Fig. 18.13 (c). Since the orbits are sun-synchronous, all parts of the Earth are viewed at the same local time on each pass. The difference between pictures of the same scene is consequently very slight making it possible to detect any feature changes that occur.

Imagery from Landsat is collected by using several different acquisition systems: (1) a three channel return beam vidicon (RBV) or television system, (2) a four channel multi-spectral scanner system (MSS), and (3) thematic mapper. RBV systems have three camera-like recorders, operating simultaneously in the green, red, and infrared frequency bands. The images are picked off light sensitive surfaces and relayed to Earth by a scanning electron beam as in a commercial television camera. The imagery is similar to conventional photography, as the ground surface is imaged instantaneously in each format.

The MSS is the primary sensor system and records imagery over a strip 185 km (115 miles) long in green, red, and two near infrared bands of the electromagnetic spectrum. Each band provides slightly different information about the Earth's surface. Band 4 (green) emphasizes sediment laden water and delineates areas of shallow water; band 5 (red) emphasizes cultural features; band 6 (near infrared) emphasizes vegetation, landforms, and land-water contrast; and band 7 (near infrared) provides good penetration of atmospheric haze for emphasizing vegetation, landforms, and land-water contrast.

The MSS imagery produced by Landsat is very good and geometric distortions are small, Fig. 18.14. The nominal ground resolution is about 80 metres (262 feet). Latitude and longitude tick marks are depicted at 30-minute intervals outside the image edge. The 15-step gray-scale tablet that appears at the bottom of the picture is used to monitor and control printing and processing functions and to provide a reference for analysis related to a particular image. Directly over the gray scale is a block of information for image identification, geographic location, and other essential information at the time the data was obtained.

The Landsat system has been upgraded with a launching of a fourth satellite in 1982. In addition to carrying the MSS imaging system, an advanced scanner called the Thematic Mapper has been added. This allows "theme type maps" that highlight farms, forests, or mineral resources to be produced. The Thematic Mapper detects seven spectral bands that will provide more information than is possible with MSS. Since the bands are narrow, it can detect finer differences of light reflected by related objects for identifying feature detail. The image produced has a sharper ground resolution, about 30 metres (98 feet), than the MSS scanner. Instead of recording information on board, the data is beamed directly to one of two Tracking and Data Relay Satellites. The satellites instantly retransmit the data to a ground station where it is again retransmitted, via a domestic communications satellite, to a main receiving station. With this system, data may now be acquired in several days instead of weeks.

New applications for Landsat imagery, Fig. 18.15, appear on a regular basis. Several uses include broad regional resource evaluations, water movement, meteorological phenomena, and investigations involving landforms, vegetation land use, and geology.

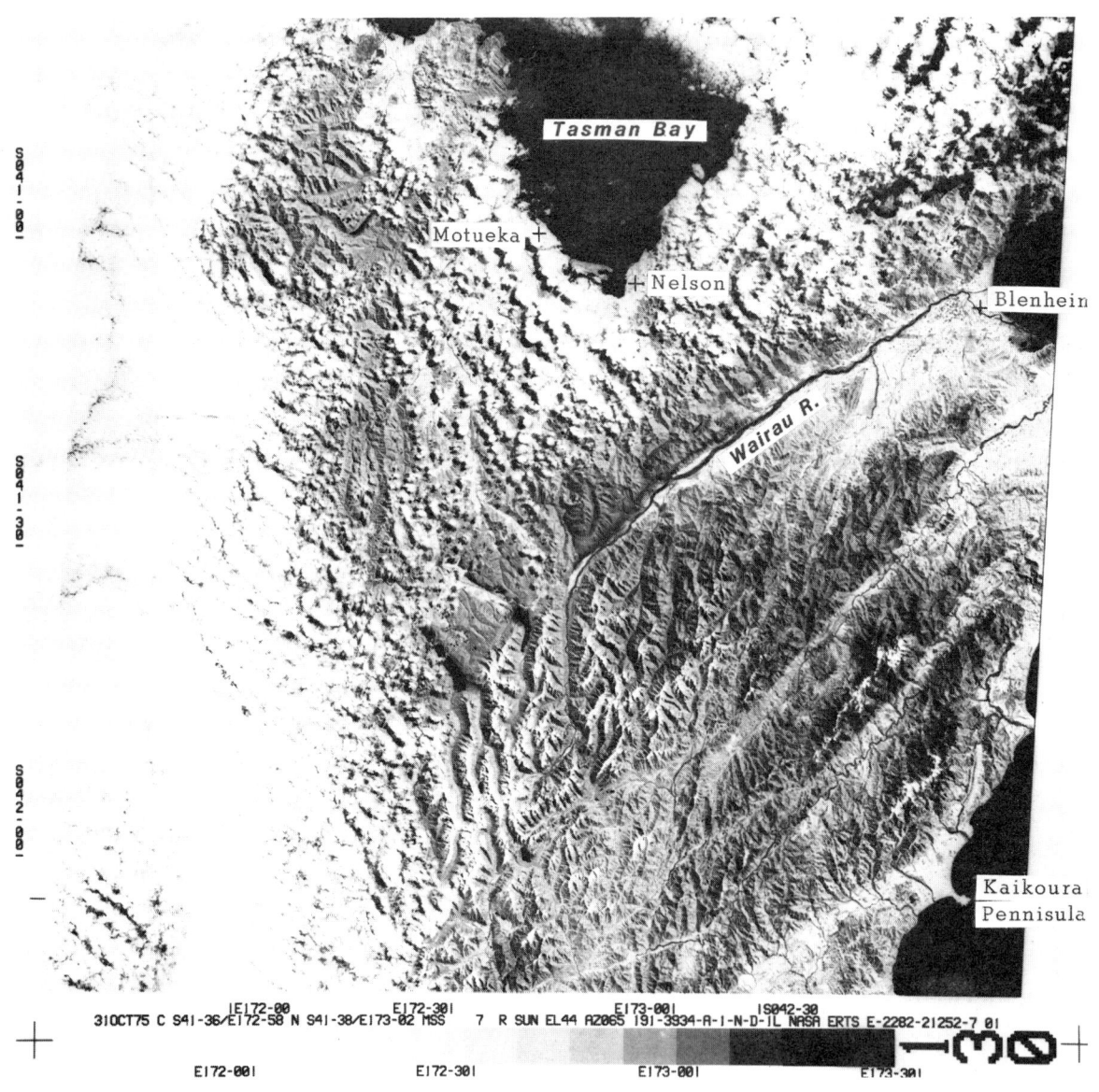

Fig. 18.14 Landsat MSS imagery for northern part of South Island, New Zealand *(Courtesy Physics Engineering Laboratory, DSIR, Lower Hutt, N.Z.)*.

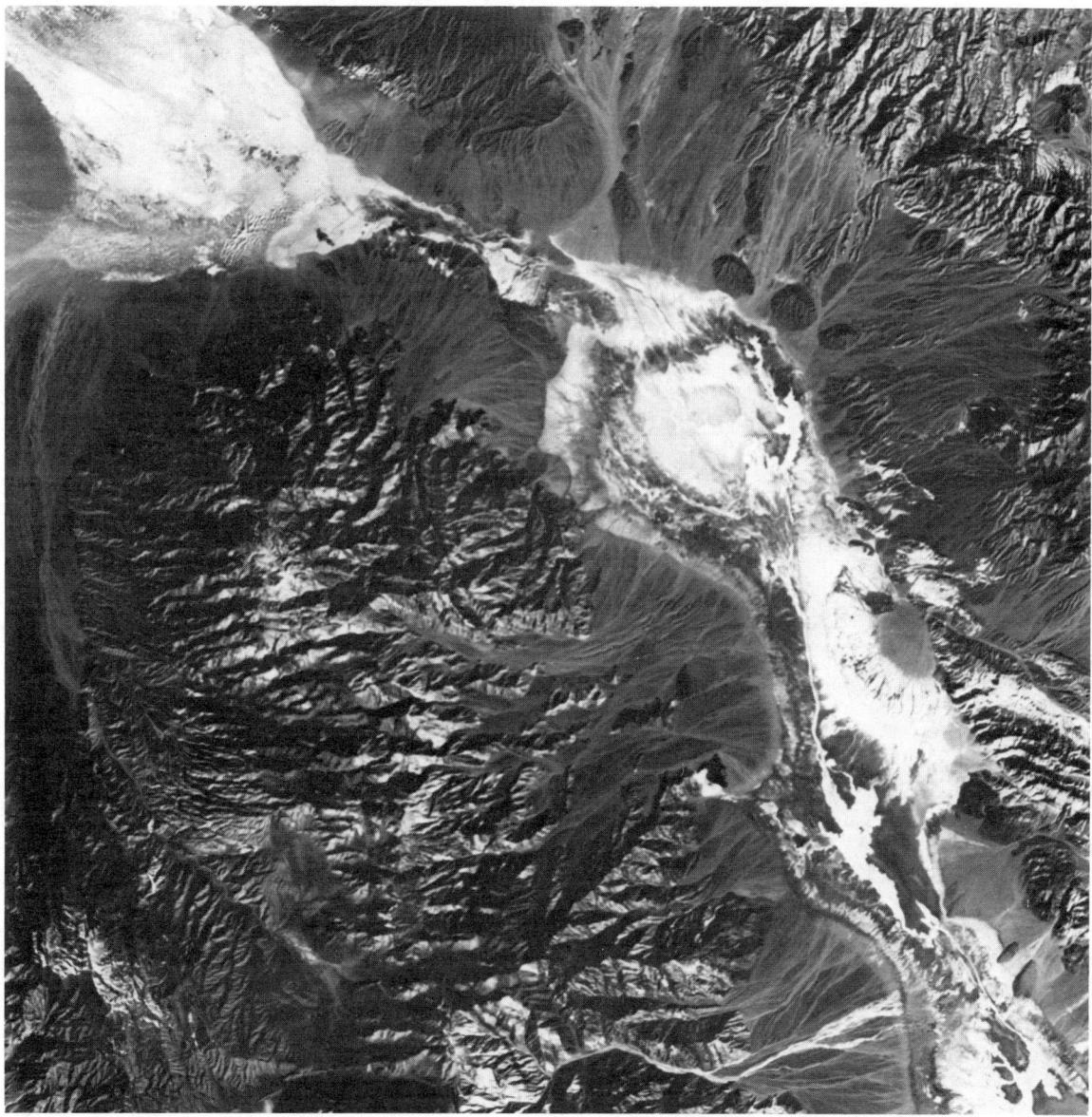

Fig. 18.15 Landsat Thematic Mapper imagery for Death Valley, California *(Courtesy USGS)*.

18.4 AERIAL PHOTOGRAPHS

Aerial photographs used for mapping generally are classified as one of three types, depending on the angle of the axis of the camera (line of sight) in relation to the ground surface. These may be classified as:

1. Vertical photographs, Fig. 18.16 (a) and 18.4, are those taken with the camera axis vertical, i.e., at right angles to the ground surface. Most photographs for mapping are of this type, since the view of the ground as seen from directly above is similar in appearance to that as seen on a map.

2. Low oblique photographs, Fig. 18.16 (b) and 18.17, are those taken with the camera axis at angles up to approximately 30 degrees from the vertical. The apparent horizon does not appear at the top of the picture.

3. High oblique photographs, Fig. 18.16 (c) and 18.18, are those taken with the camera axis approximately 60 degrees from the vertical. The apparent horizon does appear at the top of the picture.

Oblique photographs are more pictorial and more easily interpreted than vertical photographs. The main disad-

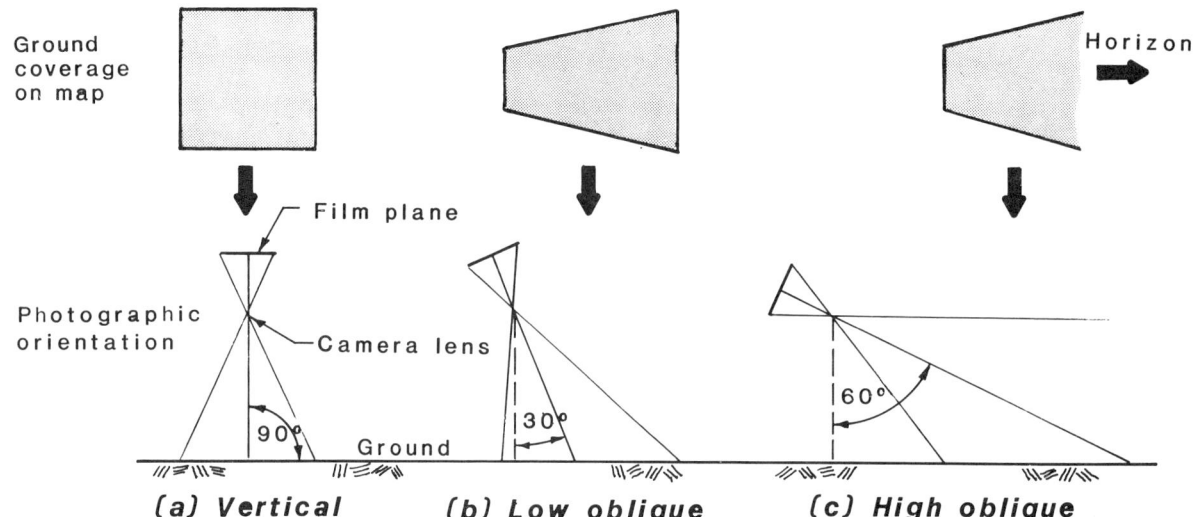

Fig. 18.16 Comparison of ground coverage for vertical and oblique aerial photographs.

Fig. 18.17 Low oblique photograph of a portion of the University of Maine at Orono campus *(Courtesy UMO)*.

Fig. 18.18 High oblique photograph of Rocky Mountain area (*Courtesy Don Comstock, USFS*).

vantage is one of scale which decreases rapidly from foreground to horizon. They are used more for reconnaissance, since a more normal view of ground objects may be seen and compared.

Aerial mapping cameras. Aerial cameras used for mapping purposes, Fig. 18.19, are precision-built with an optical distortion of less than four microns. The camera and lenses are checked for accuracy at the National Bureau of Standards at Washington, D.C. Many modern aerial camera systems also incorporate image motion compensation mechanisms to compensate for the motion of the aircraft at the instant of shutter opening. The result is a somewhat better positional accuracy of the imagery.

For topographic mapping, a six-inch focal length camera (wide angle camera) is usually preferred to provide greater relief displacement than cameras of longer focal length. Vertical distances are exaggerated, resulting in more accurate topographic maps than those from longer focal length cameras.

For planimetric mapping where minimum displacement due to topography is desired, longer focal length cameras may be used. Another way to minimize relief displacement is to use photographs taken from higher altitudes.

18.5 PHOTOGRAPHIC SCALE RELATIONSHIP

The determination and importance of map scales is discussed in chapter 3. Before a photograph can be used in drawing a map, the scale must be determined. The scale (representative fraction-RF) on a photograph may be expressed as a ratio of the focal length of the camera to the flying height above ground of the aircraft. As the flying height is increased, the scale of the photograph becomes smaller, and it becomes larger as the flying height decreases. The scale of a vertical aerial photograph is illustrated in Fig. 18.20.

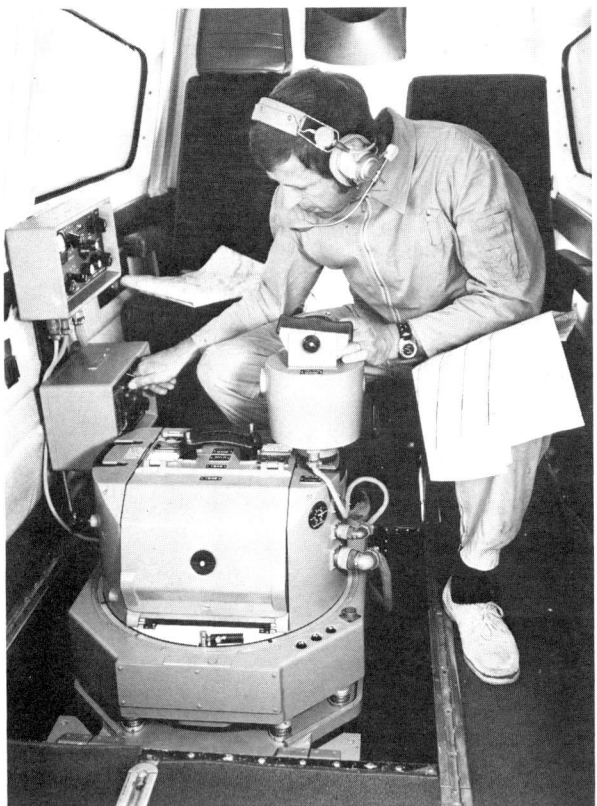

Fig. 18.19 Precision aerial camera system (*Courtesy Wild Heerbrugg Instruments, Inc*).

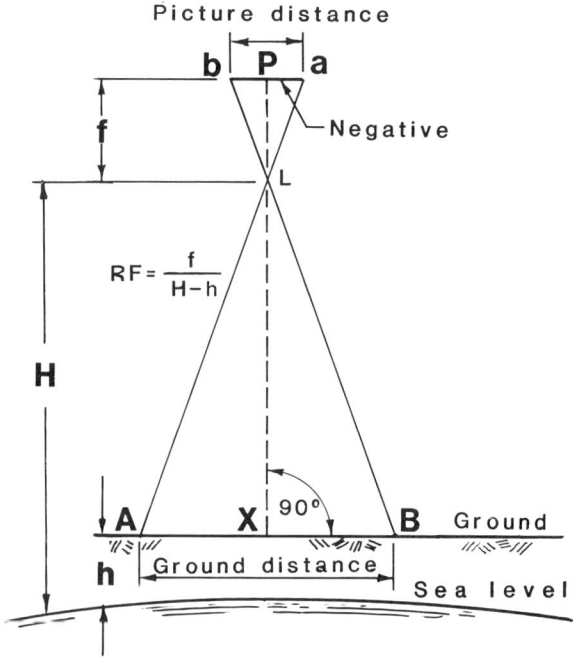

H – Camera above sea level
h – Average ground elevation
f – Focal length
X – Ground nadir
P – Picture center (principal point)
L – Camera lens
PX – Optical axis of camera

$$RF = \frac{f}{H-h}$$
where:
f = Focal length of camera
H = Altitude of aircraft (above datum, usually MSL)
h = Average ground elevation
H-h = Flying height above ground

Problem: Altitude = 4575 m (15,010'), f = 152.4 mm or 0.1524 m (6"), average ground elevation = 1525 m (5,000')

Metric: $RF = \frac{f}{H-h} = \frac{0.1524}{4575-1525} = \frac{0.1524}{3050} = \frac{1}{20\,013}$ or $\underline{1:20\,013}$ ⎤

English: $RF = \frac{f}{H-h} = \frac{6}{(15,010-5,003) \times 12} = \frac{1}{20,014}$ or $\underline{1:20,014}$ ⎦ 1:20 000

Fig. 18.20 Scale and nomenclature of a vertical photograph of flat ground.

The scale of a photograph may also be determined by scaling the distance between any two points on both a photograph and a map of the same area. It should be remembered that before the formula given below can be used, the map distance must be converted to ground distance as discussed in chapter 3.

Problem: Distance between two points on photograph = 76.2 mm or 0.0762 m (3")
Distance between same two points on map = 50.8 mm or 0.0508 m (2")
Map scale = 1:15 000

Metric: GD = 0.0508 × 15 000 = 762 m

RF (Photograph scale) $\frac{PD}{GD} = \frac{0.0762}{762} = \frac{1}{10\,000}$ or 1:10 000

English: GD = 2 × 15,000 = 30,000"

RF (Photograph scale) = $\frac{3}{30,000} = \frac{1}{10,000}$ or 1:10,000

If there is accessibility to the ground area shown on a photograph, the scale can easily be determined by measuring distances between points on the ground and relating those measurements to the photograph measurements between the same points.

18.6 DISPLACEMENT ON AERIAL PHOTOGRAPHS

Displacement on a vertical aerial photograph can best be described as the apparent movement of an object's image on the photograph in relation to its true position on the ground. In Fig. 18.21 objects at points A, B, and C when projected on a map with true scale at the datum plane will appear at points A', B, and C'. An object at point B on the ground appears at point b on the photograph. Point B is in its true datum position since it lies on the assumed datum plane where the photograph scale is true. However, points A and C are not on the datum plane and their positions on the photograph at points a and c will not show their true datum positions (that appear at A' and C' respectively).

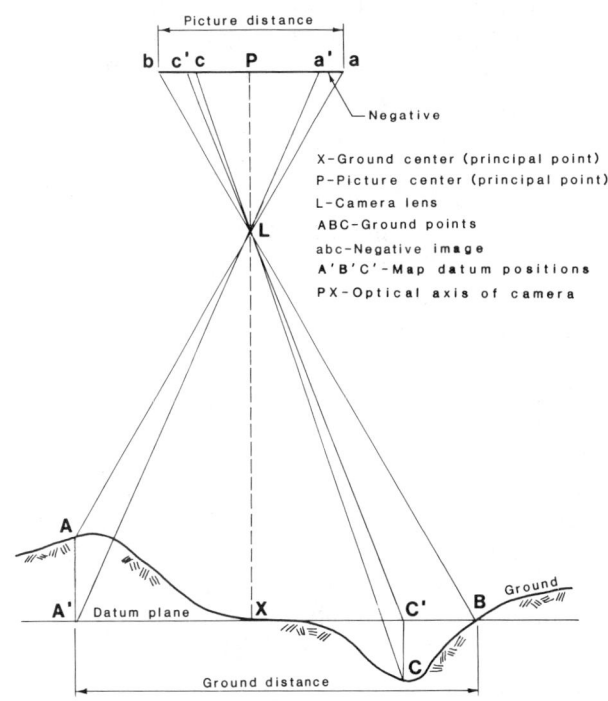

Fig 18.21 Displacement in a true vertical aerial photograph.

An object at point A is above the datum plane and is displaced outward (away from the picture center) by the distance (a' - a) on the photograph. An object at point C is below the datum plane and is displaced inward (toward the picture center) by the distance (c' - c) on the photograph.

True vertical aerial photographs rarely exist due to aircraft motions. Tilt displacement, however, is small enough in most aerial photographs so it may be ignored for most mapping purposes.

18.7 STEREOSCOPIC PRINCIPLES

Stereoscopic or binocular vision using aerial photographs is made possible by the displacement of objects on the photographs. This displacement is called *x-parallax* and is the apparent movement of one stationary object in relation to another as the viewing position changes. This can be visualized by taking a pencil and sighting along it with one eye closed so that the eye sees only the end of the pencil. If the eye is then closed and the other eye opened, without moving the pencil, it will appear to occupy a different position in space. The entire side of the pencil will also be visible. By rapidly opening and closing each eye alternately, the end of the pencil closest to the eyes appears to move in relation to the other end of the pencil, even though it is stationary. This is the parallax phenomenon.

In aerial photogrammetry parallax occurs by taking pictures of the same objects from different points in space. As the aircraft flies along the flight line, the camera shutter is timed to take pictures that have a known amount of overlap, called *endlap*. Each consecutive picture covers approximately 60 percent of the same ground area as the previous one, Fig. 18.22. Two different views of all objects on the ground are obtained in this manner. It follows then that objects at various elevations are displaced by unequal amounts (as they are different distances from the camera), giving us parallax differences that can be measured.

Determining heights of objects on vertical photographs. The displacement of objects on photographs is used to measure the height of objects and elevation differences between various points. This is also the theory in producing topographic maps from aerial photographs. If an object on a single photograph is vertical, the radial (relief) displacement of the top in relation to the bottom of the object is measured and the height calculated as illustrated in Fig. 18.23.

However, if the differences in elevation of two points on a photograph are to be measured, a stereo pair of aerial photographs is necessary. By measuring the parallax difference on the two photos, the difference in elevation of the two points may be obtained regardless of where they are located as long as they are within the 60 percent overlap of the two photographs, Fig. 18.24. When measuring parallax, the distances are always measured parallel to the line of flight (on the photo-

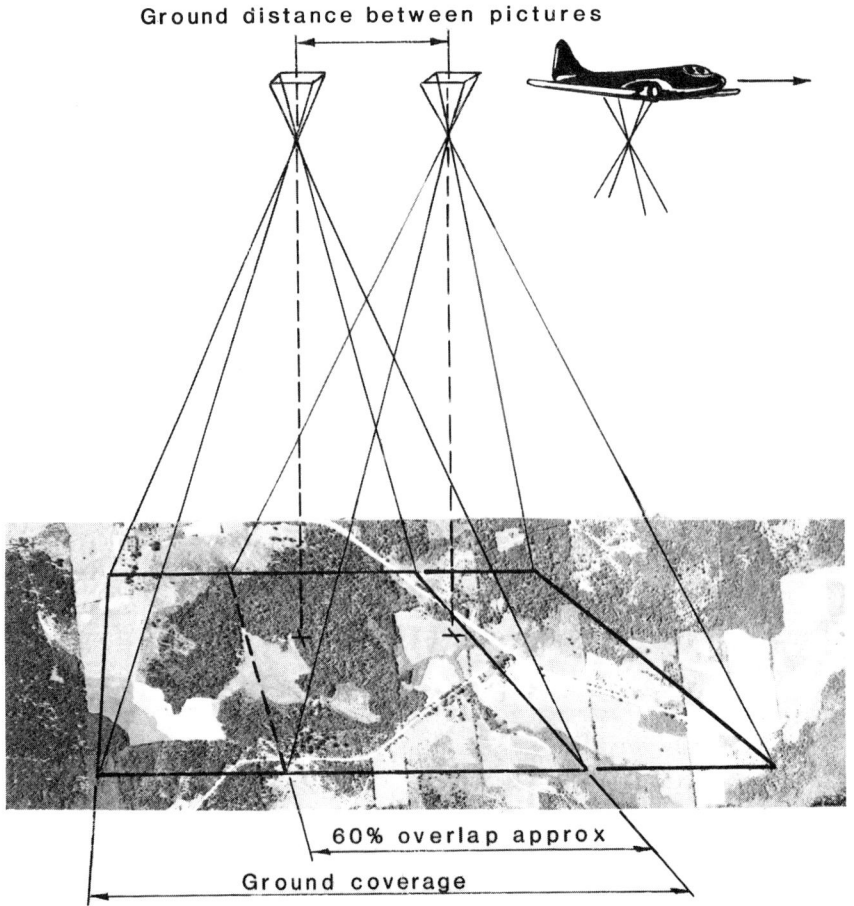

$H_o = \dfrac{d}{r}(A)$

where: H_o = height of object
 d = length of displaced image $(r - r')$
 r = radial distance from nadir (principal point to top of displaced image)
 r' = radial distance from nadir (principal point to base of displaced image)
 A = altitude (height) of camera above ground

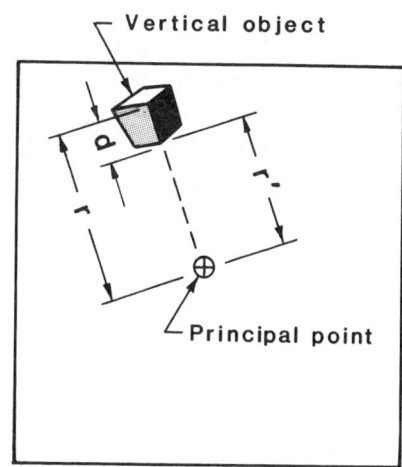

Problem: RF = 1:10 000, r = 50.8 mm (2.00"), r' = 45.2 mm (1.78"), A = 1524 m (5000') for 6" focal length camera.
What is the height (H_o) of object, Fig. 18.23?

Metric: $H_o = \dfrac{d}{r}(A) = \dfrac{5.6}{50.8}(1524) = \underline{168 \text{ m}}$

English: $H_o = \dfrac{d}{r}(A) = \dfrac{0.22}{2.00}(5000) = \underline{550'}$

Fig. 18.23 Measurement of height of an object by the radial (relief) displacement method on a single vertical aerial photograph.

graphs - lines connecting the principal points with the conjugate points).

Problem: RF = 1:10 000, H = 1524 m (5000') for 6" focal length camera,
a = 32.94 mm (1.30"), a' = 43.18 mm (1.70")
b = 38.10 mm (1.50"), b' = 50.80 mm (2.00")
What is the difference in elevation (H_o) between points A and B, Fig. 18.24?

$H_o = \dfrac{H \times dp}{p + dp}$ where:

H_o = height of object or difference in elevation between points
H = height of camera above ground datum
p = absolute stereoscopic parallax (ASP) at base of object being measured
dp = differential parallax (difference between ASP at base and ASP at top)

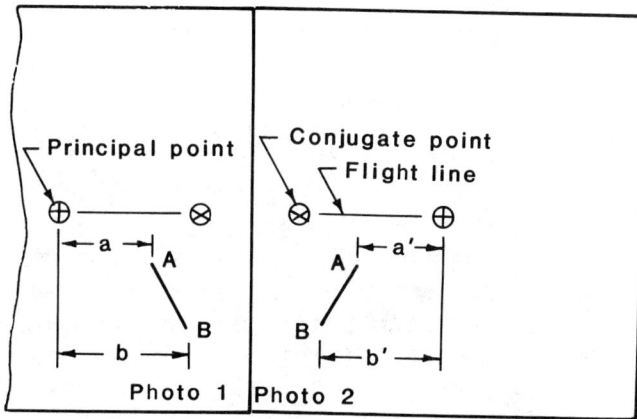

Metric: p = a + a' = 32.94 + 43.18 = 76.12
 b + b' = 38.10 + 50.80 = 88.90
 dp = 88.90 - 76.12 = 12.78
 $H_o = \dfrac{H \times dp}{p + dp} = \dfrac{1524 \times 12.78}{76.12 + 12.78} = \underline{219.1}$ m

English: p = a + a' = 1.29 + 1.70 = b + b' = 1.50 + 2.00 = 3.50
 dp = 3.50 - 3.00 = .5
 $H_o = \dfrac{H \times dp}{p + dp} = \dfrac{5000 \times .5}{3.00 + .5} = \underline{714.3'}$

Fig. 18.24 Calculation of difference in elevation of two points by means of parallax measurements.

Three-dimensional viewing of aerial photographs. It should now be clear that stereoscopic viewing of vertical aerial photographs with a stereoscope or stereomapping device is based on parallax differences in a pair of overlapping photographs to form the three-dimensional image seen by the viewer, Fig. 18.25. The two overlapping photographs correspond to the images that are seen with each eye. The three-dimensional image is formed in the mind of the observer and will appear in space.

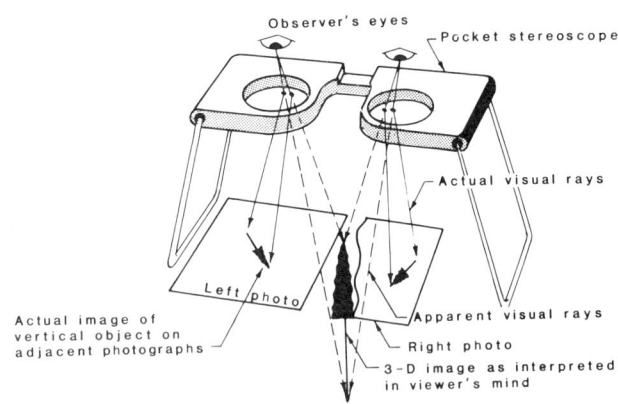

Fig. 18.25 Stereoscopic vision with the pocket stereoscope.

18.8 MAP CONTROL

An important step that must be undertaken before any mapping can be done using aerial photographs is to establish some sort of ground control. Both horizontal and vertical positions on the ground must be determined. The horizontal control is required to establish correct scale, position, and orientation. Horizontal control points that have been exactly located using surveying methods are clearly marked on the ground to be visible from the air when photographed. Vertical control is needed to establish the level datum for plotting contour lines. The elevations of selected points must be determined in the field. For each stereomodel a minimum of two horizontal and three vertical control points is required. These map-control procedures are widely used for natural resource mapping, i.e., forestry, soils, geology, etc., with aerial photographs taken as scales of 1:10 000, 1:15 840, 1:20 000, etc. For precise mapping such as engineering surveys, more accurate methods utilizing photographs taken as scales of 1:300, 1:600, 1:1200, etc., are required.

Highway survey stations markers, or USGS bench marks, if their positions can be located accurately on the photographs, are excellent control points. Survey lines which have been measured accurately, and which have identifiable stations that can be pin-pointed on the photographs, are also suitable for control. In addition to control points, other points located and marked on the photograph when using radial line plotting, Fig. 18.26, are:

1. *Principal Point*: This is the optical center of the photograph and is located by drawing a line between opposite fiducial (collimating) marks. Fiducial marks appear as ticks, notches, lines or half arrows on each of the four sides and/or the four corners of the photograph. The intersection of the lines between opposite pairs of fiducial marks locates the principal point on each aerial photograph. Each principal point must also be located on adjacent photographs.

2. *Conjugate Points*: These are the images of the principal points of adjacent photographs. Great care should be exercised to transfer the location of these points accurately.

3. *Wing Points*: A single flight line is often not adequate to give the necessary coverage of an area. Adjacent parallel flights with an overlay (sidelap) of approximately 30 percent is necessary to "tie" together the photographs taken along each flight line, Fig. 18.27. Wing points (tie points) are located along the sidelap of the photograph. Normally six of these points will be located on each picture. Wing points may be any object or point which can be clearly identified on all photographs that cover the same area. Due to parallax, an object may not appear exactly the same on all photographs,

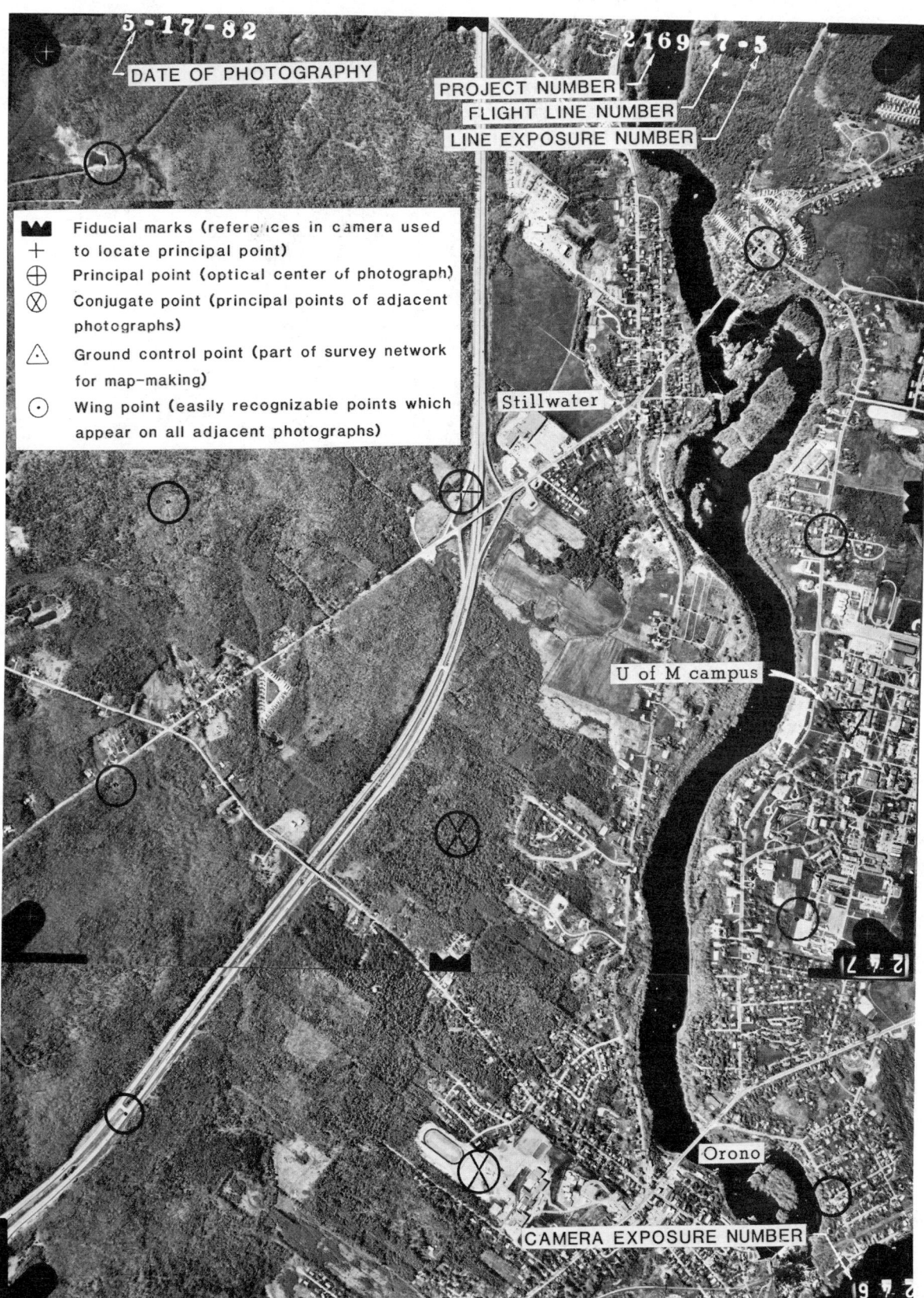

Fig. 18.26 Mosaic of two vertical aerial photographs of University of Maine at Orono campus and surrounding area illustrating nomenclature (*Courtesy James W. Sewall Co*).

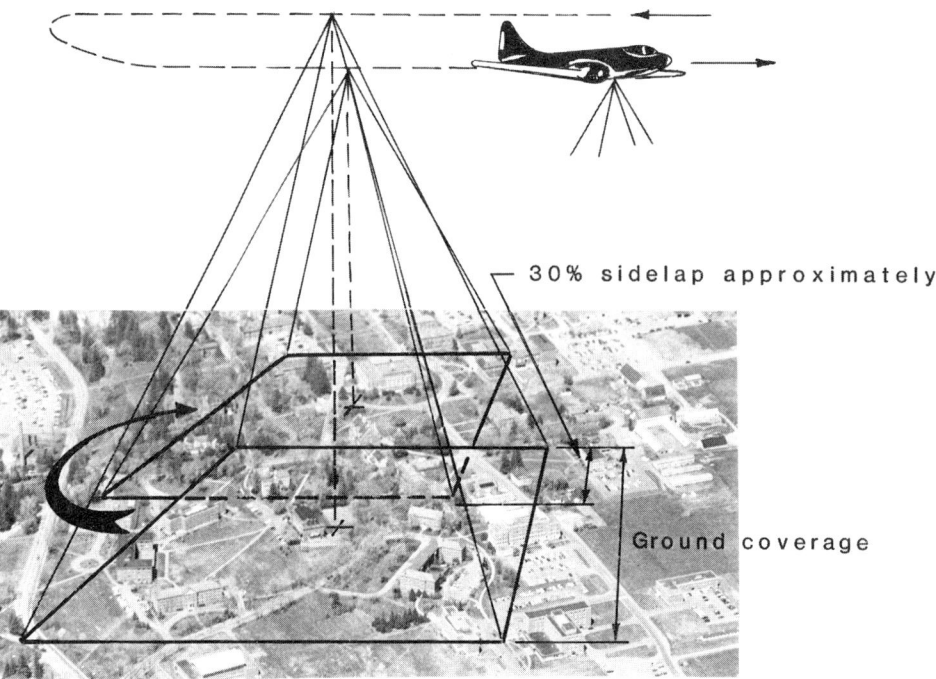

Fig. 18.27 Schematic diagram showing photographic sidelap between adjacent flight lines.

and great care must be taken to ensure that the same point (i.e., the same corner of a building or highway intersection, etc.) is located on each photograph.

Other information pertinent to the photograph will be found printed along the top or bottom. This may include date, location, time of day, focal length of camera, flying height, and job designation.

Mosaics. Several overlapping photographs may be assembled or joined together to form a *mosaic* representing a continuous picture of a larger area, Fig. 18.28. Mosaics can be classified as controlled, uncontrolled, and index. The controlled mosaic is fitted to a control plot of horizontal control points on the ground by rephotographing each vertical photograph and compensating for scale variations due to tilt and flight altitude variations. Since the center areas of the photographs are mounted directly over the control points, distances and directions may be reasonably measured. The ground features, however, where the photographs are joined may not match due to distortions and scale changes. An uncontrolled mosaic does not utilize any ground control and will have the

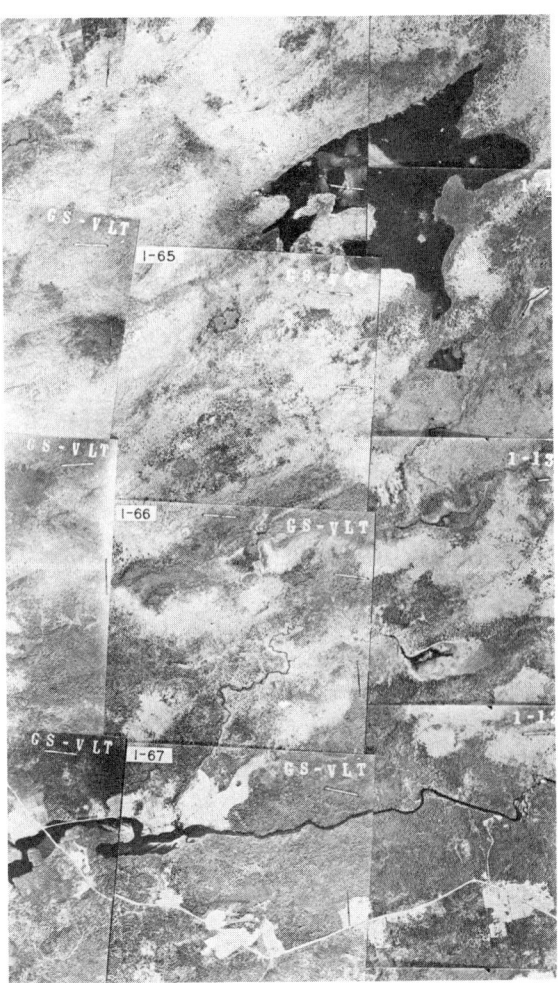

Fig. 18.28 Uncontrolled index mosaic.

border detail match. Uncontrolled mosaics indicating flight line and photo information are referred to as index mosaics. These are helpful in selecting individual photographs of specific areas for additional study.

18.9 TRANSFER OF DETAIL

Transferring detail from vertical aerial photographs to base maps can be accomplished by the use of either single photographs or stereo-pairs of photographs.

Transfer of detail using single photographs. The simplest method of transferring detail is by use of a vertical sketchmaster which uses one photograph, Fig. 18.29. It is not as precise or accurate as stereo-viewing methods, but is fast, easy, and satisfactory for filling in detail that does not need to be precisely located on the map. Prior to using the photograph in the instrument for transferring the detail to the map, it is necessary to interpret and delineate some information on the photo under a stereoscope.

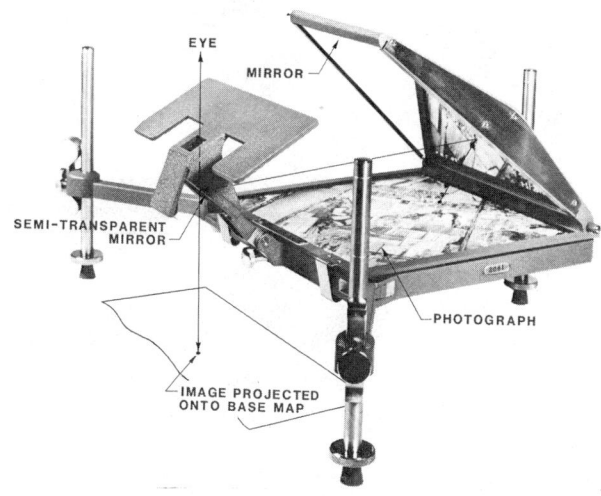

Fig. 18.29 Optical principles of vertical sketchmaster (*Courtesy Keuffel and Esser Co*).

The vertical sketchmaster makes use of a semi-transparent mirror to give the operator the impression of projecting the photographic image onto the base map. The desired information is then sketched on a 1:1 basis, a small area at a time onto the base map. Any three points of the control (including control, principal, conjugate, and wing points) are lined up by adjusting the legs of the sketchmaster up or down so the photographic image coincides with the map image. The detail within the control points is sketched, and the procedure is repeated until all the necessary detail on that photograph has been transferred to the base map.

Variable magnification transfer instruments. Instruments such as the Zoom Transfer Scope, Fig. 18.30, are used in a similar manner to the sketchmaster. The variable magnification capability of the instrument allows the photo image to be enlarged or reduced to match a base map of significantly different scale. Other advantages include built-in variable lighting, a back-lighting box for using transparencies, and the ability, through lenses and prisms, to stretch and rotate the photo image to fit the photo detail to the base map.

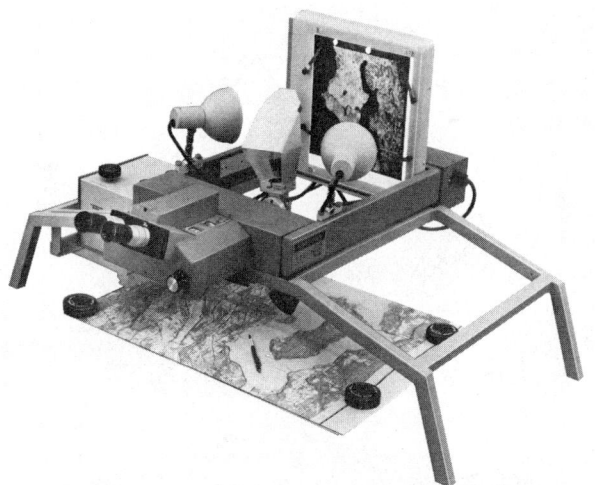

Fig. 18.30 Zoom Transfer Scope (*Courtesy Keuffel and Esser Co*).

The Stereo Zoom Transfer Scope, Fig. 18.31, utilizes stereo-pairs of photographs that allow interpretation of some detail to be done at the same time detail is transferred to the map.

For precise mapping, as in many engineering projects where large scale photographs are required, the ground control points for both horizontal and vertical control are located on a grid-system, and the photographic information transferred directly using stereo-plotting devices or storing the information on tape and creating the maps by the use of computers.

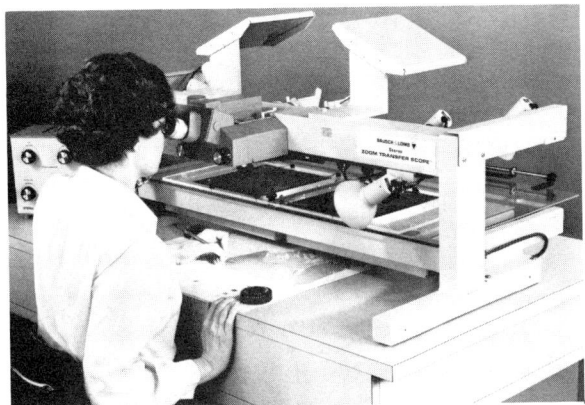

Fig. 18.31 Stereo Zoom Transfer Scope *(Courtesy Bausch and Lomb Inc)*.

Transfer of detail using stereo-pairs of photographs. Modern stereo-mapping devices are much more complex than the sketchmaster. All transfer of detail is done with stereo-pairs of photographs which are either projected onto a platen (small circular tracing table, Fig. 18.32) or viewed optically through lenses and prisms.

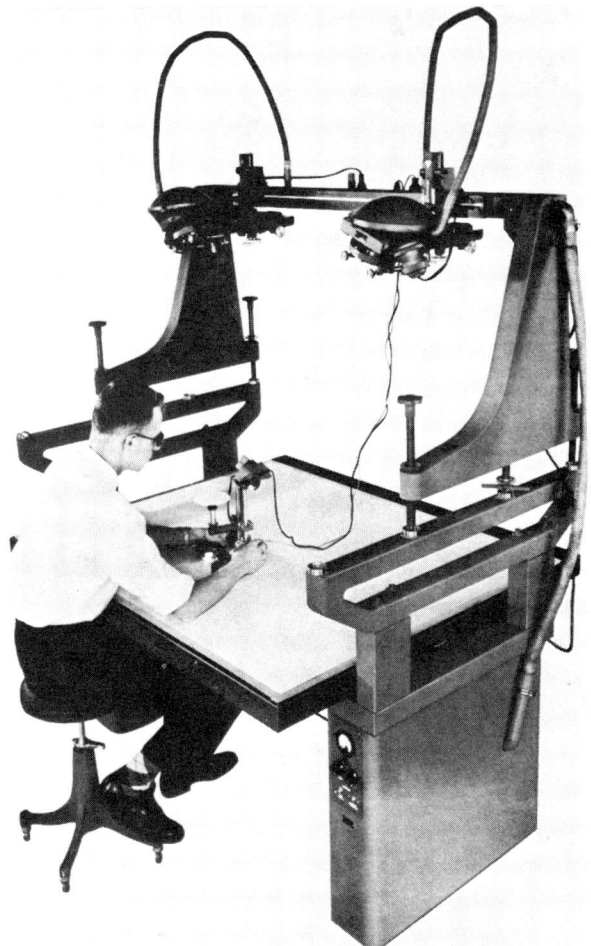

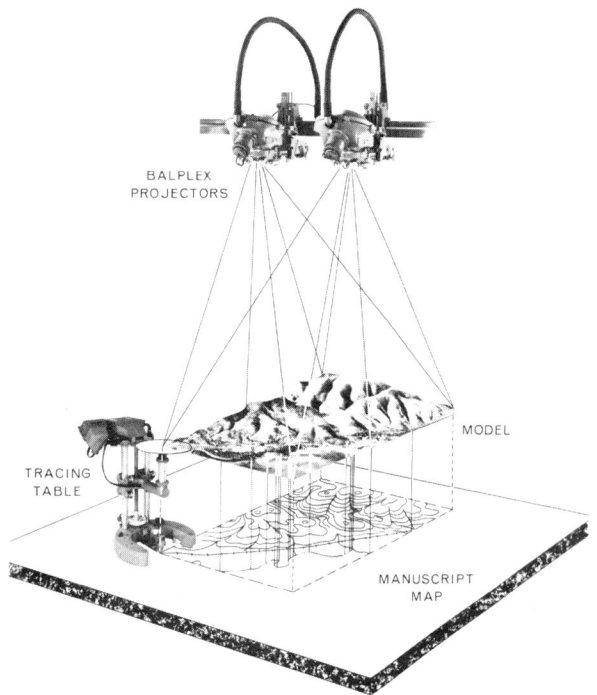

Fig. 18.32 Illustration showing principles involved in drawing manuscript map from vertical aerial photographs *(Courtesy Bausch and Lomb Inc)*.

Instruments like the Balplex and Kelsh Plotter, Fig. 18.33, project an image onto a platen by means of transparencies of the aerial photographs.

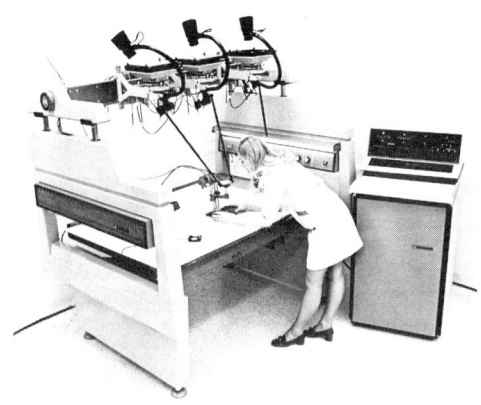

Fig. 18.33 Balplex and Kelsh plotter *(Courtesy Bausch and Lomb Inc and Keuffel and Esser Co)*.

Equipment of this type is generally manufactured in the United States. These are generally called analog plotters. Some models of these plotters are made with several projectors, but only two can be used at a time.

Aerial photos are reproduced on glass plates (diapositives) to provide the transparencies which are mounted in the projectors. One image is projected through a red filter and the other through a blue-green filter, with the operator using a pair of glasses with corresponding colored lenses to allow seeing the three-dimensional image on the platen. Newer models use polarized filters instead of colored filters, which allow the use of color or color-IR transparencies in the projectors.

In the center of the platen, a reference mark or floating dot (usually a bright point of light) appears to float in the stereoscopic model. The "dot" is kept in contact with the detail to be traced by moving the platen left-right, forward-back, and up-down. A pencil or scribe point which is located under the center of the platen traces the detail on the map manuscript (usually a polyester base drafting film) lying directly underneath, Fig. 18.34. A digitizing system can be added to permit the operator to locate a point or trace a line while digitizing simultaneously.

Photogrammetry is currently in an era where an increased use of digital techniques is involved. The analytical plotter is a revolutionary device that is renovating conventional photogrammetric practices. It is basically a digital computer that works with a stereoviewer. The digital data generated can be stored, retrieved, and displayed or analyzed using software that has been developed for a variety of applications. The system configuration illustrated in Fig. 18.35 includes the following:

1. Host computer for handling all data collection, processing, and communications with the operator.

2. CRT digital and graphical data collector with interactive graphic editing.

3. Digital stereorestitution instrument for data collection.

4. Graphics plotter for producing multicolor drafting or scribing of alphanumerics and graphics.

An analog stereoplotter either optically or mechanically reconstructs a rendition of the camera interior and exterior geometry as well as the objects being observed. It should be pointed

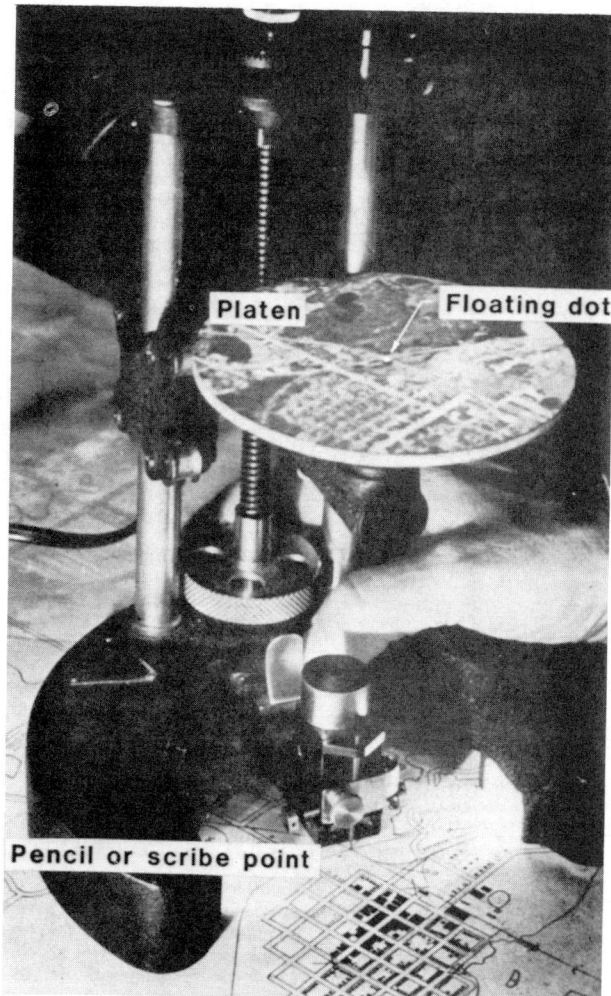

Fig. 18.34 Platten detail (Courtesy USGS).

out that plotters of this type are not obsolete, however, the transition to analytical plotters is underway. The use of an analytical plotter involves measuring picture coordinates of well-distributed ground control points. The measured picture coordinates are related to their known ground survey coordinates through a mathematical function (model). The function contains parameters to define the position and orientation of the photograph at the time it was taken. These parameters include elements of interior and exterior orientation, lens distortion, refraction, comparator errors, and emulsion (film) distortion. After comparing the survey and picture coordinates of each control point, the parameters can be determined from an adjustment computation. The survey coordinates of unknown stations can now be calculated from mathematical relationships using the orientation and other

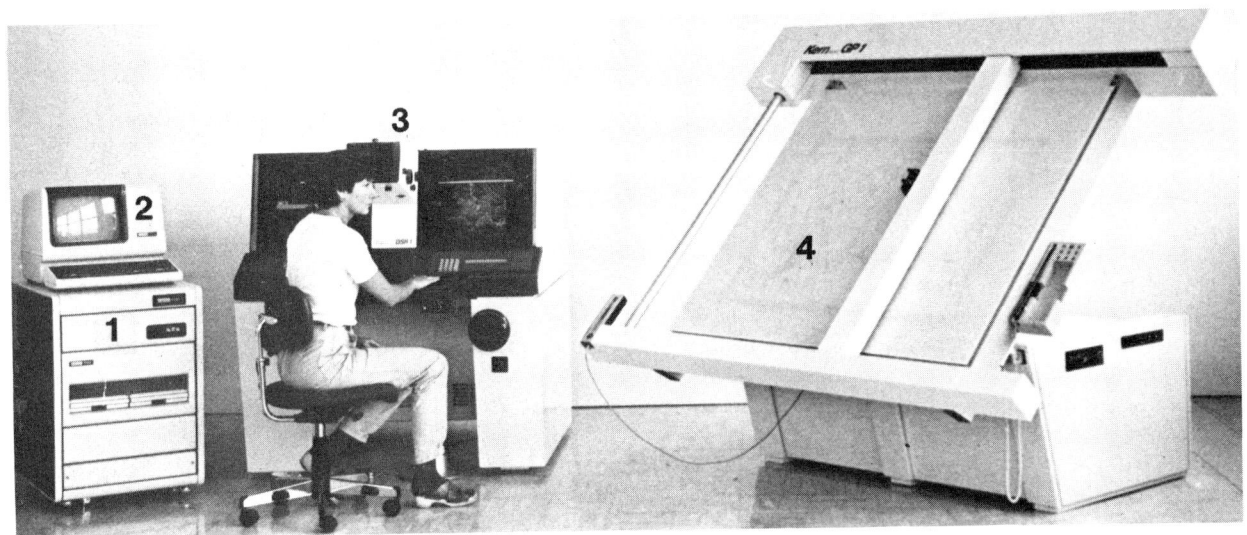

Fig. 18.35 Typical analytical stereoplotter configuration (*Courtesy Kern Instruments, Inc*).

distortion parameters just determined and the measured picture coordinates of the unknown stations.

At the present time it is economical to incorporate an analytical plotter into any mapping system. The main advantage is their adaptability to automation and the flexibility of using a wide range of input data types and combinations. Devices such as mass storage media and computer graphic devices make it possible for the user to do a considerable amount of adaptation to meet particular problems and requirements.

The analytical plotter can be used to perform any traditional stereoplotting operation. In addition it can be used as a stereocomparator (a stereoinstrument to measure x, y image coordinates) or as a stereoplotter in aerial triangulation (the precise adjustment of stereoscopic models to ground control). Since it is extremely accurate, photographs can be taken at higher altitudes. They can be from different cameras since the interior orientation is reconstructed for any calibrated camera and corrections made for all systematic errors mathematically. In map revision and collection of different types of information, a model can be quickly reset from digital data resulting in cost effectiveness. Other uses include digital terrain data acquisition, digitizing, cadastral applications, and remote sensing.

A skilled photogrammetrist or photo-interpreter can obtain a great deal of information from aerial photographs, utilizing sophisticated equipment and computers developed in the last few years. Many types of maps to satisfy almost any need are possible, as illustrated in the example of Fig. 18.36.

18.10 ORTHOPHOTOGRAPHY

In addition to conventional maps, a variety of photomaps is being produced. These are maps made by adding information such as a reference system, marginal information, and descriptive data to a photograph or assembly or photographs. Since scale distortions exist, the pictures are often altered to produce a number of different kinds of orthographic products. These are:

1. <u>Orthophotograph</u>. A vertical aerial photograph having the properties of an orthographic projection, Fig. 18.37. The photograph is derived optically from a conventional perspective photograph through a rectification or adjustment process so that image displacements caused by camera tilt and terrain relief are removed. A computer integrated system utilizing an orthoprojector, Fig. 18.38, accomplishes this by determining the mathematical relationships between ground coordi-

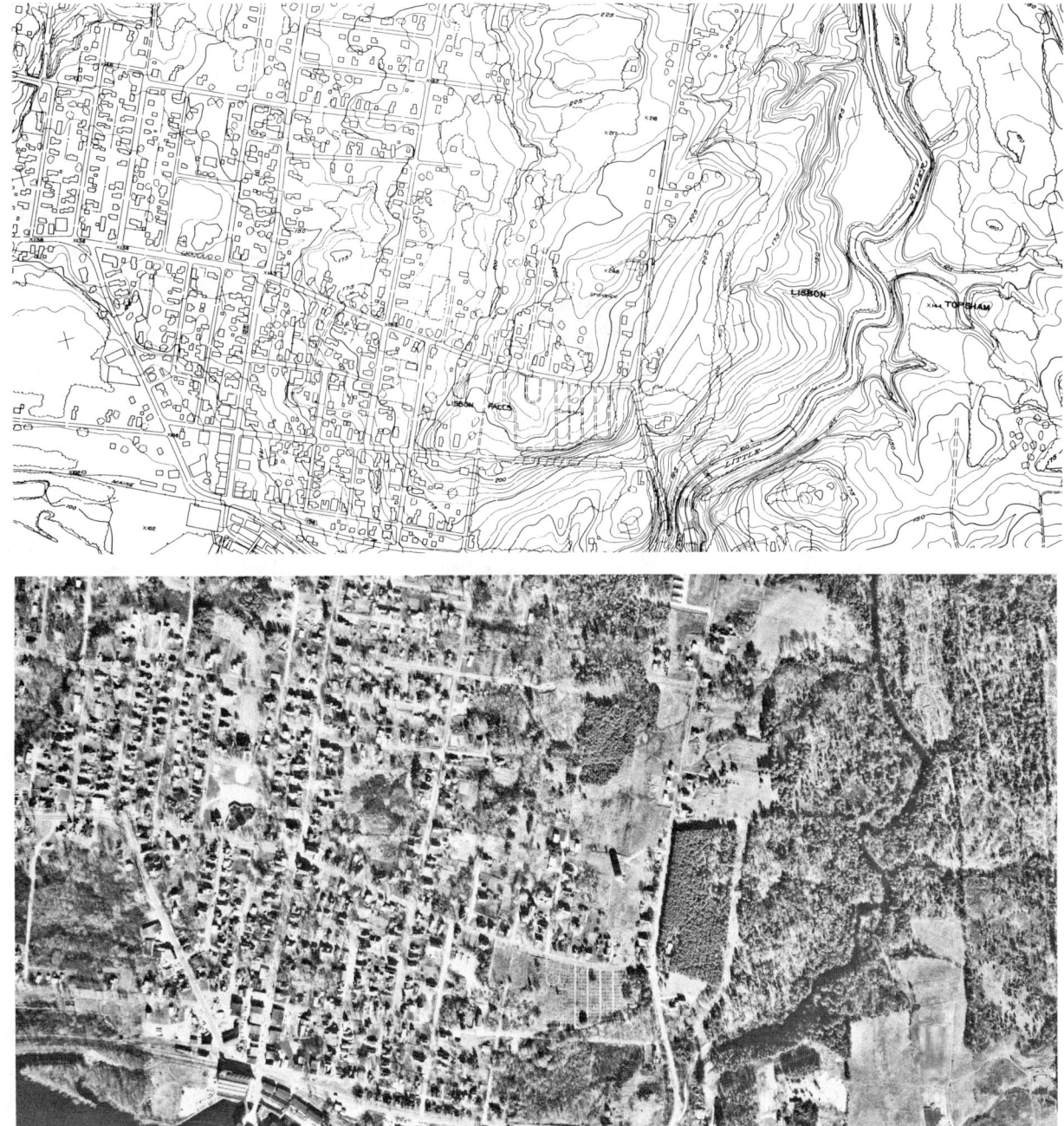

Fig. 18.36 Portion of a topographic map and aerial photograph of the same area (*Courtesy James W. Sewall Co*).

nates and coordinates in the unrectified photograph.

The orthophotograph combines the advantages of true map position along with a wealth of information that can be seen on a photograph, much of which is normally not shown on conventional line maps. The vast amount of detail in the aerial photograph lends itself to multi-purpose uses such as cadastral mapping, resource inventory, and route location planning.

2. <u>Orthophotomap</u>. A photomap made from an assembly of orthophotographs, Fig. 18.39. It may appear in standard quadrangle format and incorporate special cartographic treatment through the addition of contour lines, roads, and boundaries along with colors for features such as water and vegetation. The orthophotomap is ideal for showing complete ground detail for areas where the terrain is relatively flat, sparse vegetation

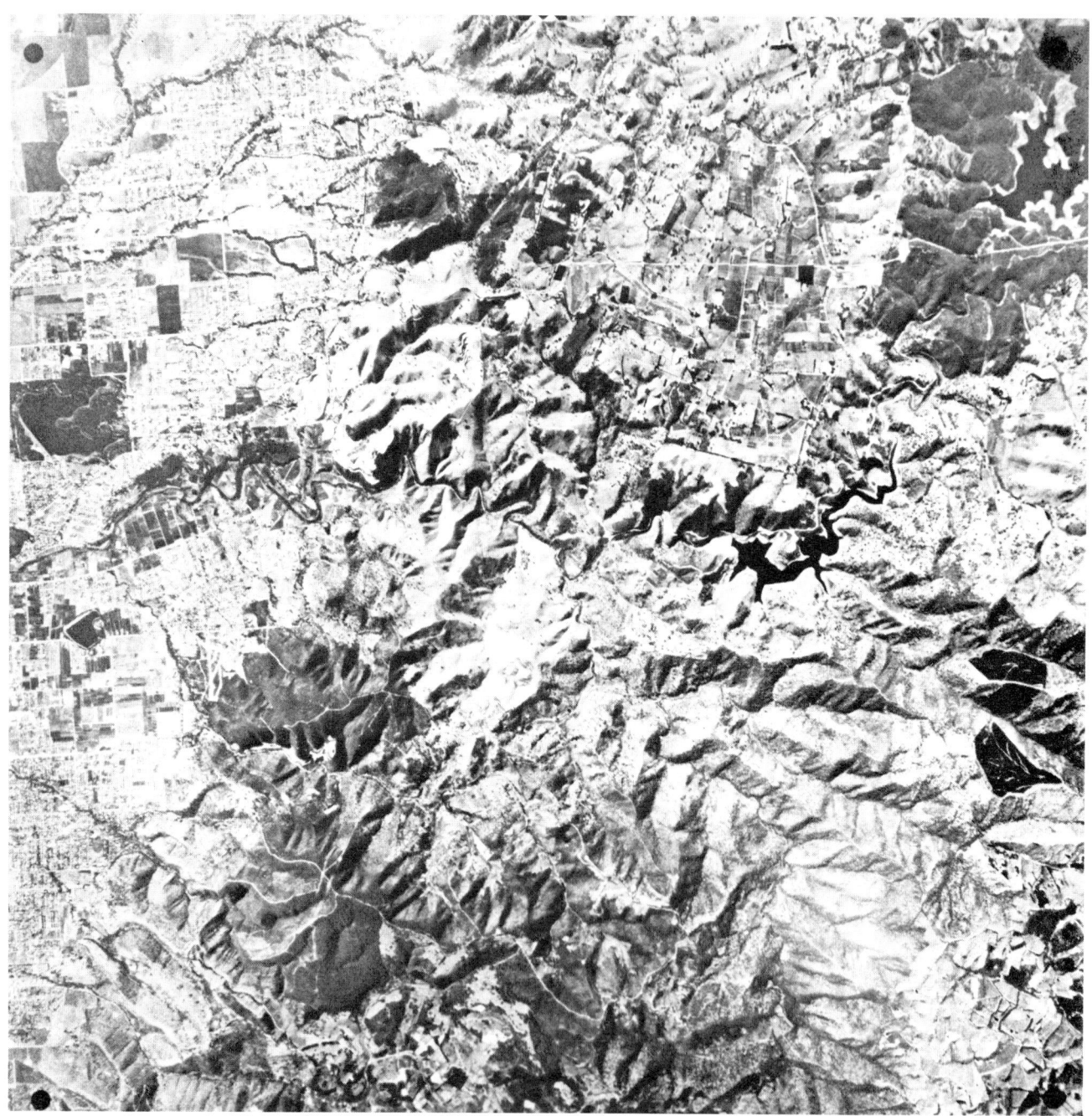
Fig. 18.37 Orthophotograph *(Courtesy Carl Zeiss, Inc)*.

347

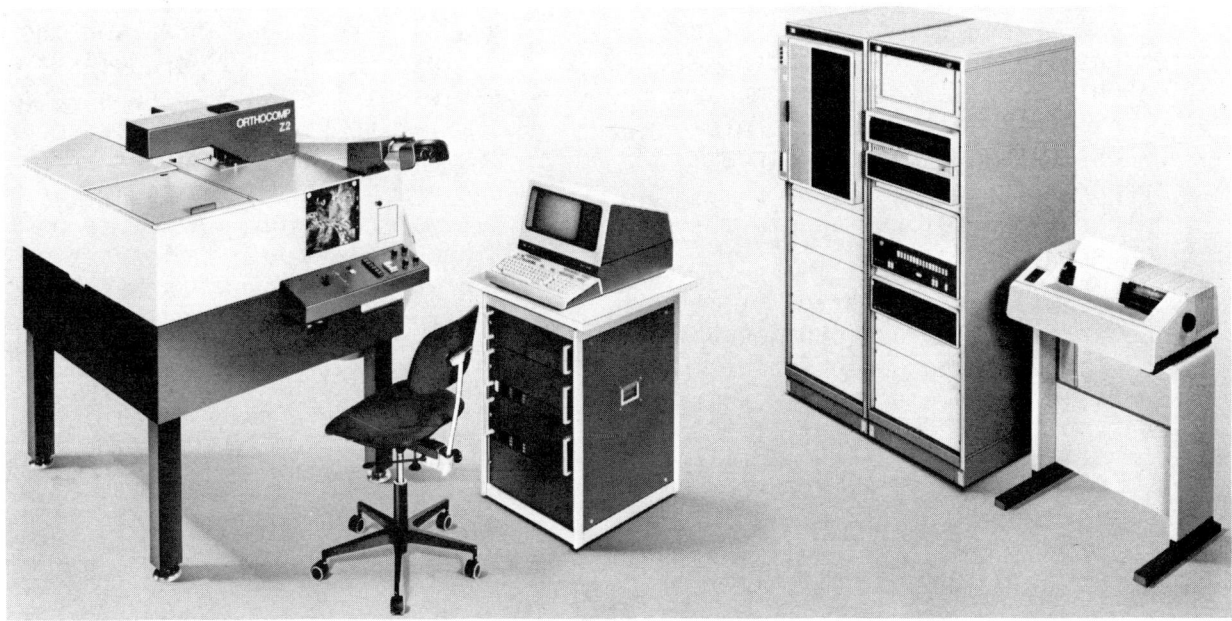

Fig. 18.38 Analytical orthoprojector basic system components *(Courtesy Carl Zeiss, Inc)*.

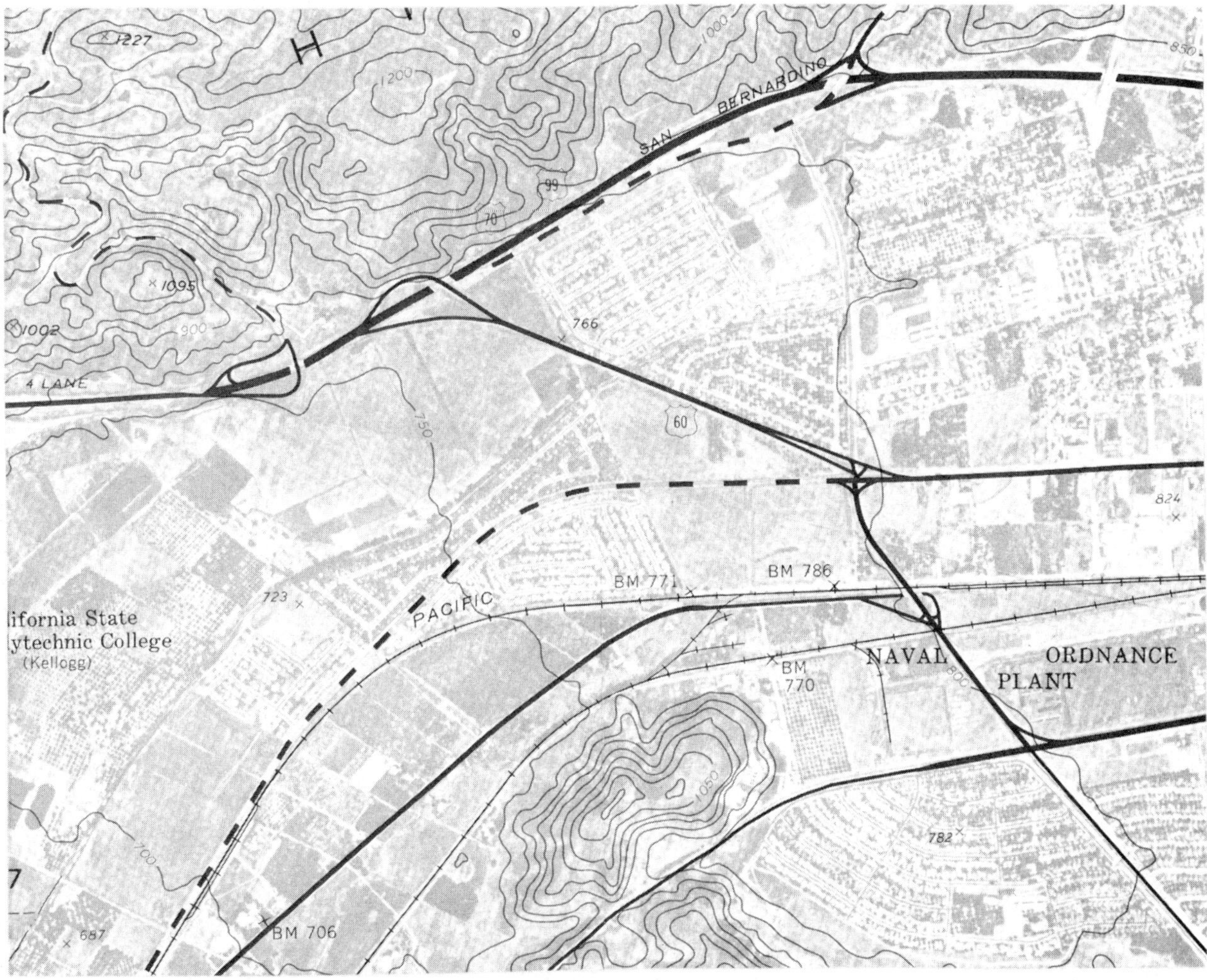

Fig. 18.39 Portion of 1:24 000 scale orthophotomap of San Dimas quadrangle *(Courtesy USGS)*.

occurs, and few cultural features appear. In areas of high relief, extensive vegetation, concentrated cultural features, and dense planimetric detail, orthophotographs are not suitable since overprinting of lines, words, numbers, and colors tend to obscure the photoimage.

3. <u>Orthophotoquad</u>. A series of monocolor rectified orthophotographs in standard quadrangle format with grid lines and place names added. No contour lines or other cartographic treatment appear on the pictures. The USGS produces a series of 1:24 000 orthophotoquads similar to Fig. 18.40. These are useful for unmapped areas and locations where line maps are in need of revision. A number of State and other agencies are using orthophotoquads as base maps for informational studies related to utilities, development and conservation of natural resources, pollution, and crop inventories.

Fig. 18.40 Portion of orthophotoquad for Mount St Helens quadrangle, WA *(Courtesy USGS)*.

COMPUTER-ASSISTED CARTOGRAPHY

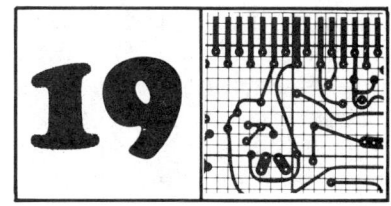

19.1 INTRODUCTION

Computer-assisted cartography (CAC) refers to the generation of information that is required for a computer to create, store, and manipulate a drawing or picture. The nation's increased interest in land and resources has created a rising demand for maps and survey information. In response to this increased need, the application of computer technology to cartographic and geographic information in a variety of modeling and analytical systems is growing.

The use of the computer in man's activities is very diverse and its application is as varied as the imagination. In cartographic related activities the outcome has been an explosion of creative new map forms. It is being used in making management decisions, estimating resource inventories, environmental protection, energy related activities, air pollution mapping, land-use studies, theoretical geography, transportation, dynamic weather patterns, demographic distributions, and the list goes on.

Traditionally, maps have been drawn basically by hand, requiring a great amount of time and expense. This is rapidly giving way to new technological developments employing the use of computers and other computer associated devices. The collection of data, equipment, and programs to produce cartographic products can be expensive. In this regard, it should be emphasized that small non-repetitive kinds of maps can still be produced by hand at less cost than by computer-assistance.

Computer-assisted mapping has, however, minimized the effort and time to produce map products through the use of a number of powerful programs that have been developed. Large amounts of data can be stored in the system, processed as rapidly as required, and a map of any form containing the required elements can be generated in seconds. Functions such as scaling and rotation may be applied to the map. Three-dimensional projections may be accomplished using a single command. Any repetitive operation can be performed automatically through the use of operator-defined groups of commands. Accuracy is greatly improved since the operator is able to examine and make corrections using the computer system's extensive edit capability before a product is drafted, rather than after it has been completed. After a map has been created, it can be stored within the system for later use, transferred to other storage devices for subsequent use, or drawn by computer devices for immediate use.

The use of computer mapping programs is advantageous for:

1. Producing working maps in a hurry by simply inputting data values into existing mapping routines.

2. Generating a series of maps depicting spatial patterns or regional variations for a large number of variables in a single set of areal values.

3. Displaying maps in a form that would be prohibitive in time, expensive to produce, and extremely difficult to design and draw by conventional means.

4. Drawing maps having repeated proportional symbols that must be calculated.

Even though the computer and its ancillary equipment relieve the need to carry out many repetitive tasks and make it easy to retrieve data and drawings, it is still the individual's creativity and experience that bring about the programs that allow the computer to do its job. The quality of the map can be no better than the quality of the data and how it is input into the computer.

This chapter is presented for individuals to gain an awareness of computer-assisted mapping developments along with the basic principles for its use.

19.2 COMPUTER-ASSISTED MAPPING SYSTEMS

The physical equipment in computer systems is called *hardware*. At the heart of a graphics hardware system is the digital computer. The computer receives (input) data from different sources. The data is then stored in memory banks for the computer to process the data and retain the final results (output) for future use. A careful study of Fig. 19.1 will help to understand the flow of operations in a typical computer-assisted mapping system.

Every year more sophisticated equipment is becoming available. New manufacturers are developing equipment in anticipation of future market sales. In many instances much of this equipment has not received user approval or is not compatible with existing equipment. It is recommended before purchasing and investing in any system that individual requirements be carefully ascertained, that the experience of other users be obtained, and that the manufacturer is in a position to provide com-

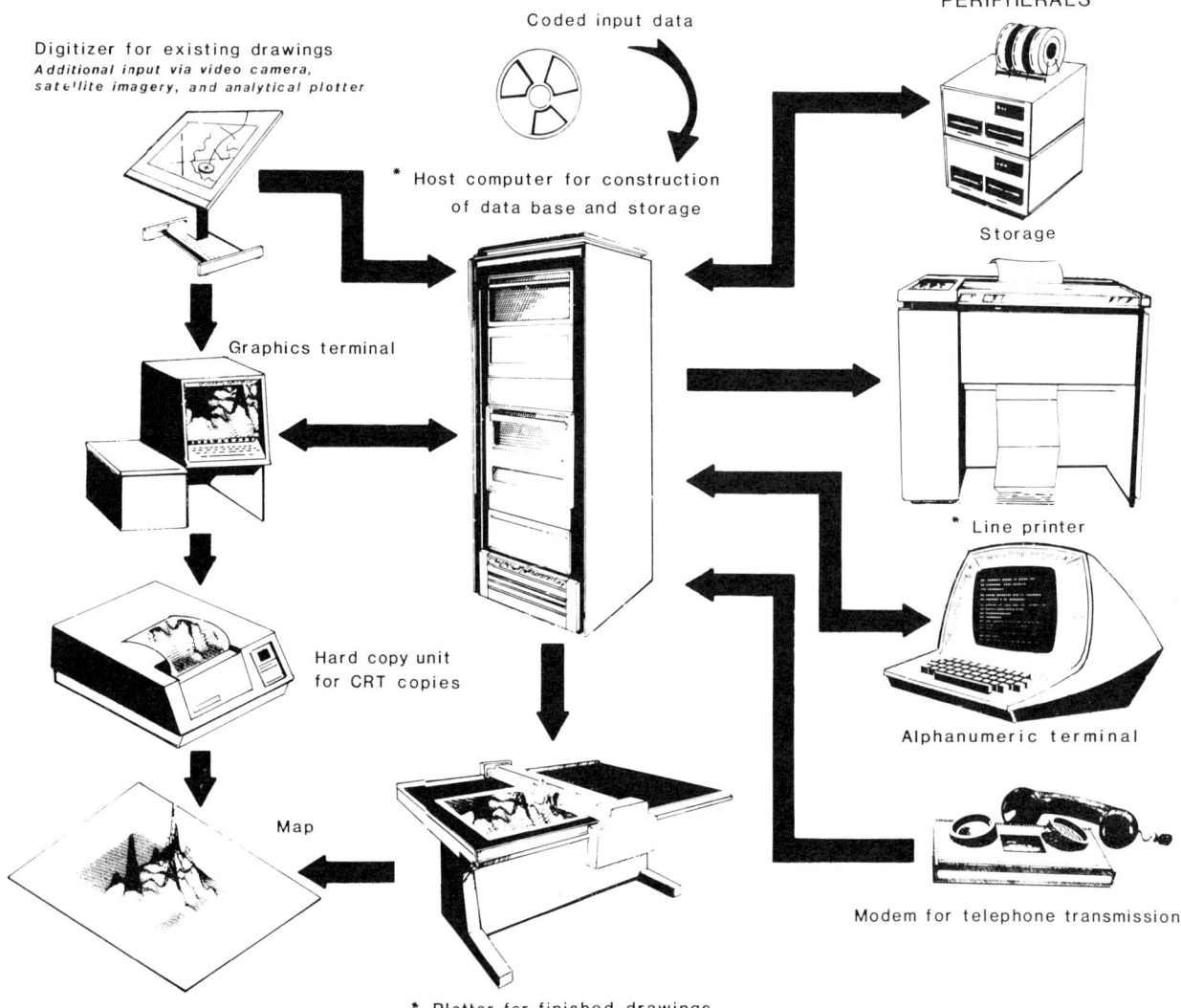

Fig. 19.1 Computer-assisted mapping system (*Components Courtesy Technical Advisors, Inc*).

plete support back-up. An excellent source containing a comprehensive listing of computer graphics supplies is *The S. Klein Directory of Computer Graphics Suppliers* - Hardware, Software, Systems and Services, Technology and Business Communications, Inc., Sudbury, MA. This directory is updated on an annual basis.

A number of manufacturers market and support a complete line of interactive graphics systems that allows for man-machine communication to enter data and to direct the course of a program. These include graphics workstations, communication interfaces, high performance peripherals, and special purpose vector graphics processors. The graphic workstations offer high resolution color and monochromatic raster display units, screen hardcopy devices, and data capture interfaces, including stereoplotter interfaces. Peripherals include tape drives, large mass storage devices, and high resolution pen and electrostatic plotters.

The main components in a stand-alone computerized stereo graphics system shown in Fig. 19.2 are as follows:

1. Line printer for alphanumeric output.
2. Computer and supporting software for processing data.
3. Color graphics terminals for image display.
4. Alphanumeric terminal for input-output.
5. Stereoplotter for map compilation using aerial photography.
6. Flatbed plotter for drawing finished product.

A number of manufacturers produce turnkey interactive graphics systems. These are ready-to-use systems offering an on-site survey, installation, training, documentation, and a hardware maintenance and software update service.

Although hardware can be expensive, costs have plummeted and a number of mini-computers, plotters, and other units are now available at relatively modest prices. For many years the development of programs, procedures, and techniques (software) for directing the hardware to perform desired functions has trailed hardware innovation. This is no longer true and many software packages are available for mapping and surveying tasks. Some companies tailor hardware systems and software packages to meet individual needs.

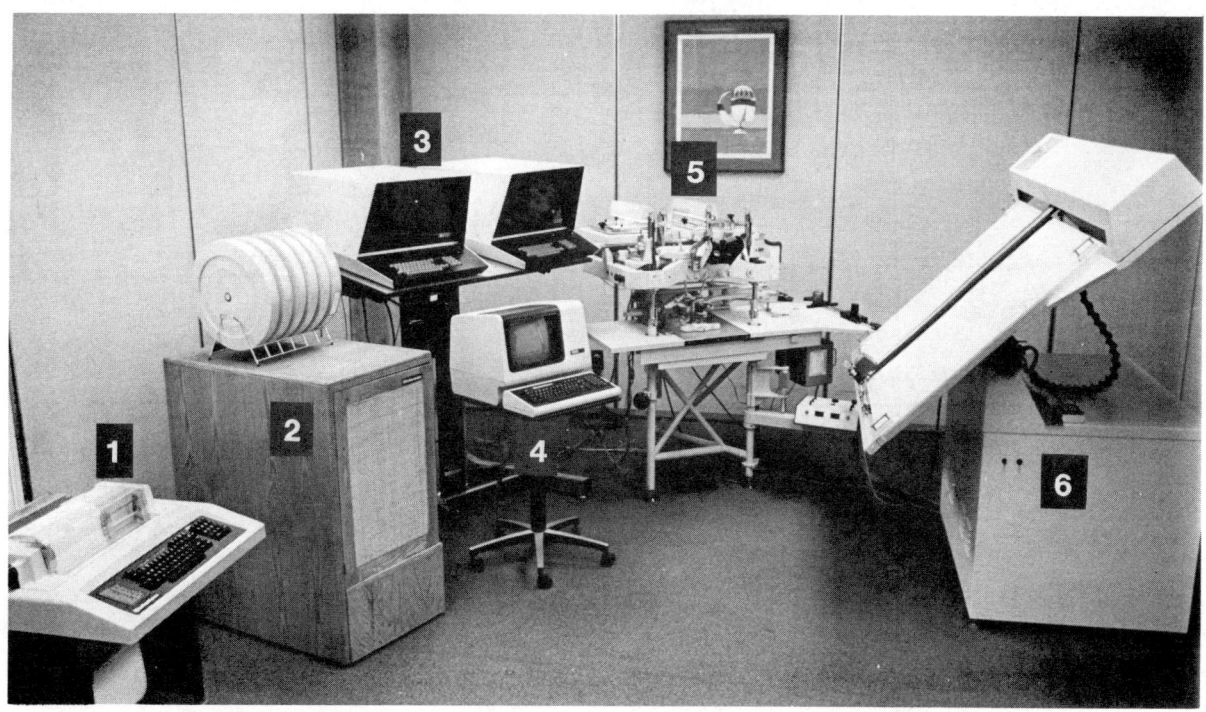

Fig. 19.2 Integrated computer graphics processing components *(Courtesy James W. Sewall Co and Kork Systems, Inc).*

19.3 COMPUTER INPUT DEVICES

Input refers to the compilation devices used to transfer information to the computer. The principal method for the input of cartographic information is to use a digitizer, Fig. 18.3 (a). Other input methods include decwriters and modems or acoustic couplers for telephone line transmission of graphic data, Fig. 19.4. Additional input information to the computer may be effectively accomplished in an interactive mode with graphics or alphanumeric terminals and peripheral devices such as light pens, joysticks, graphic tablets, video cameras, laser scanners, microdensitometers, and digital imaging sensors.

Digitizers. A digitizer consists of a work surface, a remote electronics unit (controller), and a transducer (pen or multi-button cursor), Fig. 19.3 (b). The work surface may be a solid table or a backlighted or rear-projected surface. The cursor is used to electronically encode data to a digital form. Features and functions may be directly controlled using the keys for initiation of programs, entering labels, controlling output device selection, and data recording. The cursor is basically a tracing head with cross hairs or a bulls-eye, under a magnifying lens, that may be placed over a point or moved to follow a line. The position of the points along the line to be traced is translated to digits (X-Y coordinate values) for the computer. The location of points is sensed electrically by horizontal and vertical wires in a grid under the drawing table's hard surface, or by acoustic or electromagnetic sensors.

A tablet menu, that consists of a list of features or functions commonly in symbolic form, is used for recording the nature of the feature or function at the time it is encoded, Fig. 19.3 (c). Menus vary widely in format with many organizations generating their own menus to meet specifications. The menu may be placed anywhere on the working surface. When the cursor is placed over any menu block, an appropriate button is pressed and the control unit will then translate the coordinates of that point into the correct feature or function code. A coded record of point features may include a symbol and location. Line features may be completely described in coded form to include such items as thickness, color, line type (solid, dashed, double, etc.), and function (river, boundary, timber line, etc.). Digitized information coded in this man-

(a) Digitizer

(b) Cursor

(c) Menu

Fig. 19.3 Digitizer *(a-Courtesy Hitachi, b-Courtesy UMO Forest Resources Graphics Laboratory, c-Courtesy Great Northern Paper Co).*

ner may be reproduced in the prescribed form on a plotter by subsequent processing in the computer.

Digital map data falls into three basic categories; point data, line data, and areal data. Point data includes any feature that can be located by a point and recorded by coordinates such as control stations and boundary monuments. Line data involves linear features that include transportation networks and streams. These are recorded by sequences of closely spaced coordinates. Areal data is that which requires a tint or pattern, such as vegetation and urban areas. The boundaries of areas can be defined by lines. Each data type represented, e.g., boundaries, water, buildings, etc., is digitized and can be stored as a separate layer or file in the data base.

In digitizing, three methods are employed, Fig. 19.5. Point of manual digitizing is satisfactory for most types of graphic display. In digitizing points, lines, or areas using the point by point mode, the cursor is positioned over a point and its coordinates are electronically measured and recorded directly into the computer. A continuous mode may be used that will allow points to be recorded in incre-

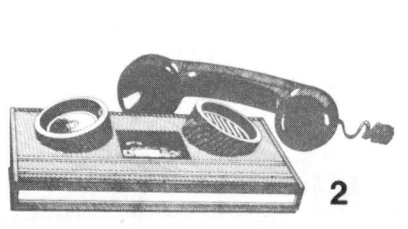
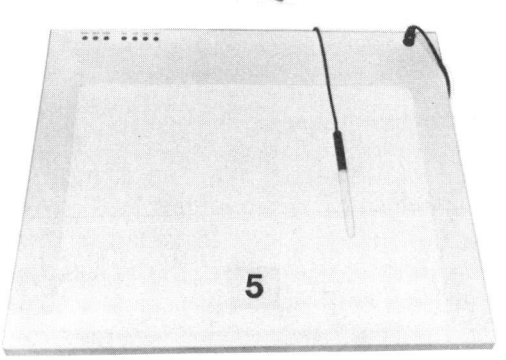

Fig. 19.4 Computer input devices: 1-decwriter *(Courtesy Digital Equipment Corp)*, 2-modem, 3-CRT graphics terminal *(Courtesy UMO Forest Resources Graphics Laboratory)*, 4-joystick, and 5-graphic tablet *(Courtesy Hitachi)*.

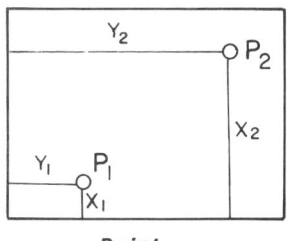 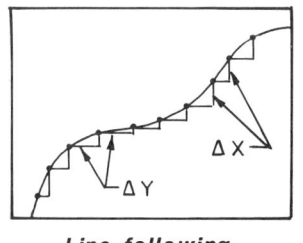 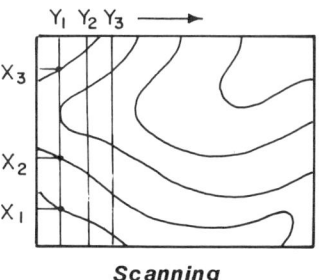

Fig. 19.5 Digitizing methods.

ments of time or distance that the cursor is moved. Accuracies of ± 0.0 - 0.2 mm (± 0.004 - 0.008 inch) using line-following or lock-on techniques and light-sensing scanning methods are finding an increased use.

Lock-on devices identify and follow the graphic image recording coordinate values at regular intervals. A photocell responds to the differences between the image element and the background or blank space to allow the detector to automatically hold onto the image element. Problems frequently occur with lines of different widths and at line intersections since the detector cannot decide which line direction to follow. In some systems the digitizer pauses when an intersection is reached. The operator may then reposition the cursor to track another line automatically until another intersection point is reached.

Scanning type digitizers show more promise than lock-on digitizers. The image is scanned on a strip basis as small as 0.25 mm (0.0001 inch) in width. As the scan line moves across the image parallel to one of the coordinate axes, all points where the scan line crosses a line or point are recorded.

Today as in pre-computer days, maps are prepared from photo-pairs using a stereoplotter. The stereoplotter operator views the three-dimensional image of the aerial photograph on a tracing table. Instead of drawing lines, the operator now moves a digitizer over the projected image, and this automatically transfers the X-Y coordinate values of all photograph features traced to computer files. Elevation information can be recorded as a third coordinate or Z value. The information on magnetic tape or disc storage is processed by the computer, creating a geographic data base.

From this data base, any desired map can be plotted automatically by the computer.

Graphics terminal. The graphics terminal, Fig. 19.6, is an interactive device that may be used for both input and output (see also section 19.4 on cathode ray tube terminals--CRT's). Data may be typed from the keyboard, verified, edited if necessary, and then transferred on demand to the computer's main memory. Input to the computer may also be achieved using a light pen, joystick, or graphic tablet, Fig. 19.7. These devices enable the user to position, enter, or delete information directly into the computer without having to use the keyboard.

Fig. 19.6 Graphics terminal (*Courtesy IBM*).

The light pen may be used to create a drawing by touching the face of a cathode ray tube with the pen. A photosensor in the light pen detects the exact time at which the raster scan electron beam of the CRT passes the location on the screen pointed to by the operator. Since the starting time and the rate of travel of the electron beam are known, the elapsed time from the start of the scan can be converted to the plane coordinates of the indicated point.

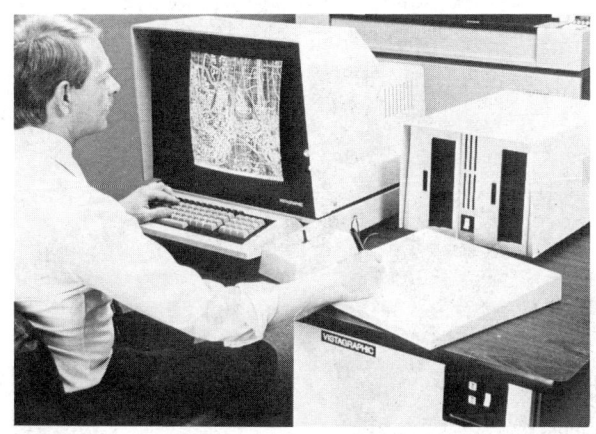

Fig. 19.7 Graphics terminal input via keyboard, joystick, or graphic tablet (*Courtesy Calcomp*).

Fig. 19.8 Voice recognition input-output system (*Courtesy Bartlett Associates, Inc*).

The movement of the pen across the face of the terminal allows the lines to be drawn. A menu containing symbols and instructions from the computer may be displayed on the terminal screen. Any item on the menu can be pointed to with the light pen and the computer will respond to those commands. For example, with the light pen a signal could be given to either enlarge, reduce, or delete a portion of the programmed image displayed. Since the drawing is always being previewed, portions can be erased, changed, or additions made. The drawing may also be stored in the computer for later use.

The joystick is a graphic input device that can be moved in any direction. During operation, a graphic cursor is displayed on the terminal's screen by manipulating the joystick for indicating point coordinates. Various features or functions are available with these devices to allow the operator at the terminal to choose the required function for displaying items.

Graphic tablets are digitizing devices normally used in conjunction with graphics terminals. A pen stylus is used with a gridded table to translate coordinate data from the position of the pen to the graphics terminal or processor.

Voice recognition and voice output systems. A wide array of voice recognition and voice output systems for man-machine communications is growing in use, Fig. 19.8. Systems of this kind are integrated with a host computer using software that allows custom designing voice applications. Each phrase is entered into a speech file. A reference code or label is assigned each message which the computer can use to access the message. In using the system, program instructions can easily be given to other devices such as digitizers and terminals by using assigned labels that incorporate menus and voice help messages.

19.4 COMPUTER OUTPUT DEVICES

The final computer results, referred to as output, are made available for use and analysis through devices such as line printers, plotters, cathode ray tubes (CRT's), and computer output microfilm (COM) units.

Line printers. Computer maps are commonly produced by line printers since this piece of equipment is an integral part of any computer system for printing output in readable form Fig. 19.9. By using the alphanumeric characters or symbols contained in the line printer, the computer instructs the line printer the position for printing each character. Printers are available with different font styles and character sets that allow characters to be condensed, expanded, or printed at different sizes. It can print a full line at a time (up to 180 characters) at a rate in excess of 1200 lines a minute.

Maps of different kinds, e.g., boundary, distribution, shaded, may be

Fig. 19.9 Line printer *(Courtesy NEC Information Systems, Inc).*

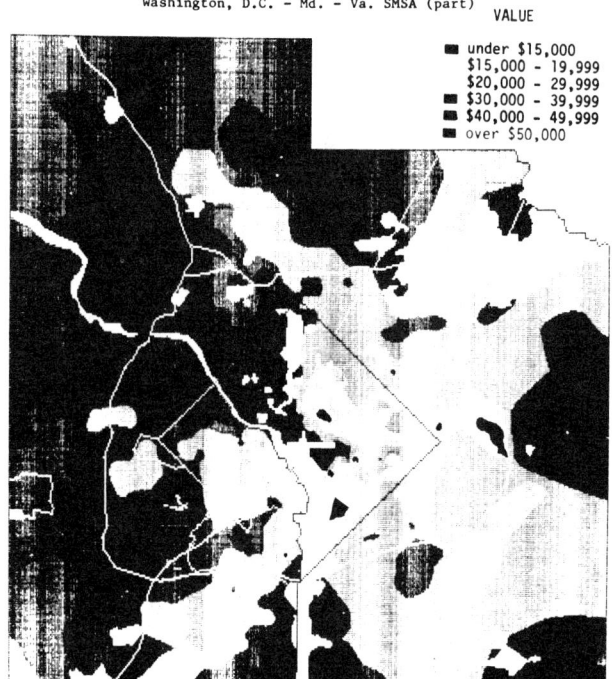

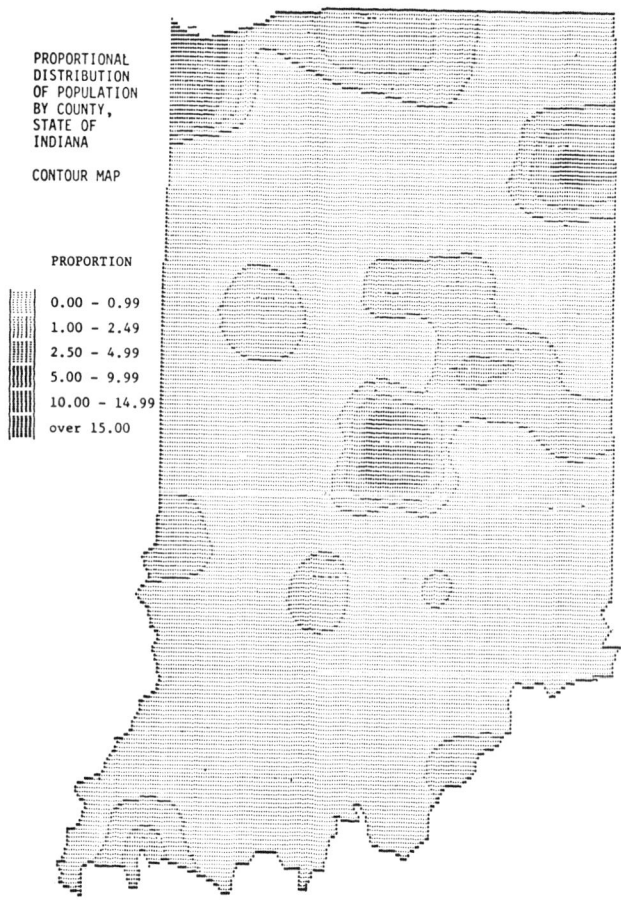

produced in this manner, Fig. 19.10. A map of any length and up to about 30 mm (13 inches) in width can be printed. For wider maps, successive strips must be joined together.

In shading a map, the line printer can be instructed to print layers of varying density or strings of letters or symbols in each character space to build up a map of the area. The line printer can over-print, making it possible to build up a range of seven or eight symbols with distinctive tones of grayness. A good printing ribbon is necessary to assure a uniform printing across the printing area.

Line printer maps are best suited for use in thematic mapping and for the representation of areal data. Since linear features are not adequately represented, the use of transparent overlays showing transportation networks, boundaries, etc. is effective to improve the maps use. Generally speaking, line printer maps are inferior to non-computer assisted creations and are not suitable for most cartographic purposes due to their sometimes sketchy or unfinished appearance. The appearance can, however, be improved by photographic reduction. Its use consequently has been by individuals more concerned with usable results than with the quality of the product.

A more recent method of producing line printer maps is through the use of matrix line printers. These devices create characters from a matrix of finely spaced dots as opposed to the many symbols and characters produced by conventional line printers. Individual dots in the matrix can be turned on or off to create recognizable symbols or characters, Fig. 19.11. The patterns seen in matrix printer maps are created by clustering a number of dots near the center of each matrix, Fig. 19.12. Matrix line printers offer several advantages over conventional line printers. A wider range of gray tones is possible, symbol configuration and orientation is more uniform, and the print quality is improved.

Fig. 19.11 Character creation from 9x6 dot matrix.

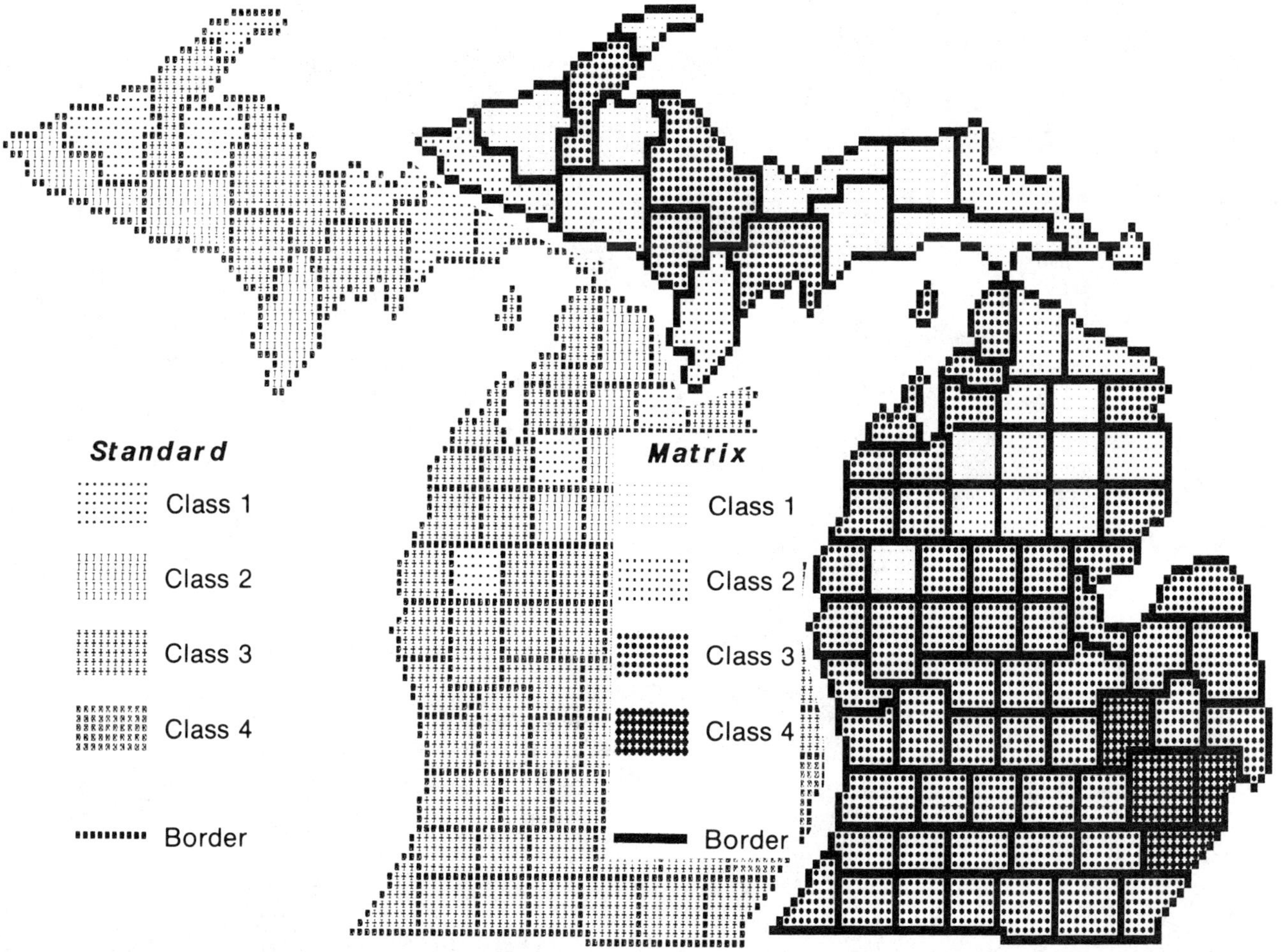

Fig. 19.12 Standard and matrix line printer map (*Courtesy R.E. Groop, Michigan State University and R.M. Smith, University of Arkansas, "The American Cartographer", Vol. 9, No. 1, 1982*).

Plotters. For graphic output, a computer-driven plotter of some type is normally used. These devices draw much faster than a drafter and vary in size, type, media used, precision, and cost. Plotters have a drawing head which may have a pencil, pen, scribe point, or fine beam of light (for drawing on photosensitive surfaces), Fig. 19.13. Depending on the model, the drawing head may have from two to eight stations to accommodate different pens or scribe points. Ordinary technical pen-type pens are used, but ball-point type pens work better for drawing at high speeds. As the drawing head moves to X and Y positions, straight lines, circles, or curves are drawn. In addition, alphanumeric characters and symbols of any shape can be drawn.

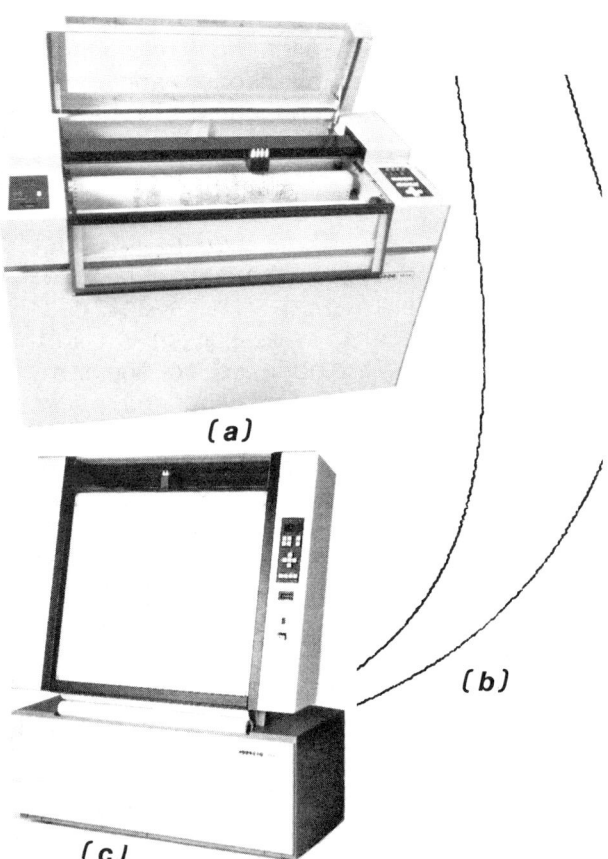

Fig. 19.14 Drum and beltbed plotter.

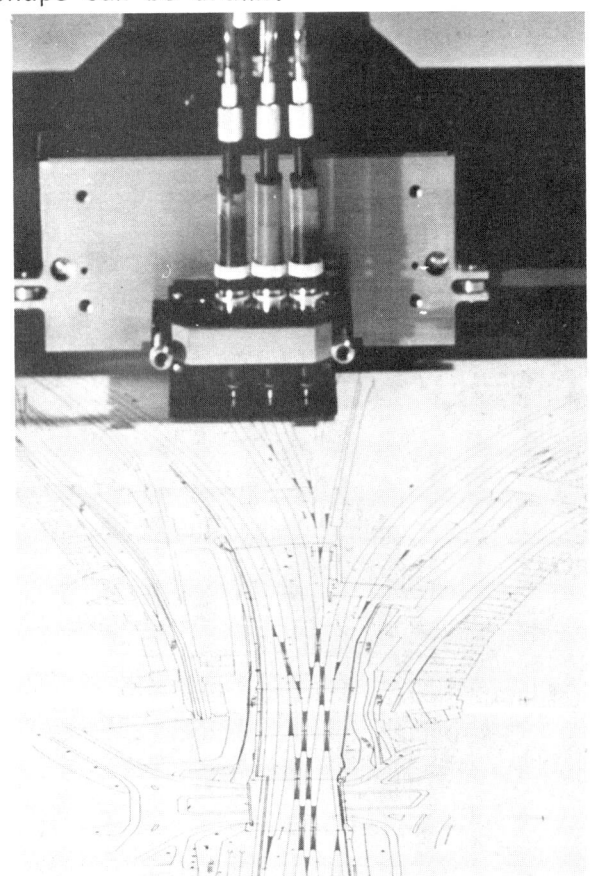

Fig. 19.13 Plotter drawing head (*Courtesy Calcomp*).

Drum type plotters are compact, inexpensive, and relatively fast, Fig. 19.14 (a). A continuous roll of paper is affixed to a rotating drum. The plotting head moves back and forth at right angles to the drum's movement. Since most drum plotters draw lines by increments, a steplike line is produced when generating diagonal lines, Fig. 19.14 (b). The resolution of this type of plotter varies with the more expensive models allowing smoother lines to be obtained. Beltbed plotters are similar to drum plotters and offer the flexibility for using cut sheet media, Fig. 19.14 (c).

Flatbed plotters have a large horizontal table, up to four or five feet, on which the drawing material is positioned, Fig. 19.15. The plotting head can be moved in any direction in a horizontal plane. For cartographic work, the finer resolution and high accuracy of flatbed plotters is recommended. Other advantages include the ability to see the entire plot and the opportunity to plot on any material or register new information to existing graphic information.

Interactive digital plotters, Fig. 19.16, provide the opportunity for communicating back and forth with the host. A number of processing operations is possible that include the drawing of alphanumerics that can be scaled, slant-

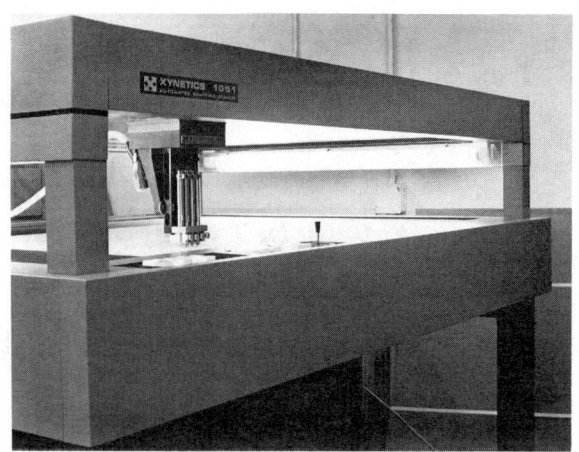

Fig. 19.15 Flatbed plotter *(Courtesy Xynetics, Inc)*.

Fig. 19.16 Interactive digital plotter *(Courtesy Tektronix, Inc)*.

ed and rotated. A joystick allows for digitizing points to a graphics terminal or processor.

Electrostatic type plotters, similar in appearance to flatbed plotters, offer a number of advantages over pen type plotters. Drawings and maps may be produced in minutes, not hours, to extend output capabilities. Plotters of this type can draw up to 34 square feet per minute with a resolution of 40,000 points per square inch. Their resolution is generally lower than line plotters and multicolored drawings are not possible. Copy can be produced on paper up to six feet in width as well as on film or translucent paper. Plots can be produced that are more complex. Some plotters will allow the addition of shading, tone patterns, grids, and variable line widths for conveying more information.

Cathode ray tube terminals (CRTs). A cathode ray tube resembles a conventional TV screen and is used for displaying alphanumeric or graphic information in black and white and color, Fig. 19.17. Two basic types of CRTs include refresh tubes and storage tubes. On refresh units the image is produced, usually 30 or 60 times a second to give a flickerless image. Storage tube devices, on the other hand, store the image until erased by producing it once, with a relatively slow moving electron beam. Information may be displayed and portions erased selectively. This process, however, requires that the whole screen be erased and the image regenerated again without the unwanted item.

The two major kinds of display processing units to create a drawing are vector and raster scan systems. In vector refresh displays, the image is traced on the screen in the form of lines or vectors that are repeatedly retraced. The display on this type of screen is bright with good resolution and provides the opportunity for the user to update. Where dynamic picture manipulation is a requirement, raster displays that are based on TV technology are used. A series of horizontal lines (raster lines) each made up of individual pixels or picture elements appear on the screen. The image is formed by a series of dots that are electronically illuminated at various light intensities on the screen. The quality of the picture, which is not as good as vector displays, will depend on the number of raster lines per inch.

Color graphics terminals are becoming more popular than monochrome

Fig. 19.17 Raster graphics terminals *(Courtesy Synercom Technology Inc)*.

units. The color image is created using raster-scan techniques to create the display by illuminated dots or pixels that map an electronic pattern in a frame-buffer memory. Each raster line is scanned sequentially from top to bottom 30 times per second by varying only the intensity of the electron beam for each pixel on a line. A large range of colors can be handled simultaneously. For monochrome raster display units, a single binary digit (bit) occurring in the memory as a one (1) or zero (0) can represent whether a pixel is activated or not for creating the image.

Terminals are now available that can display true three-dimensional images. The Space Graph, Fig. 19.18, produces three-dimensional images and graphics by the reflection of a standard terminal image in a vibrating, variable focal length mirror. The timing of the image on the terminal screen is coordinated with the position of the mirror so that individual planes of the object or scene are reflected at progressively deeper positions of the mirror image. The retentive characteristics of the human eye bind the planes together and, by frequently refreshing each plane, the graphic image in black and white or continuous gray scale format appears to be suspended in space. By moving the head, the viewer can see some side view of the object with no glasses or other special viewing aids required. Using the keyboard, the viewer can interact with the display by enhancing areas of the object, rotating the display, and scrolling through the display in the Z axis.

The development of 3-D display systems will be valuable in cartography and earth sciences such as geology. Data where movement is involved such as meterology and seismology can be effectively displayed to allow the viewer to grasp changes in the data more quickly. With 3-D, land forms can be better studied and analyzed, road layouts facilitated, and earthwork effects studied.

Output viewed on CRTs is rapid and offers the opportunity for operator interaction via the keyboard, light pen or joystick. Symbols, character sets, and basic picture segments may be developed that are pertinent to the applica-

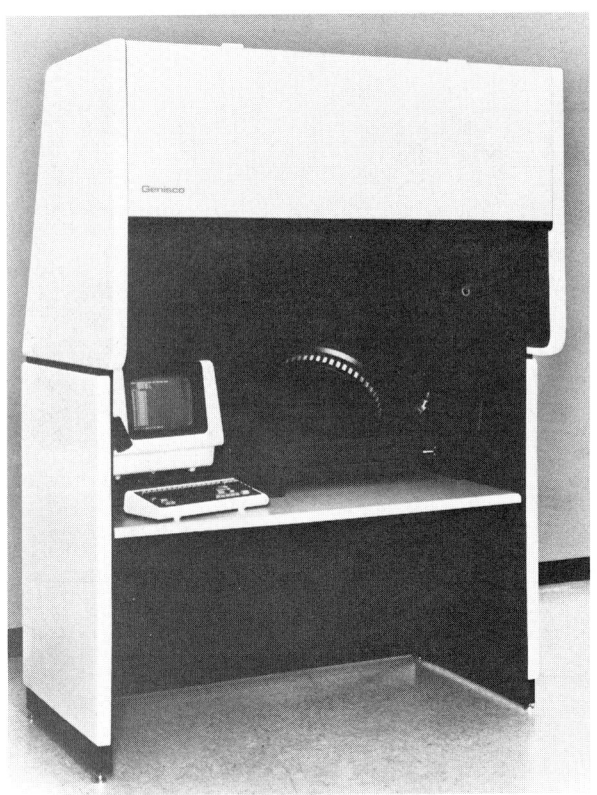

Fig. 19.18 Space Graph 3-D system configuration (*Courtesy Genisco Computers Corp*).

tion. These may be stored, manipulated, and redisplayed. A map can be shown in its entirety, panned to a selected portion, and then zoomed in for sharper detail.

Hardcopy may be obtained with a camera or with units that generate a one to one paper display from the terminal display, Fig. 19.19. These units are capable of producing hard copy with color shading and image resolution that is virtually indistinguishable from the original screen display. The problems are mainly one of size limitation and distortion which may not be acceptable for publication purposes. CRTs are most useful for data entry, check plots, manipulation and feedback, and other operations not involving high resolution output.

Computer output microfilm (COM). Output devices of this type rely on electronic equipment to plot from computer tape to 35 mm microfilm or 105 mm microfiche using a laser (light) beam. The high cost of the basic equipment and the associated photographic processing and hard copy preparation equipment

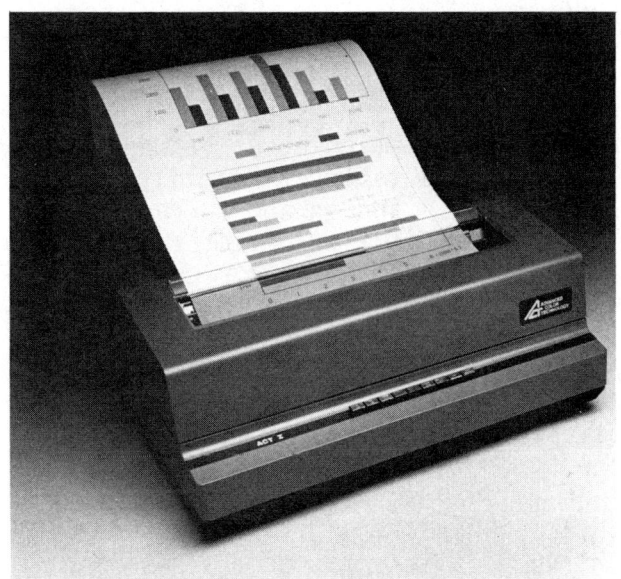

Fig. 19.19 Hard copy unit (*Courtesy Advanced Electronic Design*).

makes this system economical only when turning out large volumes of drawings. It offers, however, the best output capability available in terms of high resolution and color.

Lettering. The automation of map lettering is improving and continued work in this area is necessary. The primary problem involves the storage of multiple type fonts of different sizes, and the complexity of the proper position and location for letters so that a minimum amount of visual confusion results. In most instances it is still easier for the individual to be involved in finding the best location and size for lettering and name placement. For these reasons a combination of manual and automatic methods is still used for high quality map lettering.

19.5 COMPUTER DATA BASES

Data information is being collected and stored in digital form in computers at many levels and at great expense. One prevailing problem is that no standardization exists in compiling this data, making it difficult to translate the information or to use it on the many computer systems currently marketed. Digital cartographic data bases (DCDB) along with map products are also difficult to locate if an individual does not know where to look or who to contact for the information necessary to meet a specific need. The answer to this dilemma is to establish national standards and to create at the national or regional level a repository to act as a clearing house for receiving and distributing cartographic data. The preparation and location or exchange of data bases is consequently the primary obstacle to a more widespread use of computer-assisted cartographic techniques. Some progress in this area has been made by the U.S. Bureau of Standards assigning the USGS to serve as the lead agency in developing and managing a national cartographic data base. As part of the National Mapping Program, the USGS has as one of its primary missions the development of a DCDB that initially will contain 11 types of base map data. These are to include:

1. Reference systems--geographic and other coordinate systems except the public land survey network.

2. Hypsography--contours, elevations, and slopes.

3. Hydrography--streams and rivers, lakes and ponds, wetlands, reservoirs, and shorelines.

4. Surface cover--woodland, orchards, vineyards.

5. Nonvegetative features--lava rock, playas, dunes, slide rock, barren waste areas.

6. Boundaries--portrayal of political jurisdiction, national parks and forests, military reservations. This category does not fully set forth land ownership or use.

7. Transportation systems--roads, railroads, trails, canals, pipelines, transmission lines, bridges, tunnels.

8. Other significant manmade structures such as buildings, airports, and dams.

9. Identification and portrayal of geodetic control, survey monuments, other survey markers, and landmark structures and objects.

10. Geographic names.

11. Orthophotographic imagery.

One estimate has placed the cost as high as 200 million dollars over the next two decades for the Federal government alone to develop, organize, and manage a DCDB. The American Congress on Surveying and Mapping will play an important role in the development of digital data base standards. A National Committee for Digitial Cartographic Data Standards has been formed and will involve individuals from all segments of the mapping community to participate in their development.

Eventually, fully digital data bases in a standardized form will contain all information needed to produce a map. National Standards will be established for data formats, codes, and accuracies to allow individuals to easily use the information and contribute to its growth. The use of the computer will then allow this information, on command, to be quickly retrieved for producing instant maps in forms that would not otherwise be possible.

Data acquisition. Data from many sources must be collected to build up a useful data base. Federal, state, and private mapping agencies will be the primary sources for obtaining the many classes of information that will be needed. Other sources for obtaining information include state and county road departments, water resources agencies, natural resources departments, and environmental quality control organizations. The success of a massive undertaking of this kind requires that a free flow of information occur between these organizations for providing data on a regular basis and in the correct format.

The digitizing of information from existing basic maps is currently the primary method for building data bases. In compiling this data for use in the computer, the information should be encoded at the largest scale that shows the most detail and then reduced. Examples of the type of data that may be collected in a file and stored in the computer data bank in layers at specific levels or types of information are positions and elevations, transportation routes, lakes, streams, shorelines, terrain slopes, land use, cadastral and political boundaries, population distribution, soils, geology, and hydrology, Fig. 19.20.

The data can be recalled for viewing separately or in combination with other data, which may be readily edited, updated, changed and restored. This procedure results in lower maintenance costs, eliminates duplication, and provides ready access to precise, up-to-date information.

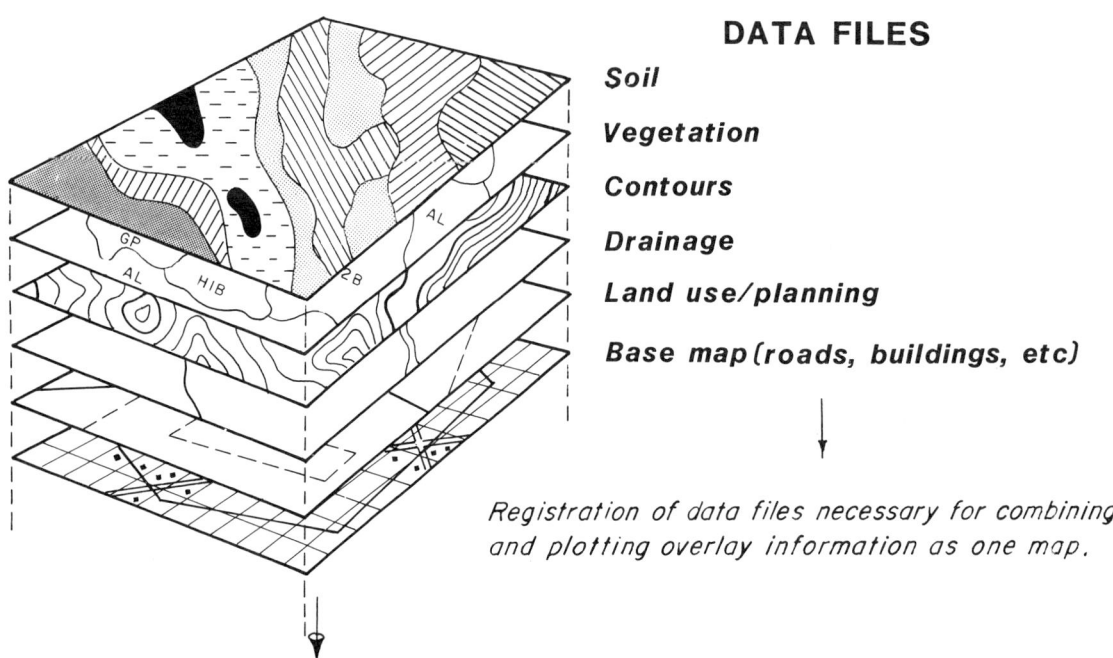

Fig. 19.20 Digitized data files.

The creation of comprehensive data banks will allow new computer generated products to be developed. One application will be the preparation of press plates for map printing using automated techniques. As new products evolve, it will be necessary to provide more technical assistance to help users find and apply cartographic information. Several services that will require strengthening are digital data distribution, applications software development, and special product preparation.

19.6 COMPUTER SOFTWARE

The instructions (algorithms) and programs that tell the computer what to do and how to do it are called *software*. The program will either contain the necessary data for execution or provide the additional information whereby this data may be obtained. The additional information may be available by means of a library subprogram or listed data stored on tape or a disc. Computer instructions for special symbols and line types (firmware) to be created are generally provided by the hardware manufacturers. Generally, the development of software is expensive and is recommended only if the program can be used many times. Several programming languages have been developed to permit the programmer to give instructions to the computer by using statements and symbols. FORTRAN (FORmula TRANslation) is the most popular, however, others such as BASIC (Beginner's All-purpose Symbolic Instruction Code) and Pascal are used.

The development of programs to prepare complex cartographic information can be a long and costly process. It requires a competent working knowledge of a programming language, an understanding of hardware operation, and the experience to take full advantage of the computer systems capability. For this reason the portability of software to enable moving to new display devices is important since it is too expensive and time-consuming to rewrite working programs. Federal agencies such as the USGS and the DMA along with other commercial organizations are in a position to employ a number of programmers for developing costly software for their own internal use. At the present time other users rely on various commercial organizations and educational institutions as the primary source for obtaining major software packages. A more recent source of obtaining software resides in the numerous new businesses emerging nationwide that provide services in software development and hardware systems design. Many available programs have the flexibility to accept data from a variety of sources, including existing maps, stereo plotters, electronic theodolite instruments, and remotely sensed data. The software products make it possible to capture and convert data, including coordinate transformations, coordinate gridding, thematic displays, and to produce maps with scale-dependent symbology. In addition, software is available for special analysis of area features and manipulation of three-dimensional surface models.

A persistent problem for cartographers is in the area of generalization as the scale of the map is reduced. Well-developed reduction and generalization software is available for performing this operation with a high degree of consistency. The degree of computer involvement in this process will vary depending on the system. Manual overrides may be employed to allow the operator to interrupt the plotting and to create the desired effects.

The available software packages fall basically into three categories; single purpose programs, general purpose programs, and special purpose subroutines. Single purpose software packages are designed to satisfy a specific objective and are restrictive in the types of maps that can be produced. General purpose mapping packages may be used for producing several types of maps. Special purpose subroutines exist for the user to call on demand to perform a specific operation or produce a certain kind of map. Eventually, software costs will be reduced and complete packages that are user oriented will become available to meet most mapping requirements.

19.7 TYPICAL APPLICATIONS

Many programs have been developed for creating a graphic display of cartographic information in a wide range of formats. The algorithms are constantly being improved and new capabilities are being explored each year for creating different types of maps such as three-dimensional surface displays. Computer-assisted methods have allowed maps to be used more for analyzing vast amounts of data as opposed to just its display. An example of this is the use of mapping satellites for obtaining information on major storms. The data received can be displayed, continually updated, and the imagery combined with other data such as rainfall measurements for predicting flooding. All of these developments have brought about the need for the user to be able to analyze maps in new forms. This will become increasingly important if the full potential of utilizing maps produced in this manner is to be realized. A number of examples of applications follow to help suggest the wide range of map forms that may be generated by computer-assisted methods.

<u>Map projections</u>. In producing a map, a projection or coordinate system is generally required to serve as a base for the addition of other types of information. Since projection or coordinate systems can be specified by equations, the graticule lines that are difficult to draw by hand can be easily drawn using computer assistance. A greater flexibility now exists by utilizing the computer to draw any type of projection having any scale, orientation, and origin, Fig. 19.21.

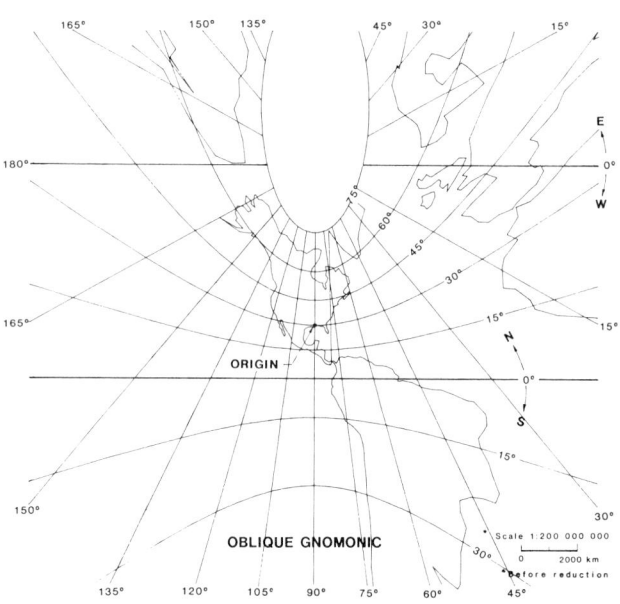

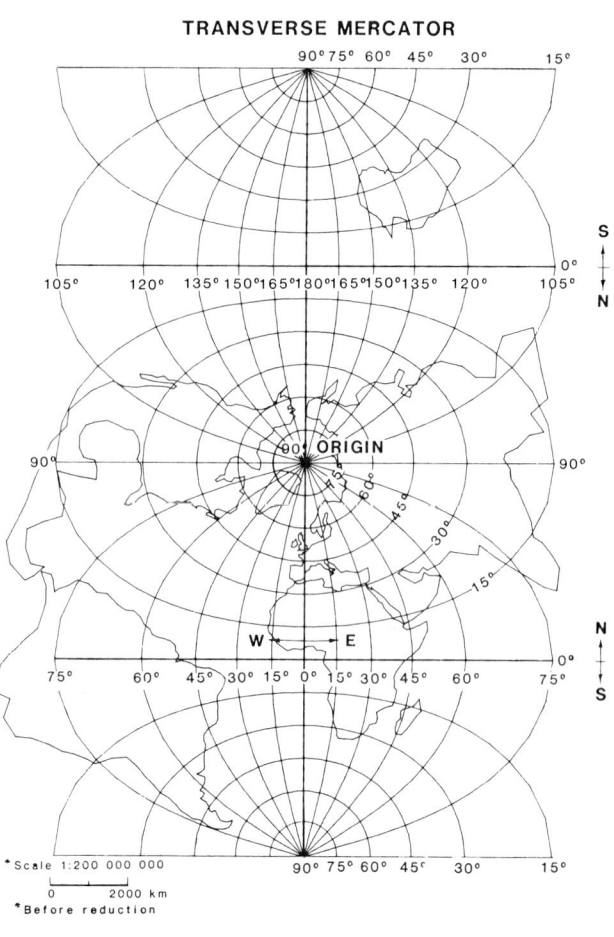

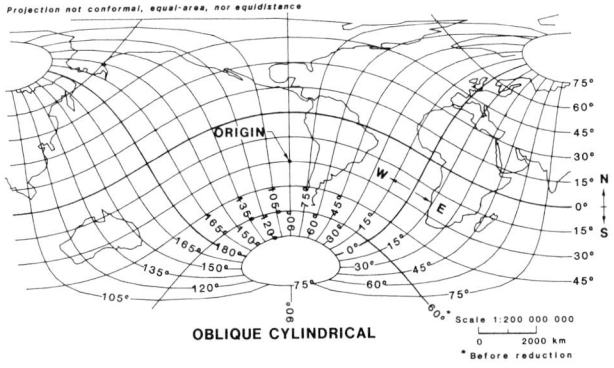

Fig. 19.21 Computer generated map projections.

Detail selection. A map can be produced from file data stored in the DCDB that will show single features or any combination of features, Fig. 19.22. Base maps containing area boundaries and other selected information, e.g., contours, hydrography, etc., may be created and used for developing new maps through the addition of current information. In the production of color separates, the computer can be programmed to redraw lines to occur in a particular color that will be in perfect registration. Since maps are continually undergoing revision, the digitizing of information allows items to be deleted from the stored tape and updated or new information for features added on a continual basis.

Enlarging-reducing. Digital methods can be used to enlarge, reduce, or change the geometry of a map, Fig. 19.23. Absolute coordinates for a map feature

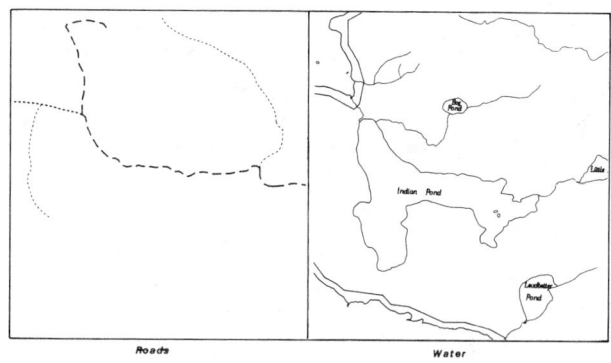
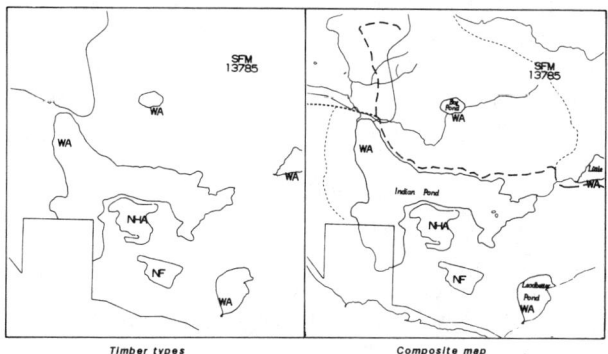

Fig. 19.22 Digitized multi-layer maps.

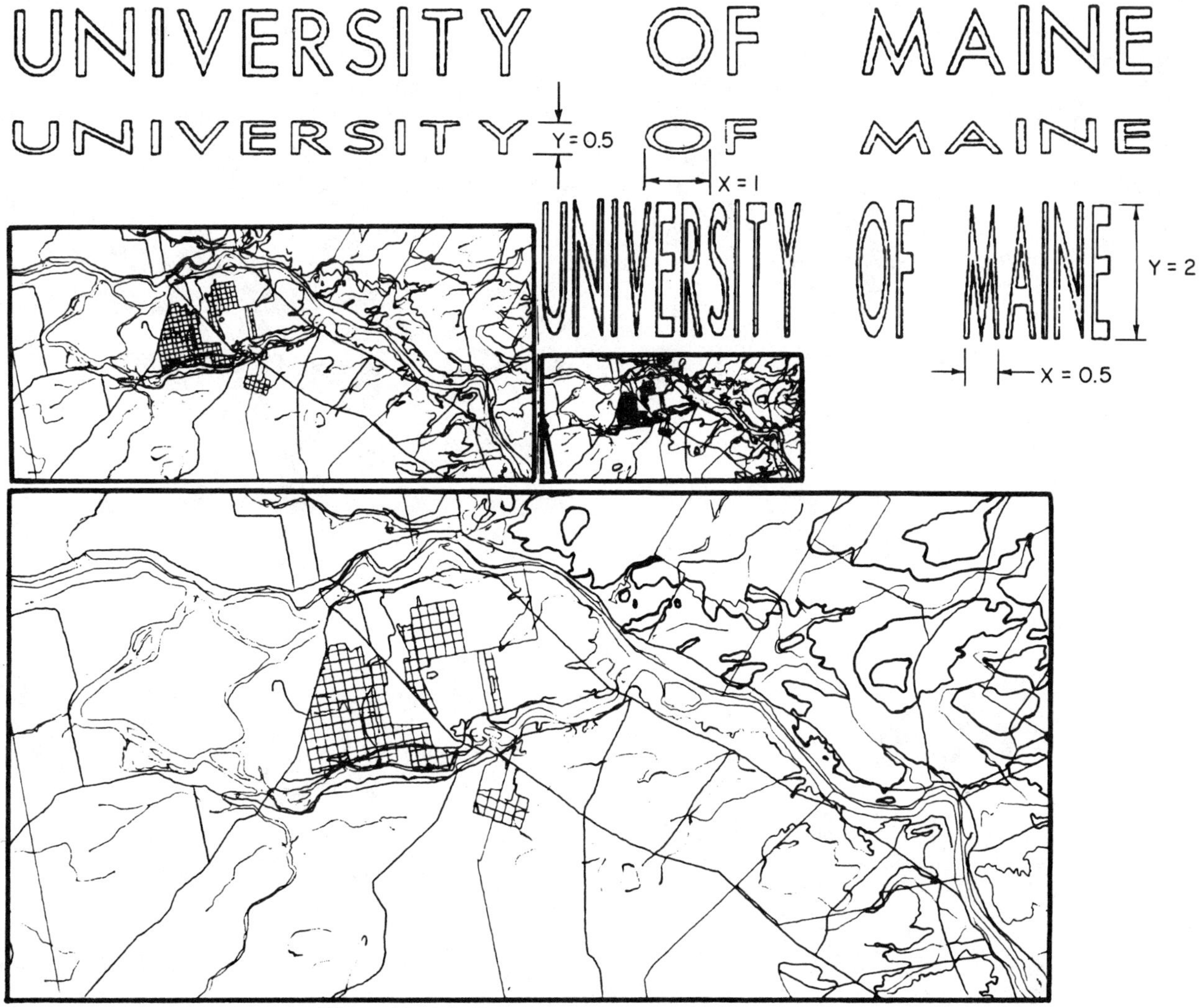

Fig. 19.23 Enlarging and reducing.

can be obtained using a digitizer. Data is recorded on tape or some other form for automated data processing. Scale change or some geometric distortion is obtained by using a numerical factor. A shape can be doubled in size by multiplying both its X and Y coordinates by a factor of 2 (2X,2Y). It could also be shortened in the X direction and lengthened in the Y direction by using a double factor such as 0.5X and 2Y. Plotting can be done using mechanical plotters on different kinds of drawing media or electronically displayed on a CRT.

Three-dimensional surface displays. A number of unique programs exist for generating three-dimensional surface displays of continuous surfaces from digital data files. Drawings in this form compliment the usual two-dimensional presentation and help to analyze and better visualize the surface. In representing the surface of the earth, a conventional contour map is generally employed, Fig. 19.23 (a). For quantitative analysis this traditional type of map remains the best way for representing the land surface since all areas are visible and accurate distances and angles can be measured. The main disadvantage is that complex surfaces cannot be easily visualized.

A realistic visual impression of the land surface is best obtained using perspective views. The mesh perspective view, Fig. 19.23 (b), represents the surface through the use of a mesh or grid of lines at right angles that are deformed to follow the surface shape. This form of surface representation is very useful for non-technical illustrations and general orientation where quantitative measurements are not necessary.

The contour perspective view utilizes contour lines to delineate the shape of the surface, Fig. 19.23 (c). Some numerical information about the relief is possible, but, precise elevations at exact coordinate locations are difficult to obtain. Since the scale changes over the perspective, the contour lines make it possible to obtain elevation information for prominent features over the entire map.

A contour key in the form of a flat contour map is sometimes used in conjunction with perspective views of a surface, Fig. 19.23 (d). The advan-

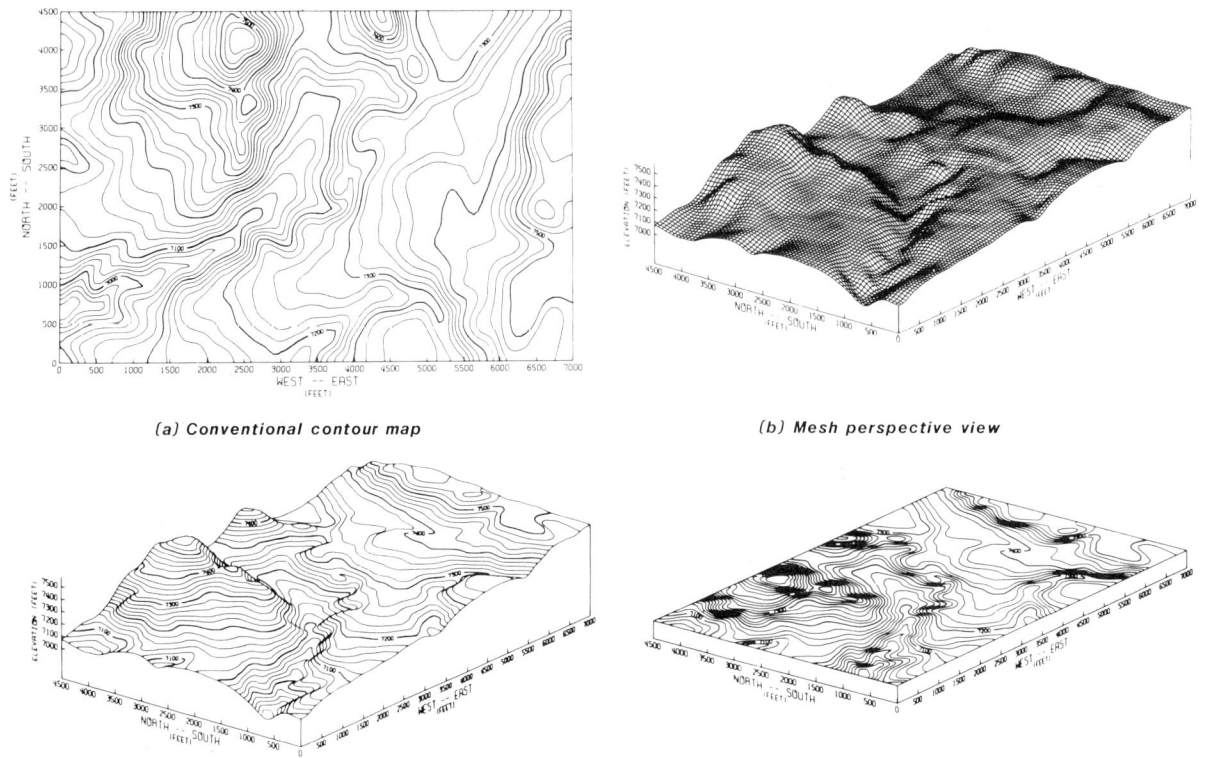

(a) Conventional contour map

(b) Mesh perspective view

(c) Contour perspective view

(d) Flat contour perspective view

Fig. 19.23 Three-dimensional surface display (*Courtesy software by Dynamic Graphics, Inc*).

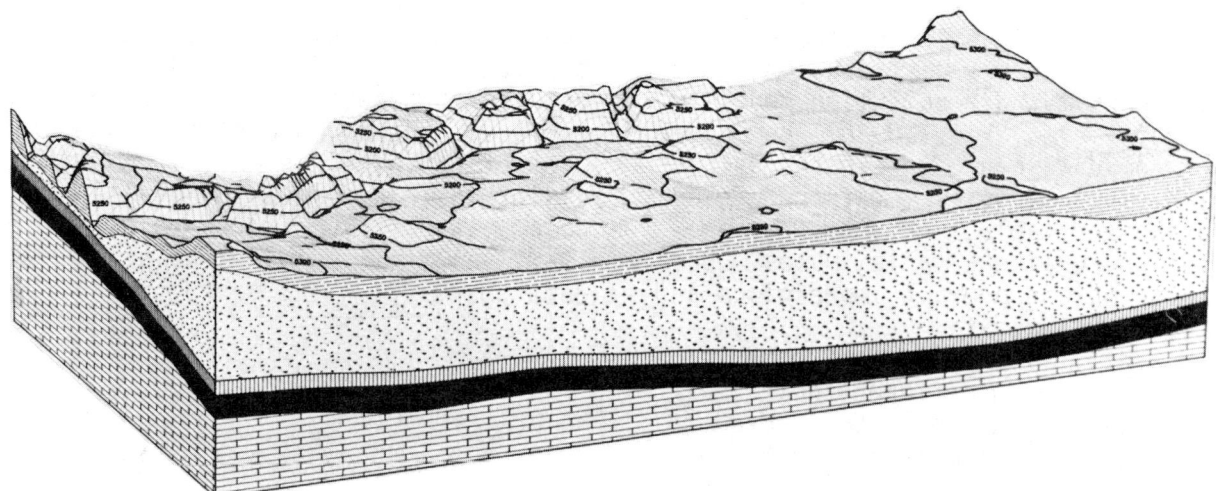

Fig. 19.24 Coal seam profile *(Courtesy Dynamic Graphics, Inc)*.

tage of this format is that the two drawings can be related to each other readily for both orientation and quantitative details.

In addition to representing the land surface, graphic information such as soils, geology, boundaries, and roads can be placed on the perspective, permitting the viewer to relate specific information to the terrain, Fig. 19.24. Other applications include site selection and development, land-use, and environmental planning.

Computer processing of digital information allows topography for an area to be displayed in new ways, Fig. 19.25. The area can be viewed from any azimuth or elevation and the relief exaggerated to any degree required for visual effect. By changing these parameters, all sides of features such as peaks and valleys can be easily inspected.

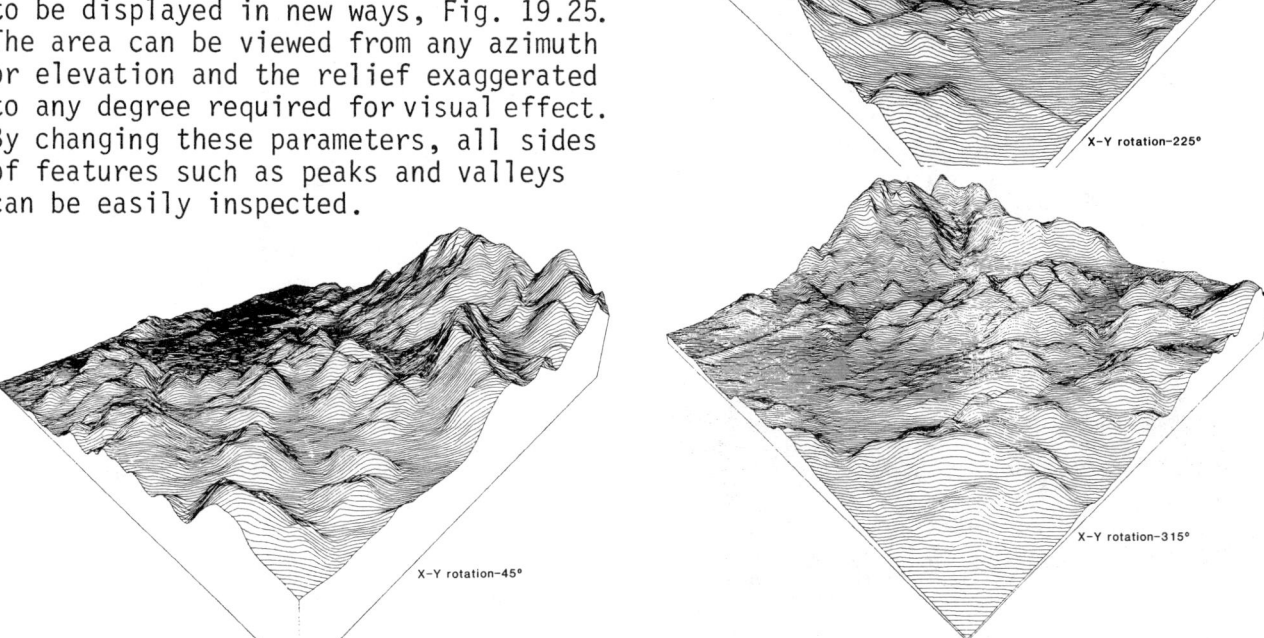

Fig. 19.25 Electrostatic plots of digital elevation or terrain variations. Horizontal angle = 20°, view point elevation = 5997, horizontal distance from view point to focal point = 15,000 feet *(Courtesy Versatec)*.

Other applications of plotting numerical data in a three-dimensional form include income distribution, population density, air pollution levels, and crop yields, Fig. 19.26. Any given matrix of data can be presented as a variable height surface rather than as an organized table. This type of presentation, unlike a table, makes it possible to obtain overall impressions and patterns of the data.

Isarithmic maps. Many computer programs based on different numerical algorithms are currently being used to produce isarithmic maps involving contours, mineral deposits, magnetic intensity values, etc. Information in the form of random or gridded data can be used to generate the character of the surface. The advantage of using a gridded set of values is that it is much easier to interpolate for additional points than for scattered points. Fig. 19.27 (a) typifies a set of scattered data points (X, Y, Z coordinates) for

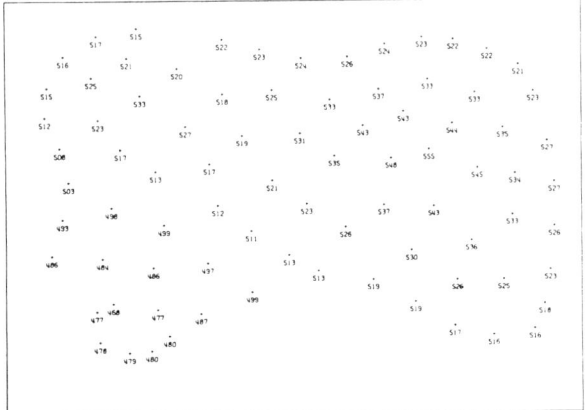

(a) Scattered data values (X,Y,Z coordinates)

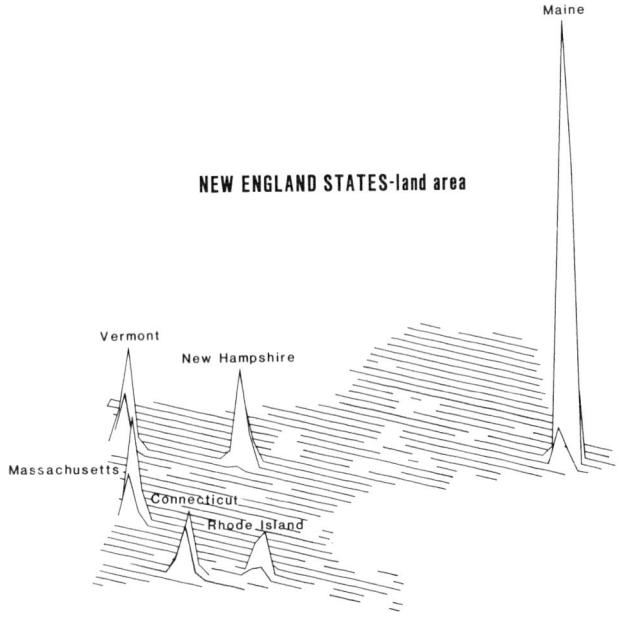

NEW ENGLAND STATES-land area

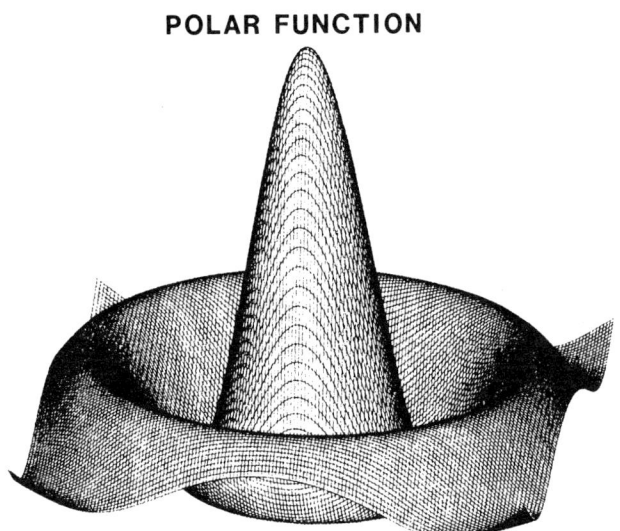

POLAR FUNCTION

A=6.0*SQRT((X-1.5)*(X-1.5)+(Y-1.5)*(Y-1.5))
Z=2.0*SIN(A)/A

where X and Y vary from 0.0 to 3.0

Fig. 19.26 Typical 3-D displays.

(b) Calculated grid values

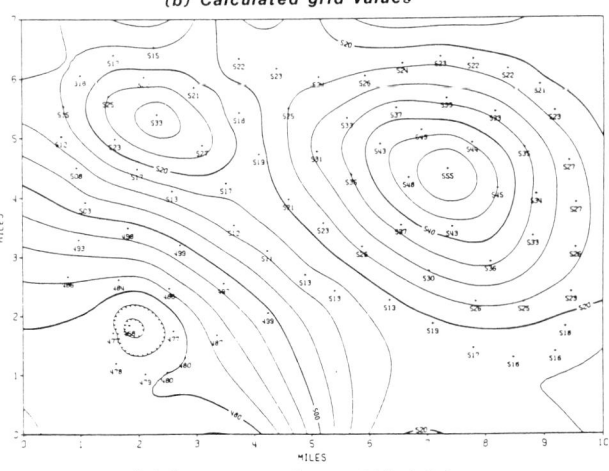

(c) Contour map from gridded data

Fig. 19.27 Isarithmic mapping *(Courtesy Dynamic Graphics, Inc)*.

producing a contour map. A new set of calculated grid point values at regular intervals that represent the same surface as the scattered points is shown in Fig. 19.27 (b). The final contour map, Fig. 19.27 (c), is produced from this gridded result.

Thematic map display. The computer has made it possible to generate many forms of thematic maps, Fig. 19.28, and to manipulate large files of geographic data. Maps of this type are increasing in numbers with applications found in most areas, e.g., medicine, crime statistics, recreation, housing, natural resources and the environment, education, etc. A number of mapping options or software packages are available to enable the user to tailor the output as required. The maps can be produced having different kinds of symbols or shaded using different line patterns or gray shading symbolism.

Terrain shading. A unique example of graphic output using computer techniques is the shaded relief map. The average map user finds that this type of presentation adds to the conventional map linework. It is helpful in determining relative elevations, terrain ruggedness, and smoothness. Computer generated shading methods utilize digitized terrain data (X, Y, Z coordinates) to find the direction of slope and slope inclination. The brightness values of slopes are calculated that will best portray various surface characteristics and a composite image is produced that simulates shaded relief.

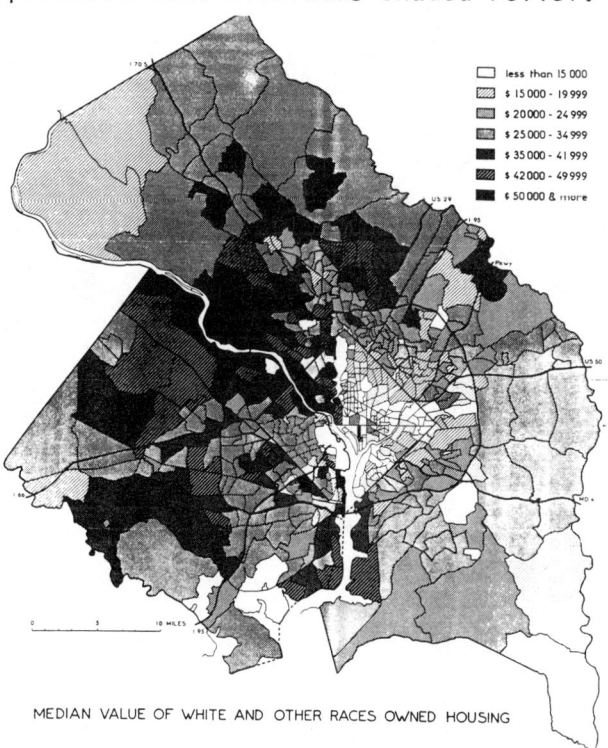

MEDIAN VALUE OF WHITE AND OTHER RACES OWNED HOUSING

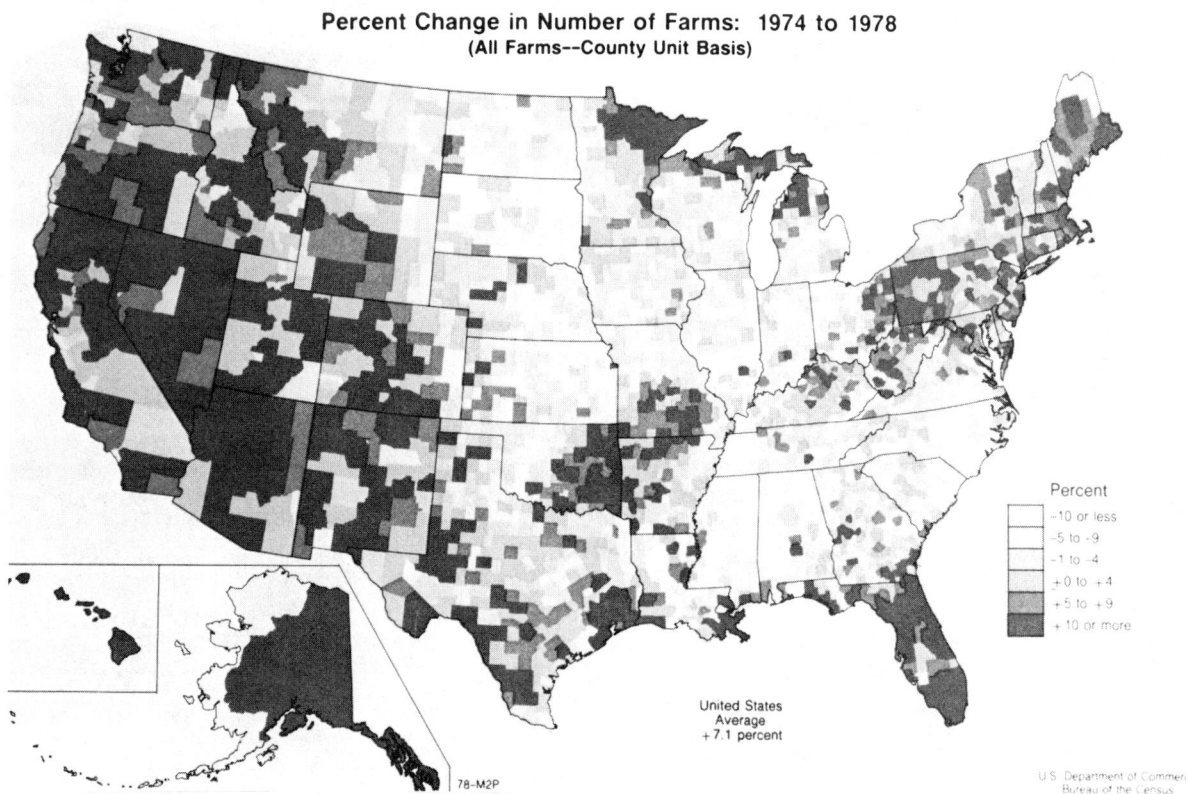

Fig. 19.28 Thematic maps *(Courtesy Bureau of the Census)*.

PRINTING PROCESSES AND REPRODUCTION SYSTEMS

20.1 INTRODUCTION

Multiple copies of printed or other graphic materials are frequently required and it is important to have a knowledge of several ways to duplicate this information. All printing and reproduction methods have inherent limitations related to different requirements and a number of considerations must be contemplated prior to selecting the final method to be used. Several factors to consider are available equipment cost, type of original artwork, number of copies required, future use of the copy, and how good the finished product should be. Regardless of the method used for reproducing material, the quality will depend directly on the care with which the original copy is prepared. The preparation of artwork is taken up in chapter 5. This information should be reviewed and understood in taking up printing processes and reproduction systems. It is difficult to separate these two topics since both activities are related and dependent on each other. In all instances it is advisable to check the process for duplicating copy and then to plan the artwork to fit the process.

The development of reproduction equipment has progressed a long way and machines are available that will reproduce any material including pencil or ink originals and photographs. A brief description of several printing processes and ways to reproduce copies of other printed or graphic materials is presented in the following sections. For each of the categories discussed there is a broad range of equipment available for general and special applications. In choosing a process, all factors should be examined and manufacturers consulted for the latest developments in materials and systems for printing and reproducing graphic information. The following information, although brief, should help individuals to choose and work effectively with the principal processes used in the printing or copying of materials.

A table summarizing several printing and reproduction processes appears at the end of the chapter. This table should be used as a guide in choosing a process that is economical to achieve the desired results.

20.2 PRINTING OPERATIONS

In any printing process a number of basic operations must be carried out prior to the actual printing. These involve photographing the original copy, processing the film, making required plates, and the presswork.

<u>Photography</u>. Photographic techniques are widely used for obtaining high quality image detail. In areas involving graphics, precision enlargements and reductions are required for printing operations to be carried out. Other operations that may also be necessary include the transferring of opaque image information onto transparent film, changing negative material to positive form and vise-versa, the conversion of continuous tone to halftone, the conversion of solids to tint screens, and color separation.

Photographic prints may be made with or without a camera. If a change in image scale is not desired, contact copies of the finished art work may be made economically. The image to be reproduced is held against the photographic film emulsion under vacuum pressure and exposed to a bright light source. Copies may be made from positive, negative or translucent art work on both film and paper. In the printing industry a process camera is used for enlarging-reducing work Fig. 20.1. Equipment of this type is quite large and expensive with the ability to copy images measuring as large as 125 x 150 cm (50 x 60 inches). The main components of a process camera are the movable copy frame for holding the artwork, an adjustable holder for the lens, and some form of artificial illumination. Since it is imperative that no distortion occur, the copy and unexposed photographic film must be held flat in parallel planes. This is achieved by holding both under vacuum pressure in the camera.

Fig. 20.1 Process camera *(Courtesy Brown Camera, Inc)*.

Films and film processing. Many kinds of film and paper may be used for contact and process work to meet a wide range of requirements. Since so many specialized products are available and constantly being introduced, manufacturers need to be consulted regarding specifications and applications of their products. For photographic work, films are coated with an emulsion that consists principally of light sensitive silver halides distributed in a colloid gelatin. The composition and size of the silver halides affect their sensitivity to light. In cartographic work where fine images or good resolution is necessary, a fine grain emulsion film is used. A high contrast emulsion type film is also needed for reproducing line and halftone images to obtain a maximum density difference between the opaque lines of the original and the clear background. Film with less contrast is required for the reproduction of continuous tones since all of the tonal values in the continuous tone negative should be maintained. In preparing artwork for photographing, preprinted shading films may be used instead of screening the negative. Shading films may also be applied to areas of negative or positive film. The effective use of this material requires that it not be reduced too much to avoid the loss of pattern detail.

The silver halides used in films are sensitive to different colors. This fact allows original copy to be prepared using a blue line guide. It is sometimes convenient to have additional lines as detail appearing on a drawing in blue as a guide and then to draw the required detail in black. If the image is photographed on ordinary black and white emulsion film (achromatic), the blue lines will not photograph and only the black lines will be copied. Other dyes are used in film emulsions to increase color sensitivity. Orthochromatic emulsions are sensitive to the green and blue portions of the color spectrum. This type of emulsion will also allow copy to be reproduced where the blue lines will be opaque and black or red lines copied as black. The film may be used in the darkroom under a red safelight for handling and processing. Panchromatic emulsions are used for color reproduction or for copying color originals in tones of gray. Films of this kind are sensitive to the entire visible spectrum.

In exposing and developing film, great care must be exercised to control these variables. If too much exposure or development occurs, background areas may be pronounced and line widths changed. After development, it will usually be necessary to opaque out undesired clear spots of the negative on a light table to prevent light passage in areas of lighter density. A black water soluble opaquing liquid or lithographer's tape is used for this purpose. The liquid flows smoothly and dries quickly without chipping, cracking or peeling. Other detail that is not needed may be deleted in the same way. If minor changes are necessary, it may be possible to cut the negative and to strip in another corrected piece.

Line and halftone. The printing of line drawings produces solid black lines of uniform ink density without graduation of tonal values. A line negative in the printing industry is made by exposing the copy to a sheet of high-contrast film in a process camera. The non-image, or white, areas of the copy reflect the most light through the camera lens to the film, producing solid black areas when the negative is processed. The image areas on the copy will reflect very little light to the film, producing the clear areas seen on a negative.

In the printing industry halftone photography must be used to copy photographs and illustrations to obtain the wide range of continuous tone values that occur. A halftone negative is made by photographing the copy through a halftone screen, Fig. 20.2. For screening black, the screen is usually at 45 degrees to the horizontal to give the most pleasing visual effect. In multicolor work, the different hues must have some angular separation to avoid undesirable patterns or moiré effects. The angles normally used are 45 degrees for black, 75 degrees for magenta (red), 90 degrees for yellow, and 105 degrees for cyan (blue). The screen is placed between the copy and a piece of high-contrast film. The film negative breaks up the continuous graduations of tones into corresponding image areas of halftone dots of various sizes and uniform spacing. The size or image area of each halftone dot is determined by the amount of light transmitted from the copy, through the screen, onto the film. Since the dots are small and closely spaced, an optical illusion is created which makes the printed halftone image look as though it were printed with varying ink densities ranging from white to black.

Tint screens allow tone percentages from 10 through 90 percent of black and screen densities from 27.5 to 200 lines per inch, Fig. 20.3. The finer the screen, the sharper the detail in the reproduction.

Duotones. A duotone is the halftone reproduction of a black and white photograph using two runs of ink, one run for detail and one run for shadow contrast and shadow density. In some instances the duotones are black halftones overprinted on a solid second color. The printing of a duotone requires two different halftone images of the original art. In discussing halftones it was pointed out that undesirable screen or moiré patterns may result by the use of improper screen angles in photographing the halftones. Screen angles are usually 30 degrees apart to reduce this undesirable affect. The halftone for carrying the important detail or for printing the darkest color is made with the screen angled at 45 degrees. The halftone for carrying the shadow and lower-middletones is made with a screen angle of 75 degrees, Fig. 20.4.

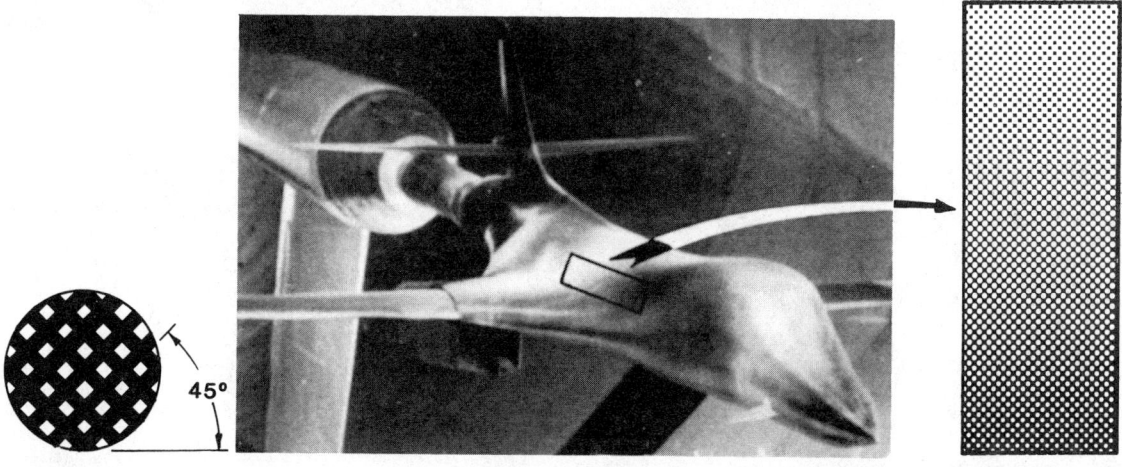

Enlarged halftone screen   *Halftone print*   *Enlarged halftone pattern*

Fig. 20.2 Halftones.

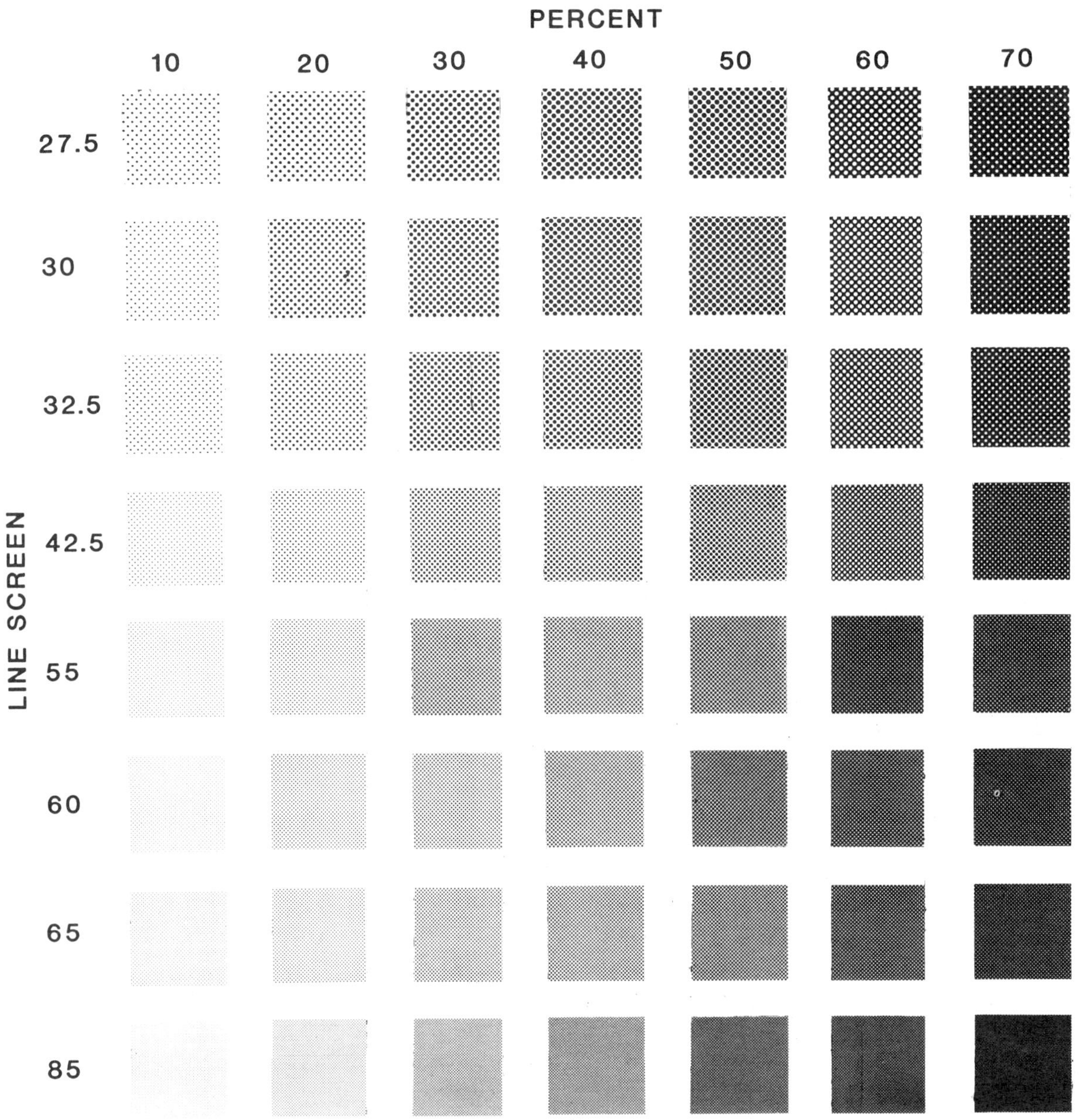

PERCENT
10
20
30
40
50
60
70
LINE SCREEN
27.5
30
32.5
42.5
55
60
65
85

Fig. 20.3 Halftone screens and values.

Screen angles 30° apart to reduce moirē patterns
75°
30°
45°

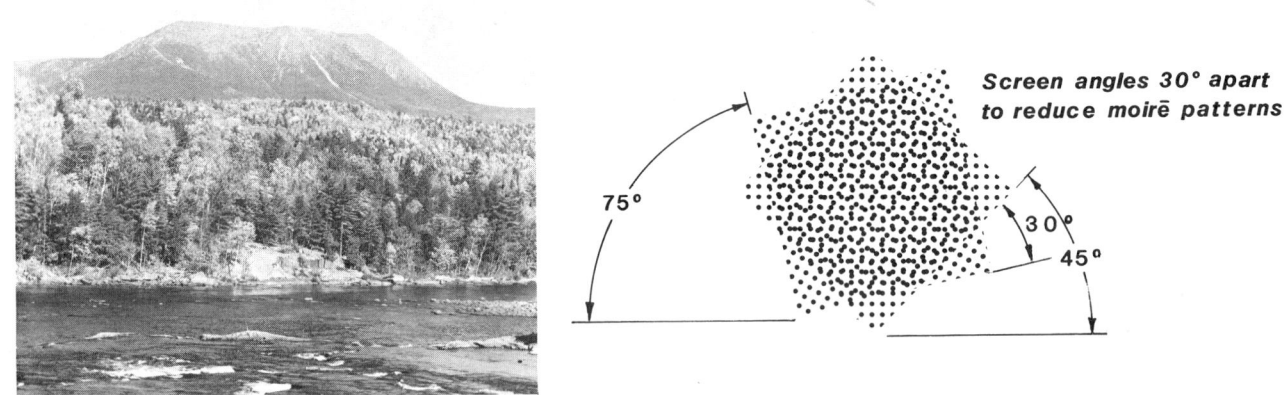

Fig. 20.4 Conventional halftone and screen angles.

Duotones are commonly used to lend depth, color, and mood to a subject. The color used frequently bears a relationship to the subject in the photograph, e.g., blue or green for a landscape. Many variations are possible by utilizing different color inks to obtain maximum highlight and shadow control. When printed, the result is an optical blending of both impressions to produce the effect of a third color, with greater intensity in the dark and middle tone areas.

20.3 PRINTING PLATES

Printing methods require an intermediate image carrier or plate before material goes on press. Metal, paper, or plastic materials having different kinds of coatings are used to make the plates. The type of material used is determined by many factors that include the printing process, type of press, paper, quality of image reproduction required, and the length and speed of the run.

Letterpress plates. In letterpress all line drawings and pictures require photoengraved line or halftone plates. The original copy is photographed to produce a film negative that is placed on pre-coated zinc, copper, or magnesium sheets. After exposing the negative covered sheets to light, the unexposed non-image areas are chemically etched. The emulsion in the image areas exposed to the light hardens and remains in relief to produce the printing surface.

Duplicate plates are often necessary for use on presses requiring curved instead of flat plates. Stereotypes are produced from paper machē or plastic molds of photoengravings. The molds are sometimes referred to as matrices or mats. Molten metal, when injected against the mat, hardens and forms the duplicate plate. Electrotypes are produced from plastic, soft metal, and wax molds made from original engravings. The mold is treated with a film of silver or graphite to make it electrically conductive. When placed in an electrolytic solution of copper, nickel or iron, a coating of metal is formed to produce the plate. The plate is removed from the mold and made ready for the press by backing it with lead.

Lithographic plates. In offset lithography camera-ready copy of line drawings and halftone illustrations is necessary for making the plates. All line copy is prepared and positioned on good quality white drawing media as it should appear on the page. Rubber cement or a hot wax machine is generally used for paste-up work. All photographs or other illustrations must be made as halftones. The paste-up line copy, minus illustrations, is photographed to produce a negative. The halftone and line negatives are then placed and taped into position on sheets of opaque ruled paper usually called flats or goldenrods.

The plates may be made on sensitized metal, plastic or paper. A wide array of automatic plate processors is available for this purpose. The flat is fastened to the plate and exposed to light through the negatives in the flat. Image areas exposed to the light become hardened while the unexposed areas remain soluble. The plate is developed to remove the emulsion in the unexposed areas. If halftones are in a separate flat that have been registered with the line flat, the plate must be "double burned" by exposing one flat onto the plate after the other.

Surface plates where the image and the surface of the plate are in the same plane are generally made from film negatives that are right-reading. These may be additive or subtractive. Additive plates are diazo presensitized by the lithographer and used for shorter press runs. Subtractive plates are diazo presensitized and lacquered by the manufacturer for use in long press runs. Both types must be exposed to a light source with the negative in vacuum contact with the coated plate and processed to develop the image areas.

Deep-etch aluminum plates with good wear qualities are used for long color work press runs. A film positive instead of negative is required for making the plate. The plate is exposed to light through the film positive that hardens the non-image areas. In developing the plate, the soft image areas are removed by chemical washing, leaving the printing areas slightly recessed in the plate.

For extremely long press runs of 500,000 or more, bimetal plates containing two metals are used. The two types of material most commonly used are copper chromium plates and steel or aluminum copper plated plates. These are processed similar to the deep-etch plates.

Gravure plates. In preparing gravure plates, the original line and continuous tone copy is photographed and screened. The negatives are then combined and photographed to form a continuous tone film positive.

A sensitized gelatin transfer sheet called carbon tissue that has been pre-screened in a contact printer is used to transfer the image from the film to the plate. The carbon tissue is exposed to screened light through the film positive that leaves gelatin areas of varying hardness according to the amount of light reaching the tissue through the film positive. After exposure, the gelatin deposits on the carbon tissue are transferred face down to a copper plate leaving the image wrong-reading. The plate is developed using hot water and chemically etched with acids that etch the metal at various depths depending on the hardness of the gelatin. Where the ink cells are shallow, the tones will be lighter and darker in areas where they are deeper.

20.4 PROOFING METHODS

To avoid errors and omissions, a drawing that is to be reproduced in quantity should be edited using a proof that will closely resemble the final printed copy. Many drawings consist of composite artwork for lines, letter type, screens, and halftones that are necessary for black and white or color reproduction. All of this work must be correct and a final proof will help to insure that all components are in their proper position and that all colors are separate and can be registered. Several processes and materials for proofing are presented in the following sections.

Photomechanical processes. A number of photomechanical systems are used for proofing. The use of diazo prints is widespread since a positive image may be obtained from positive copy. Other proofing systems, for full color or multicolor work, utilize clear polyester films coated with presensitized, light sensitive dyes. These materials may be used in daylight conditions and require very little equipment. A color, like kwik-proof sensitized colors, is wiped onto a clear acetate sheet, Fig. 20.5, and exposed to a negative in a vacuum frame. The negatives may be scribed sheets discussed in chapter 5. The film emulsion is rubbed gently using special developers to develop the image, and tap water is used to rinse the film. A single sheet of film is required for each color. All of the colored film sheets are placed one over the other in register to form the composite multicolor proof. A sheet of white backing paper is fastened to the back of the composite proof to allow for easy editing. Other products like 3M Color-Key Proofing Film may be used in a similar way for preparing color proofs.

Additional methods for producing full color photomechanical proofs involve the transferring of color pigments to a multi-layered white substrate. The coating of the sheets may be done by hand or by use of a whirler. Each layer of the substrate is exposed with one of the color negatives or positives, washed, and dried. The operation is repeated

Fig. 20.5 Kwik-proof color proofs (Courtesy Direct Reproduction Corp).

for each color, and any number of colored images may be built up on the surface to obtain a final full color proof.

Press proofs. The use of press proofs will give the best results since an opportunity exists for seeing the intended production run on the actual job paper and ink color. Proof presses operate on the same principles as offset-lithography presses.

Production press proofs may be pulled under actual production conditions. The final results are excellent, but the costs on production presses are expensive while proofs are being reviewed.

20.5 PRINTING PROCESSES

In printing, an image is reproduced in ink by transferring the image from one surface to another. A number of different printing processes and machines of different types, styles, models, and sizes may be used. The major processes used commercially are letterpress (relief), offset lithography (planographic), gravure, (intaglio), and screen printing (stencil).

Letterpress (relief) printing. In letterpress printing the image is printed from metal type engravings. The image areas are raised in relief above the non-image areas, Fig. 20.6 (a). The process involves the direct printing from the relief surfaces using cylinder or web-fed rotary presses. In a cylinder press the printing plate is mounted on a flat bed that moves back and forth under an impression cylinder, Fig. 20.6 (b). Ink is applied by rollers to the raised surfaces. The sheets of paper on which the actual printing is done enwrap the impression cylinder and are held in position by grippers. The cylinder and paper revolve together in contact with the plate.

For very high speeds, web or sheet fed rotary presses that will handle up to six colors are used. A curved metal printing plate is placed around a cylinder instead of on a flat bed. Rolls of paper provide a continuous web of paper to be fed into the machine. The paper passes between the impression cylinder and the plate cylinder, both of which rotate. Almost all magazine material is reproduced this way.

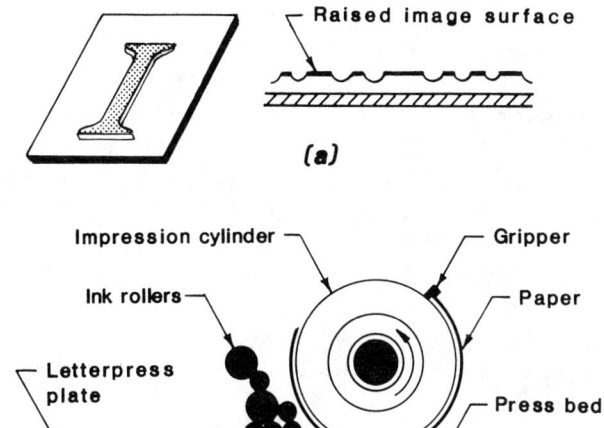

Fig. 20.6 Letterpress (relief printing).

Offset lithography (planographic) printing. In offset lithography the printed image and plate surface are in a single plane (level), Fig. 20.7 (a). The process for printing is based on the principle that oil and water do not mix. The printing image is formed on thin metal plates, using photographic processes (sometimes called photo-offset), that have been chemically sensitized to accept ink and repel water in the image areas.

Offset printing is done primarily on single or multicolor web or sheet fed presses. For printing a single color, the press has three rotating cylinders, Fig. 20.7 (b). The printing plate is

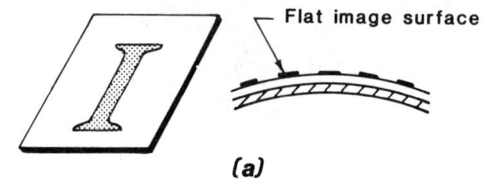
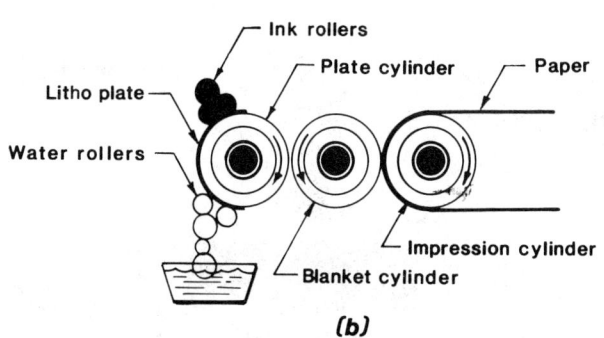

Fig. 20.7 Offset lithography (planographic) printing.

wrapped around the plate cylinder, the blanket cylinder is covered by a sheet of rubber, and the impression cylinder presses the paper against the blanket cylinder. Water rollers apply a film of water to the non-printing image areas of the plate, which then comes into contact with the ink rollers. The inked image which is right-reading is transferred to a rubber blanket where the image is reversed and then transferred or offset to the paper, where it becomes right-reading again.

A great advantage in using offset is that the flexibility of the rubber blanket allows the image to be transferred to a wide range and size of surface materials. The process is used for printing large runs of text copy and pictures in books and magazines as well as most maps. By using fine grained printing plates, fine detail is possible involving different hues and tones.

Gravure (intaglio) printing. In gravure printing, sometimes called photogravure and rotogravure, the image areas to be printed are etched into the surface of a metal plate, Fig. 20.8 (a). Reservoirs or wells are formed in the plate for ink. The depths of the wells control the amount of ink transferred and the density of tone on the paper. Halftone illustrations, like photographs, have to be screened. The printing plate is wrapped around a cylinder on printing presses similar to those used for letterpress printing, Fig. 20.8 (b).

The plate cylinder rotates in an ink resevoir which fills the depressions. A steel blade, called a doctor, wipes all ink from the non-recessed, non-image areas of the plate. Paper is pressed against the plate by the impression cylinder and picks ink out of the depressed areas coating the image in relief on the paper.

Gravure printing may be used for a wide variety of purposes such as catalogs, newsprint supplements, and packaging materials. It is used for fine line drawings and for producing quality color photographs requiring large runs.

Screen (stencil) printing. In screen printing, material such as silk, nylon or stainless steel mesh is stretched tightly over a screen frame, Fig. 20.9. The non-printing areas of the screen are filled with a pigment, and ink is forced through the image areas of the screen with a rubber squeegee. The stencils may be easily prepared by hand cutting or by photographic development. Although almost any surface, shape, and size may be screened, stencils are not suitable for long commercial runs. Good quality stencils will allow copies to be printed having a relatively high quality with the exception of halftones.

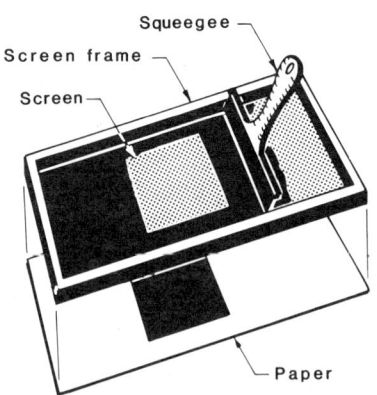

Fig. 20.9 Screen (stencil) printing.

20.6 REPRODUCTION SYSTEMS

The printing processes presented in the earlier sections require photographic methods and generally are used for obtaining a large number of copies. There is a frequent need, however, for limited copies of information that can be duplicated simply and cheaply using

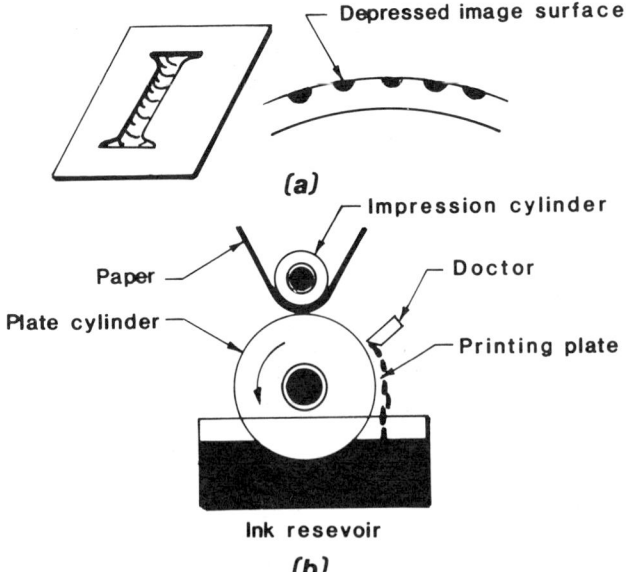

Fig. 20.8 Gravure (intaglio) printing.

other kinds of reproduction systems. Some of these systems cannot be classified as actual printing processes and some methods are non-photographic. This section presents several of the other commonly used reproduction processes.

Mimeograph. Mimeographing is an inexpensive duplicating system requiring a stencil usually made of a porous, fibrous tissue with a wax coating on both sides. The stencil is cut with a stylus or by the keys of a typewriter. The cut stencil is mounted on the cylinder of a mimeograph machine and ink, which may be of any color, is squeezed through the open areas to produce an image on the copy paper. Photomechanical processes or electronically cut stencils should be used to prepare stencils of complicated original drawings since the stencils are difficult to draw upon. Stencils are available in a wide variety of grades that have been designed to give runs up to 5,000 impressions from a single master.

Spirit. Spirit duplicating is a cheap process for reproducing simple drawings or typed pages. A spirit master is prepared on a special heavy paper with a sheet of aniline-dye carbon paper face up underneath. As the image is typed or drawn on the face of the carbon master, the dye carbon simultaneously deposits a reverse image on the back. Spirit masters may also be prepared from almost any type of printed material using office copier machines. Copies having a maximum size of $8\frac{1}{2}$ x 13 inches are made by transferring small amounts of dye from the carbon image to copy paper moistened with an alcohol base liquid. Several colors are available and may be used separately or in any combination with approximately 300 to 500 copies printed from a single master.

Electrophotography. A number of machines are available for printing based on electrostatic principles (the natural attraction of opposite electrical charges). A primary use of this concept is seen in many office copying machines. Xerox (xerography) is an excellent example of this revolutionary process. The system is completely dry and uses no ink, pressure or chemicals for reproducing on different kinds of paper or other surfaces. The image which may be either an opaque or transparent original is recorded with a special camera on a reusable plate comprised of a thin coating of selenium or zinc oxide on a metal substrate which holds an electrostatic charge. The image is developed by cascading or brushing particles of toner or powder, which is oppositely charged, over the image area and fused to the paper by heat. A wide range of high speed equipment is available which will carry out the entire process automatically in less than a minute. Some machines will permit reduced or enlarged reproductions to be made while others are designed specifically for making blowbacks from microfilm. Master plates that are ready for offset printing without further processing may be prepared directly from original material using this process.

The widespread and popular use of this type of equipment is for copying letters, documents, plans, charts, and other materials. Because of its versatality, it may also be used to imprint irregular shaped objects since no direct contact is required.

Office copiers. The market is currently inundated with many types of office copy machines used to duplicate line copy, photographs, and typed information from original material. Their popularity may be attributed to the fact that good quality copies are obtained immediately. Machines of this type should be used primarily for making a few copies and other processes should be considered where many copies are required. The machines vary in their ability to copy certain kinds of information. Some are excellent for copying drawings while others are better suited for duplicating photographs where a wide range of tone values occurs. Certain models will reproduce appliqué or adhesive materials used for eliminating repetitive drawing. Other differences include sheet size, enlargement or reduction capability, and type of copy paper required. Since so many different office copiers are available, the best advice is to select the machine that will suit individual needs.

In addition to electrostatic and diazo copy machine systems, several other methods are used. These are diffusion transfer reversal, dye transfer, and thermographic. In diffusion transfer reversal, a paper negative is exposed in contact with the original copy. This is placed face down on special copy paper and run through photocopy developer. The negative is peeled away to leave a metallic silver image on the copy paper. Dye transfer is a similar process except that the original copy is exposed to an intermediate coated with gelatin containing silver halides and dye. The dye image is subsequently transferred to a permanent media.

Thermographic processes use translucent papers or films with a plastic coating containing materials that are reactive to heat. This material is exposed in contact with the original copy and infrared (heat) radiation is used to expose and develop the image simultaneously.

<u>Diazo</u>. Diazo or ozalid reproduction equipment, combining a light source and developer, is found in most offices producing graphic kinds of information. This equipment is frequently referred to as white printers. The process is a one-to-one dry contact method that is widely used in place of older methods such as the blueprint process. Original drawings to be copied must be prepared on translucent or semi-translucent material. The drawing is placed face-up on specially prepared paper and exposed to a light source such as ultraviolet light. The light sensitive coating of the paper decomposes except where it is protected by lines that appear on the original. Exposed prints are then developed by the alkaline medium produced by ammonia vapors. A number of self-contained automatic units are available for obtaining diazo prints of any size and length, Fig. 20.10.

Diazo papers are available that will produce positive copy having black, sepia, red or blue images on a white background. Heavy papers coated on both the front and back may be obtained for two-sided reproduction. It is also possible to purchase special diazo paper which will reproduce good copies from a

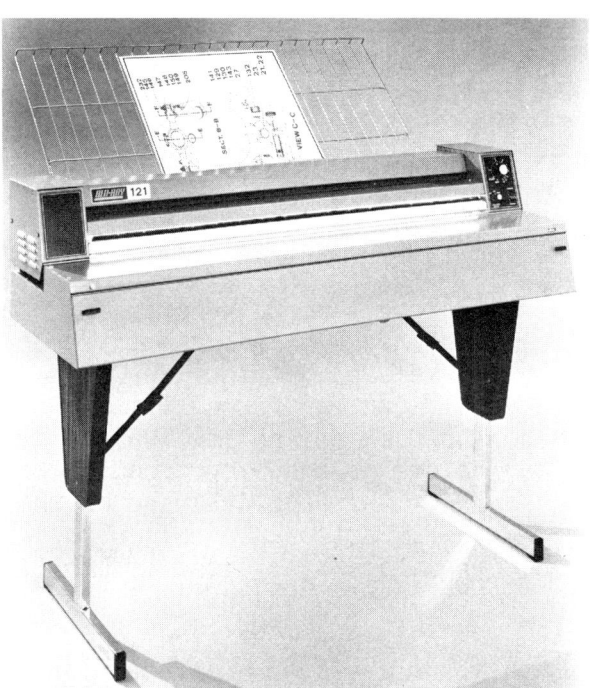

Fig. 20.10 Typical diazo machine (*Courtesy Blu-ray, Inc*).

film positive with a continuous tone image. The effects of color may be included by the use of multiple overlays made on diazo color films of the chosen colors. If additional translucent masters are needed, copies may be made on cloth and film or sepia paper intermediates. The intermediates are used to generate required copies or as a second original, to protect an original drawing from excessive handling. Some intermediate materials may be erased to allow for revision or updating of pencil, ink or typewritten information. Where any typed information appears, orange carbon backing should be used on the reverse side of the drawing material to obtain a dense image for obtaining good quality copy.

In some instances a drawing is created by combining portions of existing drawings, scissors drafting. A print of a part or portion of an existing drawing may be made on diazo adhesive backed films. The peel apart backing is removed and the film is then positioned where required. The film may also be used for printing title blocks and other repetitive type information in the same way commercial appliqué materials are used.

Micrographics. Drawings of all kinds may be drastically reduced because of the remarkable improvements that have been made in films, cameras, and other related equipment, Fig. 20.11. Micrographics is widely used since the tremendous amount of space normally needed for the storage of drawings may be reduced. Record keeping is more efficient, orders may be processed more quickly, and original drawings protected for unnecessary wear-and-tear. With high resolution materials, reductions in the magnitude of 20:1, 40:1, and greater are possible. Microfilms in general use for making microcopies are 16, 35, 70 and 105 mm. Microfilm is available in rolls, film mounted aperture cards, and microfiche. For dead storage, roll film is generally used. In active file and retrieval systems, aperture cards or other flat forms of film are preferred. An advantage in using aperture cards is that data concerning the microcopy may be keypunched on the card. Microfiche is a sheet of transparent film, approximately 100 x 150 mm, that contains many small microfilm strips set in rows. Microfiche systems are especially suited to the needs for publications of high page volume that are frequently assessed or referred to, and also require frequent updating or revision.

A variety of readers and scanning devices is available for projecting microimages in eye-legible size. Many machines are capable of producing low cost full or reduced size prints by electrostatic or photocopy methods. Transparencies may also be made which can be used for subsequent reproduction on diazo machines in addition to offset masters from which copies may be printed at a minimum cost.

20.7 OPTICAL PROJECTORS

To help meet graphic preparation needs, a number of optical projectors are available for enlarging or reducing, Fig. 20.12. Projectors of this type, sometimes referred to as opaque projectors, vary in size and accuracy. Any flat or rolled document, photograph, transparency, or negative may be pro-

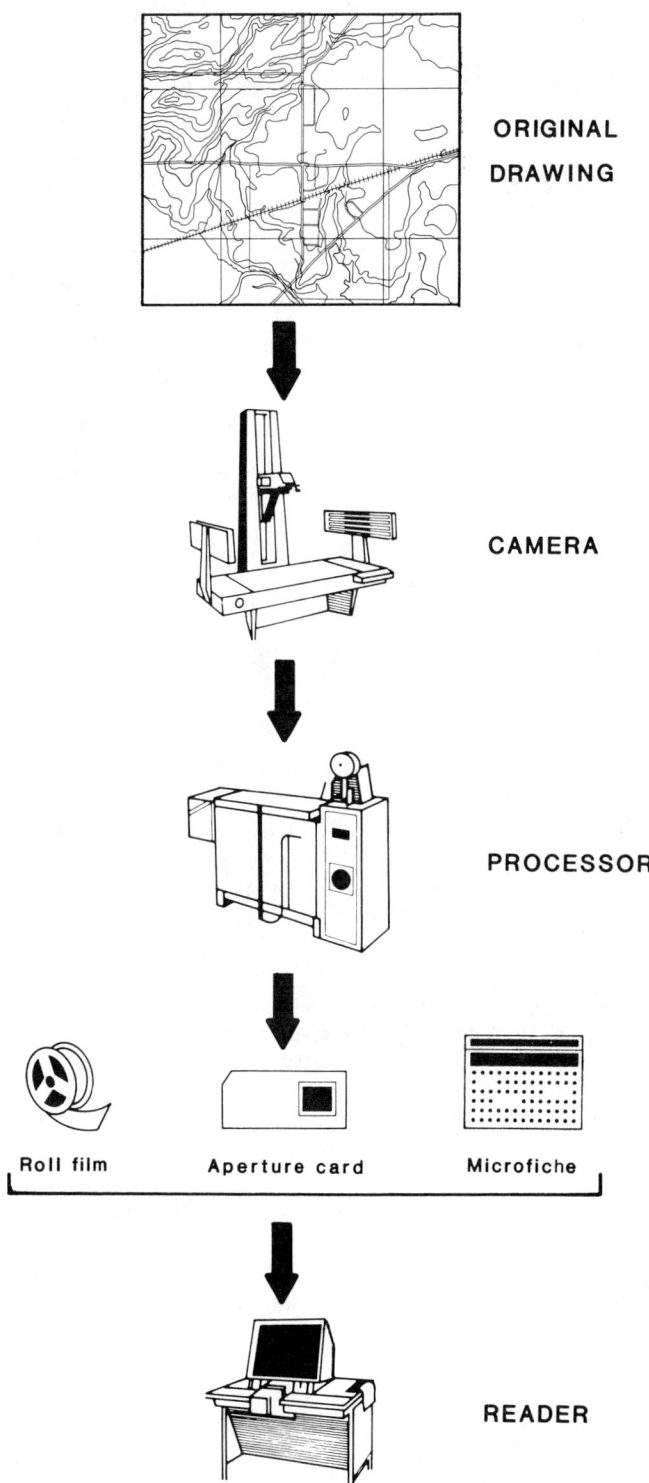

Fig. 20.11 Typical micrographic process.

jected to the drawing surface. Copy may be projected to a 1:1 scale or to any scale from about 5X plus to 5X minus. The projected image is then traced with very little distortion.

In photogrammetry aerial photographs are commonly used as photo maps for engineering projects and for large scale regional and municipal planning. Where enlarged and rectified to scale aerial photographs are required, more sophisticated equipment is required, Fig. 20.13. Rectification may be carried out rapidly since the easel or bed of the enlarger can be tilted about the diagonals. All the adjustment controls are equipped with scales so that if the rectification conditions are known, they may be preset on the instrument.

Since reflectance characteristics in infrared may be quite different than in visible light for the same object, tonal values of objects photographed in infrared may be more sharply defined and will appear differently than on panchromatic film. Water absorbs near infrared energy rather than reflecting it

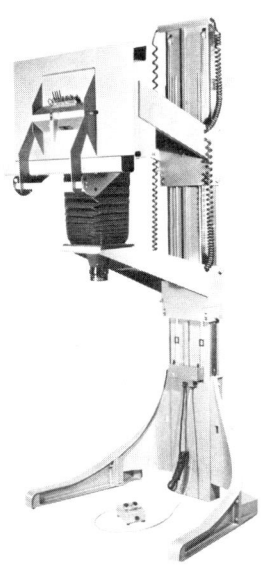

Fig. 20.12 Opaque projector (*Courtesy Artograph*).

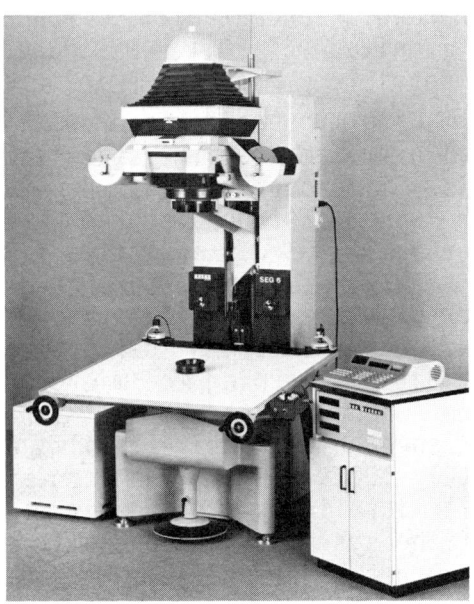

Fig 20.13 Rectifier-enlarger (*Courtesy Wild Heerbrugg*).

Table 20.1 Printing processes and reproduction systems.

Process	Number of Copies	Appearance of Print	Original Required	Description of Process	Advantages and Disadvantages
Photocopy	Unlimited. High cost	Negative or positive prints	Anything that will photograph	1. Expose negative paper 2. Process negative paper in contact with positive paper	Photographic accuracy and speed. Direct copy from open book, etc.
Offset Printing	Up to 500,000 depending on master. Low cost for runs of 500 or more copies	Black or colored lines on any background	Anything that will photograph	Master is produced photographically on paper, aluminum or zinc sheets and mounted in a cylinder press. Ink is transferred from a rubber roller to the image on the master and then to the final copy	Economical and versatile. Any number of colors. Accurate reproduction of fine lines. Enlargements or reductions possible
Mimeograph	1-5,000 from one master. Low cost	Multicolored on any color background	Master stencil	Ink transferred through master stencil to paper. Stencil may be typed, drawn or die cut	Inexpensive, permanent and rapid. Dimensional accuracy not dependable. Choice of ink and paper. Stencil difficult to draw on
Spirit	1 - 500 from one master. Low cost	Purple line on white background. Other colors available	Master carbon	Aniline dye transferred from master to spirit moistened paper to produce copy	Very cheap and rapid. Poorer contrast than mimeograph. Used for small drawings and typed copy. Copies tend to fade. Can not enlarge or reduce
Electrophotography (xerography)	Unlimited. Medium cost up to 50 copies	Black lines on white background	Any copy that can be photographed. Large solid areas or halftone images do not copy very well	Camera used to project original on a selenium coated electrostatically charged plate. Negative charged powder spread across plate. Powder image is transferred to paper and baked onto surface	Prints on almost any smooth surface. Masters can be prepared for offset printing. Must use frequently to warrant rental charges on large equipment. Enlargements or reductions possible
Diazo	Unlimited. Medium cost. Up to 50 copies	Positive print having black, blue or red lines. Paper, cloth or film may be used for printing	Any translucent material	Dry contact process: 1. Expose to light 2. Develop in ammonia fumes	Rapid. Reproducible master intermediates may be made to save wear and tear on originals. Can not enlarge or reduce

MAP INFORMATION AND DISTRIBUTION

21.1 INTRODUCTION

Surveys, photography, and mapping activities are carried out by many Federal, State, and local government agencies as well as private and business organizations. The amount of cartographic data that has or is being generated each year is staggering. Map products numbering in the thousands are prepared and released each year. These products vary in size, area covered, scale, and purpose. The problem for the map user is to locate the appropriate product to meet a specific need. This task is sometimes not easy and can be confusing when not knowing where to look for additional information. Many maps produced by different agencies are similar and considerable duplication of time and other resources is sometimes apparent in mapping activities carried out. A Federal Mapping Task Force study in 1973 reported nearly 40 Federal agencies alone as having surveying and mapping functions as producers, contractors, or grantors (see Table 21.8). Without an official overall structure for the management and coordination of these activities, the confusion in the types of data produced and their availability for the average map user will probably continue.

The three Federal agencies whose primary function is surveying and mapping include the U.S. Geological Survey (USGS), the National Ocean Survey (NOS), and the Defense Mapping Agency (DMA). Other Federal agencies also create map products to support specific responsibilities, administration of public programs, and scientific development. The work carried out by many of the agencies is frequently a cooperative effort with programs complementing and supplementing each other.

The entire spectrum of cartographic information produced by the Federal government is indispensable. The source of information is relied upon as a base for further cartographic development by private and commercial organizations. Without access to this data, it is doubtful that the plethora of information made available to the public could be produced. The equipment and expertise to generate efficiently all the data required is generally not found in all private and commercial organizations. For a map to be produced it is often necessary to conduct good research on available data as a starting place.

This chapter is presented to give some idea of the vast amount of cartographic data that is available. Major sources of information only are discussed since it is virtually impossible to describe all products or their origin. Details concerning the map products presented can be obtained by writing the agency concerned to request additional information.

21.2 LIBRARY OF CONGRESS

The Library of Congress (LC), Geography and Map Division has in its collection alone over 2,500,000 maps. Inquiries concerning its many maps and publications can be obtained by writing:

Library of Congress
Geography and Map Division
845 S. Pickett Street
Alexandria, Virginia 22304

Numerous local, college, and university libraries also have excellent map collections for reference purposes.

21.3 NATIONAL CARTOGRAPHIC INFORMATION CENTER

The National Cartographic Information Center (NCIC) was established by the U.S. Geological Survey, Department of the Interior, in 1974 to serve the public and various Federal agencies in obtaining cartographic information, Table 21.1. NCIC collects, sorts, and describes all types of cartographic information from Federal (products from more than 30 different agencies), State, and local government agencies and, where possible, from private companies in the mapping business. Cataloges, browse files, ordering facilities, and expert assistance is available so that a user can find out what is available by writing:

> National Cartographic and
> Information Center
> U.S. Geological Survey
> 507 National Center
> Reston, Virginia 22092

21.4 DEPARTMENT OF THE INTERIOR

The Department of the Interior functions as the principal conservation agency for the nation. Within this department a number of divisions and bureaus produce cartographic data both for in-house use, interagency use, and public dissemination.

<u>United States Geological Survey</u>. Since 1879, the USGS has served the public and Federal, State, and local governments by collecting, analyzing, and publishing detailed information about the Nation's mineral, land, and water resources. This work is performed primarily by the Topographic Division, the Geologic Division, and the Water Resources Division. Cartographic products produced by these divisions are in a variety of map, book, and other formats, Table 21.2.

Table 21.1 NCIC map data products.

• Advance prints • Maps • Color separates • Feature separates • Land-use and land-cover associated maps • Slope maps • Digital terrain tapes • Maps on microfilm • Orthophotoquads • Aircraft block photos • Aircraft irregular photos • Microfiche indexes of aerial and space images • Geographic computer searches • Reproductions of microfiche of State place names • Microfiche of NCIC's map catalog	• Out of print reproductions • Manned spacecraft images • Landsat images • Computer-enhanced Landsat scenes • Computer-compatible tapes of Landsat data • 35 mm viewing slides • Transformed prints • Kelsh plates • ER-55 plates • Photoindexes • Aerial Photography Summary Record System State-based graphics • Auto positives • Geodetic control data • County maps • Topographic maps on rolls of 35 mm microfilm • Geographic coordinates of various U.S. and selected world names

Table 21.2 USGS products.

- Topographic map series
- State base map series
- Anartic maps
- Photoimage maps
- Shaded relief maps
- Space-age maps
- U.S. GeoData tapes

- Slope maps
- Geologic maps
- Hydrologic maps
- Land use maps
- Digital cartographic data
- National Atlas of the United States
- Multispectral Scanner (MSS) image maps

The USGS is responsible for the National Mapping Program and publishes the popular topographic series maps on a quadrangle basis (see chapter 1). Features of color separates are available for some topographic maps. These are useful for individuals desiring to design a specific map. Only those features, such as roads and drainage, that are required can be combined to serve as a base map with other detail added as needed by the map user.

As part of the National Mapping Program the USGS now offers U.S. GeoData tapes. These are computer tapes which contain cartographic data (boundaries, transportation, and hydrographic information) in digital form. The tapes in graphic form can be used to generate computer-plotted maps. A map can be plotted to include only selected layers of information or combined as desired. Appropriate software is required, however, to allow plotting at various scales and on selected projections. For many map users the tapes offer convenience, accuracy, flexibility, and cost effectiveness.

The USGS maintains several distribution offices for obtaining maps. To order maps of areas east of the Mississippi (including Minnesota, Puerto Rico, and the Virgin Islands) write to:

Distribution Branch
U.S. Geological Survey
1200 S. Eads St.
Arlington, Virginia 22202

For areas west of the Mississippi (including Alaska, Hawaii, Louisiana, Guam, and Samoa) write to:

Distribution Branch
U.S. Geological Survey
Box 25286
Federal Center
Denver, Colorado 80225

The USGS also maintains one of the largest earth science libraries in the world. These holdings contain nearly 900,000 bound and unbound monographs, serials, and government publications; 350,000 pamphlets and reprints; 325,000 maps; 120,000 field record notebooks and manuscripts; and 200,000 album prints, transparencies, lantern slides and negatives, as well as doctoral dissertations on microfilm, and report literature on microfiche. The main library is in Reston, Virginia; three branch libraries are located in Denver, Colorado, Menlo Park, California, and Flagstaff, Arizona. The main library can be contacted as follows:

Library
U.S. Geological Survey
950 National Center
Reston, Virginia 22092

The Earth Resources Observation Systems (EROS) Program is administered by the Geological Survey. This data center, located in Sioux Falls, South Dakota, was established in 1966 to apply remote-sensing techniques to the inventory, monitoring, and management of natural resources. The center provides access primarily to multispectral imagery of the Earth from Landsat satellites, acquired by the National Aeronautics and Space Administration (NASA); aerial photographs acquired by NASA from research aircraft and Skylab, Apollo, and Gemini spacecraft. The primary functions of the center are data storage, retrieval, and reproduction and user assistance and training.

The center's computerized data storage and retrieval system is based on a geographic system of latitude and longitude, supplemented by information about image quality, cloud cover, and type of data. An inquiry can be made about the availability of remotely sensed data about a geographic point location or a rectangular area specified by latitude and longitude corner coordinates. The computer will then print out a listing of available imagery and photography for the inquirer to make a final selection. Assistance in selecting imagery or in placing an order is available from:

> User Services Unit
> EROS Data Center
> U.S. Geological Survey
> Sioux Falls, South Dakota 57198

The description of other forms of data and its availability can be obtained from:

> EROS Data Center
> U.S. Geological Survey
> Sioux Falls, South Dakota 57198

Bureau of Land Management. The Bureau of Land Management is the principal cadastral surveyor (surveys which give land ownership information) of public lands for the nation (see chapter 14). Special maps are also compiled (e.g., mineral ownership) to aid the Bureau in meeting its land and resource management responsibilities. For further information write to:

> Bureau of Land Management
> Division of Cadastral Survey
> Department of the Interior
> 18th and C Streets, N.W.
> Washington, D.C. 20240

National Park Service. The National Park Service issues a small scale map of the United States which give the locations of national parks, sea shores, monuments, historic sites, historial parks, military parks, memorial parks, battlefield sites, scenic riverways, and recreation areas. Additional maps and leaflets of the national parks and other areas are issued that contain specific information related to the facility. Information on these maps and leaflets can be obtained from:

> National Park Service
> Department of the Interior
> Washington, D.C. 20240

21.5 DEPARTMENT OF COMMERCE

The Department of Commerce is responsible for a broad range of functions designed to foster and promote the domestic and foreign commerce of the United States and to serve the needs of business and industry. Within the department, several agencies and offices produce cartographic products. The National Oceanic and Atmospheric Administration (NOAA) is the largest of these agencies with responsibility for the use of earth resources and the monitoring of the environment; air, water, and space. NOAA serves as a coordinator for the activities carried out through other offices such as the National Ocean Survey (NOS), the National Weather Service (NWS), and the National Environmental Satellite Service (NESS).

National Ocean Survey. The National Ocean Survey (formerly United States Coast and Geodetic Survey) produces nautical charts and ocean surveys of the waters of the United States and its territories. It also compiles and publishes aeronautical charts of the nations airways. Information for these items can be obtained by writing:

> Distribution Division, C44
> NOS/NOAA
> Riverdale, Maryland 20840

A major division of NOS is the National Geodetic Survey (NGS) with responsibility for establishing and maintaining the Nation's horizontal and vertical control networks (see chapter 17). A major effort now in progress is the redefinition of the horizontal and vertical control networks for the entire North American Continent. When the revised coordinates are published, the new datum will be known as the North American Datum 1983 (NAD83). Assistance in acquiring geodetic data may be obtained from:

> National Geodetic Information Center
> National Geodetic Survey, C 18
> NOS/NOAA
> Rockville, Maryland 20852

An NGS reference library is maintained by the National Geodetic Information Center (NGIC). Examples of reference materials or services available are indicated in Table 21.3. Information in obtaining these materials or in locating other geodesy data is available by writing:

> Director, National Geodetic Information Center, OA/C18X2
> NOAA
> Rockville, Maryland 20852

National Environmental Satellite Service. The weather and environment satellite program carried out by NOAA involves the application of satellites for monitoring the environment. The satellite system provides almost continuous observations of weather patterns that make possible predictions on a daily basis. These pictures can usually be seen on most television stations providing weather information.

Table 21.3 NGIC Reference Materials.

- American Congress on Surveying and Mapping Journals, 1941 to the present.
- Catalog file of the Association Internationale De Geodesie.
- Geophysical research journals, 1960 to the present.
- Reference set of International Boundary Commission Reports.
- Complete set of USCGS special publications.
- Partial set of Defense Mapping Agency special publications pertaining to geodesy.
- NOS Director's Annual Reports, which include the Superintendent of Coast survey reports dating back to 1807.
- A list of recent publications/presentations on geodesy written by NGS.
- Hundreds of historical and recent publications, presentations, texts, studies, reports, sketches, photographs, correspondence, and miscallaneous reference materials.

National Weather Service. The National Weather Service furnishes weather forecasts on a nationwide basis. Weather maps are prepared each day along with other related weather materials such as storm evacuation maps. Additional information can be obtained through:

> National Weather Service
> World Weather Building
> NOAA
> Department of Commerce
> Washington, D.C. 20233

Bureau of the Census. The Bureau of the Census, the chief statisitical agency of the United States, takes periodic censuses of population. It also collects other demographic information, business data, and tabulates statistics on a very wide range of subjects. This information in thematic map form is published in several series. A partial listing of maps published is given in Table 21.4. Other map products available from this bureau may be obtained from:

> Bureau of the Census
> Data Users Services Division
> Department of Commerce
> Washington, D.C. 20233

Table 21.4 Bureau of the Census Series Maps.

U.S. Geography Series 50(GE-50)

- Population Distribution, Urban and Rural, in the U.S., 1970.
- Number of Negro Persons, by Counties of the U.S., 1970.
- Number of American Indians, by Counties of the U.S., 1970.
- Number of Chinese, by Counties of the U.S., 1970.
- Number of Japanese, by Counties of the U.S., 1970.
- Number of Persons of Spanish Origin, by Counties of the U.S., 1970.
- Median Family Income for 1969, by Counties of the U.S., 1970.
- Number of Owner-Occupied Housing Units by Counties of the U.S., 1970.
- Number of Renter-Occupied Housing Units by Counties of the U.S., 1970.
- Congressional Districts for the 94th Congress.
- Median Gross Rent by Counties of the U.S., 1970.
- Median Value of Owner-Occupied Housing Units by Counties of the U.S., 1970.

Geography Series 70(GE-70)

- Population Distribution, Urban and Rural, in the U.S., 1970 (night-time view).
- Additional topical maps similar to those listed in the GE-50 series.

Urban Atlas Series (GE-80)

- Maps displaying selected data characteristics from the 1970 Census of population for each data item.

21.6 DEPARTMENT OF AGRICULTURE

The Department of Agriculture is a large department that consists of many agencies that have diverse responsibilities. A number of mapping programs are carried out that support its forests, soils, and crop management responsibilities. These activities are centered principally in the Soil Conservation Service (SCS) the U.S. Forest Service (FS), and the Agricultural Stabilization and Conservation Service (ASCS).

Soil Conservation Service. The SCS produces maps involving soil surveys, farmlands, watersheds, floods, and other activities. A strong cooperative working relationship exists between SCS and all levels of government; Federal, State, and local. Each of the state's have several water conservation districts that provide map products along with aerial photographs. For additional information write:

Soil Conservation Service
Education and Publication Branch
Information Division
USDA
Washington, D.C. 20250

U.S. Forest Service. The USFS issues several series of forestry and travel related maps, photography, and survey data of national forests and grasslands. To obtain information on the availability of these products write to:

U.S. Forest Service
Office of Information
Department of Agriculture
P.O. Box 2417
Washington, D.C. 20013

Agricultural Stabilization and Conservation Service. The ASCS administers programs related to agricultural concerns involving price supports, income, supplies, and resource conservation. Since aerial photographs are heavily relied upon to administer these programs, the agency serves as a major source for obtaining aerial photographs, aerial photomaps, copy negatives, glass plates, and satellite imagery. The State Agricultural Service Centers and county offices provide some of these products. For information concerning aerial photographs write to:

U.S. Department of Agriculture
ASCS
Aerial Photography Field Office
P.O. Box 30010
Salt Lake City, Utah 84125

21.7 DEFENSE MAPPING AGENCY

The Defense Mapping Agency (DMA), Department of Defense, supports U.S. forces world wide by providing cartographic information. Products include nautical charts, aeronautical charts, photomaps, and topographic series maps that encompass foreign and world wide coverage. Many of the topographic maps are in plastic relief model form. General information on map products available to the public can be obtained from:

    DMA Topographic Center
    6500 Brooks Lane
    Washington, D.C. 20315

A listing of available nautical charts can be obtained from:

    Defense Mapping Agency Depot
    5801 Tabor Avenue
    Philadelphia, Pennsylvania 19120

21.8 U.S. ARMY CORPS OF ENGINEERS

The Corps of Engineers prepares a number of maps related to the navigation of inland waters, flood control, hydroelectric power development, water and shore stabilization, water supply and storage, and recreation. Several district offices are involved in the administration of this work and should be contacted for maps related to the area. The address of district offices can be obtained from:

    U.S. Army Corp of Engineers
    Public Affairs Office
    Office of the Chief of Engineers
    Department of the Army
    James Forrestal Building
    Washington, D.C. 20314

21.9 TENNESSEE VALLEY AUTHORITY

The Tennessee Valley Authority (TVA) produces a variety of products that includes survey data, aerial photography, topographic, navigation, and recreation maps throughout its service area. This information is available from:

    Tennessee Valley Authority
    Map Information and Records Unit
    Chattanooga, Tennessee 37401

21.10 STATE AND LOCAL MAPPING AGENCIES

The individual state's conduct surveying and mapping activities that encompass many wide ranging projects involving public works, natural resources, and transportation. These programs are carried out at the State, county, and municipal level and are often cooperative efforts funded by the Federal government.

The state's mapping capability frequently resides in departments of highways (or transportation), geological units, or departments of planning and natural resources. These organizations should be contacted for cartographic products prepared at these levels. In most states, mapping activities are concentrated in the highway departments. The general highway maps that show the State and Federal highway system, county road system, drainage, railroads, pipelines, transmission lines, and some culture are best known.

The State geological maps provide a wealth of information related to earth science. Other maps at the State level include recreation, business, and agricultural type products.

At the local level different kinds of map products are produced by planning departments or engineering offices for carrying out numerous projects. The local government may also sponsor or have work done for tax mapping, utility, land use, and engineering requirements. Maps in many diversified forms are also available from Chambers of Commerce, business enterprises, historical societies, and public utilities.

Since it is not feasible to discuss and describe all the products prepared by State and local agencies, the interested map user should contact the appropriate offices for additional information.

21.11 PRIVATE MAPPING FIRMS

A number of private firms carry out surveying, aerial photography, and mapping activities. The services rendered by these firms are very diverse in meeting the needs of the map user. Detailed information on the kinds of cartographic products or services available can be obtained directly from the firm. The yellow pages of the telephone directory should be consulted for a listing of area firms. The names and addresses of other firms can be obtained from:

    American Congress on Surveying
      and Mapping
    210 Little Falls Road
    Falls Church, Virginia 22046

    American Society of Photogrammetry
    105 North Virginia Avenue
    Falls Church, Virginia 22046

21.12 COMMERCIAL MAPPING ORGANIZATIONS OR ASSOCIATIONS AND PROFESSIONAL SOCIETIES

A number of commercial organizations or associations and professional societies produce a variety of maps and related materials. These products include maps exibiting transportation networks, statistical information, marketing and advertising, real estate, mining and energy resources, and communications. The products for some organizations are well known while the firm name of other organizations will help to convey the type of products produced. A partial listing of several of these mapping organizations or associations and societies is given in Table 21.5 and Table 21.6. The journals of many societies also publish maps and articles concerned with new developments in the field of cartography. Many journals regularly carry lists and reviews of cartographic information. The principal journals are listed in Table 21.7.

Table 21.5 Commercial mapping organizations or associations.*

American Association of Petroleum
  Geologists
P.O. Box 979
Tulsa, Oklahoma 74101

American Automobile Association
811 Gatehouse Road
Falls Church, Virginia 22042

American Geographic, Inc.
3109 Thompson Rd.
Fenton, Michigan 48430

American Map Company
1926 Broadway
New York, New York 10023

American Petroleum Institute
2102 L St., N.W.
Washington, D.C. 20036

American Waterways Operators
1600 Wilson Blvd.
Suite 1101
Arlington, Virginia 22209

Arrow Publishing Company
P.O. Box 252
Canton, Massachusetts 02021

Association of American Railroads
Office of Information and Public
  Affairs
1920 L St., N.W.
Washington, D.C. 20036

Berge Exploration, Inc.
7100 N. Broadway
Denver, Colorado 80221

Bill Communications, Inc.
633 3rd Ave.
New York, New York 10017

Champion Maps
4863 N.E., 12th Ave
Oakland Park, Flordia 33334

Denoyer-Geppert Company
5235 North Ravenswood Avenue
Chicago, Illinois 60690

Dolph Map Co., Inc.
430 North Federal Hwy.
Fort Lauderdale, Florida 33301

General Drafting Company, Inc.
Convent Station, New Jersey 07940

Geographic Map Company, Inc.
P.O. Box 688
Times Square Station
New York, New York 10036

Geological Society of America
3300 Penrose Place
Boulder, Colorado 80301

Geomap Company
P.O. Box 30008
Dallas, Texas 75230

George F. Cram Company
301 South LaSalle Street
P.O. Box 426
Indianapolis, Indiana 42606

Geoscience Information Society
⁒ American Geological Institute
5205 Leesburg Pike
Falls Church, Virginia 22041

Hagstrom Co., Inc.
450 W. 33rd St.
New York, New York 10001

Hammond, Inc.
515 Valley St.
Maplewood, New Jersey 07040

Hearne Brothers
25th Floor
First National Building
Detroit, Michigan

H. M. Gousha Publications
P.O. Box 6227
San Jose, California 95114

Hudson Map Company, Inc.
2510 Nicollet Ave.
Minneapolis, Minnesota 55404

International Map Company
595 Board Ave.
Ridgefield, New Jersey 07657

Morgan-Grompion, Inc.
2 Park Ave.
New York, New York 10016

A.J. Nystrom and Company
333 N. Elston Ave.
Chicago, Illinois 60618

Pitman Maps
Oregon Blue Print Co.
930 S.E. Sandy Blvd.
Portland, Oregon 97214

Rand McNally and Company
P.O. Box 7600
Chicago, Illinois 60630

Real Estate Data, Inc.
2398 N.W. 119th St.
Miami, Florida 33167

Rockford Map Publishers, Inc
P.O. Box 6126
Rockford, Illinois 61108

Thomas Brothers Maps
550 Jackson
San Francisco, California 94133

T.N. Hubbard Scientific Co.
2855 Shermer Road
Northbrook, Illinois 60062

TV Digest, Inc.
1836 Thomas Jefferson Place, N.W.
Washington, D.C. 20036

UNESCO Publications Center
Division UNIPUB Inc.
345 Park Ave., S.
New York, New York 10016

Western Map Co.
217 S. Orange
Suite 4
Glendale, California 91204

---

* Source: Schupe, G. and O'Connell, C., Maps for Business: Accessing an Untapped Information Source: New York, Special Libraries Association, 1982.

Table 21.6 Professional mapping societies.

American Congress on Surveying and Mapping 210 Little Falls Street Falls Church, Virginia 22046 (Publisher of <u>Surveying and Mapping</u> and <u>The American Cartographer</u>) American Geographical Society Broadway at 156th Street New York, New York 10032 (Publisher of <u>National Geographic</u>) American Society of Cartographers P.O. Box 1493 Louisville, Kentucky 40201 (Publisher of <u>American Society of Cartographers Bulletin</u>) American Society of Civil Engineers 345 East 4th Street New York, New York 10017 (Publisher of <u>Journal of Surveying Engineering</u>)	American Society of Photogrammetry 105 North Virginia Avenue Falls Church, Virginia 22046 (Publisher of <u>Photogrammetric Engineering and Remote Sensing</u>) Association of American Geographers 1710 16th Street, N.W. Washington, D.C. 20009 (Publisher of <u>Professional Geographer</u>) National Geographic Society 17th and M Streets, N.W. Washington, D.C. 20036 (Publisher of <u>Geographical Review</u>)

Table 21.7 Selected cartographic resource journals.

● American Cartographer ● Canadian Cartographer ● Canadian Geographer ● Canadian Geographic ● Canadian Surveyor ● Cartographic Journal ● Cartographics ● Cartography ● Computer Graphics ● Computer Graphics and Image Processing ● Computers and Geosciences ● Ergonomics ● Geodesy, Mapping, and Photogrammetry ● Geographic Analysis ● Geographical Journal ● Geographical Review	● Geography ● Geo-Processing ● Imago Mundi ● Information Display ● International Yearbook of Cartography ● Journal of Geography ● Journal of Surveying Engineering ● New Zealand Cartographic Journal ● Photogrammetric Engineering & Remote Sensing ● Photogrammetric Record ● Professional Geographer ● Progress in Physical Geography ● Remote Sensing of Environment ● Surveying and Mapping ● Surveying Review

21.13 SUMMARY OF GOVERNMENT AGENCIES PRODUCING CARTOGRAPHIC INFORMATION

The principal government cartographic products and data sources are outlined in Table 21.8. Catalogs, ordering instructions, and order forms can be obtained from the addresses of agencies listed in Table 21.9.

Table 21.8 Map products and sources*

Products	Producing agency	Available from	Products	Producing agency	Available from
Aeronautical charts	NOS	NOS	**Geographic maps:**	NOS	NOS
Boundary information:			Land use	USGS	USGS
United States and Canada	IBC	IBC	**Highway maps:**		
United States and Mexico	IBWC	IBWC	Indian lands	BIA	BIA
Boundary and annexation surveys of incorporated places with 2,500 or more inhabitants	BC	GPO	Federal lands	FHWA	FHWA
			Federally funded roads	FHWA	GPO
			Federal primary and secondary	FHWA	GPO
Civil subdivisions and reservations	BLM	BLM	Interstate	FHWA	FHWA
State/Federal	DOS	DOS	Federal highway maps of the U.S.	FHWA	GPO
Census data (social and economic)	BC	GPO	**Historical maps and charts**	LC	LC
Climatic maps	NWS	NWS		All Federal agencies	NARS
Earthquake hazard maps	USGS	USGS	**Hydrographic charts and bathymetric maps:**	NOS	NOS
Federal property maps:				USCE	USCE
Bureau of Reclamation	BR	BR		USGS	USGS
Fish and Wildlife Service	FWS	FWS	Hydrographic surveys	NOS	NOS
National Aeronautics and Space Administration	NASA	NASA		USGS	USGS
			Nautical charts	NOS	NOS
National forests	FS	FS		USCE	USCE
National Park Service	NPS	NPS	Navigable waterways maps	USCE	USCE
Military reservations:			River and stream surveys	MRC	MRC
Air Force	USAF	USAF	River basin/watershed studies	ERC	ERC
Army	USA	USA		SCS	SCS
Coast Guard	USCG	USCG		USGS	USGS
Marines	USMC	USMC	River surveys	BR	BR
Navy	USN	USN		USGS	USGS
State maps of lands administered by Bureau of Land Management	BLM	BLM	Wildlife and scenic river jurisdiction	BLM	BLM
U.S. maps of lands administered by Bureau of Land Management	BLM	BLM	**Hydrologic investigations atlases**	USGS	USGS
			Indian reservations:		
Flood-plain maps	DRBC	DRBC	Land surveys	BIA	GPO
	FIA	FIA	U.S. maps of Indian lands	BIA	GPO
	MRC	MRC	**Land plats**	BLM	BLM
	NOS	FIA		BLM	NARS
	SCS	SCS		NPS	NPS
	USCE	USCE		USCE	USCE
	USGS	USGS	**National Atlas of the U.S.**	USGS	USGS
Geodetic control data	NOS	NOS	**Photographic products:**		
	USCE	USCE	Aerial photographs	ASCS	ASCS
	USGS	NOS/NCIC		BLM	BLM
Geologic maps:				BLM	EDC
Coal investigations	USGS	USGS		BPA	BPA
General geologic	SGA	SGA		DMA	DMA
	USGS	USGS		NASA	EDC
Geophysical investigations	NOAA	EDS		FHWA	FHWA
	NOAA	ERL		FS	EDC
	USGS	USGS		FS	NCIC
Mineral investigations	USGS	USGS		FWS	EDC
Mines	BM	BM		FWS	NCIC
Oil and gas investigations	USGS	USGS		NOS	NOS
				NPS	NPS

Products	Producing agency	Available from	Products	Producing agency	Available from
	SCS	SCS	**Utilities:**		
	USCE	USCE	Ground conductivity maps of the U.S.	FCC	GPO
	FS	FS	Principal electric-facilities maps of the U.S.	ERC	GPO
	USGS	NCIC/EDC	Principal natural-gas-pipelines maps of the U.S.	ERC	GPO
Orthophotomaps	BIA	BIA			
	NOS	NOS	**Water resources development data**	USGS	USGS
	USGS	USGS			
Space imagery:			**Miscellaneous data:**		
Landsat (ERTS)	NASA	ASCS	Clinometric (slope) maps	USGS	USGS
	NASA	EDC	Gravity survey charts	EDS	EDS
	NASA	EDS		NOS	NOS
NASA manned spacecraft	NASA	EDC		USGS	USGS
Nimbus	NWS	NWS	Income distribution maps	BC	GPO
Skylab	NASA	ASCS	Isogonic charts	USGS	USGS
	NASA	EDC	Isomagnetic charts	NOS	NOS
Tiros	NWS	NWS	Magnetic charts	EDS	EDS
Recreation maps	BLM	BLM	National science trail maps	SCS	SCS
	HCRC	HCRC	State indexes of fish hatcheries and national wildlife refuges	FWS	FWS
Seismicity maps and charts	ERL	ERL	Storm evacuation maps	NOS	NOS
	USGS	USGS	Tree danger (to powerlines) detection maps	BPA	BPA
Soils	SCS	SCS	U.S. location maps of fish hatcheries and national wildlife refuges	FWS	FWS
Soils—substation quality	BPA	BPA			
Topographic maps	USGS	USGS			
	MRC	MRC			
	NASA	NASA			

Source: Thompson, M. M., Maps for America: USGS, 1979.
*See Table 21.9 for addresses of agencies identified by acronyms in this table.

Table 21.9 Addresses of agencies.

Agricultural Stabilization and Conservation Service (ASCS)
Aerial Photography Field Office
Agricultural Stabilization and Conservation Service
Department of Agriculture
(2222 West, 2300 South)
P.O. Box 30010
Salt Lake City, Utah 84125

Bonneville Power Administration (BPA)
Bonneville Power Administration
Department of Energy
(1002 NE. Holladay Street)
P.O. Box 3621
Portland, Oreg. 97208

Bureau of the Census (BC)
Users Service Staff
Data Users Services Division
Bureau of the Census
Department of Commerce
Washington, D.C. 20233

Bureau of Indian Affairs (BIA)
Bureau of Indian Affairs
Department of the Interior
18th and C Streets, NW.
Washington, D.C. 20240

Bureau of Land Management (BLM)
Bureau of Land Management
Department of the Interior
18th and C Streets, NW.
Washington, D.C. 20240

Bureau of Mines (BM)
Environmental Affairs Field Office
Bureau of Mines
Department of the Interior
Room 3323
Penn Place
20 North Pennsylvania Avenue
Wilkes-Barre, Pa. 18701

Mine Map Repository
Bureau of Mines
Department of the Interior
Building 20
Denver Federal Center
Denver, Colo. 80225

Mine Map Repository
Bureau of Mines
Department of the Interior
4800 Forbes Avenue
Pittsburgh, Pa. 15213

Bureau of Outdoor Recreation
Deputy Director for Planning
Bureau of Outdoor Recreation
4415 Interior Building
18th and C Streets, NW.
Washington, D.C. 20240

Bureau of Reclamation (BR)
Chief, Publications and Photography Branch
General Services Division
Bureau of Reclamation
7442 Interior Building
18th and C Streets, NW.
Washington, D.C. 20240

Defense Mapping Agency (DMA)
Defense Mapping Agency
Building 56
U.S. Naval Observatory
Washington, D.C. 20305

Delaware River Basin Commission (DRBC)
Executive Director
Delaware River Basin Commission
(25 State Police Drive)
Post Office Box 7360
West Trenton, N.J. 08628

Department of Energy (DOE)
Public Affairs Director
Department of Energy
1000 Independence Avenue, SW.
Washington, D.C. 20585

Department of State (DOS)
Office of the Geographer
Bureau of Intelligence and Research
Department of State
8742 NS INR/RGE
Washington, D.C. 20520

Environmental Protection Agency (EPA)
Office of Public Awareness
Environmental Protection Agency
401 M Street, SW.
Washington, D.C. 20460

Federal Energy Regulatory Commission (ERC)
Office of Public Information
Federal Energy Regulatory Commission
825 North Capital Street, NE.
Washington, D.C. 20426

Federal Highway Administration (FHWA)
Office of Public Affairs
Federal Highway Administration
Department of Transportation
Room 4208
400 7th Street, SW.
Washington, D.C. 20590

Federal Communications Commission
Public Information Office
Federal Communications Commission
Room 202
1919 M Street, NW.
Washington, D.C. 20554

Federal Highway Administration (continued)
Aerial Surveys Branch
Highway Design Division
Federal Highway Administration
Department of Transportation
Room 3130A
400 7th Street, SW.
Washington, D.C. 20590

Federal Insurance Administration (FIA)
National Flood Insurance Program
Federal Insurance Administration
Department of Housing and Urban Development
P.O. Box 34294
Bethesda, Md. 20034

Heritage Conservation and Recreation Service (HCRC)
(formerly Bureau of Outdoor Recreation)
Federal Land Aquisition Division
Heritage Conservation and Recreation Service
130 Interior South Building
18th and C Streets, NW.
Washington, D.C. 20240

International Boundary and Water Commission, United States and Mexico (IBWC)
U.S. Commissioner
International Boundary and Water Commission, United States and Mexico
United States Section
(4110 Rio Bravo, Executive Center)
P.O. Box 20003
El Paso, Tex. 79998

International Boundary Commission, United States and Canada (IBC)
U.S. Commissioner
International Boundary Commission, United States and Canada
United States Section
Room 150
425 I Street, NW.
Washington, D.C. 20548

Library of Congress (LC)
Geography and Map Division
Library of Congress
845 S. Pickett Street
Alexandria, Va. 22304

Mississippi River Commission (MRC)
Executive Assistant
Mississippi River Commission
(Mississippi River Commission Building)
U.S. Army Corps of Engineers
P.O. Box 80
Vicksburg, Miss. 39180

National Aeronautics and Space Administration (NASA)
User Affairs Office
Office of Applications
National Aeronautics and Space Administration
247 Federal Office Building
600 Independence Avenue, SW.
Washington, D.C. 20546
Telephone: 202-755-2070

National Archives and Records Service (NARS)
Cartographic Archives Division
National Archives and Records Service
General Services Administration
Archives Building
Pennsylvania Avenue at 8th Street, NW.
Washington, D.C. 20408

National Oceanic and Atmospheric Administration (NOAA)

Environmental Data Service (EDS)
National Climatic Center
Environmental Data Service
National Oceanic and Atmospheric Administration
Department of Commerce
Federal Building
Asheville, N.C. 28801

Environmental Data Service (continued)
National Oceanographic Data Center
Environmental Data Service
National Oceanic and Atmospheric Administration
Department of Commerce
3300 Whitehaven Street, NW.
Washington, D.C. 20235

Environmental Research Laboratories (ERL)
Environmental Research Laboratories
National Oceanic and Atmospheric Administration
Department of Commerce
Boulder, Colo. 80302

National Ocean Survey (NOS)

Aerial photographs and shoreline maps:
Coastal Mapping Division, C3415
National Ocean Survey
National Oceanic and Atmospheric Administration
Department of Commerce
Rockville, Md. 20852

Chart sales:
Washington Science Center 1, C5131
National Oceanic and Atmospheric Administration
Department of Commerce
Rockville, Md. 20852

Charts:
Distribution Division, C44
National Ocean Survey
National Oceanic and Atmospheric Administration
Department of Commerce
Riverdale, Md. 20840

General cartographic information:
Physical Science Services Branch, C513
National Ocean Survey
National Oceanic and Atmospheric Administration
Department of Commerce
Rockville, Md. 20852

Geodetic control data:
National Geodetic Survey (NGS)
National Ocean Survey
National Oceanic and Atmospheric Administration
Department of Commerce
Rockville, Md. 20852

National Weather Service (NWS)
World Weather Building
National Weather Service
National Oceanic and Atmospheric Administration
Department of Commerce
Washington, D.C. 20233

National Park Service (NPS)
Office of Communications
National Park Service
3043 Interior Building
18th and C Streets, NW.
Washington, D.C. 20240

Soil Conservation Service (SCS)
Cartographic Staff
Soil Conservation Service
Department of Agriculture
Federal Building
6505 Belcrest Road
Hyattsville, Md. 20782

STATE GEOLOGIC AGENCIES (SGA)
Contact the State Geologist or
other cognizant official in each State.

Tennessee Valley Authority (TVA)
TVA Map Information & Records Unit
Chattanooga, Tenn. 37401

U.S. Air Force (USAF)
Office of Information
Office of the Secretary
U.S. Air Force
The Pentagon
Washington, D.C. 20330

U.S. Army (USA)
Public Affairs Office
Office of the Chief of Engineers
Department of the Army
James Forrestal Building
Washington, D.C. 20314

U.S. Army Corps of Engineers (USCE)
Office of Chief of Engineers
U.S. Army Corps of Engineers
Washington, D.C. 20314

U.S. Coast Guard (USCG)
Oceanographic Unit
U.S. Coast Guard
Building 159E, Washington Navy Yard Annex
Washington, D.C. 20390

U.S. Fish and Wildlife Service (FWS)
Division of Realty
U.S. Fish and Wildlife Service
Department of the Interior
555 Matomic Building
1717 H Street, NW.
Washington, D.C. 20240

U.S. Forest Service (FS)
U.S. Forest Service
Office of Information
Department of Agriculture
P.O. Box 2417
Washington, D.C. 20013

U.S. Geological Survey (USGS)

All cartographic data:
User Services Section
National Cartographic Information Center (NCIC)
U.S. Geological Survey
Department of the Interior
MS 507, National Center
(12201 Sunrise Valley Drive)
Reston, Va. 22092

Photographs and remote sensor imagery:
User Services Unit
EROS Data Center (EDC)
U.S. Geological Survey
Department of the Interior
Sioux Falls, S. Dak. 57198

Maps, aerial photographs, and control data by mail:

Alaska
Distribution Section
U.S. Geological Survey
Department of the Interior
101 12th Avenue
Fairbanks, Alaska 99701

States east of Mississippi River plus Puerto Rico
Branch of Distribution, Eastern Region
Publications Division
U.S. Geological Survey
Department of the Interior
1200 South Eads Street
Arlington, Va. 22202

States west of Mississippi River plus Hawaii, Guam, and American Samoa
Branch of Distribution, Central Region
Publications Division
U.S. Geological Survey
Department of the Interior
MS 306, Box 25286
Denver Federal Center
Denver, Colo. 80225

Commercial dealers are listed on sales indexes which can be obtained from any of the above three offices.

U.S. Government Printing Office (GPO)
Assistant Public Printer
(Superintendent of Documents)
U.S. Government Printing Office
North Capitol and H Streets, NW.
Washington, D.C. 20402

U.S. Marine Corps (USMC)
Information Branch
Information Division
U.S. Marine Corps
Arlington Annex
Columbia Pike and Arlington Ridge Rd.
Arlington, Virginia 20370

U.S. Navy (USN)
Research and Public Inquiries Office
Public Information Division
Office of Information
U.S. Navy
The Pentagon
Washington, D.C. 20350

Source: Thompson, M.M., Maps for America: USGS, 1979.

APPENDIXES

contents

1	Selected bibliography	402
2	Decimal and millimetre equivalents	405
3	Metric-English length units and conversions	406
4	Metric-English area units and conversions	407
5	Metric-English volume units and conversions	408
6	Angular units and conversions	409
7	Selected values useful in mapping	410
8	Spelling and symbols for Engish-Metric units	411

APPENDIX 1. SELECTED BIBLIOGRAPHY.

A number of books and other publications have been published about graphics, maps, and mapping. The bibliography, although not complete, serves as an additional source of information for individuals interested in exploring subject areas in greater detail. The professional journals listed in Table 21.7, chapter 21 publish articles concerned with cartographic information.

Analytical Plotter and Workshop Proceedings: American Society of Photogrammetry, Falls Church, VA., 1980.

Avery, T.E., *Interpretation of Aerial Photographs:* Burgess Publishing Company, Minneapolis, MN., 1977.

Bagrow, L. and Skelton, R.A., *History of Cartography:* Harvard University Press, Cambridge, MA., 1964.

Barrett, E.C. and Curtis, L.F., *Environmental Remote Sensing:* Crane, Russak and Company, Inc., NY., 1974.

Barrett, E.C. and Curtis, L.F., *Introduction to Environmental Remote Sensing:* John Wiley & Sons, NY., 1976.

Bertin, J., *Graphics and Graphic Information Processing:* Walter de Gruyter, NY., 1981.

Birch, T.W., *Maps: Topographical and Statistical:* Oxford University Press, London, 1964.

Brinker, R.C. and Wolf, P.R., *Elementary Surveying:* Harper & Row Publishers, NY., 1977.

Brown, L.A., *The Story of Maps:* Little, Brown and Company, Boston, MA., 1949.

Carrington, D.K., *Map Collections in the United States and Canada: A Directory:* New York Special Libraries Association, 1978.

Castner, H.W. and Robinson, A.H., *Dot Area Symbols in Cartography: The Influence of Pattern on Their Perception:* American Congress on Surveying and Mapping, Falls Church, VA., 1969.

Crane, G.R., *Maps and Their Makers:* Hutchinson's University Library, Hutchinson House, London, 1953.

Curran, H.A., Justus, P.S., Perdew, E.L., and Prothero, M.D., *Atlas of Landforms:* John Wiley & Sons, Inc., NY., 1974.

D'Agapeyeff, A. and Hadfield, E.C.R., *Maps:* Oxford University Press, London, 1950.

Davis, J.C. and McCullagh, M.J., *Display and Analysis of Special Data:* John Wiley and Sons, NY., 1975.

Deetz, C.H. and Adams, O.S., *Elements of Map Projections:* Special Publication No. 68, U.S. Department of Commerce, Coast and Geodetic Survey, Washington, D.C., 1945.

Dickinson, G.C., *Maps and Air Photographs:* John Wiley & Sons, NY., 1979.

Dickinson, G.C., *Statistical Mapping and the Presentation of Statistics:* Edward Arnold (Publishers) Ltd., London, 1973.

Drazniowsky, R., *Map Librarianship:* The Scarecrow Press, Inc., Metuchen, NJ., 1975.

Ellis, M.Y., *Coastal Mapping Handbook:* USGS and NOS, U.S. Government Printing Office, Washington, D.C., 1978.

Evans, R.M., *An Introduction to Color:* John Wiley & Sons, NY., 1948.

Foley, J.D. and Van Dam, A., *Fundamentals of Interactive Computer Graphics:* Addison-Wesley Publishing Company, 1982.

Giesecke, F.E., Mitchell, A., Spencer, H.C., and Hill I.L., *Technical Drawing:* The MacMillan Company, NY., 1930.

Greenhood, D., *Mapping:* University of Chicago Press, Chicago, IL., 1971.

Halftone Methods for the Graphic Arts: Eastman Kodak Company, Rochester, NY., 1976.

Harvard Library of Computer Graphics: Harvard University Laboratory for Computer Graphics and Spatial Analysis, 1979.

Herubin, C.A., *Principles of Surveying:* Reston Publishing Company, Reston, VA., 1978.

Holz, R.K., *The Surveillant Science: Remote Sensing of the Environment:* Houghton Mifflin Company, Boston, MA., 1973.

Hopkins, R.A., *International (SI) Metric System & How It Works:* Polymetric Services, Tarzana, CA., 1974.

International Symposium Commission IV: New Technology for Mapping Proceedings: Canadian Institute of Surveying, Ottawa, Canada, 1978.

International Symposium on Computer-Assisted Cartography Proceedings: American Congress on Surveying and Mapping, Falls Church, VA.

    Auto Carto II, 1975
    Auto Carto III, 1978
    Auto Carto IV, 1979
    Auto Carto V, 1982
    Auto Carto VI, 1983

Inhof, E., *Cartographic Relief Representation:* Walter de Gruyter, NY., 1982.

Keates, J.S., *Cartographic Design and Production:* Longman Group Limited, London, 1976.

Keates, J.S., *Understanding Maps:* John Wiley & Sons, NY., 1982.

Larsgaard, M., *Map Librarianship: An Introduction:* Libraries Unlimited, Littleton, CO., 1978.

Kuchler, A.W., *Vegetation Mapping:* The Ronald Press Co., NY., 1967.

Lawrence, G.R.P., *Cartographic Methods:* Methuen & Company, NY., 1979.

LeMaraic, A.L. and Ciaramella, J.P., *Complete Metric System with the International System of Units:* Abbey Books, Somers, NY., 1973.

Lem, D.P., *Graphics Master:* Dean Lem Associates, Los Angeles, CA., 1977.

Lillesand, T.M. and Kiefer, R.W., *Remote Sensing and Image Interpretation:* John Wiley & Sons, NY., 1979.

Lister, R., *Antique Maps and Their Cartographers:* G. Bell and Sons Ltd., London, 1970.

Lobeck, A.K., *Things Maps Don't Tell Us:* The MacMillan Company, NY., 1958.

Loxton, J., *Practical Map Production:* John Wiley & Sons, NY., 1980.

Maling, D.H., *Coordinate Systems and Map Projections:* George Philip and Son Ltd., London, 1973.

Manual of Color Aerial Photography: American Society of Photogrammetry, Falls Church, VA., 1968.

Manual of Instructions for the Survey of the Public Lands of the United States: Bureau of Land Management, U.S. Department of the Interior, Washington, D.C., 1947.

Manual of Photogrammetry: American Society of Photogrammetry, Falls Church, VA., 1966.

Manual of Remote Sensing: American Society of Photogrammetry, Falls Church, VA., 1975.

McCormac, J.C., *Surveying:* Prentice Hall, Inc., Englewood Cliffs, NJ., 1976.

McDonnell, P.W., Jr., *Introduction to Map Projections:* Marcel Dekker, Inc., NY., 1979.

Mitchell, H.C. and Simmons, L.G., *The State Coordinate Systems: A Manual for Surveyors:* Special Publication No. 235, U.S. Department of Commerce, Coast and Geodetic Survey, Washington, D.C., 1945.

Monkhouse, F.J. and Wilkinson, H.R., *Maps and Diagrams: Their Compilation and Construction:* Methuen and Company Ltd., London, 1971.

Monmonier, M.S., *Computer-Assisted Cartography Principles and Prospects:* Prentice Hall, Inc., NJ., 1982.

Moore, L.C., *Cartographic Scribing Materials, Instruments, and Techniques:* American Congress on Surveying and Mapping, Falls Church, VA., 1975.

Morrison, J.L., *Method-Produced Error in Isarithmic Mapping:* American Congress on Surveying and Mapping, Falls Church, VA., 1971.

Muehrcke, P.C., *Map Use:* J.P. Publications, Madison, WI., 1980.

Raisz, E., *Principles of Cartography:* McGraw-Hill Book Company, NY., 1962.

Remote Sensing: National Academy of Sciences, Washington, D.C., 1970.

Richardus, P. and Adler, R.K., *Map Projections:* North-Holland Publishing Company, Amsterdam-London, 1972.

Ristow, W.W., *Guide to the History of Cartography: An Annotated List of References on the History of Maps and Mapmaking:* Library of Congress, Washington, D.C., 1973.

Robinson, A., Sale, R., and Morrison, J., *Elements of Cartography:* John Wiley & Sons, NY., 1978.

Ryan, Daniel L., *Computer-Aided Graphics and Design:* Marcel Dekker, Inc., NY., 1979.

Schwartz, S.I. and Ehrenberg, R.E., *The Mapping of America:* Harry N. Abrams, Inc., NY., 1980.

Skelton, R.A., *Decorative Printed Maps of the 15th to 18th Centuries:* Spring Books, London, 1965.

Special Effects for Photomechanical Reproduction: Eastman Kodak Company, Rochester, NY., 1974.

Steers, J.A., *An Introduction to the Study of Map Projections:* University of London Press, London, 1965.

Taylor, D.R. Fraser, *The Computer in Contemporary Cartography:* John Wiley & Sons, NY., 1980.

Technical Papers of the American Society of Photogrammetry: American Society of Photogrammetry, Falls Church, VA., 1982.

Technical Papers of the American Congress on Surveying and Mapping: American Congress on Surveying and Mapping, Falls Church, VA., 1982.

Teicholz, E. and Dorfman, J., *Computer Cartography World-Wide Technology and Markets:* International Technology Marketing, Newton, MA., 1976.

Thompson, M.M., *Maps for America:* U.S. Geological Survey, U.S. Government Printing Office, Washington, D.C., 1979.

Thrower, J.W., *Maps & Man:* Prentice Hall, Inc., NJ., 1972.

Tooley, R.V., *Maps and Map-Makers:* Crown Publishers, Inc., U.S.A., 1978.

Turnbull, A.T. and Baird, R.N., *The Graphics of Communication:* Holt, Rinehart, and Winston, Inc., NY., 1968.

Wattles, G.H., *Survey Drafting:* G.H. Wattles Publications, Orange, CA., 1977.

Wilford, J.N., *The Mapmakers:* Alfred A. Knopf, NY., 1981.

APPENDIX 2. DECIMAL AND MILLIMETRE EQUIVALENTS.

1/2	1/4	1/8	1/16	1/32	1/64	Decimals	millimetres	1/2	1/4	1/8	1/16	1/32	1/64	Decimals	millimetres
					1	.015625	.396875						33	.515625	13.096875
				1		.031250	.793750					17		.531250	13.493750
					3	.046875	1.190625						35	.546875	13.890625
			1			.062500	1.587500				9			.562500	14.287500
					5	.078125	1.984375						37	.578125	14.684375
				3		.093750	2.381250					19		.593750	15.081250
					7	.109375	2.778125						39	.609375	15.478125
		1				.125000	3.175000			5				.625000	15.875000
					9	.140625	3.571875						41	.640625	16.271875
				5		.156250	3.968750					21		.656250	16.668750
					11	.171875	4.365625						43	.671875	17.065625
			3			.187500	4.762500				11			.687500	17.462500
					13	.203125	5.159375						45	.703125	17.859375
				7		.218750	5.556250					23		.718750	18.256250
					15	.234375	5.953125						47	.734375	18.653125
	1					.250000	6.350000		3					.750000	19.050000
					17	.265625	6.746875						49	.765625	19.446875
				9		.281250	7.143750					25		.781250	19.843750
					19	.296875	7.540625						51	.796875	20.240625
			5			.312500	7.937500				13			.812500	20.637500
					21	.328125	8.334375						53	.828125	21.034375
				11		.343750	8.731250					27		.843750	21.431250
					23	.359375	9.128125						55	.859375	21.828125
		3				.375000	9.525000			7				.875000	22.225000
					25	.390625	9.921875						57	.890625	22.621875
				13		.406250	10.318750					29		.906250	23.018750
					27	.421875	10.715625						59	.921875	23.415625
			7			.437500	11.112500				15			.937500	23.812500
					29	.453125	11.509375						61	.953125	24.209375
				15		.468750	11.906250					31		.968750	24.606250
					31	.484375	12.303125						63	.984375	25.003125
1						.500000	12.700000	2	4	8	16	32	64	1.000000	25.400000

APPENDIX 3. METRIC-ENGLISH LENGTH UNITS AND CONVERSIONS.

Approximate Common Units	
1 inch	= 25.4 millimetres (0.083 33 feet)
1 foot	= 0.3 metre
1 yard	= 0.9 metre
1 chain ⎤ Surveyors	= 66 feet (20.12 metres)
80 chains ⎦	= 1 statute mile
1 chain ⎤ Engineers	= 100 feet
52.80 chains ⎦	= 1 statute mile
1 nautical mile	= 6 076.103 feet (1852 metres)
1 statute mile	= 63 360 inches (5280 feet, 1.6 kilometres)
1 nautical mile	= 1.15 statute miles (6069.85 feet, 1.85 kilometres)
60 nautical miles	= 76 statute miles
1 millimetre	= 0.04 inch
1 centimetre	= 0.393 7 inch (10 millimetres)
1 metre	= 39.37 inches (1.1 yards, 3.280 8 feet, 100 centimetres)
1 kilometre	= 0.6 mile (1000 metres)

Conversions		
inches	X 0.083 33	= feet
"	X 25.4	= millimetres
"	X 2.54	= centimetres
"	X 0.025 4	= metres
feet	X 0.333 3	= yards
"	X 0.000 189 4	= miles
"	X 0.000 164 6	= nautical miles
"	X 3048	= centimetres
"	X 0.304 8	= metres
yards	X 36	= inches
"	X 3	= feet
"	X 0.914 4	= metres
miles	X 5280	= feet
"	X 1760	= yards
"	X 1 609.34	= metres
"	X 0.869 0	= nautical miles
"	X 1.609 34	= kilometres
millimetres	X 0.039 370 1	= inches
"	X 0.10	= centimetres
centimetres	X 0.393 701	= inches
"	X 328.084	= feet
"	X 10	= millimetres
"	X 0.01	= metres
metres	X 39.37	= inches
"	X 3.280 84	= feet
"	X 1.093 61	= yards
"	X 0.000 621 4	= miles
metres	X 10 000	= millimetres
"	X 100	= centimetres
"	X 0.010	= kilometres
kilometres	X 3281	= feet
"	X 1 093.6	= yards
"	X 0.621 371	= miles
"	X 0.54	= nautical miles
"	X 1 000 000	= centimetres

APPENDIX 4. METRIC-ENGLISH AREA UNITS AND CONVERSIONS.

Approximate Common Units

1 square inch	= 6.5 square centimetres
1 square foot	= 144 square inches (0.111 1 square yards, 0.093 square metre)
1 square yard	= 1296 square inches (9 square feet, 0.8 square metre)
1 square mile	= 640 acres (2.590 square kilometres)
1 acre	= 43 560 square feet (10 square chains, 0.001 562 5 square mile, 0.404 68 hectares)
1 square acre	= 4 040.873 square metres
1 square millimetre	= 0.001 55 square inch
1 square centimetre	= 0.16 square inch (100 square millimetres)
1 square metre	= 11 square feet (1.2 square yards, 10 000 square centimetres)
1 square kilometre	= 0.386 1 square miles (247.04 square acres 1 000 000 square metres, 100 hectares)
1 hectare	= 2.471 acres (100 ares, 10 000 square metres)
1 are	= 100 square metres

Conversions

Square inches	X 0.006 944	= square feet
" "	X 645.2	= " millimetres
" "	X 6.452	= " centimetres
" "	X 0.000 654 16	= " metres
square feet	X 144	= square inches
" "	X 0.111 1	= " yards
" "	X 3.587X10^{-8}	= " miles
" "	X 0.000 022 96	= acres
" "	X 929 030	= square centimetres
" "	X 0.092 903 04	= " metres
" "	X 3.587X10^{-8}	= " miles
square yards	X 9	= square feet
" "	X 3.28X10^{-7}	= " miles
" "	X 0.000 206 6	= acres
" "	X 0.836 127	= square metres
square miles	X 27 880 000	= square feet
" "	X 3 098 000	= " yards
" "	X 640	= acres
" "	X 2 589 988	= square metres
" "	X 2.589 988	= " kilometres
acres	X 43 560	= square feet
"	X 4840	= " yards
"	X 0.001 562	= " miles
"	X 4 046.856	= " metres
"	X 0.4	= hectares
hectares	X 107 639	= square feet
"	X 2.471	= acres
"	X 100 000	= square metres
square millimetres	X 0.001 55	= square inches
" "	X 0.01	= " centimetres
square centimetres	X 0.155 000	= " inches
" "	X 0.001 076	= " feet
square metres	X 10.763 9	= square feet
" "	X 1.195 99	= " yards
" "	X 0.000 247 1	= acres
" "	X 3.861X10^{-7}	= square miles
square kilometres	X 10.76X10^{6}	= square feet
" "	X 0.386 138	= " miles
" "	X 247.1	= acres
section	X 2 589 988	= square metres
township	X 93 239 570	= square metres

APPENDIX 5. METRIC-ENGLISH VOLUME UNITS AND CONVERSIONS.

Approximate Common Units	
1 cubic inch	= 16 cubic centimetres
1 " foot	= 1728 cubic inches (0.03 cubic metres)
1 " yard	= 27 cubic feet (0.8 cubic metres)
1 " centimetre	= 0.06 cubic inch (1000 cubic millimetres)
1 " metre	= 35 cubic feet (1.307 cubic yards, 1 000 000 cubic centimetres)

Conversions		
cubic inches	X 0.000 578 7	= cubic feet
" "	X 0.000 021 43	= " yards
" "	X 16.387 1	= " centimetres
" "	X 0.000 016 39	= " metres
" feet	X 1728	= cubic inches
" "	X 0.037 04	= " yards
" "	X 283.168	= " centimetres
" "	X 0.028 316 8	= " metres
cubic yards	X 46 656	= " inches
" "	X 27	= " feet
" "	X 764 555	= " centimetres
" "	X 0.764 555	= " metres
cubic centimetres	X 0.061 023 7	= cubic inches
" "	X 0.000 035 31	= " feet
" "	X 0.000 001 308	= " yards
" "	X 10 000 000	= " metres
cubic metres	X 61.023	= cubic inches
" "	X 35.314 7	= " feet
" "	X 1.307 95	= " yards

APPENDIX 6. ANGULAR UNITS AND CONVERSIONS.

Approximate Common Units	
1 mil	= $\frac{1}{6400}$ circle (0.056 25 degrees, 0.062 5 grad)
1 grad	= $\frac{1}{400}$ circle (16.2 mils, 0°54', 0.9 degree)
1 degree	= $\frac{1}{60}$ circle (17.8 mils, 1.1 grad)
360 degrees	= 400 grads (6400 mils)
1 radian	= 57°17'44.8" (57.295 779 51°)
1 degree great circle	= 60 nautical miles
1 degree longitude (at equator)	= 69.172 miles (111.321 kilometres)
1 minute longitude (at equator)	= 1.15 statute miles or approximately 1 nautical mile (1.85 kilometres)
1 second longitude (at equator)	= 0.019 statute mile or 101 feet (30 metres)
1 degree latitude (at equator)	= 68.703 statute miles (110.566 kilometres)
1 degree latitude (at poles)	= 69.407 statute miles (111.699 kilometres)
15 degrees longitude	= width of one time zone (360°/24 hr.)
360 degrees	= 24 hours

Conversions		
degrees (angle)	X 60	= minutes
" "	X 0.017 45	= radians
" "	X 3600	= seconds
minutes (angle)	X 0.000 290 9	= radians
seconds (angle)	X 4.848X10^{-6}	= radians
radians	X 57.30	= degrees (angle)
"	X 3438	= minutes (angle)
"	X 0.637	= quadrants

APPENDIX 7. SELECTED VALUES USEFUL IN MAPPING.

	Values
pie (π)	= 3.141 592 654
10 000 kilometres	= distance from equator to pole (basis for length of metre)
6 356 583.8 metres	= earth's polar semi-axis (Clarke ellipsoid 1866)
6 378 206.4 metres	= earth's equatorial semi-axis (Clarke ellipsoid 1866)
20 906 000 feet	= mean radius of earth (3,960 miles)
1 acre foot	= 1 acre 1' deep or 43,560 cu. ft.
6 miles or 480 chains	= length, width, of normal township
36	= number of sections in normal township
640 acres	= one normal section of one sqaure mile
1 pole	= 16.5 feet
1 rod	= 16.5 feet
1 link*	= 7.92 inches
1 chain*	= 4 rods, 66 feet, or 100 links
1 mile	= 320 rods, 80 chains, or 5,280 feet
1 square rod	= 272.25 square feet
1 acre	= 43,560 square feet or 160 square rods
1 acre	= 208.75 feet sqaure (approximate)
1 acre	= 8 x 20 rods or any two numbers whose product is 160
1 vara	= 33.33 inches
varas/1.08	= yards
yards x 1.08	= varas
varas x 100/36	= feet
square varas/5,645	= acres
1 labor	= 1,000,000 square varas or 177.1 acres
1900.8 varas	= 1 mile
5,645 square varas	= 1 acre
arpent	= a tract 192 feet square

* Measurements in government surveys are in chains and links.

APPENDIX 8. SPELLING AND SYMBOLS FOR ENGLISH-METRIC UNITS.

Unit	Symbol	Unit	Symbol
acre	acre	hectometre	hm
angle (plane)	∠	inch	in.
angle (solid)	$\varnothing, \emptyset$	kilometre	km
area	A	length	l
are	a	metre	m
centimetre	cm	mile	mi
chain	ch	millimetre	mm
cubic centimetre	cm^3	radian	rad
" decimetre	dm^3	second	S
" dekametre	dam^3	square centimetre	cm^2
" foot	ft^3	" decimetre	dm^2
" hectometre	hm^3	" dekametre	dam^2
" inch	in^3	" foot	ft^2
" kilometre	km^3	" hectometre	hm^2
" metre	m^3	" inch	in^2
" millimetre	mm^3	" kilometre	km^2
" yard	yd^3	" metre	m^2
decimetre	dm	" mile	mi^2
dekametre	dam	" millimetre	mm^2
foot	ft	" yard	yd^2
hectare	ha	volume	V
		yard	yd

INDEX

Abbreviations, 97
Abscissa, 112
Aerial photography
  cameras, 333
  classification, 331
  determination of heights, 336-337
  displacement, 335
  map control, 338-340
  parallax, 335
  scale of, 334-335
  stereoscope, 338
  stereoscopic principles, 335
  transfer of detail, 341-344
Agonic line, 244
Agricultural Stabilization and Conservation Service (ASCS), 390
Agriculture, Department of, 390
Aligned dimensions, 158
Alber's equal-area conic projection, 229
American Congress on Surveying and Mapping (ACSM), 392
American Geographical Society, 15, 394
American National Standards Institute, (ANSI), 26
Ames lettering guide, 93, 95
Analemma, 251
Angles
  deflection, 242
  dihedral, 253
  dimensioning, 166
  in isometric, 171-172
  in oblique, 179
  interior, 242
  to bisect, 51
  to draw equal angles, 51
  to draw-tangent method, 52
Angular perspective, 185
Appliqué, 66-70, 202
Arc
  tangent to arcs, 57
  tangent to line and arc, 56
  tangent to lines, 56
Architect's scale, 40
Arcs
  dimensioning, 165
  in isometric, 173-175
  in oblique, 179-180
  sketching, 195

Area
  determination of
    by coordinates, 278
    by double meridian distance (DMD), 279
    dot grid, 274
    planimeter, 277
    Simpson's rule, 276
    strip method, 275
    trapezoidal rule, 276
    triangle method, 275
  of cross sections, 280
Area charts, 124
Area symbols, 199
Army Corps of Engineers, 391
Arrowheads, 158
Auxiliary views, 153-155
Axes
  graphs, 112
  isometric, 169, 171
  oblique, 178
Axonometric projection, 169
  dimetric, 176
  isometric, 169
  trimetric, 176
Azimuth, 241
Azimuthal projections, 231
  azimuthal equal-area, 235
  azimuthal equidistant, 234
  globular, 237
  gnomonic, 231-234
  orthographic, 231
  stereographic, 235-237

Bar charts, 121-123
Bearing, 241
Bench marks, 293
Block diagrams, 188-191
Bolt circle, 165
Bonne projection, 227
Bored hole, 145
Bowditch rule, 262
Breaks, conventional, 147
Broken-out section, 151
Bureau of Land Management (BLM), 15, 258, 388
Bureau of the Census (BOC), 389

Cabinet drawing, 178
Cadastral maps, 3

Cadastral survey, 257
Calligraphy, 89
Cameras, 320, 324, 333, 372
Cartographic Information Center, National (NCIC), 386
Cartography
  computer-assisted, 349-369
  definition, 5
  worldwide, 15
Cathode ray tube (CRT), 352, 354, 359-360
Cavalier drawing, 178
Centerlines, 27, 136
Center of vision, 183
Chain dimensions, 163
Chamfers, 166
Chord, 60
Charts, 112, 121-122, 124-125
  map, 211
Choropleths, 203
Circles
  geometry of, 60
  great, 252
  isometric, 173-175
  oblique, 179-180
  sketching, 195
  through three points, 52
  to draw, 22
Color
  chroma, 87
  functions of, 87
  hue, 82
  in map symbols, 88
  primary, 82
  processes, 87-88
  secondary, tertiary, 82
  separation drawings, 85-87
  shade, 83
  tints, 83
  value, 82
Compass
  bearing of a line, 241
  conversions, 246-247
  declination, 243-244
  lead points, 22
  types of, 22-24
  use of, 22-24
Computer-assisted cartography
  data bases, 361-363
  defined, 349
  digitizing, 353-354
  hardware, 350
  input devices
    digitizers, 352
    graphics terminals, 345-355
  output devices
    cathode ray tube terminals (CRTs), 359-360
    computer output microfilm (COM), 360
    line printers, 355-357
    plotters, 358-359
  software, 363
  systems, 350-351
  typical applications, 364-369
Computer output microfilm (COM), 360
Condensed lettering, 95
Conformality, 220
Conic sections, 59
Conical projections, 227
  Albers equal-area, 229
  Bonne, 228
  Lambert conformal, 227
  perspective conic, 227
  polyconic, 230
Conjugate principal points, 338
Continuous tone, 373
Contour interval, 289
Contour lines
  interpolation, 303-305
  types of, 288-289
Coordinate axes, 112
Conventional breaks, 147
Conventional practices, 145
Conversions
  direction, 246-247
  scale, 46-47
  tables, 405-410
Coordinates
  for drawing contours, 302
  geographical, 251
  plotting, 261-263
  rectangular, 112, 260
  spherical, 251
  State Plane, 263-265
  Universal Transverse Mercator (UTM), 265-266
Counterbored hole, 145
Countersunk hole, 145
Cross-sections, 311
CRT, 352, 354, 359-360
Curve fitting, 116
Curved lines, 140
Curves
  dimensioning, 165
  graph, 116
  in isometric, 175
  in oblique, 179-180
  irregular, 31-32
Cut and fill, 297, 314
Cutting plane line, 147
Cylinders-dimensions, 165

Cylindrical projections, 223
  Gall's, 226
  Mercator, 223
  Miller, 226
  oblique Mercator, 225
  perspective cylindrical, 223
  transverse Mercator, 225

Datum, 293
Decimal
  dimensions, 159-160
  scale, 39
Declination
  diagram, 244-245
  grid, 244
  magnetic, 243
Defense Mapping Agency (DMA), 391
Deflection angles, 242
Degrees, slope, 306
Department of Agriculture, 390
Department of Commerce, 388
Department of the Interior, 386
Depression, 294
Diagram, 112, 126
Diazo, 380
Digitizer, 352
Dihedral angle, 253
Dimension line, 27
Dimensioning
  angles, 166
  arcs, 165
  chamfers, 166
  circular features, 164-165
  curved surfaces, 165
  datum, 165
  finished surfaces, 160
  holes, 164-165
  keyways, 166
  notes, 160, 164
  pictorial, 167-168
  slots, 166
Dimensions
  aligned, 158
  arrowheads, 158
  English units, 159
  intermediate, 163
  lines, 157
  location, 161, 163
  metric units, 159
  on or off views, 162
  overall, 162-163
  placement, 161-166
  principal, 131
  reference, 163
  repetitive, 162
  size, 161
  unidirectional, 159
Dimetric, 176
Direction
  azimuth, 241
  bearing, 241
  conversions, 246-247
  true directional, 220
Distortion-map, 221-222
Dividers, 24
Dot maps, 204-206
Drafting machine, 29
Drawing
  boards, 29
  compasses, 22-24
  dividers, 24
  equipment, 21
  ink, 35
  instruments, 21
  lines, 30
  media, 70-72
  pencils, 24
  reading a, 144-145
  reproduction, 381
  scales, 39
  types of, 2
Drawings
  one-view, 134
  two-view, 134
  three-view, 135
Drilled hole, 145
Duotones, 373
Dusting brush, 26

Earth
  coordinates, 251
  figure of, 249
  latitude-longitude, 253
  meridians-parallels, 251
  statistics, 250
Earth Resources Observation System (EROS), 387
Eckert projection, 239
Edge view of a plane, 142, 154-155
Electromagnetic spectrum, 319-320
Elevation, 285
Ellipse
  alternate four-center, 174, 179
  axes, 60
  concentric circle, 61
  definition, 60
  foci, 60
  four-center, 173
  parallelogram, 61
  plotting, 60-61
  sketching, 195
  templates, 33

Engineer's scale, 39
Equal area, 220
Equator, 251-252
Equidistant, 220
Equilateral triangle, 49
Erasers, 26
Erasing
  ink, 35
  machine, 28
  shield, 27
Error of closure, 261
Exploded assembly, 181
Extended lettering, 95
Extension line, 27, 157

False color, 321
FAO, 161
Federal mapping agencies, 385-391, 395-399
Fiducial marks, 338
Fillets and rounds, 146, 175
Film
  peelable, 78-79
  photographic, 321, 372
  polyester, 74-75
  sensitized, 76
Finish marks, 160
First-angle projection, 131
Flaps, 72
Flow diagrams, 126
Foci, 60
Folding lines, 130
Four-center ellipse
  isometric, 173-174
  oblique, 179
Fractions
  dimensioning, 159
  guide lines, 93-95
  lettering, 91
Freehand sketching, 193-197
French curve, 31-32
Frontal
  line, 140
  plane, 142
  projection plane, 129
Full section, 151

Gall's stereographic cylindrical projection, 226
Generalization, 210-217
Geodesy, 249
Geodetic control data, 290-293
Geoid, 249
Geological Survey, U.S., 15, 386-387
Geometric constructions, 49-64
Geometric figures, 49-50

Glass box, 129
Globe, 219
Globular projection, 237
Gnomonic projection, 231-234
Gores, 219
Goode's interrupted homolosine projection, 239
Gothic lettering, 89-93
Grade, 306
Gradient, 307
Grads, 307
Graduated symbols, 208-211
Graphic, 1
Graphic scale, 44
Graphs
  area, 124
  axes, 112
  bar, 121-123
  classification of, 111-112
  curve fitting, 116
  grids, 113
  lettering of, 116-117
  location and selection of axes, 113
  logarithmic, 119-120
  organization and flow, 126
  pictorial, 125
  points and curves, 115-116
  rectangular coordinate, 113
  scale units, 114-115
  semi-logarithmic, 119
  surface, 125
  titles, 117-118
Graticule, 219, 220
Gravers, 75
Gravure printing, 378
Great circle, 220, 225, 251
Grid
  declination, 244
  magnetic angle, 245
  map, 260
  military grid reference system, 266-267
  north, 243
  State Plane Coordinates, 263-265
  Universal Polar Stereographic (UPS), 267-270
  Universal Transverse Mercator (UTM), 265-266
Guide lines, 93-95

Hachures, 285
Half sections, 151
Half views, 138
Halftones, 373-374
Hammer-Aitoff projection, 238
Hexagon, 54

Hidden lines, 27, 137
Hieroglyphics, 1
Histogram, 122
Holes
   dimensioning, 164-165
   representation of, 145
Horizon, 183
Horizontal lines, 140
Horizontal planes, 142
Horizontal projection plane, 129
Hue, 82
Hydrography, 88, 97, 215
Hyperbola, 59, 64
Hypsography, 88, 98

Infrared
   false color, 321
   near-infrared, 319
   thermal, 319
Inch marks, 159
Inch/millimetre conversions, 405
Ink
   drawing, 32-36
   erasing, 35
   instruments, 33-34
   lines, 36
Inking, order of, 36-37
Interior angle, 242
International date line, 252
International foot, 43
International Standards Organization (ISO), 131
Irregular curve, 31-32
Isarithms, 287, 368
Isogonic lines, 244
Isolines, 205
Isometric
   angles, 171-172
   assemblies, 180-181
   arcs, 173-175
   axes, 169-171
   centering of, 172
   circles and arcs, 173-174
   construction of, 171
   curves, 175
   dimensioning, 167-168
   drawing, 170
   projection, 169
   sections, 180-181
   sketching, 196
Isopleths, 203

Keyways, 166

Lambert's conformal conic projection, 227

Landforms, 285, 297-300
Land survey systems
   metes and bounds, 256-257
   U.S. Public Land Survey, 257-260
Land Management Bureau (BLM), 15, 258, 388
Landsat, 328-329
Latitude, 253
Layer (altitude) tinting, 287
Leaders, 27, 158
Leads, 24
Legend, 108
Leroy, 101
Lettering
   basic strokes for, 90-91
   computer, 361
   consistency, 96
   dry transfer, 69, 99-100
   equipment, 101-103
   guide lines, 93-95
   inclined (italic), 91
   lower-case, 90-91
   map-features, 97-99
   marginal data, 106-108
   microfont, 109
   on maps, 97-99
   pica point, 104
   positions for, 96-97
   proportion, 95
   reduction of, 94
   reservoir pens, 102
   single-stroke Gothic, 90
   spacing, 95-96
   stability, 95
   styles, 89, 92-93
   templates, 101-103
   vertical, 90
Letters
   extended and condensed, 95
   Gothic, 90-91
   map, 93
Letterpress, 377
Level
   lines, 140
   planes, 142
Library of Congress, 385
Line
   divided into equal parts, 50
   to bisect, 50
   true length, 139
Line printers, 355-357
Lines
   center, 27, 136
   contour, 287-289
   cutting plane, 147
   drawing of, 30

hidden, 27, 137
inclined, 30
parallel, 30
perpendicular, 30-31
precedence of, 137
section, 148
sketching, 194
types of, 140
vertical, 30
visible, 27
Location dimensions, 161, 163
Logarithmic graphs, 119-120
Longitude, 253
Lower-case letters, 90-91
Loxodrome, 223, 254

Magnetic declination, 243
Magnetic north, 243
Major axes, diameter, 60
Map drawing
   appliqué, 66-70
   negative artwork, 65
   open window negatives, 79-81
   positive artwork, 65
   registration, 72
   scribing, 74-81
Map information and distribution, 385-399
Map scale, 43-44
Mapping milestones, 9
Maps
   cadastral, 3
   charts, 211
   choropleth, 203
   classification of, 3
   computer, 4
   definition, 3
   dot, 205-207
   extraterrestrial, 18-19
   historical evolution, 7-15
   land location and description, 249-271
   lettering, 93, 97-99
   orientation, 241-247
   photomaps, 4, 345
   planimetric, 3
   sources of, 385-399
   symbols, 199-218
   thematic, 4, 203
   topographic, 3, 17, 285-317
   use of, 6-7
Masks, 74
Materials, sectioning, 149
Mechanical engineers scale, 39
Mercator projection, 223, 225
Meridians, 251
   guide, 258
   prime, 252
Metes and bounds, 256-257
Metric
   dimensioning, 159
   equivalents, scale, 48
   scale, 40
Microfilm, 381
Mile, types of, 406
Military grid reference system, 266-267
Miller's cylindrical projection, 226
Mils, 307
Mimeograph, 379
Mining and geology terms, 311-313
Minor axes, diameter, 60
Modem, 350, 352
Moiré, 373
Mollweide's projection, 238
Mosaics, 340
Multispectral scanner, 324-327
Multiview
   arrangement of views, 129
   dimensions, 131
   first angle, 131
   glass box, 129
   projection, 127-129
   projection planes, 129
   third angle, 131
   selection of views, 133
   sketching, 196
   spacing views, 135

National Aeronautics and Space Administration (NASA), 387
National Cartographic Information Center (NCIC), 386
National Forest Service (NFS), 390
National Geographic Society, 15
National Mapping Program, 20
National Oceanic and Atmospheric Administration (NOAA), 388
National Ocean Survey (NOS), 15, 388
National map accuracy standards, 47-48
National Park Service (NPS), 388
National Weather Service, 389
Near-infrared, 319
Non-isometric lines, 172
North
   grid, 243
   magnetic, 243
   true, geographical, 242
Notes, dimensioning, 160-164
Numerals
   guide lines, 93-95
   inclined (italic), 91
   vertical, 90

Oblique
  angles, 179
  assemblies, 180-181
  dimensioning, 167-168
  drawing, 177
  lines, 140
  planes, 142
  projection, 177
  sketching, 196
  types of, 178
Octagon, 55
Offset printing, 377
Offset section, 151
Ogee curve, 57-58
One-point perspective, 187
One-view drawing, 134
Open window negative, 79-81
Ordinate, 112
Organization diagrams, 126
Orientation, map, 241-247
Orthodrome, 225, 251
Orthographic projection, 127, 129-133
Orthomorphic, 220
Orthophotograph, 4, 344
Orthophotomap, 345
Orthophotoquads, 348
Overlays, 72-74

Panchromatic film, 321
Parabola, 59, 62-63
Parallel
  lines, 30
  perspective, 187
  standard, 223
Parallels, 251
Parallax, 335
Parallelogram, 49
Partial auxiliary views, 155
Partial views, 138
Paste-up artwork, 65-66
Patterns, 67-70, 203
Peelable film, 78-81
Pencil
  grades, 24
  lines, types of, 26-27
  points, 25
  sharpeners, 25-26
  sketching, 193-197
Pens
  Leroy, 101
  reservoir, 33
Pentagon, 53
Percent grade, 306
Perpendicular lines, 30
Perspective, 128, 169
  angular, two-point, 188
  considerations, 183-185
  of a line, 186-187
  nomenclature, 182-183
  parallel, one-point, 187
  types of, 185
Phantom section, 152
Photogrammetry, 305
Photography, 371
  black and white, 321
  color, 321
  films, 372
  infrared, 321
Photomaps, 4, 345
Pica point, 104
Pictorial, 169
  assemblies, 180-181
  charts, 125
  dimensioning, 167-168
  dimetric, 176
  isometric, 169-175
  oblique, 177-180
  perspective, 182-188
  sections, 181
  sketching, 196
  trimetric, 176
Picture plane, 127, 183
Pie charts, 124
Pins, registry, 72
Place names, 93
Plane coordinates, 263-265
Plane, true size, 141
Planes
  frontal, 129
  horizontal, 129
  profile, 129
  projection, 129
  reference, 130
  representation of, 141
  types of, 142
Planimeter, 277-278
Planimetric maps, 3
Plates, printing, 375-376
Plotters, 342, 358-359
Points
  and lines, 139-140
  on graphs, 115
  multiview projection of, 139
  numbering points on views, 144
Polar planimeter, 277-278
Polyconic projection, 230
Polyester film, 74-75
Polygons, 50
Precedence of lines, 137
Principal dimensions, 131
Principal points, 338
Principal projection planes, 129

Primary colors, 82
Prime meridian, 252
Printing, 383
Plates, 375-376
Processes
　gravure, 378
　letterpress, 377
　offset, 377
　screen, 378
Proofing methods, 376
Private mapping companies, 392
Process camera, 372
Profile
　lines, 140
　maps, 308, 310
Projection
　axonometric, 169
　cabinet, 178
　cavalier, 178
　dimetric, 176
　first-angle, 131
　isometric, 169-175
　multiview, 127, 129
　oblique, 177-180
　of objects, 143
　of lines, 139
　of planes, 141
　of points, 139
　orthographic, 127, 129, 133
　parallel, 128
　perspective, 182-188
　planes of, 129
　theory, 127
　third-angle, 130-131
　trimetric, 176
　types of, 128
Projection planes, 129, 142
Projections, map, 219-239, 364
　azimuthal (zenithal), 231-237
　classification, 221
　conformal, 220
　conical, 227-231
　cylindrical, 223-226
　equal-area, 220
　equidistant, 220
　equivalent, 220
　special, 237-239
　summary of, 222
Projectors, 129
Proportional dividers, 24
Protractor, 31
Public Land Survey, U.S., 257-260

Quadrangle maps, 15, 17
Quadrant, 130
Quadrilaterals, 49

Qualitative data, 114, 205
Quantitative data, 114, 205

Radar, 320, 327
Range lines, 358
Ratio of precision, 262
Reading a drawing, 144-145
Reamed hole, 145
Rectangle, 49
Rectangular coordinate graphs, 113
Rectangular map coordinates, 260-263
Reference dimensions, 163
Registration, 72
Relief
　block diagrams, 188-191
　contour characteristics, 294-297
　contour interval, 289
　contour map applications, 308-317
　contours, types of, 287-289
　datum, 293
　defined, 285
　hachures, 285
　land forms, 297-300
　layer (altitude) tinting, 287
　models, 285
　shading, 286
Remote sensing
　cameras, 320, 324
　defined, 319
　electromagnetic spectrum, 319-320
　multispectral, 324-327
　photography, 320-324
　satellite imaging systems, 327-329
　scanning, 324-327
　sensors, 320
　side-looking airborne radar (SLAR), 327
　thematic mapper, 329
Removed section, 152
Representative fraction (RF), 43
Reproduction systems, 383
　diazo, 380
　electrophotography, 379
　micrographics, 379
　mimeograph, 379
　office copiers, 379
　spirit, 379
Revolved section, 152
Rhombus, 49
Rhumb line, 220, 223, 254
Runouts, 146

Sandpaper pad, 25
Satellite images, 327-329
Scale
　aerial photographs, 334-335

area relationship, 45
large vs. small, 45
transformation, 46-47
Scales
architect's, 40
civil, engineers, 39
mechanical engineers, 39
metric, 40
map
graphic, 44
representative fraction (RF), 44
verbal, 44
Scalene triangle, 49
Secant, 223
Screens, 67, 373-374
Scribing
how to, 75-77
materials, 74, 76
peelable film, 78-81
sensitized film, 76
Section
lines, 27, 148
symbols, 149
Sections
conic, 59
pictorial, 181
theory of, 147
types of, 151-152
Sector, 60
Segment, 60
Semi-logarithmic graphs, 119
Semiotics, 199
Shading relief, 286, 369
Sheet sizes, 70
Side Looking Airborne Radar (SLAR), 327
Simplification, 212
Simpson's Rule, 276
Sinusoidal projection, 237
Size dimensions, 161
Sketching
blocking-in, 196
isometric, 196
maps, 197
materials, 193
multiview, 196
oblique, 196
proportion, 196
techniques, 193-195
Sketchmaster, 341
Slope, determination of, 306-308
Slope ratio, 307
Software, 363
Soil Conservation Service (SCS), 390
Sources of maps
commercial, 392-393
federal, 385-391, 395-399

local, 391
private, 392
professional societies, 392, 394
state, 391
Spacing
letters, 95-96
views, 135
Spectrum, electromagnetic, 319-320
Spherical coordinates, 251
Spherical triangles, 253-256
Spheroid, 249
Spot elevations, 293
Spot facing, 145
Stability, of letters, 95
Standard parallels, 258
Standard sheet sizes, 70
State plane coordinates, 263-265
Station point, 183
Stereographic projection, 235-237
Stereoplotters, 342-344
Stereoscopes, 338
Stick-up, 66-70
Strike and dip, 311
Successive auxiliary views, 153-154
Surface charts, 125
Symbols
map
considerations, 201-202
types of, 199
variations, 201
sectioning, 149
Système International d'Unités (SI), 40
linear units, 41
prefixes, 41
symbols, 41

Tangencies, 147
Tangent points, 56-57
Tapes, 68
Templates, 32
Tennessee Valley Authority (TVA), 391
Third-angle projection, 130-131
Three-point perspective, 185
Three-view drawing, 135
Thematic Mapper, 329
Thematic maps, 4, 203, 369
Time, 252
Title block, 107
Titles of graphs, 117
Tone, continuous, 373
Topographic
map series, 17
symbols, 200
types of maps, 3
Topography
defined, 285

methods of obtaining, 301-305
Townships, 258
Tracing
  cloth, 71
  paper, 70
Transverse Mercator projection, 225
Trapezium, 49
Trapezoid, 49
Triangles
  to draw, 53
  types of, 30, 49
Trimetric projection, 176
True length of line, 139
True north, 242
True size of plane, 141
T-square, 29
Two-point perspective, 188
Two-view drawings, 134

Ultraviolet waves, 319-320
Unidirectional dimensions, 159
Universal Polar Stereographic (UPS), 267-270
Universal Transverse Mercator (UTM), 265-266
U.S. Army Corps of Engineers, 391
U.S. Census Bureau, 389
U.S. Department of Agriculture (USDA), 390
U.S. Forest Service, 390
U.S. Geological Survey (UGS), 15, 386-387
U.S. Public Land Survey, 257-260
U.S. Survey Foot, 42

Value, color, 82
Vanishing points, 183
Varigraph, 101
Vertical
  lines, 140
  planes, 142
Vertical exaggeration, 310
Views
  alignment, 129-130
  alternate position, 136
  auxiliary, 153-155
  choice of, 133-135
  half, 138
  orientation, 133
  partial, 138
  removed, 152
  sectional, 147-152
  sketching, 196
  spacing, 135
  visualization, 144-145
Visible line, 27

Vision, center of, 183
Visual rays, 129
Visualization, 144-145
Volume
  by contours, 283
  by prisms, 282
  by profiles and cross sections, 281

Waxer, 66
World Geographic Reference System (GEOREF), 270-271

X-axes, 112
X-rays, 319-320
Xerography, 379

Y-axes, 112

Zenithal projections, 231